MW00800612

Fluorescence Imaging
Spectroscopy and Microscopy

CHEMICAL ANALYSIS

A SERIES OF MONOGRAPHS ON
ANALYTICAL CHEMISTRY AND ITS APPLICATIONS

Editor
J. D. WINEFORDNER

VOLUME 137

A WILEY-INTERSCIENCE PUBLICATION

JOHN WILEY & SONS, INC.

New York / Chichester / Brisbane / Toronto / Singapore

Fluorescence Imaging Spectroscopy and Microscopy

Edited By

XUE FENG WANG and BRIAN HERMAN

Department of Cell Biology and Anatomy
University of North Carolina at Chapel Hill
Chapel Hill, North Carolina

A WILEY-INTERSCIENCE PUBLICATION

JOHN WILEY & SONS, INC.

New York / Chichester / Brisbane / Toronto / Singapore

Copyright © 1996 by John Wiley & Sons, Inc.

All rights reserved. Published simultaneously in Canada.

Library of Congress Cataloging in Publication Data:
Fluorescence imaging spectroscopy and microscopy / edited by X. F. Wang
and Brian Herman.
    p. cm.—(Chemical analysis ; v. 137)
    "A Wiley-Interscience publication."
    Includes bibliographical references and index.
    ISBN 0-471-01527-X (cl : alk. paper)
    1. Fluorescence spectroscopy. 2. Fluorescence microscopy.
I. Wang, X. F. (Xue Feng), 1963 –   . II. Herman, Brian.
III. Series.
QD96.F56F57 1996
502'.8'2—dc20   95-31993

Printed in the United States of America

10 9 8 7 6 5 4 3 2 1

CONTENTS

CONTRIBUTORS

Peter L. Becker, Department of Physiology, Emory University School of Medicine, Atlanta, Georgia

K. M. Berland, Laboratory for Fluorescence Dynamics, Department of Physics, University of Illinois at Urbana-Champaign, Urbana, Illinois

Robert A. Buckwald, Applied Spectral Imaging Ltd., Industrial Park, Migdal Haemek, Israel

Dario Cabib, Applied Spectral Imaging Ltd., Industrial Park, Migdal Haemek, Israel

Robert M. Clegg, Max Planck Institute for Biophysical Chemistry, Molecular Biology Department, Göttingen, Germany

C. Y. Dong, Laboratory for Fluorescence Dynamics, Department of Physics, University of Illinois at Urbana-Champaign, Urbana, Illinois

T. French, Laboratory for Fluorescence Dynamics, Department of Physics, University of Illinois at Urbana-Champaign, Urbana, Illinois

Yuval Garini, Applied Spectral Imaging Ltd., Industrial Park, Migdal Haemek, Israel

Gerald W. Gordon, Department of Cell Biology and Anatomy, University of North Carolina at Chapel Hill, Chapel Hill, North Carolina

E. Gratton, Laboratory for Fluorescence Dynamics, Department of Physics, University of Illinois at Urbana-Champaign, Urbana, Illinois

Susan Halabi, Tulane University, School of Public Health, New Orleans, Louisiana

Brian Herman, Laboratories for Cell Biology, Department of Cell Biology and Anatomy, University of North Carolina at Chapel Hill, Chapel Hill, North Carolina

Nir Katzir, Applied Spectral Imaging Ltd., Industrial Park, Migdal Haemek, Israel

Satoshi Kawata, Department of Applied Physics, Osaka University, Osaka, Japan

Raoul Kopelman, Department of Chemistry, University of Michigan, Ann Arbor, Michigan

Joseph R. Lakowicz, Center for Fluorescence Spectroscopy, Department of Biochemistry, University of Maryland at Baltimore, School of Medicine, Baltimore, Maryland

John J. Lemasters, Laboratories for Cell Biology, Department of Cell Biology and Anatomy, University of North Carolina at Chapel Hill, Chapel Hill, North Carolina

E. Neil Lewis, Laboratory of Chemical Physics, National Institute of Diabetes, Digestive and Kidney Diseases, National Institutes of Health, Bethesda, Maryland

Stephen Lockett, Lawrence Berkeley Laboratory, University of California, Berkeley, California

Zvi Malik, Life Sciences Department, Bar Ilan University, Ramat-Gan, Israel

Hiroshi Masuhara, Department of Applied Physics, Osaka University, Osaka, Japan

Ann McKalip, Laboratories for Cell Biology, Department of Cell Biology and Anatomy, University of North Carolina at Chapel Hill, Chapel Hill, North Carolina

Ammasi Periasamy, Department of Cell Biology and Anatomy, University of North Carolina at Chapel Hill, Chapel Hill, North Carolina

M. Rhadhakrishna Pillai, Regional Cancer Center, Kerala State, India

David W. Piston, Department of Molecular Physiology and Biophysics, Vanderbilt University, Nashville, Tennessee

Keiji Sasaki, Department of Applied Physics, Osaka University, Osaka, Japan

P. T. C. So, Laboratory for Fluorescence Dynamics, Department of Physics, University of Illinois at Urbana-Champaign, Urbana, Illinois

Dirk G. Soenksen, Applied Spectral Imaging, Inc., Carlsbad, California

Henryk Szmacinski, Center for Fluorescence Spectroscopy, Department of Biochemistry, University of Maryland at Baltimore, School of Medicine, Baltimore, Maryland

Weihong Tan, Department of Chemistry and Brain Institute, University of Florida, Gainesville, Florida

Xiaolu Wang, Brimrose Corporation of America, Baltimore, Maryland

Xue Feng Wang, Department of Cell Biology and Anatomy, University of North Carolina at Chapel Hill, Chapel Hill, North Carolina

Pawel Wodnicki, Laboratories for Cell Biology, Department of Cell Biology and Anatomy, University of North Carolina at Chapel Hill, Chapel Hill, North Carolina

W. M. Yu, Laboratory for Fluorescence Dynamics, Department of Physics, University of Illinois at Urbana-Champaign, Urbana, Illinois

PREFACE

Fluorescence spectroscopy has historically been used to learn about the molecular structure and function of proteins and lipids on a nanosecond to picosecond timescale. This technique has been very useful, as it allows details of molecular activity to be obtained quantitatively. Unfortunately, these approaches required most studies to be performed on isolated proteins, artificial lipid membranes, and often in cuvettes, precluding understanding of the function of these molecules *in situ*. Recently, a number of developments in digitized video microscopy have made possible the imaging of the dynamics of molecular, chemical, structural, and functional environments at the subcellular level; the ability to combine digitized video microscopy with fluorescence spectroscopy promises to provide unparalleled information about protein and lipid structure and function at the single intact living cell level. In the past two years, the marriage of these two technologies has begun to become a reality.

In an attempt to communicate the usefulness of these techniques and broaden the knowledge of these exciting advances, we have organized a collection of detailed monographs on fluorescence imaging spectroscopy and microscopy written by authors at the forefront of the field. A major purpose of the book is to bring together scientists of distinct disciplines and to foster interactive integration of instrumentation and applications. This volume critically evaluates the principles and new ideas, methods, instruments, and applications of fluorescence imaging spectroscopy and microscopy in a variety of scientific disciplines. Biochemists, cell biologists, biophysicists, physiologists, neuroscientists, applied physicists, and materials scientists, as well as scientists in industrial, agricultural, basic and clinical medical research, and the pharmaceutical sciences, should find these monographs useful.

The monographs in this book emphasize the *multifunctional* and *multidimensional* uses of the combination of fluorescence spectroscopy and digital imaging microscopy. Imaging of fluorescence excitation and emission spectra, fluorescence intensity, fluorescence polarization, fluorescence resonance energy transfer, fluorescence lifetime, and two-photon spectroscopy can yield distinct but complementary information; different fluorescence spectroscopic techniques can also be combined with different microscopic techniques (e.g., conventional, confocal, and near-field) for two- and three-dimensional measurements. Laser trapping (or tweezers) techniques allow manipulation of

xxiii

the external and internal milieus of the sample under study. The goal of research on and applications of fluorescence imaging spectroscopy and microscopy is to *watch* in greater detail (i.e., improve spatial resolution); to *measure* more specifically, sensitively, and accurately (i.e., improve spectroscopic information); and to *manipulate* more precisely (i.e., improve control) simultaneously at the intact single-cell level. This important philosophy was introduced by a former supervisor of one of the editors, Dr. Shigeo Minami (President, The Japan Society of Applied Physics and Professor Emeritus, Department of Applied Physics, Osaka University, Osaka, Japan), several of whose students have contributed to this book. We respectfully dedicate this volume to him.

We hope that this book reveals the excitement, diversity of techniques, applications, and potential use of the combination of fluorescence spectroscopy and microscopy in a broad spectrum of scientific fields, and that it will prompt scientists to use the approaches described here in their studies.

XUE FENG WANG
BRIAN HERMAN

Chapel Hill, North Carolina
March 1996

CHEMICAL ANALYSIS

A SERIES OF MONOGRAPHS ON
ANALYTICAL CHEMISTRY AND ITS APPLICATIONS

J. D. Winefordner, *Series Editor*

Fluorescence Imaging
Spectroscopy and Microscopy

CHAPTER

1

QUANTITATIVE FLUORESCENCE MEASUREMENTS

PETER L. BECKER

Department of Physiology
Emory University School of Medicine
Atlanta, Georgia 30322

1.1. INTRODUCTION

Over the last decade there has been a dramatic increase in the range of questions that can be addressed with quantitative fluorescence microscopy (QFM) techniques. This expansion has been fueled by technological advances in virtually every major area. Perhaps the most significant advances have been in the development of new fluorescent probes, allowing a much wider range of processes to be studied. In addition, the improvements in imaging hardware, the continual increase in the performance/cost ratio of computers, and the progressive evolution of software tools have contributed to the field. These latter developments have greatly reduced the time and energy investment new users must make to assemble and operate a functional system. As a result, there has been a similarly dramatic increase in the number of investigators using QFM systems.

A common aim of QFM applications is to obtain an individual measurement or set of measurements of the fluorescence intensity from a specimen that is sufficiently *reliable* to answer some question. By that standard, a QFM application is not necessarily defined by the aims of the investigation. Given the general tendency to push the limits of resolution, it is inevitable that at some point every investigator will confront the issue of reliability. A rational approach to understanding and improving reliability requires an understanding of the sources of measurement uncertainty and the ability to assess their relative contributions. This chapter will summarize the basic principles and practical considerations involved in making quantitative fluorescence measurements on microscope-based systems. I will focus almost exclusively on

Fluorescence Imaging Spectroscopy and Microscopy, edited by Xue Feng Wang and Brian Herman. Chemical Analysis Series, Vol. 137.
ISBN 0-471-01527-X © 1996 John Wiley & Sons, Inc.

those aspects of the measurement system and the measurement strategy that influence the reliability of fluorescence intensity measurements and of the estimates of the specimen characteristics that are based on them. My goal is to help quantitative fluorescence users improve measurement reliability by examining the sources of uncertainty and by describing how to quantitatively assess them.

1.2. THE QFM PROCESS

The goal of a QFM application is to produce a reliable quantitative estimate of some biologically relevant property or characteristic of a specimen: the concentration and/or location of some molecular species, the rate of some biochemical process, the properties of a local, subcellular milieu, etc. Figure 1.1 is a flow diagram that outlines the general process involved in a typical QFM application. The path along the upper part of the diagram is directed at generating a detector signal that is indicative, in some way, of the specimen characteristic of interest. The specimen is prepared and stained with a fluorescent dye, with the aim of having the physical characteristics of the probe represent the specimen characteristic. The specimen is then illuminated with light at an appropriate wavelength to excite the dye, generating fluorescence that is dependent on the physical state of the dye. Finally, the fluorescent light is collected by the microscope and detected, producing a raw signal that is sampled and permanently stored.

To have meaning, this raw detector signal needs to be processed and interpreted through a sequence of signal-processing steps. I have broken this down into a three-stage process along the bottom part of the figure that conceptually aims to reverse each of the processing steps taken along the upper path. The final outcome is a quantitative estimate of the specimen characteristic. Along the way, estimates of each of the intermediate characteristics are produced. A variety of offsets and scaling factor corrections may be applied to the raw signal to account for such things as background light, specimen autofluorescence, nonlinearity of the detector, etc., to produce an estimate of the probe's fluorescence. This, in turn, must generally be calibrated by mathematically combining it with other experimentally derived constants to produce an estimate of the physical state of the probe; for example, the signal from ion-selective dyes must be compared to maximum and minimum values to estimate the fraction of the dye bound to the ion. Finally, this estimate of the physical characteristics of the probe will likely be processed further by calibration factors that estimate the characteristics of the dye in the specimen (e.g., affinity for the target molecule) to generate the estimate of the specimen characteristic.

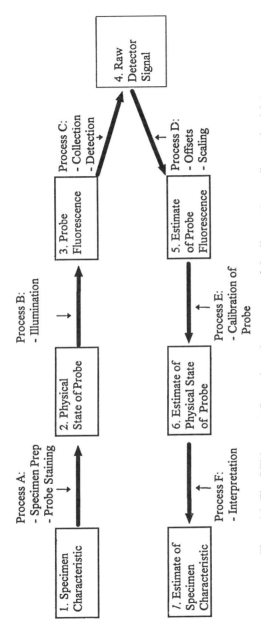

Figure. 1.1. The QFM process. Steps along the upper part of the diagram (proceeding to the right) are directed at producing a detector signal that is indicative, in some way, of the specimen characteristic of interest. Steps along the lower part to the diagram (proceeding to the left) are directed at producing an estimate of the specimen characteristic from the raw detector signal.

For reasons that will be outlined below, any particular estimate produced by the above process will differ from the actual magnitude of the parameter by some amount, an amount I refer to as the *error*. Except in trivial cases, the error is always unknown. Thus, the estimate has a degree of *uncertainty*, defined as the range of possible error between an estimate and the actual value. Although usually the uncertainty is also unknown, many factors that contribute to it can be quantitatively estimated, and thus one often can specify a lower limit to the range of possible error. One source of uncertainty is *noise*, defined as the random, or seemingly random, variations in the signal magnitude about its mean value. When quantifiable, uncertainty is expressed as the standard deviation (SD) of a representative sample of the parameter.

1.2.1. Detectors and the Detection Process

Quantitative fluorescence measurement requires a device to detect the photons emitted by the specimen. For imaging applications, common detectors include silicon intensified target (SIT) cameras, charge-coupled device (CCD) cameras, and intensified CCD cameras. For nonimaging applications and for confocal microscopes, photomultiplier tubes (PMTs) are the common detector of choice. Although the characteristics and attributes of these detectors differ greatly, the photon detection process is based on the same fundamental principle. Every detector has a photosensitive surface that can absorb incident light. When a photon strikes this surface, there is some probability that it will liberate an electron. This electron is termed a *photoelectron* (PE). The probability that a PE is created is specified by the *quantum efficiency* of the detector (QE_{det}). The PEs are then captured and quantified by some strategy that is characteristic of the detector type. In the practical implementation of most detectors, the signal is integrated over some time interval before it is sampled, the duration of which may or may not be adjustable. For the purposes of this discussion, a *detector* will be considered to include the hardware where the photoelectron is created, along with any and all hardware used to quantify and digitize the signal. Thus, for example, if the output of a video camera is recorded on analog videotape and then later digitized by a computer, the videotape recorder and the computer digitizing hardware are to be considered part of the detector. By this convention, the detector output is always a digital value, and no additional noise can be added to the signal beyond this stage. The measured signal will be in detector units (DU) that are proportional to some number of PE. For PMTs operating in photon counting mode, 1 DU can equal 1 PE, but for most other detectors, 1 DU will usually (but not always) correspond to more than 1 PE.

1.3. SOURCES OF UNCERTAINTY

The raw detector signal is a sample of the detector output at a particular time, and it is natural to think of its value as a concrete entity, something that by definition has no error. However, in a very real sense, this measurement is an estimate: it is an estimate of the average magnitude of the detector signal that would be obtained from an infinite number of samples made under exactly the same conditions (moment in time, lamp intensity, probe concentration, etc.). As such, even the raw detector signal has an element of uncertainty associated with its value. As the interpretation of the signal is refined with each step in the signal processing chain, this uncertainty will be passed along and additional uncertainty will be added.

Most of the uncertainty in the magnitude of the raw detector output is due to noise. Uncertainty due to noise comes from two general sources: shot noise and measurement system noise. All subsequent processing steps will add additional uncertainty. Most of the offsets and scaling factors used in these processing steps will have been derived from measurements using the same or similar hardware, and thus will have an uncertainty due to the noise associated with their measurement. Typically, the experimenter can minimize the noise in these factors by repeated sampling or by integrating the signal for a longer time. However, additional uncertainty will exist in these factors because they necessarily cannot be measured at the same time and under the same exact conditions as when the measurements from the specimen were made. I'll refer to these as *circumstantial uncertainties*.

Circumstantial uncertainties and the uncertainty that results from noise (in the signal and in scaling factors) can be estimated. Thus, one can estimate the range of potential error due to them. Less tractable are what might be termed *interpretive uncertainties*. The validity of the number of assumptions made in carrying out the final two processing steps (E and F) in Figure 1.1 generally cannot be proved. One assumes that the signal processing strategy accounts for (1) all changes in the probe that result from its presence in the specimen milieu and (2) all changes in the specimen characteristic that result from the presence of the probe. These uncertainties will be specific to the particular preparation and application, and the effect they have on reliability will depend greatly on the specific question being addressed. Because it is difficult to quantitatively estimate these types of uncertainties, they won't be considered further. However, one should be aware of their existence and recognize that in some situations they may be paramount in any subjective assessment of reliability.

1.4. NOISE

At the level of the raw detector output, the major (and often exclusive) source of uncertainty is noise. For any real measurement system, this noise has two components: shot noise and measurement system noise. In low light level applications, noise can be the major source of the overall uncertainty in the final estimate of the specimen characteristic. Like uncertainty, noise is expressed quantitatively as the SD of the process or measurement.

1.4.1. Shot Noise

Shot noise is the statistically random variations in signal intensity due to the stochastic nature of the events that compose the signal. A stochastic process consists of uncorrelated, random, discrete events. The detection of photons coming from a reasonably large number of illuminated fluorescent molecules is a stochastic process that obeys Poisson statistics. A stochastically occurring event cannot be described with *any* degree of certainty of its occurrence in advance. The best one can do is to assign a probability to its occurrence. This probability can be expressed in several ways, one of which is to specify the mean number of events that will occur during a given time interval. The actual number of events recorded in successive time intervals will vary in a statistically characteristic manner (Papoulis, 1965).

Assume that a beam of light with a mean intensity of I photons per second is aimed at a *noiseless* photon detector having a QE of 1.0 (i.e., every photon is detected). The mean number of photons that strike the detector during a time interval Δt is $A = (I \, \Delta t)$. For any given Δt interval, there is some probability, p_n, that *exactly* n photons will be detected; p_n varies with A:

$$p_n = (A^n/n!)e^{-A} \tag{1}$$

The relation p_n vs. n has a Poisson distribution. For A greater than about 50, this relation approaches the continuous Gaussian function:

$$p(n) = (1/(2\pi A)^{0.5})e^{-[(n-A)^2/(2A)]} \tag{2}$$

This function is normally distributed about a peak value at $n = A$ and has a variance equal to A. This is a hallmark of a Poisson process: the variance equals the mean.

If we continually sampled the detector output for successive Δt intervals, its value would fluctuate randomly (i.e., the signal would be noisy). Since the magnitude of noise is defined as the SD of the signal, for a purely stochastic process the noise is equal to the square root of the signal. Because of this relation, an individual measurement can be used to calculate a reasonably

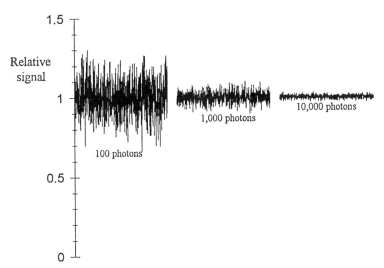

Figure 1.2. Dependence of relative noise on signal intensity. For each mean signal intensity indicated, 1000 randomly generated numbers were produced having population statistics equivalent to shot noise (i.e., SD = the square root of the mean). These data points were normalized to the mean value and then displayed consecutively.

good estimate of the shot noise. Note, however, that the signal must be expressed in units of the elementary events that compose the signal; in this case it would be the number of photoelectrons *detected*. Since the noise increases only with the square root of the signal intensity, the relative noise declines. This is illustrated in Figure 1.2, which displays the normalized signal that might be observed if the mean detected intensity per sampling interval was 100, 1000, or 10,000 photons. Each data set is a consecutive plot of 1000 data points. Note the decrease in the relative noise level at higher signal intensities.

The *signal-to-noise ratio* (SNR) is a commonly used estimate of the quality of a measured signal, S. For a purely Poisson process, the SNR increases with the square root of the signal (since $SNR = S/\sqrt{S} = \sqrt{S}$). Most real measurement systems will add additional noise to the signal, so the overall noise will be greater than $\sqrt{S}$. Thus, the SNR can never be higher than $\sqrt{S}$, but it can be, and often is, lower. Thus, shot noise sets an upper limit to the SNR that can be achieved at any given signal intensity.

1.4.2. Measurement System Noise

In addition to shot noise, the raw detector signal will very likely have a noise component due to the measurement system. This consists principally of detector noise and lamp noise. For a reasonably well engineered instrument, the detector noise will be constant for a particular integration time and should be independent of signal intensity. In contrast, lamp noise will generally always remain a fixed percentage of the signal. That is, if the lamp output varies by $\pm 1\%$, the mean signal will also vary by this amount.

1.4.2.1. Detector Noise

All detectors will add noise to the signal, noise that arises from one or more qualitatively different processes, the presence and magnitude of which will depend on the particular detector and detector type. From a practical standpoint, it is sufficient to consider these noises as one of two types, dark signal noise and readout noise. All detectors have a *dark signal*; that is, they will accumulate PEs in the absence of light due to a variety of processes, such as spontaneous thermal creation at the photocathode, cosmic rays, etc. The dark signal will be linearly proportional to the integration time, and the noise due to the dark signal will be equal to the shot noise that would be associated with a signal of equivalent magnitude. The dark signal is temperature dependent, hence the rationale for cooling some detectors. Most detectors also have *readout* (RO) *noise*, noise that becomes apparent only when the detector signal is read. True RO noise is due to various types of electronic noise (Burgess, 1955; Pierce, 1956) in the circuits that quantify the PEs and convert them to a digital value. Strictly speaking, RO noise is independent of both the integration time and the signal, but it may vary with the RO rate in those detectors where this is adjustable.

Detectors with variable integration times will have the RO noise and the dark signal noise specified separately. This allows one to estimate the detector noise that will exist for different exposure times. However, for detectors with fixed integration times, the dark signal noise is commonly lumped together with the RO noise; since integration time is fixed, the dark signal contribution to the noise will be fixed as well. The usual convention for specifying the magnitude of detector noise (either the overall noise or a specific component) is the SD of the signal in units of PE. Thus, if an RO noise is specified as 20 PE, the noise would be equivalent to the shot noise of a signal having a mean value of 400 PE. Occasionally the dark noise is specified as a dark signal of some number of PEs per unit time, so the actual noise would be estimated as the square root of the calculated number of PEs that would be accumulated in a given integration time. Note that the dark noise is not affected by camera "black level" or other offset adjustments that allow one to subtract off the dark signal. A procedure to estimate these noise characteristics of any particular detector will be presented later.

1.4.2.2. Lamp Noise

If signal shot noise can be said to provide an upper limit to the SNR that can be achieved for a *given* signal strength, lamp noise sets an upper limit to the SNR that can be achieved at *any* signal strength. Unlike other forms of noise, lamp noise is linearly dependent on the signal. If the lamp output varies with, say, an SD of 1% when sampled at some frequency, the fluorescence intensity will also vary by at least this much, and so the SNR can never exceed 100. Lamp noise is generally due to instability of the lamp power supply. Because these instabilities tend to occur at high frequencies, lamp noise is generally a more important concern for studies with signal integration times on the order of a few milliseconds or less (although I know of one lamp power supply on the market that uses rather poorly filtered rectified AC power, and it is unsuitable for even much longer integration times). Short of replacing the lamp with a better version, true lamp noise can only be minimized either by simultaneously measuring the output intensity (using the result to scale the fluorescence measurement) or by increasing the integration time. A number of manufacturers produce low noise power supplies, with output variations less than 0.05% measured at 1 kHz.

The output intensity of most lamps will decrease during the life of the bulb. Further, there can be drifts in the output level of as much as a few percent over the course of an hour of operation. Thus, if the application requires the calibration of the raw data to an absolute intensity standard, frequent calibrations or simultaneous sampling of the lamp output may be necessary to minimize this uncertainty. Other types of lamp instabilities, such as arc "dancing," should be corrected by replacing the bulb. Most recent bulb designs from the major manufacturers have minimized this tendency, at least for xenon bulbs.

1.5. ESTIMATING NOISE AND UNCERTAINTY

In virtually every fluorescence microscopy application the investigator will need to judge the reliability of individual estimates, and in truly quantitative applications the investigator will often need to assess the significance of differences between multiple estimates. Uncertainty will be present in the raw detector signal in the form of noise, and more uncertainty will be added as the signal is sequentially processed to produce estimates of more germane parameters. Thus, it is necessary to be able to calculate the uncertainty to properly analyze the data. For each estimate, one is interested in the uncertainty relative to its magnitude, i.e., the estimate-to-uncertainty ratio (EUR), a term analogous to the SNR.

The raw detector signal noise (N_{Ds}) can be estimated from the signal shot noise and from knowledge of the noise characteristics of the measurement system. Since the magnitude of the shot noise depends on the mean signal intensity, we can estimate the signal shot noise from the magnitude of the detector signal. Quantitatively, it can be expressed as

$$N_{DS} = (N_S^2 + N_{MS}^2)^{0.5} \qquad (3)$$

where N_S is the signal shot noise and N_{MS} is the measurement system noise. Note that to estimate the shot noise, you need to know the number of PEs that compose the signal, which means you *must* know the number of PEs per detector unit. In a later section I will describe a procedure used to estimate this parameter, as well as the noise characteristics of the measurement system.

The postacquisition processing of data will generally involve sequentially applying offsets and scaling factors to convert the raw detector signal to an estimate of the specimen characteristic. Most of these factors are themselves measurements made with the system, and thus each will also have an uncertainty due to noise. Usually, this noise uncertainty can be greatly minimized by averaging multiple measurements or by increasing the signal integration time to increase the number of PEs that compose the measurement. However, these factors will also have circumstantial uncertainties due to the fact that they were not measured at the same time and under exactly the same conditions as those prevailing when the specimen was measured. This type of uncertainty can be minimized by carefully evaluating the conditions that affect the measurement and reproducing them as best you can when these offsets are estimated. However, it is inevitable that some conditions will not be faithfully reproduced, resulting in errors. For example, most applications require the subtraction of specimen autofluorescence, but in practice it is often not possible to estimate this parameter in the same specimen that is stained with dye. In this situation, one will generally use the average autofluorescence of a representative population of nonstained specimens. The SD of the mean value of this sample can be used as an estimate of its uncertainty.

1.5.1. Propagation of Uncertainties

The first postacquisition processing step aims to estimate the dye-specific signal from the total specimen signal. This usually consists of subtracting background and specimen autofluorescence intensity estimates (but may also include factors that correct for nonlinearities of the detector):

$$FI_{dye} = I_{DS} - I_{bk} - I_{af} \qquad (4)$$

where FI_{dye} is the estimated dye-specific fluorescence intensity; I_{DS} is the detector signal intensity; and I_{bk} and I_{af} are the estimated intensity of the background and autofluorescence, respectively. The uncertainty of the FI_{dye}

estimate will depend on the uncertainty of all three components. The rules for calculating the combined uncertainty depend on the mathematical operation applied, as noted by Moore et al. (1990):

For offsets:

$$X \pm Y; \qquad U = (U_Y^2 + U_X^2)^{0.5} \tag{5}$$

For scaling factors:

$$XY; \qquad U = (X^2 U_Y^2 + Y^2 U_X^2)^{0.5} \tag{6}$$

$$X/Y; \qquad U = (X^2 U_Y^2 + Y^2 U_X^2)^{0.5}/Y^2 \tag{7}$$

Thus, the formula for calculating the uncertainty in FI_{dye} [determined in Eq. (4)] would be

$$U_{dye} = (N_{DS}^2 + U_{bk}^2 + U_{af}^2)^{0.5} \tag{8}$$

where N_{DS} is the noise of the detector signal [from Eq. (3)], and U_{bk} and U_{af} are the uncertainty in background and autofluorescence estimates, respectively. Further processing will generally be performed to calibrate the signal against some set of standard measurements. For example, the dye-specific signal might be multiplied by a factor K that converts FI_{dye} to dye concentration:

$$[dye] = FI_{dye} \cdot K \tag{9}$$

The uncertainty of the dye concentration estimation would then be

$$U_{[dye]} = (FI_{dye}^2 \cdot U_K^2 + K^2 U_{FI_{dye}}^2)^{0.5} \tag{10}$$

Although the expression of the raw detector signal in PE units was necessary to estimate the shot noise, all other parameters and their uncertainty can be expressed in any appropriate unit. Note that when a scaling factor is applied, the EUR of the result will always be less than the smallest EUR of any multiplier or divisor.

The above equations on combining uncertainty hold only if all parameters are uncorrelated. Otherwise, the assessment becomes mathematically quite complex. Consider the calibration of the signal from the calcium-sensitive dye fluo-3 (Minta et al., 1989). When excited at 490 nm, fluo-3 has a fluorescence at 520 nm that is low when not bound to calcium (F_{min}) and high when saturated with calcium (F_{max}). The $[Ca^{2+}]$ is determined by measuring the FI at a particular time and then applying the following formula (Grynkiewicz, et al., 1985):

$$[Ca^{2+}] = K_d(FI - F_{min})/(F_{max} - FI) \qquad (11)$$

Note that FI appears twice in this equation; thus, the uncertainty in the numerator is partially correlated with the uncertainty in the denominator. The simplest way I have found to assess the accumulation of uncertainty in these situations is to model it with a large number of cases using a spreadsheet program. Calculate the uncertainty of all parameters that are not correlated using the above guidelines. Fill a spreadsheet column with a randomly generated population of about 1000 values having a mean and an SD equal to that of each parameter,[1] and have a final column be the formula that calculates the desired result based on the individual values in each row. Calculate the mean and SD for the values in this result column.

1.5.2. Using Uncertainty Estimates

When assessing the accumulation of uncertainty, it is important to keep in mind the question being addressed. Although the above guidelines will allow one to assess the quantifiable uncertainty of the estimate of the specimen characteristic, the uncertainty of this estimate may not be appropriate for judging the validity of your conclusions. Consider the situation where the effect of some agent on the cytosolic $[Ca^{2+}]$ is studied with a specimen loaded with the calcium-sensitive dye fluo-3. If the question is "Does agent X increase the $[Ca^{2+}]$?" and the experiment is to measure the FI before and after rapidly applying X to a given specimen, then a comparison of the raw detector signals, assessed in light of *their* uncertainty, may be sufficient to answer the question. The uncertainty in the background and autofluorescence signals and in the calibration parameters F_{max}, F_{min}, and k_d are not relevant (assuming control experiments show that X does not affect the autofluorescence) because the error in each of these will be the same for both measurements (i.e., these parameters will be correlated). Estimating the uncertainties in the $[Ca^{2+}]$ estimate produced by Eq. (11) by considering the error in these parameters would lead to an underestimate of the reliability and possibly an invalid conclusion. On the other hand, if the measurements were from two different specimens, one exposed to agent X and one not, then accounting for the

---

[1]Some spreadsheet programs can create this population for you given a mean and a SD. Otherwise, a random number generator function that produces a value from 0 to 1 can be used (e.g., *a rand* in Lotus 1-2-3 or Quattro Pro). Individual members of a population of numbers having a reasonably accurate gaussian distribution can be obtained by subtracting the sum of six random numbers from the sum of six more random numbers, multiplying the result by the desired SD, and adding the mean value.

uncertainty of the background and autofluorescence would be necessary. If the aim was to specify as precisely as possible the $[Ca^{2+}]$ achieved after treatment with X, then the additional inclusion of the uncertainty of the calibration parameters would be required.

1.6. ASSESSING MEASUREMENT SYSTEM NOISE CHARACTERISTICS

It is important to have a reasonably good assessment of the noise characteristics of the measurement system. Such information is essential to accurately estimate the uncertainty of the raw detector signals and of subsequent estimates based on them. Further, with this information in hand, it is easier to rationally diagnose equipment problems, identify components for improvement, and predict the feasibility of new applications. A protocol for assessing the noise characteristics of a measurement system is presented below; it is based on the strategy used by Moore et al. (1990). It is suitable for just about any type of setup, requires no prior knowledge about the characteristics of the system, and assumes only that the detector does not add a signal-dependent noise (a reasonable assumption for a quality detector). Performing and analyzing the results of this procedure should take no more than an afternoon.

Assessing the noise characteristics of the detector requires a stable light source. A reasonably inexpensive one can be constructed with a light-emitting diode (LED), a 9 V battery, a 330 Ω resistor, and a 10 kΩ, 20 turn potentiometer (constructed with the two resistors in series, limiting the LED current). With my high-time resolution setup (a PMT and photon counter), the output of this lamp appears purely stochastic when sampled at 2 kHz. The LED should be placed so that its light will enter the microscope and reach the detector (I find that a 10 × objective works well, as does no objective at all). For imaging detectors, position the LED well out of focus and try to ensure that the image intensity is as spatially uniform as possible (regional variations less than 10%).

The initial part of this procedure is to measure the noise on the detector signal as the incident intensity is varied over a range that corresponds to those obtained under experimental conditions. This variation is best done using precise neutral density filters to attenuate the light by a known amount (which will allow an assessment of the linearity of the detector), but if a good set is not available, one can just alter the LED current. If the detector has an adjustable gain, it should be set to a level used experimentally (if the gain is normally adjusted for each experiment, then this assessment procedure should be done for a range of settings). Further, if the detector has a "black level" type of adjustment, it should be set low enough so that almost all pixels have a value greater than zero when no light strikes the detector. In general, a data set consisting of repeated samplings of the detector signal is required to calculate the noise level (i.e., the SD of the signal). The time resolution should be the fastest employed under experimental conditions. How this data set is obtained depends on the type of detector used.

For nonimaging (e.g., PMT) detectors, repeated consecutive samples should be taken (1000 or so are sufficient). For imaging detectors, the data collection strategy depends on whether there is a correlation between the signal contained in neighboring pixels (see below). For cameras that have a very low degree of correlation between adjacent pixels, two complete images acquired a short time apart will suffice, since each image contains thousands of individual samples. For cameras with significant correlation, numerous images (> 30; the more the better) will have to be taken. In any case, you *must* also make sure that the time interval between images is significantly longer than the persistence time constant of any element in the detector.

A correlation between neighboring pixels will generally exist in intensified cameras, since the light that leaves the intensifier stage can strike a larger area on the secondary detection surface. This effect can be examined by aiming the camera at a point source of light (e.g., an LED masked by a pinhole) located several meters across the room at a magnification such that the image on the detector surface will be much smaller than an individual pixel area. You may have to iteratively adjust the position of the point source to get its image located approximately in the center of a pixel, and of course the optics should be such that the diffraction-limited size of the image will fit comfortably within a pixel area on the detector surface. Take an image of the point source at camera gains equivalent to those used under experimental conditions and examine the intensity distribution of the image of the spot. A low correlation would exist if more than 95% of the intensity is within a single pixel. An alternative procedure to detect the presence of this problem (but not to describe it) is to acquire the data as suggested for a camera with a correlation problem but to analyze the data both ways. If the variance is significantly higher when analyzed as if it were correlated, then it almost certainly was.

To determine the noise, you need to calculate the mean and variance of the data. For nonimaging systems, this is a straightforward processing of the data set. For imaging systems without a correlation problem, determine the mean intensity (per pixel) for one image; then subtract one image from the other (pixel by pixel), determine the variance of this difference image, and then adjust the variance by dividing it by 2. For imaging systems with a correlation problem, you need to select a fixed region in each image to assess, such as the central 30-by-30 pixel region. Calculate the mean signal per pixel in this region in all images, determine the mean and variance of this set of mean values, and then multiply the variance by the number of pixels in the region used. Repeat

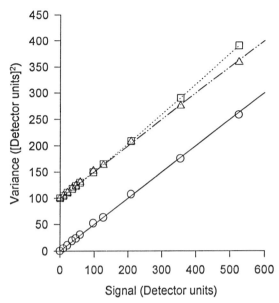

Figure 1.3. A plot of signal variance vs. detector signal intensity for a PMT system (open circles) and for two hypothetical measurement systems. Note that the detector signal is in detector units (DUs) and the variance is expressed in DU^2 units. The data plotted in circles are actual values measured with my high time-resolution microfluorimeter at a 2 kHz sampling rate, with a battery-powered LED as the light source. The line through these data is the predicted relation for a purely Poisson process. The data plotted in triangles are the same as these plotted in the open circles except that a 20 PE RO noise was added. The data plotted in squares are the same as those plotted in triangles with the addition of a 1% lamp noise.

this overall procedure at different incident light intensities, from very low levels up to the limit of the detector output (however, take care not to saturate the detector output range). A final set of measurements should be made with no light at all reaching the detector. Subtract the mean detector value of this no-light data set from the mean signal of the other samples. Construct a graph plotting the variance as a function of the signal intensity.

An example of such a plot is shown in Figure 1.3, displaying actual data from my PMT setup in the open circles. The mean value of the no-light data set was < 1. Note that the variance appears to be linearly dependent on the signal, and the Y-intercept is approximately 0. If the underlying signal was a Poisson process, the linear portion of the relation should have a slope of 1.0 if the signal was expressed in PE units and the variance in PE^2 units. If the slope is not 1.0, we assume that this is because 1 DU does not equal 1 PE. Thus, the reciprocal of the slope of the line is the number of PEs per detector unit (PE/DU). For the open circle data, the slope is 1/2, indicating that the photon counter circuit increments by one for every two photons detected (which is correct). The line through these data is the variance that would be expected if the noise was due solely to shot noise. The fit of the actual data would indicate that the noise on my detector signal is overwhelmingly due to shot noise. The Y-intercept is proportional to the sum of the RO noise and dark signal noise of the detector, which for my PMT–photon counter should be 0 at a 2 kHz sampling rate. The data plotted with the triangles is the same data with added RO noise, meant to illustrate the finite detector noise that will commonly be observed with an imaging detector. The magnitude of the RO and dark signal noise, in PEs, can be calculated from the Y-intercept by the equation

$$N_{RO + dark} = (Y_{int})^{0.5} \cdot PE/DU \tag{12}$$

which, for the hypothetical case illustrated with the triangles, corresponds to a noise of 20 PEs.

For detectors with fixed RO rates, such as video output cameras, the dark signal will be a constant, and thus this noise can be conveniently lumped into the RO noise. However, for detectors with variable RO rates, such as PMTs and slow scan CCDs, you will also want to determine the noise component due to the dark signal. The general procedure used to do this is to block all light from reaching the detector and compare the raw signal produced with different integration times. Then calculate the number of PEs detected per unit time from the slope of this relation. Remove this component from the $N_{RO + dark}$ to calculate the specific RO noise:

$$N_{RO} = [(N_{RO + dark})^2 - (PE_{dark})]^{0.5} \tag{13}$$

where PE_{dark} is the estimated number of dark signal PEs that would accumulate during the integration time used. Now you can estimate the detector noise for any integration time by applying the formula

$$N_{Det} = [(N_{RO})^2 + (PE_{dark})]^{0.5} \tag{14}$$

1.6.1.   Caveats

The detector noise will most likely vary with the gain of the detector, so it is important to repeat this process over the range of gain settings normally used experimentally. Further, the above assessment procedure assumes that the detector does not add a signal-dependent noise. If it did, your plot of the variance vs. signal might not appear linear. If the plot curves upward, that indicates a signal-dependent noise that grows faster than shot noise (i.e., the noise is greater than the square root of the signal). If the noise grows just as fast as shot noise, the plot will appear linear, but you will underestimate the number of PEs per detector output unit. Your only clue that you have this type of noise would be if your estimated PE/DU value differed significantly from what the manufacturer specified.

1.6.2. Assessing Lamp Noise

Repeating the above procedure using your normal excitation lamp as the light source will allow you to determine the noise characteristics of this component. A convenient way to do this is to illuminate a dilute solution of a fluorescent dye with the instrument and sample the fluorescence intensity. The dye's concentration should be adjusted to produce a reasonably strong signal when the detector sensitivity is set to the same level used above. It is important to ensure that the mean intensity of the signal remains constant over the sampling interval. If the dye photobleaches, one may have to reject the initial samples until the photobleaching process and diffusion of fresh dye into the illumination region reach a steady state. The signal should be varied by serially diluting the dye or by inserting neutral density filters into the excitation light path. The same assessment procedure should be employed on the data and the results plotted on the same graph (variance vs. signal). A stable lamp should produce data that fall on the same line as the data produced previously. If the lamp is noisy, the variance will be higher at any given signal and the plot will curve upward. A hypothetical example of the effect of lamp noise is shown in Figure 1.3 (square symbols). This model assumes a lamp noise of 1% on top of the data modeled with the RO noise. Note the pronounced upward curve of the plot. I should also mention that with the xenon excitation lamp used on my system (75 W Osram xenon bulb powered by a model 1600 low noise power supply from Opti Quip, Highland Mills, New York) the lamp noise is undetectable at sampling rates of 1 kHz or less when the signal is averaging several thousand photons.

1.7. FLUORESCENCE: OVERVIEW AND LIMITATIONS

In many cases, the major contributor to the overall uncertainty will be the signal shot noise. One general strategy to minimize shot noise uncertainty is to increase the fluorescence intensity so that the signal consists of more PEs. However, efforts to do this are subject to several constraints due to the nature of fluorescence and of fluorescent dyes.

The absorption of a photon (and its energy) by a fluorescent molecule causes it to make a transition from a ground state to a higher-energy excited state (singlet state). For a good fluorescent dye, one of three things will generally happen at this point: (1) the molecule will return to the ground state by emitting a photon (i.e., fluorescence); (2) it will return to the ground state by nonradiative processes; or (3) the molecule can undergo a chemical reaction, resulting in the effective loss of the fluorescent molecule (i.e., it will photodestruct or photobleach). If and when the molecule returns to the ground state, it can absorb another photon and repeat the sequence. The ratio of the number of fluorescence photons emitted (at all wavelengths) to the number of photons absorbed is referred to as the *quantum yield of fluorescence* (QY_f), and the ratio of the number of molecules that photodestruct to the number of absorption events is called the *quantum yield of photodestruction* (QY_{pd}).

The average energy of the fluorescent photons emitted from a dye is less than the average energy of the absorbed photons. The peak absorption λ is shorter than the peak emission λ, a difference that is referred to as the Stokes shift. This energy difference is due to the fact that both energy level transitions (absorption up and fluorescence down) generally are to higher energy thermal (or vibrational) levels. This excess thermal energy is rapidly lost (within picoseconds). In contrast, the time the molecule spends in the excited state is generally on the order of a couple of nanoseconds. The average time a molecule spends in the excited state is referred to as the fluorescence lifetime, symbolized by τ.

The ability of a molecule to absorb light of a particular wavelength λ is characterized by its molar extinction coefficient (ε, in units of $L\,mol^{-1}\,cm^{-1}$) at that wavelength. The absorbance of light by a solution of dye is described by Beer's law:

$$\log(I_i/I_o) = \varepsilon c d \tag{15}$$

where I_i and I_o are the input and output intensities, respectively; c is the concentration of dye (in moles/liter); and d is the path length (in centimeters) of the light beam through the sample. Beer's law in the above form is not very practical for assessing fluorescence from a sample on a microscope, since one generally is not quantifying the input and output excitation beams' intensity. Alternative equations for specifying the absorbance rate are sometimes more useful. If the excitation light intensity (I_{ex}) is known in units of photons $cm^{-2}\,s^{-1}$, the absorbance rate of one molecule is given by the expression (Mathies and Stryer, 1986)

$$AR(photon/s) = I_{ex}\varepsilon\rho \tag{16}$$

where $\rho = (3.8 \times 10^{-21}\,mol\,L^{-1}\,cm^3)$, a constant that, when multiplied by ε, yields the absorption cross section of a single molecule.

The total number of photons emitted by a single molecule in time interval t will be equal to the absorbance rate times the QY_f times t. Thus, the fluorescence emission of a large sample of molecules is

$$FI_{sample} = AR \cdot QY_f \cdot t \cdot V \cdot [dye] \cdot N_a \qquad (17)$$

where V is the volume of the sample; $[dye]$ is the dye concentration; and N_a is Avogadro's number. Expanding Eq. (17) with Eq. (16) yields

$$FI_{sample} = (I_{ex}\varepsilon\rho) \cdot QY_f \cdot t \cdot V \cdot [dye] \cdot N_a \qquad (18)$$

1.7.1. Nonlinearities in Fluorescence

Note that the fluorescence intensity (FI) calculated in Eq. (18) is assumed to be linearly proportional to the illumination intensity (I_{ex}) and the dye concentration. In many cases, this approximation is acceptable. However, it is important to note that two assumptions are made in arriving at this approximation. The first is that the illumination intensity is sufficiently low that any given molecule has a very low probability of absorbing a photon on the timescale of the fluorescent lifetime. The second is that the product ($\varepsilon \cdot [dye] \cdot$ path length) is sufficiently low that the excitation light is not significantly attenuated as it passes through the sample. If either of these assumptions is invalid, then the emitted FI will not be linearly proportional to the excitation intensity. In some applications, this can be a problem.

1.7.1.1. Ground State Depopulation

Equation (18) states that the rate of absorption (photons/s) of the dye sample is proportional to the number of dye molecules, N_t (where N_t equals $V \cdot [dye] \cdot N_a$). In actuality, it is proportional to the number of dye molecules in the ground state (N_0). If the absorption rate of a single molecule is low relative to the fluorescent lifetime, then $N_0 \cong N_t$. However, as the excitation intensity is increased, any given molecule will spend an increasingly larger fraction of time in the excited state, and $N_0 < N_t$. This phenomenon is known as *ground state depopulation*. The FI will increase linearly with I_{ex} at low intensities but will approach a constant value at higher intensities. The limit will be reached when I_{ex} is so intense that any given molecule absorbs a photon immediately after returning to the ground state. As a consequence, the maximum number of photons that can be emitted by a single dye molecule per second is equal to the QY_f divided by the fluorescent lifetime:

$$FI_{max} = QY_f / \tau \qquad (19)$$

Ground state depopulation limits our ability to increase FI by boosting I_{ex}. This type of problem is generally more likely to occur with excitation schemes that use periodic high-intensity illumination, such as flash lamps, pulsed lasers, or rapid scanning lasers (common with confocal microscopes), since although the mean excitation intensity may be low, the peak can be quite high. The impact of this problem is compounded if the quantization scheme relies on multiple measurements at different excitation wavelengths when the dye has significantly different ε values (since the magnitude of the effect will be different at the different wavelengths) or with dyes that indicate changes in some property of the specimen by a change in ε. The calcium-sensitive dye fura-2 and the pH-sensitive dye BCECF [2',7'-bis-(2- carboxyethyl)-5(6)-carboxyfluorescein] are representative of dyes that fall into both categories. The magnitude of a ground state depopulation effect can be estimated by varying the excitation intensity with precision neutral density filters and assessing the linearity of the dependence of the fluorescence intensity on the excitation intensity.

1.7.1.2.   Inner Filtering

A second phenomenon that can produce nonlinear dependence of FI on I_{ex} is inner filtering, the attenuation of either the excitation or the emission light by the dye or other absorbing molecules in the specimen. Equation (18) is based on the assumption that the illumination intensity is constant throughout the specimen, so that every molecule is illuminated at the same intensity. However, if the product (εcd) is large, significant attenuation of the excitation light may occur as the light passes through the sample, and thus dye molecules on the back side of the specimen will absorb light at a lower rate relative to dye molecules on the front side of the specimen. A qualitatively similar effect can occur if either the dye or other molecules within the specimen absorb emission light, since the attenuation is greater for light originating deeper within the specimen.

The inner filter effect limits our ability to compensate for a low FI by increasing the dye concentration. The magnitude of the problem again depends on the type of investigation and the dye/measurement scheme employed. Like ground state depopulation problems, inner filter problems are potentially more significant if the measurement strategy employs dyes that indicate specimen characteristics via an altered ε or dyes that are excited at multiple wavelengths. Thick multicellular preparations are more prone to this type of problem since the path lengths can be large. Inner filter effects are not easy to measure, but the potential magnitude of the effect resulting from the presence of the dye can be assessed from estimates of the dye concentration and the specimen thickness.

1.7.2. Limits to Total Fluorescence Detection

1.7.2.1. Photodestruction

Photodestruction, or photobleaching, generally refers to a variety of processes that result in the gradual loss fluorescence at a rate that is dependent on the degree of light exposure. The literature on this process is sparse and for the most part anecdotal, probably because the specific types of reactions that can take place are as varied as the molecule structures of the different dyes. The rate of photobleaching correlates with the local concentration of oxygen, and thus it is thought that the interaction of an excited dye molecule with oxygen and/or oxygen radicals is a common route for this process. Further, all other conditions being equal, UV light seems more effective in producing photodestruction, perhaps because UV light is more efficient in coproducing oxygen radicals within the specimen (see Rost, 1991, for a more complete summary of this process).

Photodestruction ultimately limits the total number of fluorescent photons that can be emitted by a sample of dye. Although QY_{pd} is generally quite small for good fluorescent dyes, it is finite. For example, for fluorescein it was estimated to be 2.7×10^{-5} (Mathies and Stryer, 1986), for fura-2 bound to calcium at 340 nm excitation it has been estimated to be about 6.5×10^{-4} (Becker and Fay, 1987). Thus, with continual illumination, there will be continual "consumption" of the dye. The QY_f/QY_{pd} ratio gives an estimate of the number of fluorescent photons an average molecule will emit before it is destroyed. For fluorescein, with a QY_f of 0.99, the estimate is 37,000 photons. For fura-2, with a QY_f in the presence of calcium of 0.49, it is 760 photons. Although strategies to lower the QY_{pd} (such as lowering the oxygen concentration or adding free radical scavengers to the specimens) can help, in many applications they are inappropriate, and in any event there will still be a finite limit.

1.7.2.2. Collection and Detection Efficiency

Although several hundred to several thousand fluorescent photons can be emitted from each dye molecule in the specimen, inefficiencies in the collection and detection of the light will result in a much lower number detected per molecule. Even the highest NA (numerical aperture) objective will collect at best 25% of the light emitted from an object in its field of view. Losses will also occur because of inefficiencies in the dichroic mirrors and emission filters used to select the emission wavelength band. Further, additional losses will occur in the optical elements of the microscope as the image is transmitted to the detector. Finally, detectors have quantum efficiencies less than 1.0, and thus

	PMT	CCD
Objective collection efficiency:	0.25	0.25
Dichroic filter transmission efficiency:	0.85	0.85
Emission filter transmission efficiency:	0.50	0.50
Microscopy transmission efficiency:	0.80 (?)	0.80 (?)
Detector quantum efficiency:	0.10	0.50
Photons leaving specimen:	1000	1000
Photons reaching detector:	85	85
Photons detected:	9	43

Figure 1.4. A comparison of system collection efficiencies for two typical microscope-based fluorescence detection systems that differ with respect to the quantum efficiency of their detectors (a PMT vs. a CCD camera). The objective collection is the fraction of light leaving the specimen that can enter the objective. The dichroic and emission filter differences are the fractions of the emission spectrum that can pass through these elements. Microscopy efficiency includes all other optical elements, including the transmission efficiency of the objective.

will not detect every photon that strikes them. Thus, the detector output FI will be significantly lower than the sample FI due to these losses. Expressed in the form of previous equations

$$FI_{det} = FI_{sample} \cdot CE_{obj} \cdot TE_{filters} \cdot TE_{mic} \cdot QE_{det} \qquad (20)$$

where CE_{obj} is the collection efficiency of the objective, $TE_{filters}$ is the transmission efficiency of the dichroic and emission filters, TE_{mic} is the transmission efficiency of the microscope (including the objective), and QE_{det} is the quantum efficiency of the detector.

Figure 1.4 illustrates the estimation of the efficiency of two systems having different detector QEs (one representative of a PMT, one of a CCD camera) but identical efficiencies for the other components. My intent is to illustrate approximate overall efficiencies, and so values for all of the components were chosen to be typical of commercially available good-quality products but not necessarily the best available. Note that typically less than 10% of the photons emitted by a sample will even reach the detector, and fewer still will be detected. The value chosen for the emission filter efficiency may strike some readers as too low. This value was chosen by assuming the filter had a mean transmission efficiency of about 0.80 over its bandpass and that its bandpass covered the spectral region that includes about two thirds of the emitted light. The QY_f includes all photons emitted at all wavelengths, and given that most emission spectra tail off gradually at longer wavelengths, the value chosen is actually quite realistic. Two practical considerations generally limit one from improving the signal quality by increasing the bandpass coverage of the emission spectrum beyond this level. First, at shorter wavelengths, the emission filter will generally start to encroach on the transition region of the dichroic mirror, risking the collection of scattered excitation light. Second, extending the bandpass at longer wavelengths leads to diminishing returns, as more background and autofluorescence light will reach the detector.

The most uncertain value in Figure 1.4 is the microscope's transmission efficiency, which I consider to encompass all optical elements including the objective. It is curious that the manufacturers of just about every component of the system are willing to provide quantitative information about the performance of their products except the manufacturers of microscopes and objectives. Manufacturers of optical filters will provide quantitative transmission spectra. Detector manufacturers will provide quantitative information on the noise characteristics and the quantum efficiencies of their cameras and PMTs. Manufacturers of lamps and lasers will provide quantitative information on the spectral and power outputs of their products. This information makes it easy to assess components and at least rule out designs that are inferior or inadequate for the application. In contrast, the leading microscope manufacturers seem to treat quantitative information about their products as if it were a trade secret. Although it is clear from the improvements in their most recent models that they are concerned about efficiencies, they still convey this information using qualitative, subjective criteria. Despite touting the "improved efficiency" of newer designs over the last few years (particularly relative to competitors' models), no salesperson I ever talked to could provide an estimate of what it actually was. Instead one hears a variety of ridiculous anecdotal claims (my favorite was a microscope salesperson who claimed that with the new design he only had to turn the overhead lamp power knob to position 4 to view a slide, whereas with the older model he had to dial in 6!). Information about the transmission efficiency of objectives is similarly difficult (but not impossible) to extract. This state of affairs will change only if customers start putting pressure on these companies, and thus I urge you to ask for this information when considering your next purchase.

1.8.   REDUCING UNCERTAINTY

Optimizing the measurement strategy and the hardware setup to minimize uncertainty involves a balance of several considerations, many of which are application specific and thus can only be treated generally. A rational approach to this task requires a reasonably accurate assessment of the sources of uncertainty and their relative contributions to the overall uncertainty. You need to know the noise characteristics of the measurement system, have a reasonable idea of the magnitude of the specimen signal that will be measured (in PE units), and know the typical values of the magnitude and uncertainty of the offsets and scaling factors used to process the data. You also need to have a reasonable idea of the level of uncertainty that can be tolerated in a particular application. Although you can and should target all sources of uncertainty, the biggest dividends will come from improving the worse offenders.

System improvements can have collateral benefits even if uncertainty is not intolerable. The same EUR could be obtained at a higher spatial or temporal resolution. Alternatively, one could decrease the excitation intensity to get the same EUR with a reduced rate of photobleaching (so that the specimen can be studied longer) and a reduction in nonspecific irradiation side effects (heating, photodamage, etc.).

1.8.1. Improving Fluorescence Intensity

In low-light applications (such as fluorescence), the signal shot noise will often be a major limiting factor to the EUR. In such cases, maximizing the number of PEs that compose the raw detector signal should be an important goal. Although there are many ways to do this, the trick is to accomplish it in a manner that minimizes the addition of other uncertainties. The following relation summarizes the potentially modifiable factors that influence the detector signal emitted from a particular region of the specimen during a sampling interval t (ignoring the specific factors that determine I_{ex}):

$$\text{FI}_{det} \propto (\varepsilon \cdot \text{QY}_f) \cdot [\text{dye}] \cdot I_{ex} \cdot \text{CE}_{obj} \cdot \text{TE}_{filters} \cdot \text{TE}_{mic} \cdot \text{QE}_{det} \cdot t \qquad (21)$$

Obviously, some of these factors are more easily modified than others. The first two are characteristics of the dye, the choice of which may be dictated by the application. Except for the really adventurous, TE_{mic} is probably fixed (moreover, many of the obviously inefficient aspects of microscope designs have been improved in the more recent models from the larger manufacturers). Generally QE_{det} is determined by the imaging strategy and the available budget, although one would hope that QE and noise characteristics were considered at the time of purchase. This leaves [dye], I_{ex}, CE_{obj}, and $\text{TE}_{filters}$ as the most likely targets for improving FI. In terms of the trade-offs, the most beneficial improvement can come from improving CE_{obj} and the average transmission of the dichroic and emission filters, factors that (along with TE_{mic}) determine what fraction of the emitted light reaches the detector. Improvements at these points are essentially penalty free; there will be no adverse effects on the relative autofluorescence and background signals or on the rate of photobleaching, nor will it exacerbate inner filter or ground state depopulation problems.

Increasing the dye concentration, the excitation light intensity, or the bandpass range of the filters can yield improvements, but such adjustments

should be made only with full awareness of the potentially negative consequences. Increasing I_{ex} will result in a faster rate of photobleaching and hasten
ground state depopulation. The benefits of increasing the dye concentration
must be balanced by the potential for creating an inner filter effect and for
exacerbating the deleterious effects of the dye on the specimen. Altering the
bandpass region of the emission filter to collect more emitted photons has to
be balanced by the increased collection of background and autofluoresecence
photons. These risks can be evaluated only on an individual basis, and clearly
require familiarity with the performance characteristics of the hardware and
a good assessment of the potential for encountering the fluorescence limitations described above, as well as typical magnitude and uncertainty in the
background, autofluorescence, and other factors associated with a particular
application. For example, taking in more autofluorescence photons by widen-
ing the emission filter bandpass might be worthwhile if the contribution to the
overall uncertainty due to this parameter was low relative to the signal shot
noise, but not if the autofluorescence uncertainty was large.

If photodestruction is a problem, a strategy that might be beneficial, when
one is using a dye having a short Stokes shift, is to illuminate at a shorter,
suboptimal excitation wavelength to allow the dichroic and emission filters to
capture more of the emission spectrum. Typical filter combinations provided
by microscope manufacturers generally are designed to have the dichroic
transition centered between the absorption and emission peaks, and the
excitation and emission filters spaced a safe distance away. This may optimize
the fluorescence intensity for a given lamp output, but not the photodestruction cost per detected photon.

Finally, if despite all efforts the signal intensity is still unacceptably low,
one will have to consider sacrificing temporal or spatial resolution. I have
deliberately avoided the issue of spatial and temporal resolution and the
uncertainty in specifying them; the considerations are wholly different and
much more application specific. However, several points are worth noting
about the SNR implications of achieving various spatial and temporal resolutions. By means of a variety of averaging techniques, both the temporal and
spatial resolutions can be lowered after acquisition to reduce noise. The
advantage of oversampling in both domains is that one retains the option of
sacrificing them to improve the SNR *after* the data are acquired. However,
combining multiple measurements to reduce resolution also combines the
noise of each measurement in the final result. The highest SNR is achieved by
adjusting the magnification and the detector integration time so that the data
are acquired at the desired resolution, as less measurement system noise will be
accumulated. This is best illustrated in equation form. Consider the effect of
a four-pixel box-average strategy to reduce the spatial resolution after acquisition (assume that all the pixels have approximately the same signal). Initially,

the signal and noise will be

$$\text{initial signal} = S \tag{22}$$

$$\text{initial noise} = [S + N_{ms}^2]^{0.5} \tag{23}$$

After the four neighboring pixel box average, the signal will be $4S$ and the noise will be

$$\text{"boxed" noise} = [4S + 4N_{ms}^2]^{0.5}(= 2 \times \text{initial noise}) \tag{24}$$

resulting in twofold improvement in SNR. If instead the image was demagnified so that the information from the region of the specimen that was originally focused on the four pixels was now focused on only one pixel, the signal would still be $4S$ but noise would be

$$\text{"demagnified" noise} = [4S + 1N_{ms}^2]^{0.5}(< \text{"boxed" noise}) \tag{25}$$

The measurement system noise is now factored in only once. Thus, the overall noise is lower, and the SNR will be more than twice the initial SNR.

An analogous situation applies to reducing temporal resolution by varying the detector integration time, although in this case the components of the measurement system noise (dark signal and RO noise) need to be considered separately. Initially, with the signal $= S$, the noise would be

$$\text{initial noise} = [S + N_{dark}^2 + N_{RO}^2]^{0.5} \tag{26}$$

The resulting of averaging four successive samples would be identical to the four-box average procedure above; the noise would be

$$\text{"averaged" noise} = [4S + 4N_{dark}^2 + 4N_{RO}^2]^{0.5}(= 2 \times \text{initial noise}) \tag{27}$$

and the signal would be four times larger, resulting in a twofold improvement in the SNR. If the time resolution was instead decreased by increasing the detector integration time by a factor of 4, the situation would be

$$\text{"4} \times \text{integration time" noise} = [4S + 4N_{dark}^2 + N_{RO}^2]^{0.5}(< \text{"averaged" noise}) \tag{28}$$

Because the total integration time is the same for both strategies, the dark signal noise contribution would remain the same. However, the RO noise will be factored in only once, so the net noise will be less than twice as great, and thus the SNR will be more than twofold larger.

In summary, one should avoid acquiring the data with a spatial or temporal resolution greater than needed. If the decision is made to sacrifice resolution, postacquisition processing can yield significant improvements in the SNR, although the maximum benefit is achieved by adjusting the resolution prior to acquision. The extra SNR improvement made by using this strategy depends on the magnitude of the nonduplicated noise component relative to all other sources. In general, improvement will be greatest when the signal is weak.

1.8.2. Reducing Offsets and Scaling Factor Uncertainties

The uncertainty in the magnitude of offset and scaling factors includes a noise component (since these factors are also estimated with the measurement-hardware) and a circumstance component. In addition, offsets such as the background and autofluorescence increase the signal shot noise because they are part of the raw detector signal. Thus, one should strive to minimize these offsets as much as possible prior to detection. Background signals should be minimized by preventing stray light from entering the microscope and by using field stop diaphragms to restrict illumination to the portion of the field of view containing the specimen. Certainly the room should be darkened, and all objects emitting light at the emission wavelengths should be attenuated as much as possible (red acetate sheets are reasonably effective, if using blue-green dyes, for video monitors and LEDs on instrument panels). Further, with nonimaging detectors, a mask or field stop diaphragm should be used to limit the view of the detector to the illuminated region. Minimizing autofluorescence (relative to the dye signal) generally requires improving the wavelength selectivity of the excitation and/or emission filters for the dye, a strategy that, as discussed above, may conflict with a desire to boost the dye-specific signal. The shot noise component of offsets and scaling factors should be minimized by estimating these parameters with large numbers of PEs. That is, use long integration times and/or average many successive estimates. The circumstantial component to their uncertainty should be minimized by reproducing as carefully as possible the conditions that existed when the specimen was measured.

1.8.3. Other Improvements

If the noise characteristics of the measurement system are significant relative to the signals being measured, then improvements here may be called for. This is particularly true if the assessment procedure described earlier indicates noise levels significantly greater than the rated specifications of the instrument. Replacing a lamp or detector is a more expensive step than the other strategies considered so far, and thus should not be done unless there is good evidence that it will help. Most manufactures will allow you to evaluate an instrument for a week or two. If you do not yet have a system to evaluate new components, it is worthwhile to seek out one to do so. The noise characteristics of the detector should be considered with a reasonable idea of how the instrument will be used, as different noise components can vary in their importance, depending on the measurement strategy.

SUGGESTED READINGS

Many of the topics covered here have been given fuller treatment in other sources. For additional information on the general topic of fluorescence, I would suggest Lakowicz (1983). The best survey of available fluorescent dyes I have come across is the Molecular Probes, Inc. catalog (Haugland, 1992). Inoue (1986) is an excellent reference for microscopy in general and video techniques in particular, and also discusses some of the features of video cameras. Aikens et al. (1989) describe the general characteristics and imaging strategy of CCD cameras. Two books by Rost (1991, 1992) cover a number of general topics on microscopes and quantitative fluorescence. Moore et al. (1990) cover a number of specific issues regarding uncertainties in fluorescence measurements, particularly with respect to imaging calcium; further, the issue of *Cell Calcium* in which their article appears is devoted to papers on calcium imaging techniques.

REFERENCES

Aikens, R. S., Agard, D. A., and Sedat, J. W. (1989). Solid state imagers for microscopy. *Methods Cell Biol.* **29**.

Becker, P. L., and Fay, F. S. (1987). Photobleaching of fura-2 and its effect on the determination of calcium concentrations. *Am. J. Physiol.* **253**, C613–C618.

Burgess, R. E. (1955). Electronic fluctuations in semiconductors. *Br. J. Appl. Phys.* **6**, 185–190. Reprinted in *Electrical Noise; Fundamentals & Sources* (M. S. Gupta, Ed.), pp. 59–64. IEEE Press, New York, 1977.

Grynkiewicz, G., Poenie, M., and Tsien, R. Y. (1985). A new generation of Ca^{2+} indicators with greatly improved fluorescence properties. *J. Biol. Chem.* **260**, 3440–3450.

Haugland, R. P. (1992). *Molecular Probes Handbook of Fluorescent Probes and Research Chemicals*. Molecular Probes, Eugene, OR.

Inoue, S. (1986). *Video Microscopy*. Plenum, New York.

Lakowicz, J. R. (1983). *Principles of Fluorescence Spectroscopy*. Plenum, New York.

Mathies, R. A., and Stryer, L. (1986). Single-molecular fluorescence detection: A feasibility study using phycoerythrin. In *Application of Fluorescence in the Biomedical Sciences* (D. L. Taylor, A. S. Waggoner, R. F. Murphy, F. Lanni, and R. R. Birge, Eds.), pp. 129–140. Alan R. Liss, New York.

Minta, A., Kao, J. P., and Tsien, R. Y. (1989). Fluorescent indicators for cytosolic calcium based on rhodamine and fluorescein chromophores. *J. Biol. Chem.* **264**, 8171–8178.

Moore, E. D. W., Becker, P. L., Fogarty, K. E., Williams, D. A., and Fay, F. S. (1990). Ca^{2+} imaging in single living cells: Theoretical and practical issues. *Cell Calcium* **11**, 157–179.

Papoulis, A. (1965). *Probability, Random Variables, and Stochastic Processes*. McGraw-Hill, New York.

Pierce, J. R. (1956). Physical sources of noise, *Proc. IRE* **44**, 601–608. 1956. Reprinted in *Electrical Noise: Fundamentals & Sources* (M. S. Gupta, Ed.), pp. 51–58. IEEE Press, New York, 1977.

Rost, F. W. D. (1991). *Quantitative Fluorescence Microscopy*. Cambridge University Press, New York.

Rost, F. W. D. (1992). *Fluorescence Microscopy*, Vol. 1. Cambridge University Press, New York.

CHAPTER

2

AUTOMATED IMAGE MICROSCOPY

PAWEL WODNICKI,[1] STEPHEN LOCKETT,[2] M. RHADHAKRISHNA
PILLAI,[3] SUSAN HALABI,[4] ANN McKALIP,[1] and BRIAN HERMAN[1]

[1]*Laboratories for Cell Biology
Department of Cell Biology and Anatomy
University of North Carolina at Chapel Hill
Chapel Hill, North Carolina 27599*

[2]*Lawrence Berkeley Laboratory
University of California
Berkeley, California 94720*

[3]*Regional Cancer Center
Kerala State, India*

[4]*Tulane University, School of Public Health
New Orleans, Lousiana*

2.1. INTRODUCTION

For the past three centuries the microscope has been in use as a primary tool of investigation in biological science. The continuous search for improvements in microscopy has led to development of many types of microscopes, with the guiding developmental principle being that any new microscope should provide a visual source of information about objects that are otherwise too small to see with the unaided eye and can be recorded and analyzed by the researcher. It is impossible to overlook the importance of such visual information, as our own sense of vision provides us with most of our information about the outside world. Thus, it is easy to understand why substantial efforts has been invested to make this great tool even more useful.

Following the development of the microscope, it soon became clear that the ability to preserve the image generated by the microscope was essential for

Fluorescence Imaging Spectroscopy and Microscopy, edited by Xue Feng Wang and Brian Herman. Chemical Analysis Series, Vol. 137.
ISBN 0-471-01527-X © 1996 John Wiley & Sons, Inc.

widespread dissemination of scientific findings and for analysis that cannot be achieved by visual examination alone. This goal was achieved using photographic techniques and subsequently, through the development of electronic devices capable of converting optical signals into electronic form (i.e., and electronic camera) that could be stored for later use (i.e., on a video recorder). These improvements constituted the first video microscope (1).

The increasing complexity of biological experiments performed with optical microscopy and the complexity of data that can be obtained from specimens have resulted in the generation of even more sophisticated optical microscopy systems. The coupling of computers to video microscopy systems has resulted in the development of devices that can perform complicated experiments on living cells or tissue using multiple indicators of physiological activity, providing more data, convenient storage of and access to the data, and most important substantial aid in the analysis of large volumes of data. The next logical advance in the development of such systems will be to automate as much of the data acquisition and analysis as possible. In this chapter we present an overview of what constitutes an imaging system, with special emphasis on automation of image acquisition, image analysis, and imaging techniques used in the context of biological applications.

2.2. AUTOMATED IMAGE ACQUISITION SYSTEM

The purpose of an automated image acquisition system is to perform the desired experiment and to gather and store data for later retrieval or analysis. Depending on the requirements of a particular experiment, imaging systems with various levels of complexity and automation can be used. Two basic types of imaging systems currently exist: one that includes components designed to perform specific functions and one that includes universal components designed to perform many different kinds of experiments. Automated imaging systems can be further divided into two major classes, depending on the functions they perform. In the first class, the microscope plays the central role of creating an image of the specimen that is later converted to an analog voltage signal using an image sensor and is then digitized and stored in a computer. The microscope and associated optics and detectors therefore determine the quality of the data obtained, particular care must be taken to match all components in order to get the best possible results. In the second class, the automated microscope builds on the features of the first type but, in addition, provides control over the experiment via automation, for example by automatically selecting the excitation light intensity and the wavelength of excitation, positioning the specimen in the field of view, and autofocusing. This category of microscope can also contain computer software and hardware that

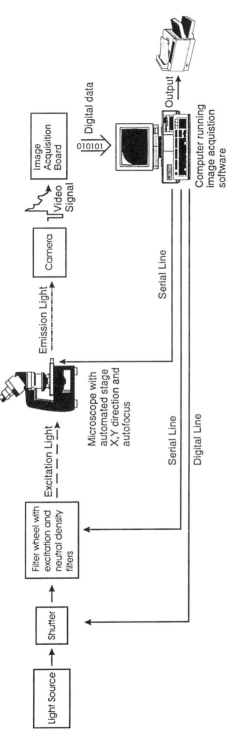

Figure 2.1. Typical automated imaging system.

can automatically identify objects in the images, classify them, and extract relevant information about each object (e.g., size, presence or absence of a diagnostic marker). Figure 2.1 illustrates a typical automated imaging system and its relevant components.

Physically, imaging systems consist of a host computer and separate hardware components that communicate with each other using some kind of digital bus, which is a collection of digital signal lines carrying electronic signals according to a defined protocol. Components that reside in the host computer itself (i.e., imaging boards) use the computer's internal bus, as this configuration provides the highest possible speed of communication, which is primarily required for the imaging boards. There are many types of digital buses. Some, such as ISA, NuBus, and SBus, are computer dependent, others, like VME and PCI, are universal buses that enable use of the same hardware components in different environments. PCI is especially well suited for imaging applications due to its high speed of operation and support from many different computer manufactures. Other components that do not need such a high speed of communication can use other types of connections, such as serial lines (RS-232, RS-485), simple digital input/output (I/O) lines, and the GPIB bus, which is often built into test and measurement equipment.

2.2.1. Excitation Light

Excitation light of some form is required to form an image of the specimen. In the case of epifluorescence microscopy, specific wavelengths of excitation are usually required. A convenient approach to providing the required wavelength(s) of excitation light is to use a mercury, xenon, or metal halide arc lamp coupled to a filter wheel. These lamps output light over a wide spectrum of wavelengths, and the specific wavelengths(s) required can be selected through bandpass excitation filters mounted on a filter wheel. The main advantages of this approach are the ability to use a multispectral lamp as a light source and the ability to control filter wheel position (and hence wavelength of excitation) remotely by a computer. By using a second filter wheel populated with neutral density filters, one can control the intensity of the excitation light in addition to the wavelength of excitation, adding further flexibility to the system. The combination of these two filter wheels thus allows the selection of specific wavelengths over the entire visible spectrum at intensity levels that do not photobleach the fluorescent probe or result in cell damage. Lasers can also be used as light sources, although available wavelengths are quite limited in low-cost units, and the cost of tunable lasers is prohibitively high for widespread use. Another way to provide the desired excitation wavelengths is to use monochromators which provide fast switching times, on the order of 2 ms, comparable full-width half maximum (FWHM) bandwidths as interference filters, but unfortunately less excitation intensity. Acousto-optic (AO) devices are becoming more available and allow microsecond variations in excitation wavelengths and intensity. Unfortunately, current AO devices require high-intensity light sources to provide sufficient light throughput and can suffer from image shifting. The liquid crystal tunable filter (LCTF) does not suffer from the image shifting of the AO devices, but care must be taken to prevent the LCTF from becoming too hot during exposure to excitation light.

It is very important, especially when working with living cells or tissue, to control the time of exposure of the specimen to the excitation light. This will minimize photodamage to the specimen and photobleaching of the fluorescent reporter molecules, as well as extending the lifetime of the interference and/or neutral density filters in the light path. This can be accomplished as demonstrated in Figure 2.1, through the use of mechanical shutters. Typical shutters come with their own computer interfaceable controllers that can be used to close and open shutters manually, as well as remotely via computer; more complicated controllers also provide control over the time the shutter remains open. The minimum opening speed of these types of shutters is on the order of 10 ms, which is usually adequate for many types of experiments. If shorter pulses are required, one can employ flash lamps or pulsed laser systems.

2.2.2. Specimen Positioning

The position of the specimen relative to the microscopic field of view can be precisely controlled ($\pm 1\mu$m) by using a stage equipped with a stepper motor and interfaced to the host computer. This is another important feature of an automated system because it can greatly enhance the speed of data collection and thus the amount of data later available for analysis. It can also facilitate experiments in which serial observations of cell structure and/or function require removal and subsequent return of cells to the microscope stage. For example, the correlation of an early event (say, changes in the level of free calcium following growth factor stimulation) with later events (say, entry of cells into DNA synthesis) requires incubation of living cells between the measurement of calcium and the entry of cells into DNA synthesis (24–36 h). Practically, this requires removing from the stage the slide or chamber containing the cells and placing it in a CO_2 incubator for 24–36 h before returning it to the microscope stage. Specimen positioning provides the ability to return to specific coordinates and thus the same region of the slide (cell) that was observed initially prior to removal of the slide. Another example is when one wishes to screen a large collection of cells for a specific marker and then correlate the presence or absence of the marker with cellular cytomorphology. It is very helpful to be able to automatically return to precise positions (cells) previously observed. Motorized stages provide large ranges of positioning in both horizontal X and Y directions. Having control over stage position in the vertical Z direction provides the ability to perform automatic focusing (essential for automated experiments) or to collect sets of images of a specimen at different planes, which can then be used to create a three-dimensional representation of the specimen through volume rendering. However, the accuracy and repeatability of repositioning can be less than satisfactory in some instances due to the design of stepping motors and errors inherent in the mechanical nature of stage movement. In addition, while use of an automated stage for specimen positioning can be quite convenient, care must also be taken to ensure that a simple method is available for manual control of the stage without losing information on current position; restoring this information would require realignment procedures.

2.2.3. Imaging Detectors

The most critical part of any imaging system and one that ultimately determines the sensitivity and quality of the signal, is the image sensor. Many types of image sensors exist, and the choice of which one to employ will be dictated by the requirements of a given experiment. In many cases, light emitted from fluorescently labeled biological specimens has very low intensity. In order to detect this emitted light, it has to be amplified and/or collected for a long time by the detector. Image intensifiers amplify light before it reaches the camera faceplate. However, amplifying weak signals can introduce nonstationary noise; thus, some means of reducing noise must be used. One way to enhance the signal-to-noise ratio is to average several video frames into one image, but at a cost of a reduction in temporal resolution. More sensitive detectors, such as cooled charge-coupled device (CCD) chips, allow integration of the signal on the CCD faceplate for periods of up to several minutes.

The most frequently utilized image detectors are based on a semiconductor (CCD). The sensor area of the CCD is an array of detectors placed on a rectangular grid. The size of the individual detector can vary (6–23 μm × 6–23 μm). Operation of the CCD is based on the actions of photons hitting the detector surface and generating an electrical charge that remains trapped in the area of the pixel. Charges accumulated in the pixels over a period of time are proportional to the light intensity and later are converted sequentially, row by row, into a voltage signal, which is then sent out of the CCD camera as a video signal. CCD cameras have many advantages over other types of image detectors, as they offer larger dynamic range, better quantum efficiency and low noise characteristics, linear response, and negligible geometrical distortion; in addition, they are smaller and more rugged. The majority of CCD cameras are compatible with standard video signals, but the high-resolution, low-light CCDs are not. A growing number of specialty cameras are particularly well suited for scientific applications in automated systems, offering unique features like large chip areas and thus higher resolution up to $4k \times 4k$ pixels, digital output, cooling to lower noise and dark current, flexible timing, and full control by a computer. Detailed discussion of CCD devices can be found in Aikens et al. (2).

2.2.4. Imaging Boards

Once the optical signal has been converted into a voltage signal, it has to be digitized for numerical analysis and permanent storage; imaging boards or frame grabber boards accomplish this task. The type of imaging board used has to be matched to the camera in terms of type of signal, resolution, and precision. In the past, imaging boards also performed a number of image processing functions, However, because computers have become more powerful, general image processing functions can now be performed more conveniently, and in many cases more rapidly using computers, although some functions that have to be done at video rate are still better done by the imaging board (e.g., averaging of video frames). The ability to capture and digitize camera signals at video rate does not mean that the data can be recorded and stored by the computer at that speed, usually due to limitations in hard disk speed and size. If there is a need to record data at video rate, short recording times are possible if the imaging board has enough memory; otherwise, other devices (optical disk, videotape, etc.) are required.

2.2.5. Image Acquisition Software

Software plays a major role in automated image microscopy. Figure 2.2 shows the multilayered structure of image acquisition software.

Low-level functions are performed by modules interacting directly with hardware to control its functions. Because information regarding complex operations is found in this low-level programming, higher level programming can focus on simple, standard ways of software interaction with a piece of hardware. By abstracting functions of similar components into a uniform set of operations, the system gains modularity and hardware independence. The advantage of this approach is the ability to exchange elements of the system that perform identical functions, such as imaging boards, without disturbing the functionality of the system.

The middle layer of software consists of instructions that implement the core functions of the automated imaging system; it is by far the most complicated portion of the system. It responds to user input via communication with upper level software to control experimental conditions such as

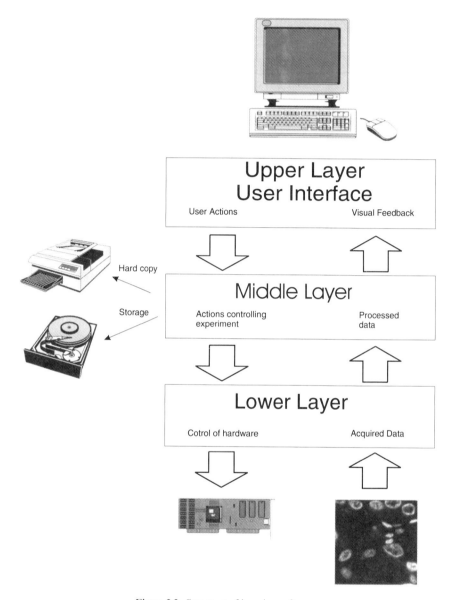

Figure 2.2. Structure of imaging software.

camera integration time and number of frames to average. Once the conditions of an experiment are defined, the middle layer executes a sequence of procedures that constitute the experiment. This layer is also responsible for collecting and storing data from imaging sensors, often together with additional information about the experiment, such as wavelength of excitation, number of frames averaged, the date, and other information pertinent to the experiment.

The upper layer of software is responsible for interacting with the user and communicating input to the lower levels of software, as well as providing feedback from the experiment. Particular implementation will depend on the computer employed, but commonly used elements include buttons for simple actions such as grabbing an image, input fields for numerical values and textual information, menus for choices, windows for displaying acquired images, and windows presenting extracted information about the current experiment.

2.3. AUTOMATED IMAGE ANALYSIS

Imaging systems help to collect data by automating the process of image collection and by controlling routine and sometimes tedious tasks. While this level of automation makes the acquisition of microscopic imaging data more convenient, it does not provide the investigator with any new information beyond what can be seen in the acquired images. However, the power of the computer system can be applied to extract useful information from the images that previously required human interpretation. The development of a system that analyzes the images with the intention of extracting information about the specimen that is not obtainable, or not easily obtainable, by visual inspection would be of great importance.

Recent technological advances in electronics that have brought considerable computing power to the average user are key factors in building systems that are capable of analyzing images in reasonable amount of time. However, as hardware limitations have been overcome, software for image analysis has become a limiting factor. The development of software that will accurately and reliably extract biological information from the images is still largely a goal for the future. Most software developed to date has dealt with images on the lowest level of abstraction, i.e., as arrays of picture elements (pixels) (Figure 2.3), without attempting to extract information. For the most part, such software processes the image in order to make visual analysis easier. Although a lot of useful information can and is obtained using his kind of approach, it is just the beginning. The next step will be to build a system that directly and automatically extracts the information desired by the investigator. An ultimate goal is to make the computer "understand" images in the same way as

164	128	100	225
145	133	175	181
80	111	93	40
0	67	16	21

Figure 2.3. Image and corresponding array of intensities.

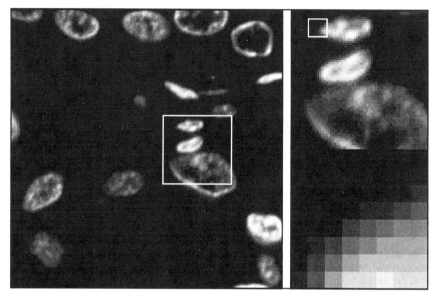

Figure 2.4. Image of a typical biological sample.

humans do. Achieving this goal will require applying new approaches that do not yet exist or are in their infancy, such as artificial intelligence, machine learning, and knowledge of the analytical mechanism of the human brain, which are subjects of intensive research.

The process of understanding an image can be viewed as a procedure of reducing the large amount of data in the image by deriving new descriptions of the data at a higher level of abstraction. The outcome of this process will ultimately lead to a description of the specimen image using terms natural for humans. To begin, we start with a raw image (Figure 2.4) that is nothing more than an organized collection of pixels represented by numbers. At this level, we look at the image by looking at single pixels or small neighborhoods of them. However, this approach does not allow us to tell if there is any structure or objects in the image. Though some information can be obtained on this level, it will be either specific to one pixel or general in nature such as statistical information.

By grouping pixels into larger regions based on chosen features (the simplest being a range of intensities), we achieve a higher level of description. The image is no longer composed of pixels; instead, it has distinct regions with their own properties, and new operations can be performed on them. This process is known as image segmentation. For many applications, knowledge of these regions and their properties is enough to describe the image. However, further information can be extracted from the image by grouping regions into clusters and characterizing them, resulting in a new level of description. It should be clear from this description that two kinds of operations can be performed on an image. On affects only the current representation of the image data, such as filtering; the second creates a new level of description.

Creating automated image analysis systems begins by specifying requirements for the system and then selecting a set of image processing functions that will meet those requirements. At present, this process is not formalized; it is done by a researcher and then coded into a computer program. In addition, these systems are often limited to one specific application. Universal systems capable of automatically learning and finding desired user information do not yet exist.

Because images contain a lot of information and because processing times can be very long, it is important to minimize the number of operations performed during analysis. Efficient implementation of image processing functions is not a trivial task. A lot of research is being done in the area of parallel computing to design optimal computer architecture as well as algorithms.

2.3.1. Introduction to Image Analysis

Low-level image processing is based on operations applied to each pixel separately (point operations), small neighborhoods of pixels (neighborhood operations), or ones that are transforms of the entire image (global operations). Some operations have mathematical roots and are therefore well understood, whereas others were discovered empirically.

2.3.1.1. Operations on Single Pixels (Point Operations)

The simplest operations are ones performed on values of single pixels and include operations with a constant image or two images. These operations can be represented using the relationship

$$R[x, y] = A[x, y] \text{ OPERATION } C$$

or

$$R[x, y] = A[x, y] \text{ OPERATION } B[x, y]$$

where $R[x, y]$, $A[x, y]$, $B[x, y]$ are corresponding pixels in the resultant image R and source images A and B; C is a constant; and OPERATION is a mathematical operator (e.g., addition, subtraction, logical AND, OR, etc.) These kinds of operations are often used as part of more complicated manipulations that enhance image quality in general or emphasize specific features of the image.

Mathematical operations have a variety of uses for image enhancement. Subtraction is probably the most often used operation, employed, for example, to implement background subtraction. Figure 2.5 shows subtraction of a constant from an image, as well as subtraction of two images. Background subtraction helps remove artifacts that are additive in nature; however, background subtraction will not correct for shading. Shading can be due to a variety of causes, such as nonuniform illumination or staining or spatial variation in the sensitivity of the sensor. The best way to correct for shading is to divide the experimental image with the image of a uniform signal taken under the same conditions.

Logical operations are very useful when dealing with multiple regions. By creating an image mask for regions of interest, one can perform operations that affect only those particular regions. To illustrate this operation, suppose that we have defined a mask covering one of the cells, as in Figure 2.6 (see Color Plates). To obtain the average intensity of this cell, we average the

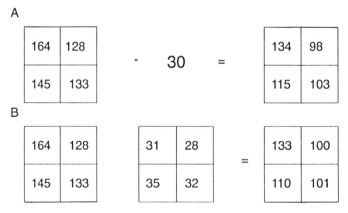

Figure 2.5. (A) Subtraction of a constant from an image; (B) subtraction of two images.

Figure 2.7. Binary image after thresholding.

intensity in the whole image, but by incorporating an AND operation with the region masking the cell, we are selecting only pixels from this region. It is easy to repeat the same operation for a different region by selecting another mask.

Thresholding, or density slicing, is another example of a point operation that is used to create regions within an image by selecting pixels whose intensities are in the range defined as valid. This is probably the simplest way of performing segmentation on an image (Figure 2.7).

2.3.1.2.   Operations Based on Convolution

An important class of operations is based on convolving the image with a small kernel that is a matrix of coefficients (3). These are neighborhood operations. Typical kernels have the same size in *x* as well as in *y* directions and are usually odd numbers. This operation derives new values of the pixel based not only on this original value in the pixel but also on the values of a set of pixels from the area surrounding the original pixel covered by the kernel. Convolution is performed for every pixel in the image by aligning the center of the kernel with the pixel being convolved and then calculating the sum of the products of the pixel values and kernel coefficients. This new value is often normalized by dividing it by the sum of the coefficients in order to preserve brightness. This then becomes the new value of the pixel. There are no restrictions on the size of the convolution kernel, although satisfactory results in most cases can be achieved with kernel sizes of 3×3, 5×5, and 7×7. Using the smallest possible kernel speeds up processing, which is often desirable in systems that are used interactively. Convolution is described by the following relationship:

$$R_{i,j} = \sum_m \sum_n A_{i-m, j-n} K_{m,n}$$

where R is the resultant image; A is the source image; and K is the convolution kernel.

For a 3×3 kernel, convolution can be written as follows:

$$R[x,y] = A[x-1, y-1] \cdot K[1,1] + A[x-1, y] \cdot K[1,2] + A[x-1, y+1] \cdot K[1,3]$$
$$+ A[x, y-1] \cdot K[2,1] + A[x,y] \cdot K[2,2] + A[x, y+1] \cdot K[2,3]$$
$$+ A[x+1, y-1] \cdot K[3,1] + A[x+1, y] \cdot K[3,2]$$
$$+ A[x+1, y+1] \cdot K[3,3]$$

Many useful operations are based on this principle, most notably digital filters. Low-pass filters are used to reduce random noise and to smooth images.

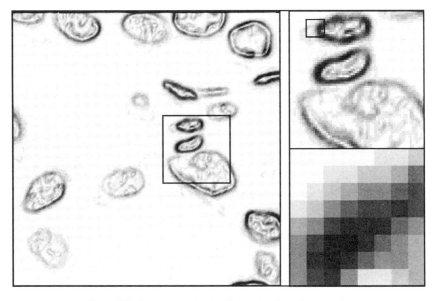

Figure 2.8. Image convolved with an edge detection kernel.

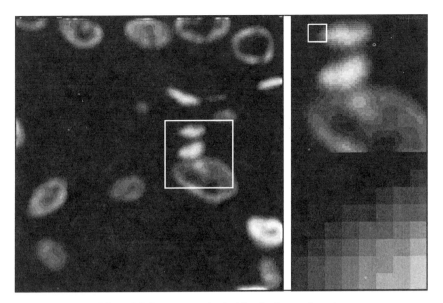

Figure 2.9. Image convolved with a 7 × 7 mean kernel.

High-pass filters enhance object boundaries and, as such, can be used in edge detection. Figure 2.8 shows the results of convolution with an edge detection kernel, and Figure 2.9 shows an image convolved with a 7 × 7 mean kernel.

2.3.1.3. Operations Based on Relations in the Neighborhood

Another type of local operations is based on the relationship between pixels in a neighborhood (i.e., their presence or absence), and as such, they work best for images that are masks of regions or boundaries. The operation of dilation works by filling gaps around existing pixels, thus developing regions or boundaries. Figure 2.10A shows a binary image after two dilations. Erosion has the opposite effect, as it removes pixels from the edges of the mask but always preserves connectivity. This operation is not the reverse of dilation, as can be seen in Figure 2.10B. Dilation and erosion, used alternately will smooth the edges of regions, fill small gaps, and connect close objects.

2.3.1.4. Fourier Transform

Image contents can be analyzed by looking at spatial information. An alternative way to described an image is in terms of its frequency contents. The

A

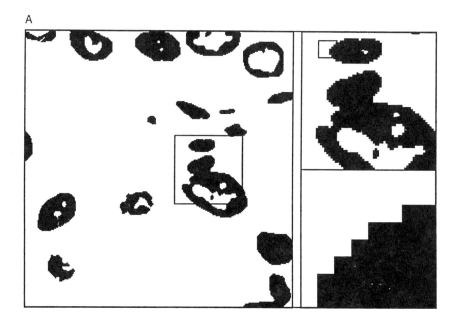

B

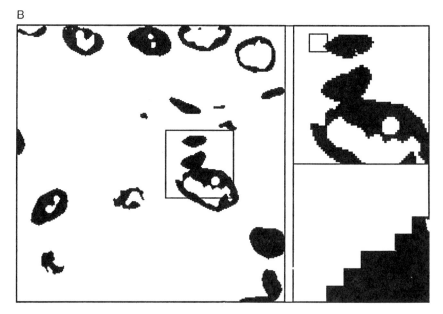

Figure 2.10. (A) Binary image after two dilations; (B) binary image after two erosions.

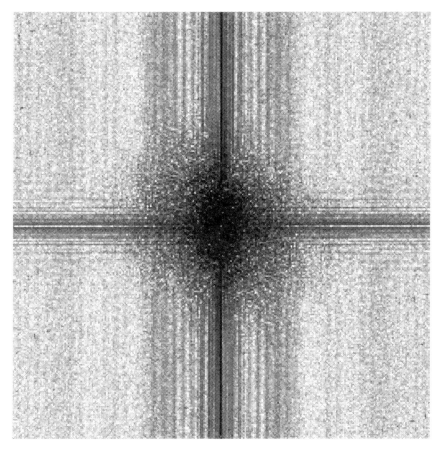

Figure 2.11. Fourier transform of a specimen image.

relation between the spatial and frequency domains is expressed by the Fourier transformation. Fourier transformation on a digital image yields an image that represents the frequency contents of the original image. Details of Fourier transformation are described in Gonzalez and Wintz (3) and Papoulis (4). Fourier transform analysis finds many uses in image processing for example, in finding regular structures in a specimen and selective removal of unwanted frequencies due to harmonic noise. Fourier transform can also be used in three-dimensional image reconstruction and is an example of a global operation. Figure 2.11 shows an image of the Fourier spectrum of the specimen image from Figure 2.4.

2.3.2. Segmentation

The purpose of segmentation is to group pixels into larger regions based on a chosen feature. Segmentation is the first step in deriving a description of image contents. Segmentation can be based on a variety of features, the simplest being the range of intensities defined by thresholding. However, such simple methods often do not work well, so other methods of segmentation are utilized. One way to increase the accuracy of segmentation is by using a combination of different methods and criteria; after all of the chosen segmentation methods are performed, the results can be compared and the best result chosen. Newer segmentation algorithms include application of neural networks that can be trained to recognize specific regions.

2.3.3. Image Contents Description

Deriving a meaningful description of image contents from the raw image is the ultimate goal of image analysis. The complexity of the description generally depends on the questions asked by the user of the system and can be as simple as counting the number of objects or plotting changes in intensity vs. time. For more complicated descriptions, properties of the specific regions can be described, including area, perimeter, width, length, minimum enclosing rectangle or circle, and other shape factors, as well as statistical properties. Other descriptions would also include mutual relations between objects, their location, and their presence in or absence from the image.

2.3.4. Image Presentation and Output Devices

Pseudocolor images are a simple and powerful technique for data presentation. In this technique, the intensity of an image is not displayed directly but serves as an index for a look-up table that translates gray scale values into colors that are then displayed. An example of a pseudocolor image is presented in Figure 2.12 (see Color Plates). Sometimes visualizing two-dimensional images as three-dimensional drawings in the form of surface plots can greatly enhance understanding of image contents. Hard copies of images can be created in many ways, ranging from photographing the computer screen to printing using printer devices. A convenient way of getting a hard copy of an image from a video signal is to use a video printer that takes the video signal as input and produces a printed copy of it. Output from the image processing computer can also be sent directly to color printers.

2.4. AUTOMATED MICROSCOPE FOR CLINICAL IMAGING

The hardware and software components we have just described have been used to assemble an automated image microscope that we have used for automated detection and quantitation of human papillomavirus in clinical Pap (Papanicolaou) smears.

Accumulating evidence strongly associates high-risk human papillomavirus (HPV) infection with the development of cervical cancers (5). A number of epidemiological studies suggest that only women with persistent HPV infection develop cervical cancer (6). In addition, recent studies indicate that the amount of high-risk HPV (viral load or HPV gene copy number) in cervicovaginal epithelial cells may be a risk factor for cervical cancer. Thus, a technique that could detect the genotype and quantity of HPV in smears of cervicovaginal epithelial cells would be of major import in assessment of patient clinical status as well as in epidemiological studies relating HPV infection to cervical cancer.

The primary screening test for cervical abnormalities is the Pap smear. Unfortunately, there are certain disadvantages to this test. These include its lack of sensitivity when used as a triage method for colposcopy, detecting only 50–80% of abnormalities subsequently found by histologic examination, and a false-negative rate of between 5 and 20%(7). Lastly, it is often difficult to identify HPV-specific viral infection by cytologic examination of a Pap smear. For these reasons, we developed a microscopic imaging system that, when coupled to flourescent *in situ* hybridization (FISH) techniques, is capable of detecting one copy of HPV per cell, and can genotype and quantitate the amount of HPV present at a single cell level in cervical Pap smears (8). This system, known as automated fluorescence image cytometry (AFIC), is able to automatically identify each individual cell nucleus in a Pap smear, and to quantitate HPV copy number, allowing assessment of cellular cytoarchitecture and HPV status at the single-cell level.

AFIC consists of an epifluorescence microscope, a low-light-level camera, and a computer (10). The images from the microscope are captured by a CCD camera, digitized, and stored in computer memory. For precise quantification, the images are corrected for background (by subtracting an image not containing stained objects) and for variations in the efficiency of the system over the imaging area (shading). Shading is assessed by measuring a standard fluorescent object at different positions over the imaging area. The nuclei are automatically detected by first labeling them with a fluorescent DNA stain (e.g., Hoechst), recording images of them, correcting the images for background, and shading and segmenting the images into regions corresponding to stained nuclei and (unstained) background. In addition, the AFIC has a computer-controlled stage that automatically records the location of the images relative to a fixed reference point to a precision of 1 μm. This feature allows the user to remove the slide from the stage (for further staining, etc.) and then reimage the same scenes as before. Also, the focus drive of the microscope is under computer control in order to acquire images at different focal depths.

The ability of our microscope system to automatically scan and identify individual nuclei (cells) in a smear in based on the fact that, when stained with the fluorescent DNA dyes Hoechst or propidium iodide, the nuclei are very bright compared to background, allowing highly reliable identification of the nucleus on a slide using image segmentation techniques (9). These techniques are based on the fact that nuclei (1) have higher intensities than the backgroud in the images; (2) are approximately circular, and (3) have a limited range of sizes. However, clustered nuclei are particularly difficult to segment into individual entities because significant background between them does not exist. We have developed algorithms that cope with this problem by segmenting both high-pass and low-pass filtered versions of the images. Following filtering, the stained nuclei in the original and filtered images are detected by calculating adaptive, gradient-weighted threshold intensities between the intensities of the nuclei and the background. Next, the algorithm finds the optimum edges of the nuclei by assuming that such edges correspond to pixels with a locally maximum slope. At this stage of the analysis, the algorithm has not distinguished between regions representing individual nuclei vs. regions representing clusters of touching or overlapping nuclei. Such cluster are recognized based on the fact that they are larger and have more irregular shapes than single nuclei. Clustered nuclei are then divided into individual nuclei by first determining their skeletons, followed by searching for the best dividing path across them. Each path (which need not be straight), crosses the skeleton once and is the path possessing the highest average contrast per pixel. After the individual nuclei in the original and filtered version are identified are three segmented regions corresponding to each nucleus. For each nucleus, one of the segmented regions representing the nucleus is chosen as the final result, based on the closeness of the regions to average nuclear morphology.

Once the individual nuclei have been detected, numerous properties of these nuclei can be directly quantified. These properties routinely include their location in the image, area, perimeter, total fluorescence intensity from the DNA stain, and variation of the DNA stain within the nucleus. The same properties measured from other fluorescent labels in the same nuclei can be quantified by mapping the regions corresponding to each nucleus over the images of each label. The output from the algorithms is an image showing the regions automatically defined as being nuclei. Additional algorithms are available for reporting the results as histograms (e.g., DNA ploidy distributions) and scattergrams for comparing the distribution of one property with respect to another.

The performance of the AFIC was determined in terms of (1) the proportion of fluorescent-stained nuclei correctly detected and (2) the precision with which the total fluorescence from the area and the area of stained nuclei could be quantitated. The AFIC correctly detected over 90% of the nuclei in typical clinical specimens (8.9). For well-separated nuclei, the AFIC detects virtually 100% of the fluorescence-stained nuclei. In specimens containing large portions of clustered cells, the performance of the AFIC drops below 100% but remains above 90%. For example, in images obtained from 2 µm thick sections of prostate carcinoma containing 296 isolated and 174 clustered epithelial cell nuclei (assessed by manual inspection of the images), 98% and 95% of the isolated and clustered nuclei, respectively, were automatically identified. In images of breast carcinoma that contained 37 isolated and 97 clustered nuclei, 100% and 91%, respectively, were correctly identified.

The precision with which AFIC can quantify the integrated fluorescence intensity from an object and its area were determined using fluorescent standard beads (i.e., fluorescent beads containing known amounts of fluorescent molecules per bead) and 4',6-diamidino-2-phenylindole (DAPI)-stained nuclei (8). The coefficient of variation (CV = standard deviation from the mean for a set of measurements on identical objects) for measurements of the total fluorescence signal from correctly identified nuclei was 2–3% and was 1–2% for the area of the nuclei. For touching objects, the precision reduces to 5% C.V. The linearity and sensitivity of AFIC were examined in two ways: (1) through the use of fluorescent beads containing different amounts of fluorescent molecules per bead and (2) by performing FISH for HPV detection in cervical cell lines that contain varying copy numbers of HPV DNA. The response of the AFIC was linear (8). The sensitivity of AFIC (the lowest detectable concentration of fluorescence) was 200 molecules of equivalent soluble fluorescein per square micrometer. Finally, photobleaching was not found to be a problem under the conditions with which our images were acquired; typically, a 1.7% reduction in fluorescence intensity was found during exposure to excitation energy during image acquisition when imaging DAPI-stained nuclei.

Following characterization, the AFIC was combined with the FISH technique and used to screen cervical specimens to investigate the prevalence of HPV-16 in women with varying grades of cervical dysplasia/carcinoma in comparison to polymerase chain reaction (PCR) and Southern blot analysis. We tested the sensitivity and specificity of the FISH procedure for the presence of HPV-16 genome by FISH in comparison to PCR and Southern blot analysis in 99 cervical specimens obtained from a cervical dysplasia clinic. Of the specimens tested, 29% were positive for HPV-16 by PCR and 14% were positive of HPV-16 by the FISH procedure. When the sensitivity and specificity of FISH were compared to that of PCR in terms of HPV detection and clinical disease status, it was found that the specificity of FISH and PCR with respect to predicting cervical intraepithelial neoplasia (CIN) 2–3 and carcinoma in situ was 90% and 75%, respectively, while the sensitivity of FISH and PCR with respect to predicting CIN 2–3 and carcinoma in situ was 21% and 36%, respectively. Because the FISH procedure preserves cellular morphology, has higher specificity than PCR yet comparable sensitivity, use of the FISH procedure for detection of high-risk HPV may serve as a useful adjunct to cytological screening for detection of high-grade cervical disease.

The ability of high-risk HPV to contribute to malignant progression appears to depend on experssion of the E6 and E7 oncogenes (10). E6 binding to p53, which we have detected by fluorescence imaging, is thought to lead to degradation of p53 through the ubiquitin-dependent proteolysis system and thus loss of its growth-suppressing activity (12). In addition, it has recently become clear that p53 plays a role in tumor suppression by both inhibiting cell growth and inducing apoptosis (programmed cell death) (12). Thus, inactivation of p53 can lead to loss of responsiveness of cells to radiotherapy and chemotherapy due to alterations in apoptosis. Recent findings indicate that nonfunctional p53 can lead to alterations in the level of bcl-2, a regulator of apoptosis (13). Therefore, using AFIC, we examined Pap smears from 94 women with varying degrees of cervical disease for the presence or absence of p53, HPV 16-E6, and bcl-2 protein using immunofluorescence microscopy. Our findings indicate that there was a statistically significant inverse association between the presence of p53 and advanced cervical disease [odds ratio (OR) = 0.28, 95%; confidence interval (CI) = 0.1–0.7]. Moreover, the odds of being diagnosed with an advanced stage of cervical cancer (≥ CIN 2) was four times higher (95% CI = 1.7–10.6) for women positive for E6 expression and 62 times higher (95% CI = 7.8–483) for women positive for bcl-2 expression compared to women negative for E6 or bcl-2. However, in multivariate analysis, the presence of bcl-2 (OR = 41, 95% CI = 4.8–345.3) was the only variable that showed significant association with advanced cervical disease after adjusting for the presence of p53 and E6. Chi square analysis also demonstrated a strong association between E6 and bcl-2 expression ($p < 0.001$), as well as between E6 or bcl-2 expression and diagnosis ($p = 0.0015$ and $p < 0.001$, respectively). These findings indicate that (1) the expression of high-risk HPV E6 protein results in the alteration of mediators of apoptosis; (2) expression of bcl-2 is strongly associated with the development of high-grade cervical disease; and (3) the pattern of expression of E6, p53, and bcl-2 proteins may be useful in identifying women at increased risk for the development of cervical cancer.

2.5. SUMMARY

Automated image microscopy provides the ability to enhance the speed of data acquisition as well as data analysis in a highly reproducible fashion.

Although great strides have been made in the development of automated image microscopy, its capabilities are still quite limited relative to its potential. The incorporation of parallel processing, artificial intelligence, neural networks, more sensitive detectors and sophisticated image analysis algorithms, promises substantial improvements and versatility in the use of automated image microscopy. Such improvements promise to reduce human operator input and hence the fatigue associated with processing micrscopic images manually. This should lead to a lower error rate, more cost-effective acquisition of information, and better understanding of basic biomedical processes.

REFERENCES

1. Inoue, S. (1987). *Video Microscopy.* Plenum, New York.

2. Aikens, R. S., Agard, D. A., and Sedat, J. W. (1989). Solid-state imagers for microscopy. *Methods Cell Biol.* **29**, 291 313

3. Gonzalez, R. C., and Wintz, P. (1987). *Digital Image Processing.* Addison-Wesley, Reading, MA.

4. Papoulis, A. (1977). *Signal Analysis.* McGraw-Hill, New York.

5. Kiviat, N. B., and Koutsky, L. A. (1993). Specific human papillomavirus types as the causal agents of most cervical intraepithelial neoplasia: Implications for current views and treatment. *J. Natl. Cancer Inst.* **85**(12), 934 935.

6. Koutsky, L. A., Galloway, D. A., and Holmes, K. K. (1988). Epidemiology of genital human papillomavirus infection. *Epidemiol. Rev.* **10**, 122 163.

7. Moscicki, A. B., Palefsky, J. M., Gonzales, J., and Schoolnik, G. K. (1991). The association between human papillomavirus deoxyribonucleic acid status and the results of cytologic rescreening tests in young, sexually active women. *Am. J. Obstet. Gynecol.* **165**(1), 67 71.

8. Siadat-Pajouh, M., Periasamy, A., Ayscue., A. H., Moscicki, A. B., Palefsky, J. M., Walton, L., DeMars, L. R., Power, J. D., Herman, B., and Lockett, S. J. (1994). Detection of human papillomavirus type 16/18 DNA in cervicovaginal cells by fluorescence based in situ hybridization and automated image cytometry. *Cytometry* **15**(3), 245 257.

9. Lockett, S. J., Orand M., Rinehart, C. A., Kaufman, D. G., Herman, B., and Jacobson, K. (1991). Automated fluorescence image cytometry. DNA quantification and detection of chlamyydial infections. *Anal. Quant. Cytol. Histol.* **13**(1), 27 44.

10. Kanda, T., Watnabe, S., and Yoshiike, K. (1988). Immortalization of primary rat cells by human papillomavirus type 16 subgenomic DNA fragments controlled by the SV40 promoter. *Virology* **165**(1), 321 325.

11. Liang, X. H., Volkmann, M., Klein, R., Herman, B., and Luckett, S. J. (1993). Colocalization of the tumor-suppressor protein p53 and human papillomavirus E6 protein in human cervical carcinoma cell lines *Oncogene* **8**(10), 2645 2652.

12. Shaw, P., Bovey, R., Tardy, S., Sahli, R., Sordat, B., and Costa, J. (1992). Induction of apoptosis by wild-type p53 in a human colon tumor-derived cell line. *Proc. Natl. Acad. Sci. USA.* **89**(10), 4495–4499.

13. Haldar, S., Negrini, M., Monne, M., Sabbioni, S., and Croce, C. M. (1994). Down-regulation of bcl-2 by p53 in breast cancer cells. *Cancer Res.* **54**(8), 2095–2097.

CHAPTER

3

MULTISPECTRAL IMAGE PROCESSING FOR COMPONENT ANALYSIS

SATOSHI KAWATA and KEIJI SASAKI

Department of Applied Physics
Osaka University
Osaka 565, Japan

3.1. INTRODUCTION

Various computerized image-processing methods have been developed for analyzing qualitative and quantitative information on spatial patterns of specific components in observed images, and applied to remote sensing in environmental sciences, medical diagnostics with X-ray images, robot vision for automatic control systems, and so forth. In pattern-recognition studies, images are classified into different components in the feature space by comparing component features such as shape, size, density, histogram, and direction (1). One example for these methods is texture analysis, in which the components are classified by their spatial stasitics or frequency distributions (2). However, if a component does not have a specific spatial feature different from the features of other components, how can it be distinguished in the image? It often happens that the components are not classifiable in the feature hyperspace because of their mutual dependency in the spatial domain.

In such a case, spectral (optical frequency) information should be used for component analysis. Different materials must have different spectral responses to electromagnetic waves of a certain energy band. In remote sensing of land and ocean images observed by airplanes and spacecraft, the spectral classification techniques with typically six or seven visible and infrared bands is commonly used for land use, agricultural resource, and surface mineral content pattern mappings (3). In diagnostic X-ray images, the distributions of bone and tissue are separately reconstructed by using two different X-ray energy levels (4). Since the absorption characteristics of bone and tissue at two

Fluorescence Imaging Spectroscopy and Microscopy, edited by Xue Feng Wang and Brian Herman. Chemical Analysis Series, Vol. 137.
ISBN 0-471-01527-X © 1996 John Wiley & Sons, Inc.

energy levels are different from each other, their spatial distributions can be reconstructed by inverting the spectral characteristic functions. Such image analyses based on spatial and spectral information are called *imaging spectroscopy*.

In this chapter, we describe several component pattern analysis methods of imaging spectroscopy, by which spatial patterns of individual components contained in the scene and in their spectral curves are analyzed from the multispectral image that is a set of images observed at plural wavelengths or spectral bands. These methods are based on linear least squares fitting, principal component analysis in multivariate analyses, nonlinear constraints given by a priori information, nonlinear optimization algorithms, entropy minimization criteria, and so forth. In addition to the mathematical theories, some examples of chemical and biological applications are introduced.

3.2. MULTISPECTRAL IMAGE

In the human visual system, a scene is sensed by three types of neurochemical sensors in the retina, and these three images are recognized in the brain as a color picture. Unless each image is black and white, those images are different from each other because the three detectors have different spectral responses in the visible region. In machine vision, the number of spectral bands is not limited to three; by using interference color filters or dispersive elements, hundreds of distinct images can be obtained in the range from UV to IR for a single scene (5). We call a set of these images a multispectral image. Figure 3.1 illustrates our definition of the multispectral image: the image data set (on the left) may be regarded as a spectral data set of mixture samples (on the right).

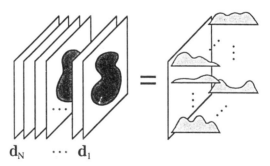

$$d_N \quad \cdots \quad d_1$$

Figure 3.1. Illustration of the multispectral image. The right-hand side of the figure is an image set measured at N wavelengths, and the left-hand side is a spectrum set at individual image pixels.

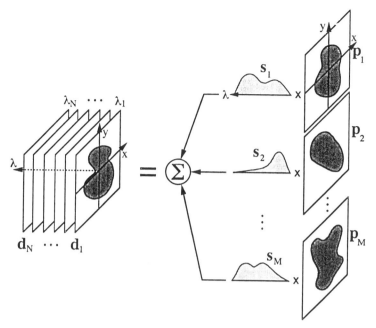

Figure 3.2. Imaging model for multispectral patterns. The multispectral image is given by linear combinations of M component patterns weighted by their corresponding spectral responses.

Unless the sample is uniformly mixed over the image, the spectra on the right vary pixel by pixel with their shape. Here we have plenty of spectral information about the sample components as a form of mixtures, although we do not know them individually.

Suppose that the observed sample is composed of M components having different spectral curves in a proper spectral range. All the spectra on the right in Figure 3.1 can then be expressed as combinations of the spectral responses of M components. Figure 3.2 shows another interpretation of the multispectral image. Each of the images taken at N wavelengths is a linear combination of M component distributions weighted by their corresponding spectral responses.

If the system is assumed to be linear, Figure 3.2 can be described by the following linear matrix equation in discrete form:

$$[D] = [S][P] \tag{1}$$

where $[D]$ is a matrix of the observed multispectral image whose row vectors

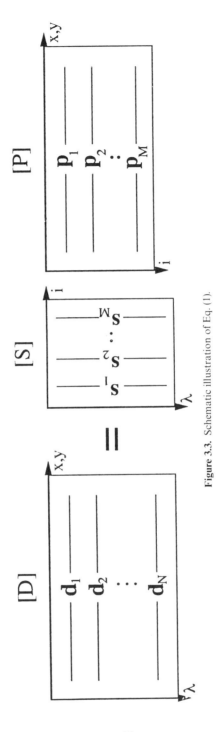

Figure 3.3. Schematic illustration of Eq. (1).

58

$\mathbf{d}_1, \ldots, \mathbf{d}_N$ represent the images of the scene at individual wavelengths with L lexicographically arranged pixels; matrix $[S]$ represents a set of spectral responses $\mathbf{s}_1, \ldots, \mathbf{s}_M$ of M constituent components in its column vectors; and $[P]$ is a matrix composed of M component spatial patterns $\mathbf{p}_1, \ldots, \mathbf{p}_M$ in its row vectors. The dimensions and elements of these three matrices are graphically shown in Figure 3.3.

For fluorescence and emission images, the light intensity distribution meets the linear combination of Eq. (1). For the absorption image, the system in an optical density scale is linear; we must take the logarithm of the observed image intensity divided by the illumination light intensity to satisfy the linear relation of Eq. (1). In this case, the column vectors of $[S]$ form absorbance spectra of components.

The purpose of the following methods is to estimate the component spatial patterns $[P]$ and/or the component spectra $[S]$ from the multispectral image $[D]$. These methods should be used appropriately for each application, depending on given a priori information such as data from preliminary experiments, spectral libraries, and known physical conditions.

3.3. COMPONENT PATTERN SEPARATION

In pattern separation, it is supposed that the number of components M is known and that the number of observed wavelengths or bands N is more than M. In addition, the spectra of all components $[S]$ are assumed to be obtained in advance by preliminary experiment or from a spectral library. Thus the purpose of this method is to separate each pattern $[P]$ from the observed data $[D]$ by using the spectral information $[S]$.

If the number of components is the same as the number of observed wavelengths ($M = N$), the matrix $[S]$ is $M \times M$ square so that each column of matrices $[D]$ and $[P]$ composes M simultaneous equations for M unknowns. Therefore, the component patterns $[P]$ can be estimated with the matrix inverse:

$$[P] = [S]^{-1}[D] \qquad (2)$$

where $[S]^{-1}$ denotes the inverse matrix of $[S]$.

In the case of $N > M$, the number of simultaneous equations exceeds that of unknowns $[P]$, so that the solution satisfying all the simultaneous equations does not exist, except for the perfectly noiseless data; that is, the solution for M selected equations does not agree with other $(N - M)$ equations due to the random noise. The linear least squares method can be applied to such a case. In this method, the optimum solution is determined so that the sum of the squared errors between the left- and right-side matrix elements in Eq. (1) should be minimized, which is given by

$$[P] = ([S]^t[S])^{-1}[S]^t[D] \tag{3}$$

where t represents the matrix transpose. The term $([S]^t[S])^{-1}[S]^t$ is called a generalized inverse matrix. Equation (3) requires the matrix inverse processing for the $M \times M$ square matrix $([S]^t[S])$, not for the large image data matrix $[D]$. The matrix $[D]$ is processed only with the multiplication by the generalized inverse matrix of the spectral library.

As one of the applications, Figure 3.4 shows the pattern separation for human blood corpuscles. Blood includes red and white corpuscles, lymphocytes, granulocytes, and so forth, and the number of each kind of corpuscle gives important information for clinical diagnoses. Corpuscle counting is usually performed by eye, which is tedious and takes a very long time; therefore, a computerized processing method has long been expected. As preliminary processing for corpuscle counting, we applied the present method to the spatial pattern separation of red and white corpuscles.

Figure 3.4a, b shows the multispectral image of a Giemsa-stained blood specimen observed with a charge-coupled device (CCD) camera and interference color filters (425 and 550 nm, respectively; spectral width = 15 nm) under a transmission microscope (Olympus BH-2) with an objective lens of $20 \times$ magnification (numerical aperture = 0.70). The digitized images were

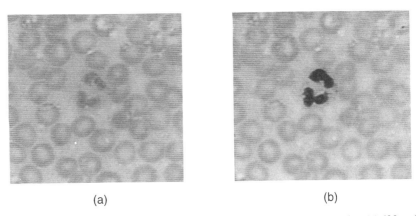

(a)   (b)

Figure 3.4. Multispectral image of a Giemsa-stained blood specimen observed at (a) 425 and (b) 550 nm.

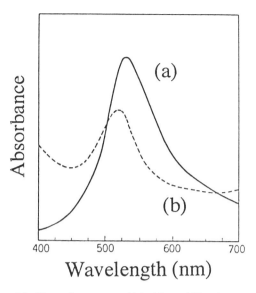

Figure 3.5. Absorption spectra of (a) white and (b) red corpuscles.

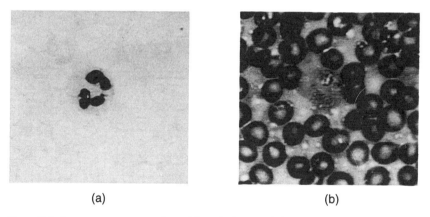

Figure 3.6. Component patterns estimated from the images of Figure 3.4 and the spectra of Figure 3.5. Patterns (a) and (b) correspond to spatial distributions of white and red corpuscles, respectively.

obtained by a frame grabber of 256×256 pixels with 8 bit resolution. The white and red corpuscles were colored reddish purple and bluish red, respectively, by the dye, so that the contrast between a white corpuscle at the center of the images and the surrounding red corpuscles was clearly different in the two observed images.

The absorption spectra of white and red corpuscles observed in the separate experiments are shown in Figure 3.5. When the values of absorbance at 425 and 550 nm are substituted in the spectral matrix $[S]$ of Eq. (3), the individual component patterns $[P]$ can be obtained from the multispectral image $[D]$, as shown in Figure 3.6. A personal computer (NEC PC-9801F, 16-bit CPU) was used for the calculation, and the computation time was $\sim 30\,s$. Figures 3.6a and 3.6b correspond to the spatial distributions of the white corpuscle (cell nucleus) and the red one, respectively. In this experiment, the matrix $[S]$ did not include the spectrum of cytoplasm in the white corpuscle, which is different from that of the nucleus; the pattern of the cytoplasm was overlapped with both of the estimated patterns in Figure 3.6.

3.4. OPTIMAL WAVELENGTH SELECTION FOR PATTERN SEPARATION

Since each data image is composed of an enormous number of pixels, the number of wavelengths at which the multispectral image is observed will be limited by the memory capacity, measurement time, and processing speed of a computer. In addition, a compact, simple apparatus required for remote sensing, robot vision, and so forth provides a limited number of spectral bands. When the number of wavelengths is fixed, what wavelengths or bands should be selected for the measurement of the multispectral image? The quality of the component patterns estimated from the multispectral image depends on the selection of wavelengths. How can the most efficient information of the sample in the pattern separation be obtained?

Wavelength selection usually depends on the experiments' intuition or experience and often lacks objectivity and reproducibility. Automatic selection by a computer is preferable; for this, one has to develop the criterion to pick up the optimal wavelength combination and the realistic computer algorithm based on the criterion to accomplish the automatic selection. We used the popular criterion of minimum mean square error U, where the expected value of the squared error is theoretically derived from the component spectra $[S]$ at the selected wavelengths and the autocorrelation matrix of noise $[W]$, as follows (6):

$$U = \mathrm{tr}[([S]'[W]^{-1}[S])^{-1}] \qquad (4)$$

where tr denotes the matrix trace. If noise obeys an uncorrelated process with a zero mean and a constant variance σ^2, Eq. (4) reduces to

$$U = \sigma^2 \operatorname{tr}[([S]^t[S])^{-1}] \qquad (5)$$

Although the optimal set may be chosen by comparing the mean square errors for all feasible combination sets of wavelengths, there is normally an enormous number of combinations, so that calculation of mean square errors for all combinations is practically impossible. For example, the number of combinations of five wavelengths among 200 candidates is 2,535,650,040. If the calculation of U for one combination takes 1 ms, the total computation time for all feasible combinations requires about 1 month. Two approaches to save computation time have been discussed (7). In one, the number of wavelength candidates is reduced by acceptance of the resolution reduction or by the use of spectroscopic knowledge. In the other, an approximation algorithm such as stepwise down regression is used. However, these methods may often miss the optimal wavelength set.

We have developed a generalized and practical algorithm utilizing the branch and bound method, which is popular in the field of combinatorial optimization studies (8, 9). The optimal wavelength set can be selected from all feasible combinations in a short computation time without any approximation. The algorithm is composed of two processes. Instead of calculating U values for the feasible combinations, the combination set is partitioned into subsets at every wavelength by either selection or nonselection (this process is called *branching*), and a lower bound L, which should be smaller than any U values for all combinations in the given subset, is calculated for comparison with the current minimum $U_{\min}$ (this process is called *bounding*). The derivation of the lower bound L and the details of the algorithm have been reported in Sasaki et al. (6).

We applied this method to optimal wavelength selection for the pattern separation of xylene–isomer mixtures using their infrared absorption spectra (6). Figure 3.7 shows library component spectra of *o*-, *m*-, and *p*-xylene sampled at every $4 \mathrm{~cm}^{-1}$ between 1200 and $964 \mathrm{~cm}^{-1}$. Each spectrum comprises 60 sampling points. Three selected wavelengths are shown by solid vertical lines in Figure 3.7. Three is the minimum number needed to resolve three components. Analytical chemists tend to believe that the optimal wavelengths to be selected in spectra are the positions where only one component has absorption (or emission) but others are in the baseline region. For example, in Figure 3.7, we might intuitively select the wavelength set $\{\lambda_a, \lambda_c, \lambda_d\}$ or $\{\lambda_a, \lambda_b, \lambda_d\}$, shown by dotted lines, rather than the solid lines automatically selected. However, the present experiment led to the fact that the U values for $\{\lambda_a, \lambda_c, \lambda_d\}$ and $\{\lambda_a, \lambda_b, \lambda_d\}$ where 6.4 and 1.5 times as large as that for the optimal set, respectively.

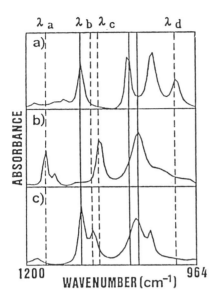

Figure 3.7. Three wavelengths optimally selected for the pattern separation of xylene–isomer mixtures: (a), (b), and (c) show component spectra of *o*-, *m*-, and *p*-xylene, respectively. Solid vertical lines indicate the optimally selected wavelengths, and dotted lines show the intuitively selected wavelengths.

The computation time needed for the present method to find the three wavelengths was 0.09 s, with the use of a large-scale computer (NEC ACOS 77 system 1000), whereas the enumerative method that calculated U values for all feasible combinations took 7.36 s, although giving the same result. The number of combinations enumerated in the new method was only 268 out of 34,220 combinations. This reduction enables one to select the wavelength by using a desktop computer attached to a commercial spectrophotometer. The same calculation by a personal computer with the BASIC language took 33 s.

Besides the advantages of data reduction and measurement time economy, wavelength selection enhances the accuracy of pattern separation, in terms of the signal-to-noise ratio, compared to the use of full spectral data. It was experimentally confirmed that the squared error of the patterns estimated from the selected spectral data was normally smaller than that from the full spectral data, where the exposure time of the former data was longer than that of the latter one, so that the total measurement time was fixed (6). This shows that wavelength selection gives higher quality pattern separation than the generalized inverse of full spectral data if the total exposure time is fixed. This conclusion might conflict with one's intuitive belief. However, it is true because only the most significant data useful for pattern separation should be selectively analyzed; insignificant data should be excluded.

3.5. COMPONENT PATTERN ESTIMATION

3.5.1. Eigenvector Analysis and Nonnegativity Constraints

If even one unknown or unpredicted component is included in the scene, it becomes impossible to estimate other component patterns or it results in misestimation. Furthermore, the component spectra often vary depending on environmental conditions, so that the spectral library cannot be utilized for pattern separation. In this section, a new spectral pattern analysis method, by which both the spatial patterns of the components contained in the multi-spectral image and their spectral responses are reconstructed without knowledge of either their spectroscopic characteristics or their spatial features, is described (10). This method will be useful for finding new materials that have never been discovered in mixed images and for analyzing spatial and spectral properties of reactive molecules under unknown physical/chemical conditions.

This method is also based on the linear assumption of Eq. (1). The number of components M is unknown, so that even the dimensions of $[P]$ and $[S]$ are not given before the analysis. The purpose of the method is to estimate both the component pattern matrix $[P]$ and the spectral matrix $[S]$ from the observed image data $[D]$ alone. Let us start by finding the number of components M, which is determined by eigenvalue analysis. Based on the linear assumption, the rank of the multispectral image matrix $[D]$ or the autocorrelation matrix $[D][D]^t$ for a rectangular matrix $[D]$ corresponds to the number M, which is given by the number of nonzero eigenvalues of those matrices if noise is absent in the observed images (11). If noise exists, we should count the number of eigenvalues greater than the threshold given by the noise variance (1). The threshold or noise variance can be found by measuring or calculating the detector noise, the nonlinearity of the detector, the quantization error of the analog-to-digital converter, and so forth. Some criteria, such as Akaike's information criterion (13), may also be employed.

The multispectral image $[D]$ can be recomposed by the nonzero eigenvalues and their corresponding eigenvectors:

$$[D] = [U][E][V] \qquad (6)$$

where $[E]$ is an $M \times M$ diagonal matrix whose diagonals are square roots of the eigenvalues, and $[U]$ and $[V]$ are matrices composed of eigenvectors $\mathbf{u}_1, \ldots, \mathbf{u}_M$ and $\mathbf{v}_1, \ldots, \mathbf{v}_M$ of $[D][D]^t$ and $[D]^t[D]$, respectively, corresponding to the eigenvalues of $[E]$. The dimensions of three matrices are shown in Figure 3.8. Note that based on Figure 3.8, $[U]$ may be called eigenvectors in a spectral space while $[V]$ is those in a spatial space.

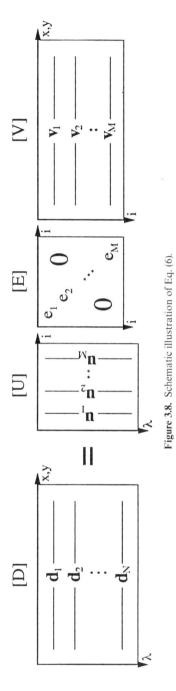

Figure 3.8. Schematic illustration of Eq. (6).

66

This method of eigenvalue–eigenvector analysis of a data set is called principal component analysis in multivariate analyses, and the matrix expansion of Eq. (6) is called singular value decomposition in signal processing. In the fields of pattern recognition and data compression, it may be called a Kahunen–Loéve expansion.

When the matrix $[V]$ is used, the component spatial pattern $[P]$ can also be expressed as the linear transformation:

$$[P] = [T][V] \qquad (7)$$

where $[T]$ is an $M \times M$ matrix composed of coefficients for eigenvectors to form $[P]$. When Eqs. (1) and (6) are used, the spectral matrix $[S]$ is also given by $[T]$ and $[U]$:

$$[S] = [U][E][T]^{-1} \qquad (8)$$

Here the problem is to determine the matrix $[T]$. However, the multispectral image data do not give any information on $[T]$. Some additional knowledge of the matrices $[P]$ and $[S]$ is required for the estimation. We utilize the physical conditions as the nonnegativity constraints for $[P]$ and $[S]$:

$$[P] \geq [0] \qquad (9)$$

$$[S] \geq [0] \qquad (10)$$

where $[0]$ denotes the matrix with all zero elements, and inequalities (9) and (10) mean that all elements of $[P]$ and $[S]$ are nonnegative. There exists no negative density at any pixel of any component pattern, and absorbance and emission intensity of the component spectra must be nonnegative at every wavelength, which restricts the matrix elements of $[P]$ and $[S]$ to a positive value or zero (10). Although these constraints may not seem powerful for finding $[P]$ and $[S]$ from $[D]$, the number of inequalities obtained by the (9) and (10) is enormous. For 256×256 pixel images of four components, constraint (9) constitutes a set of $\sim 250{,}000$ inequalities; $[P]$ and $[S]$ can be coupled by using eigenvectors, as shown in Eqs. (7) and (8):

$$[T][V] \geq [0] \qquad (11)$$

$$[U][E][T]^{-1} \geq [0] \qquad (12)$$

In relations (11) and (12), the elements of $[T]$ are relative values. If a value for

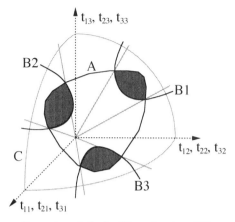

Figure 3.9. Explanatory illustration of the feasible region in a $[T]$ space determined by the nonnegativity constraints and the normalization for a three-component system. Cone A is the constraint of inequality (11), and the three concave boundaries B1, B2, and B3 are made by inequality (12). Spherical surface C is given by Eq. (13).

any nonzero elements of $[S]$ or $[P]$ is known, $[T]$ reduces to the absolute. If such an element does not exist, $[T]$ should be normalized by

$$\text{diag}([T][T^t]) = [I] \tag{13}$$

where $[I]$ is the identity matrix.

Figure 3.9 is an explanatory illustration of constraints (11) and (12) and Eq. (13) in a $[T]$ space with $M = 3$ (10). Inequality (11) limits $[T]$ within cone A, and inequality (12) makes three concave boundaries $B1$, $B2$, and $B3$. The three separated cones enclosed by $A, B1, B2$, and $B3$ are cut by a spherical surface C given by Eq. (13). In this figure, three shaded areas on the cut surface give the bounded solutions for row vectors of $[T]$, providing the solutions $[P]$ and $[S]$ by Eqs. (7) and (8), respectively.

For two-component systems, an algorithm for finding the bounded solutions is simple (10). Lawton and Sylvestre developed an algorithm for analyzing the component spectra of two-dye mixtures (14). We proposed an advanced algorithm that suppresses the influence of noise on the results. In this case, the feasible region satisfying the nonnegativity constraints is found as follows: inequality (11) is expressed by using the individual elements $\{t_{ij}\}$ of $[T]$ and $\{v_{ij}\}$ of $[V]$:

$$t_{11}v_{1i} + t_{12}v_{2i} \geq 0, \qquad i = 1, 2, \ldots, L \tag{14}$$

$$t_{21}v_{1i} + t_{22}v_{2i} \geq 0, \qquad i = 1, 2, \ldots, L \tag{15}$$

These inequalities yield the relationship between t_{11} and t_{12} and between t_{21} and t_{22}:

$$-[\max(v_{2i}/v_{1i})]^{-1} \leq t_{12}/t_{11} \leq -[\min(v_{2i}/v_{1i})]^{-1} \tag{16}$$

$$-[\max(v_{2i}/v_{1i})]^{-1} \leq t_{22}/t_{21} \leq -[\min(v_{2i}/v_{1i})]^{-1} \tag{17}$$

Similarly, inequality (12) reduces to

$$(u_{j1}e_1t_{22} - u_{j2}e_2t_{21})/(t_{11}t_{22} - t_{12}t_{21}) \geq 0, \qquad j = 1, 2, \ldots, N \tag{18}$$

$$(u_{j2}e_2t_{11} - u_{j1}e_1t_{12})/(t_{11}t_{22} - t_{12}t_{21}) \geq 0, \qquad j = 1, 2, \ldots, N \tag{19}$$

where e_1 and e_2 are diagonal elements of the matrix $[E]$. By giving the relations $t_{11} > 0$, $t_{21} > 0$, and $t_{12}/t_{11} > t_{22}/t_{21}$ for determining the arrangement of the order of component vectors in $[P]$ and $[S]$, relations (18) and (19) yield

$$t_{22}/t_{21} \leq \min(e_2 u_{j2}/e_1 u_{j1}) \tag{20}$$

$$t_{12}/t_{11} \geq \max(e_2 u_{j2}/e_1 u_{j1}) \tag{21}$$

The normalization of Eq. (13) is given as a function of $\{t_{ij}\}$, that is,

$$t_{11}^2 + t_{12}^2 = 1 \tag{22}$$

$$t_{21}^2 + t_{22}^2 = 1 \tag{23}$$

Inequalities (16), (17), (20), and (21) and Eqs. (22) and (23) give the feasible solution regions.

Although this method neglects the noise component, the actual data contain the measurement noise, which may degrade the quality of the estimates. The noise term is included in the elements of $\mathbf{v}_1$ and $\mathbf{v}_2$. Especially in inequalities (16) and (17), which provides the solution range, the noise in v_{1i} of the largest $|v_{2i}/v_{1i}|$ could severely affect the estimation and sometimes exclude the true solution from the limit. To overcome this problem, we modified the algorithm so that v_{1i} as small as the noise level is automatically removed from the candidates for bounding the solution range. The modification is as follows (15): v_{1i} in inequalities (16) and (17) is replaced by

$$v'_{1i} = (v_{1i}^2 + \alpha^2)/v_{1i} \tag{24}$$

where α is a constant. Equation (24) is used to cut out v_{1i} smaller than α and alter it to a large v'_{1i}. This modification is one of noise filtering. The value of α can be determined by the noise level of the observed data.

For systems with more than three components, the problem becomes much more difficult than with two-component systems. A point pair (the maximum and the minimum) in a $[T]$ space for the estimates of $[P]$ and $[S]$ cannot be determined, but the solutions are given as closed regions on a hypersurface for individual rows of $[T]$ (see Figure 3.9). In this case, the problem turns out to be a constrained nonlinear optimization problem that finds the maximum and minimum values for p_{ij} of $[P]$ for all i and j, subject to constraints (11) and (12) and Eq. (13) with respect to $[T]$. This optimization can be performed by the sequential unconstrained minimization technique (SUMT) by using the idea of the penalty function. The details of this algorithm will be described later.

We carried out some experiments with real multispectral image data to verify the present theory (10). Figure 3.10a–c shows microscopic images of a *Paramecium* specimen observed at 450, 550, and 650 nm under a transmission microscope. This sample was stained with hematoxylin and Fast Green. It is known that hematoxylin is deposited in the nucleus and the cytoplasm, whereas Fast Green is deposited on the cilium and the cell membrane.

We performed component analysis of spatial and spectral patterns for these images without knowledge of the sample and the dyes. We actually used 13 band images from 400 to 700 nm at every 25 nm interval, including the three images shown in Figure 3.10. Table 3.1 shows eigenvalues of $[D][D]^t$. The number of eigenvalues over the noise level, which corresponds to the number of components, was determined by the quantization error of the highest absorbance to be two. The nonnegativity constraints of the optical density and absorbance bound the feasible space of $[T]$ by inequalities (16), (17), (20), and (21). The final solutions of spectral and pattern estimates of two components are shown in Figures 3.11 and 3.12. The two component spectra should be in the shaded areas in Figure 3.11. Bands a and b include the expected spectra of hematoxylin and Fast Green, respectively. The spatial pattern of hematoxylin is estimated between Figures 3.12a and 3.12b, where only the nucleus and the cytoplasm are recognized, while the pattern of Fast Green is estimated between Figures 3.12c and 3.12d, where only the cilium and the cell membrane are observed (the nucleus is not). The computation time required to obtain these results by using the personal computer was $\sim 30\,\text{s}$.

3.5.2. Optimal Estimation with an Entropy Minimization Criterion

In practical applications, the nonnegativity constraints given by inequalities (11) and (12) are strong enough to limit the feasible solution set to a reasonably narrow band in an image space as well as in a spectral space. Although the

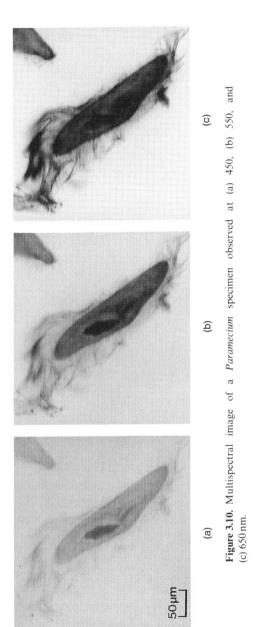

(a)

(b)

(c)

Figure 3.10. Multispectral image of a *Paramecium* specimen observed at (a) 450, (b) 550, and (c) 650 nm.

71

Table 3.1. Eigenvalues of the Multispectral Image Data Shown in Figure 3.10

Principal Components	Eigenvalues
1	7.1×10^{-2}
2	1.3×10^{-3}
Noise level	2.9×10^{-4}
3	6.3×10^{-5}
4	3.0×10^{-5}
5	8.3×10^{-6}
6	4.2×10^{-6}
7	1.9×10^{-6}
8	1.8×10^{-6}
9	9.9×10^{-7}
10	6.3×10^{-7}
11	5.3×10^{-7}
12	3.8×10^{-7}
13	1.9×10^{-7}

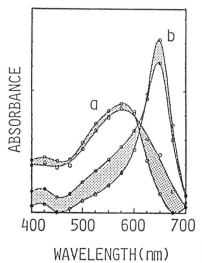

WAVELENGTH (nm)

Figure 3.11. Estimated band spectra of two components (shaded areas a and b).

experimental results resolved component patterns fairly well in observed scenes, in advanced method to find a unique solution for every component pattern and spectrum is expected for precise quantitative analyses.

If there is no particular a priori information about the components, the estimation theory to find the optimal solution within a feasible solution set

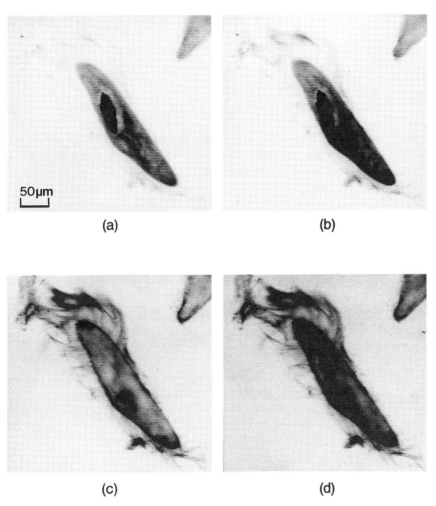

Figure 3.12. Estimated band patterns of two components. The solution bands are limited by patterns (a) and (b) and by patterns (c) and (d), which correspond to spectra a and b of Figure 3.11, respectively.

must be used. We proposed a method that estimates the components so that their spectral features are most enhanced (16). In other words, this method chooses the spectra for individual components that are most independent of other component spectra.

This estimation can be realized by minimizing the function

$$H([S]) = -\sum\sum a_{ij}\ln a_{ij} \rightarrow \min \qquad (25)$$

where

$$a_{ij} = |s''_{ij}| / \sum |s''_{ij}| \qquad (26)$$

and s''_{ij} is the second derivative of a candidate spectrum with respect to the wavelength for component i at wavelength j.

Optimization (25) is exactly equivalent to the entropy minimization if $\{a_{ij}\}$ is regarded as a probability density function of a certain process; $\{a_{ij}\}$ is, of course, not that function, but the recognition of this similarity helps us to understand this estimation method. A uniform distribution of a probability density function gives the largest entropy, while a single delta function gives the smallest one. Therefore, minimizing the entropy of relation (25) localizes peaks in spectra and smooths baselines. Peak localization emphasizes the features of the individual components and separates the features of other components in the spectral space. The reason for taking the second derivative of the spectrum is to emphasize the peak features.

If the components have some features in their spatial patterns rather than in the spectra, entropy minimization should be performed in the spatial domain such that

$$H([P]) = -\sum\sum b_{ij} \ln b_{ij} \to \min \qquad (27)$$

where

$$b_{ij} = |p''_{ij}| / \sum |p''_{ij}| \qquad (28)$$

In Eq. (28), p''_{ij} is the Laplacian of a candidate spatial pattern for component i at pixel j. Optimization (27) can be used to localize the spatial peaks and edges and to smooth the background areas.

To minimize the nonlinear function under the nonnegativity constraints, SUMT can be used (17). In this technique, the cost function to be minimized is

$$R([P],[S]) = H([P],[S]) + \gamma Q([P],[S]) \to \min \qquad (29)$$

where $H([P],[S])$ is the entropy function given by optimization (25) or (27) or their combination, and $Q([P],[S])$ denotes a penalty function with a scaling factor γ. With the nonnegativity constraints, $Q([P],[S])$ is given as

$$Q([P],[S]) = \sum\sum F(p_{ij}) + \sum\sum F(s_{ij}) \qquad (30)$$

where p_{ij} and s_{ij} are the elements of $[P]$ and $[S]$, respectively, and

$$F(x) = 0 \quad (x \geq 0) \quad \text{and} \quad F(x) = x^2 \quad (x < 0) \qquad (31)$$

With the help of this penalty function, the problem reduces to an unconstrained optimization.

Given a value for γ, optimizing R in relation (29) can be performed by the simplex method, the Davidon–Fletcher–Powell method, or a combination of the two. The minimum of H subject to the nonnegativity constraints is given by $\gamma \to \infty$ in minimization (29). However, since noise exists in the image measurement, relation (29) does not give the optimum at $\gamma \to \infty$. Some negative values should be permitted for absorbing the noise component at values near zero. In SUMT, a small γ is given first for optimization, resulting in weak constraint. With the resultant solution as the first estimate of the next step, the optimization is redone with a larger γ value. This repetition continues until γ exceeds the signal-to-noise ratio of the data.

We applied this method to various multiband image data. Here we show a practical application to the histochemical analysis of tissue sections (16, 18). Figure 3.13 shows a microscopic image of a tissue section of rat liver. This specimen was stained with nitroblue tetrazolium (NBT). It is known that the NBT dye is reduced to two chemicals, monoformazan and diformazan, by enzyme activity (dehydrogenase reaction) in liver (19). The weighted sum of the densities of the two formazans corresponds to the enzyme activity at the current pixel. However, these formazans are chemically unstable, and chemical isolation of them is difficult. We analyzed the density patterns and spectra of individual components by the present method.

We observed 13 band images. Figure 3.14 shows the images at 450, 550, and 650 nm bands. Eigenvalue analysis indicated that the number of components

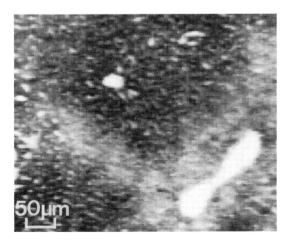

Figure 3.13. Microscopic image of a tissue section of rat liver stained with NBT.

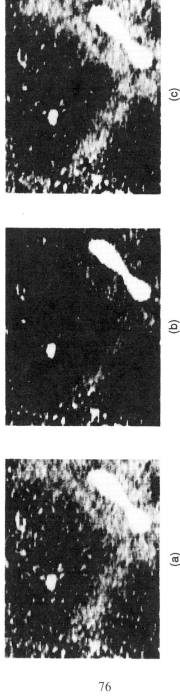

(a)

(b)

(c)

Figure 3.14. Multispectral image of the specimen shown in Figure 3.13, which was observed at (a) 450, (b) 550, and (c) 650 nm.

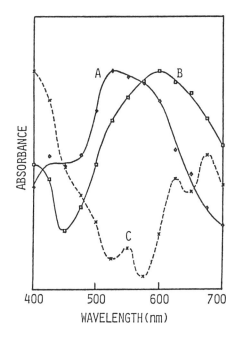

Figure 3.15. Estimated spectra of three components.

was three. Figure 3.15 shows the optimal component spectra estimated by entropy minimization with the nonnegativity constraints. From the features of the estimated spectra, we assigned spectrum A in Figure 3.15, with a peak at 525 nm and a shoulder at 575 nm, to monoformazan; spectrum B to diformazan; and spectrum C to the NBT remaining in the tissue. Figure 3.16 shows the spatial density distributions of the three components reconstructed by the present method, with profiles along a line. Figure 3.16a shows that monoformazan was distributed uniformly, while diformazan was distributed in a gradient from a central vein (a round hole in the figure) to a portal triad (a gourd-shaped hole in the figure). These distributions approximately match the theoretical expectation of enzyme activity given by Chikamori et al. (20). The computation time for the SUMT algorithm was ~ 10 s with a large-scale computer.

Here the limitation of the present method is discussed. Since this is an estimation, the solution may not be the true one, but it is the optimal one under the entropy minimization criterion. Let us consider a computer-simulated example in which the method estimates a solution different from the true one (16). Figure 3.17a shows two spectra generated by a computer that contain two peaks at the same positions but with different peak-to-height ratios. We simulated a number of mixtures of these two spectra by the computer and

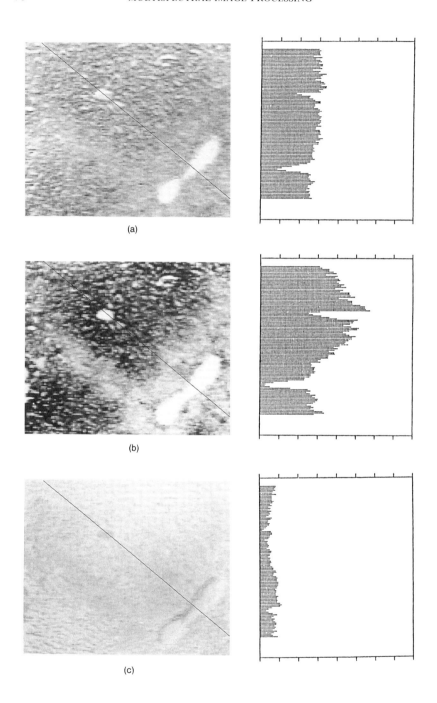

(a)

(b)

(c)

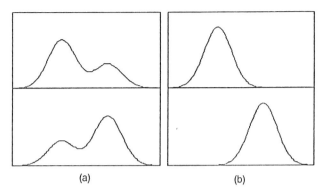

(a)   (b)

Figure 3.17. Example of the estimation giving a solution different from the true one: (a) computer-simulated component spectra; (b) estimated spectra obtained by entropy minimization.

estimated the components from them. The result is shown in Figure 3.17b, which is different from the original in Figure 3.17a. However, the spectra of Figure 3.17a can be recognized as the compositions of the spectra in Figure 3.17b with different absorbing mechanisms that correspond to the two peaks, and the entropy minimization successfully resolved them into the components. Fortunately or unfortunately, in a number of practical applications of the present method, including non-image-type spectral data separations (21), we have not experienced such a situation.

3.6. FLUORESCENT PATTERN SEPARATION BY DOUBLE EIGENANALYSIS

In this section, we describe a method by which the true solution of component patterns, as well as their spectra, can be uniquely and deterministically found without any constraint or criterion (22). This method is applicable not to general cases of multispectral images but rather to images formed by two physical processes, such as fluorescent excitation–emission images and those obtained by photoacoustic spectrometry. We introduce the mathematical theory to obtain the true solution of components from the multispectral image data for the case of fluorescent samples.

The sample is excited (illuminated) at two different spectral bands, and the fluorescent images are observed as multispectral data. The two sets of the

Figure 3.16. Estimated density patterns and profiles along a line: (a), (b), and (c) correspond to spectra A, B, and C in Figure 3.15, respectively.

multispectral images are expressed as

$$[D_1] = [S_1][P] \tag{32}$$

$$[D_2] = [S_2][P] \tag{33}$$

where $[S_1]$ and $[S_2]$ are spectral response matrices under two different excitations. The objective is to estimate the elements of the matrices $[S_1], [S_2]$ and $[P]$ from the observed matrices $[D_1]$ and $[D_2]$. Equations (32) and (33) are also unsolvable; therefore, we utilize the additional information on the physical processes of molecular fluorescence. The shape of the fluorescent spectrum of each component is normally independent of the excitation wavelength, but the total intensity varies in proportion to absorbance at the wavelength. Therefore, the component fluorescent spectrum is given by a product of absorbance and fluorescence emission intensity at the excitation and emission wavelengths, respectively. By introducing a diagonal matrix $[A]$, the emission–spectral matrices $[S_1]$ and $[S_2]$ are combined such that

$$[S_2] = [S_1][A] \tag{34}$$

where diagonals of $[A]$ are absorbance ratios of band 2 to band 1 for the individual components, which are also unknown.

By using the eigenvalue-eigenvector analysis for the multispectral image $[D_1]$, the unknown matrices $[P]$ and $[S_1]$ are decomposed to

$$[P] = [T][V_1] \tag{35}$$

$$[S_1] = [U_1][E_1][T]^{-1} \tag{36}$$

where $[E_1]$ is an eigenvalue matrix, and $[U_1]$ and $[V_1]$ are eigenvector matrices of $[D_1][D_1]^t$ and $[D_1]^t[D_1]$, respectively. Substitution of Eq. (34) into Eq. (33) and replacement of $[P]$ and $[S_1]$ by Eqs. (35) and (36) yield

$$[D_2] = [S_1][A][P] = [U_1][E_1][T]^{-1}[A][T][V_1] \tag{37}$$

or

$$[E_1]^{-1}[U_1]^t[D_2][V_1]^t = [T]^{-1}[A][T] . \tag{38}$$

Note that all the matrices on the left-hand side of Eq. (38) are known. With the definition

$$[B] = [E_1]^{-1}[U_1]^t[D_2][V_1]^t \tag{39}$$

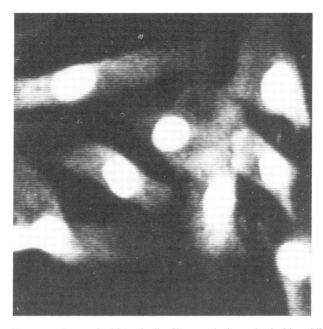

Figure 3.18. Fluorescent image of cultivated cells of human gingiva stained with acridine orange.

Eq. (38) becomes

$$[B] = [T]^{-1}[A][T] \tag{40}$$

Since $[A]$ is a diagonal matrix, Eq. (40) is again another form of the eigenvector decomposition; $[A]$ and $[T]$ can be found from $[B]$ by diagonalization with the eigenanalysis, which determines these matrices as eigenvalues and eigenvectors of $[B]$, respectively. Finally, $[P]$, $[S_1]$, and $[S_2]$ are obtained from Eqs. (35), (36), and (34), respectively, by using the determined values of $[A]$ and $[T]$. In this method, the number of pixels and emission wavelengths must be larger than the number of components, while the number of excitation bands does not depend on the sample but is fixed at only two.

Figure 3.18 shows a fluorescent image of cultivated cells of human gingiva stained with acridine orange (22). This sample was excited at two different spectral bands, and their fluorescent images were measured through four different spectral bands. Two filters used for band-selective excitation covered 455–490 and 475–490 nm, respectively, while four filters for emission band selection passed fluorescence over 515, 530, 570, and 590 nm, respectively.

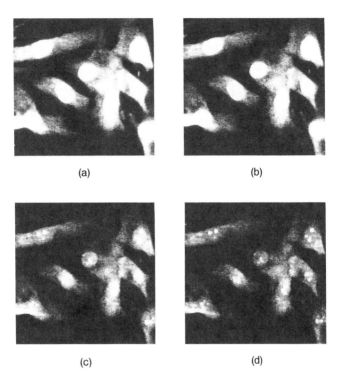

(a)   (b)

(c)   (d)

Figure 3.19. Multispectral image of the sample of Figure 3.18 excited at a 455–490 nm band and measured at the spectral bands over (a) 515, (b), 530, (c) 570, and (d) 590 nm. Parts (e)–(h) are the same but excited at a 475–490 nm band.

Figure 3.19 shows these bandpass images of the same sample excited at the different bands.

Figure 3.20 is a plot of eigenvalues calculated by using the images shown in Figure 3.19a–d. The number of components is found to be two by counting the number of eigenvalues over the noise-power level shown in Figure 3.20. We calculated the separation of spatial patterns and spectra into two components from the image data of Figure 3.19. Figure 3.21 shows the spectral responses [S] of the fluorescence intensities for the two separated components, and Figure 3.22 shows the estimated spatial patterns of the components.

We knew the components of the sample to be separated before the experiment; this experiment was performed to verify the present method.

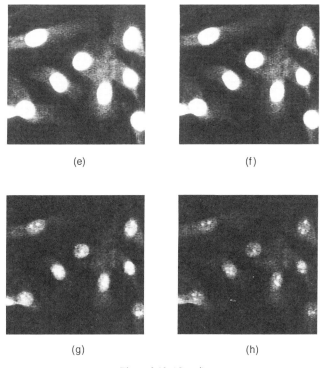

(e)   (f)

(g)   (h)

Figure 3.19 *(Contd)*

Acridine orange is known as a multimarker of DNA and RNA. It binds with DNA and forms a monomeric complex, while it forms a dimeric one by binding with RNA. Figure 3.22a represents the distribution of DNA, which exists in nuclei, while Figure 3.22b represents the distribution of RNA, which exists in cytoplasms and nucleoli in nuclei. These two components are successfully separated in the resultant images.

The present method does not require any constraint, support or any criterion; rather, it uses two sets of multispectral images. This method can be applied only to images formed by two physical processes. In addition, these two processes have to contribute independently and multiplicatively to image formation, as in Eqs. (32)–(34). For example, multiband fluorescent images are given by the multiplicative form of emission spectra and absorbance spectra, as described above.

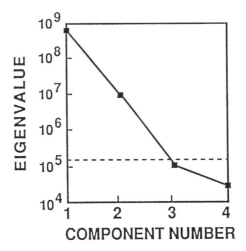

Figure 3.20. Plot of eigenvalues of the image data shown in Figure 3.19a–d. The dashed line indicates the noise level.

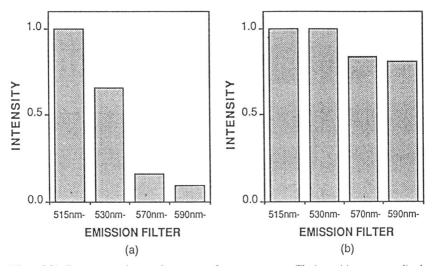

Figure 3.21. Reconstructed spectral responses of two components. The intensities are normalized by the maximum value.

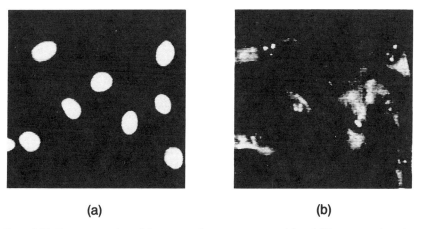

(a) **(b)**

Figure 3.22. Reconstructed spatial patterns of two components: (a) and (b) correspond to the spectral responses of Figures 3.21a and 3.21b, respectively.

3.7. CONCLUSION

We have introduced several multispectral image analysis methods that enable one to estimate spatial distribution patterns of individual components existing in the image and/or in the corresponding spectral responses. These methods are based on different assumptions, so their use depends on the obtained a priori information and the observed data. It is worth noting that the key to these component analyses is the use of two-dimensional data as a function of spatial position and wavelength. The analyses cannot be performed by independent observation of images and spectra. Furthermore, the fluorescence emission–excitation image data, which add one dimension (excitation wavelength) to the multispectral image, provide enough information to determine the unique solution for the component patterns and spectra, without any constraint or optimization cirterion. These methods are also applicable to nonimage data such as time-resolved fluorescence and absorption spectra of photophysical and photochemical processes. They can be expected to elucidate the reaction dynamics and to discover new transient species (15, 21, 23, 24).

REFERENCES

1. Sklansky, J., and Wassel, G. N. (1981). *Pattern Classifiers and Trainable Machine.* Springer-Verlag, New York.

2. Pratt, W. K. (1978). *Digital Image Processing*. Wiley, New York.

3. Nagy, G. (1972). *Proc. IEEE* **60**, 1177.

4. Lehmann, L. A., Alvarez, R. E., Macovski, A., Brody, W. R., Pelc, N. J., Riederer, S. J., and Hall, A. L. (1981). *Med. Phys.* **8**, 659.

5. Chiou, W. C. Sr. (1985). *Appl. Opt.* **24**, 2085.

6. Sasaki, K., Kawata, S., and Minami, S. (1986). *Appl. Spectrosc.* **40**, 185.

7. Charalambous, G. (Ed.) (1984). *Analysis of Foods and Beverages*. Academic Press, New York.

8. Paradimitriou, C. H., and Steiglitz, K. (1982). *Combinatorial Optimization: Algorithms and Complexity*. Prentice-Hall, Englewood Cliffs, Nd.

9. Salkin, H. M. (1975). *Integer Programming*. Addison-Wesley, Menlo Park, CA.

10. Kawata, S., Sasaki, K., and Minami, S. (1987). *J. Opt. Soc. Am. A* **4**, 2101.

11. Lawson, C. L., and Hanson, R. J. (1974). *Solving Least Squares Problems*. Prentice-Hall, Englewood Cliffs, Nd.

12. Malinowski, R. R. (1977). *Anal. Chem.* **49**, 612.

13. Akaike, H. (1974). *IEEE Trans. Autom. Control.* **AC-19**, 716.

14. Lawton, W. H., and Sylvestre, E. A. (1971). *Technometrics* **13**, 617.

15. Kawata, S., Komeda, H., Sasaki, K., and Minami, S. (1985). *Appl. Spectrosc.* **39**, 610.

16. Sasaki, K., Kawata, S., and Minami, S. (1989). *J. Opt. Soc. Am. A* **6**, 73.

17. Kowalik, J., and Osborne, M. R. (1968). *Methods for Unconstrained Optimization Problems*. Elsevier, New York.

18. Araki, T., Chikamori, K., Sasaki, K., Kawata, S., Minami, S., and Yamada, M. (1987). *Histochemistry* **86**, 567.

19. Altman, F. P. (1974). *Histochemie* **38**, 155.

20. Chikamori, K., Araki, T., and Yamada, M. (1985). *Cell. Mol. Biol.* **31**, 217.

21. Sasaki, K., Kawata, S., and Minami, S. (1984). *Appl. Opt.* **23**, 1955.

22. Sasaki, K., and Kawata, S. (1990). *J. Opt. Soc. Am. A* **7**, 513.

23. Sasaki, K., Kawata, S., and Minami, S. (1983). *Appl. Opt.* **22**, 3599.

24. Sakai, H., Itaya, A., Masuhara, H., Sasaki, K., and Kawata, S. (1993). *Chem. Phys. Lett.* **208**, 283.

CHAPTER

4

SPECTRAL BIO-IMAGING

YUVAL GARINI, NIR KATZIR, DARIO CABIB,
and ROBERT A. BUCKWALD

Applied Spectral Imaging Ltd.
Industrial Park
Migdal Haemek, Israel

DIRK G. SOENKSEN

Applied Spectral Imaging Inc.
Carlsbad, California 92009

ZVI MALIK

Life Sciences Department
Bar Ilan University
Ramat-Gan 52900, Israel

4.1. INTRODUCTION

Spectroscopy is a well-known analytical tool that has been used for decades to characterize the spectral signatures of chemical constituents. The physical basis of spectroscopy is the interaction of light with matter. Traditionally, spectroscopy is the measurement of the light intensity (I) that is emitted, transmitted, scattered, or reflected from a sample, as a function of wavelength, at high spectral resolution—but without any spatial information. Imaging, on the other hand, is primarily concerned with obtaining high spatial resolution information from a two-dimensional sample—but imaging usually provides only limited spectral information (e.g., by imaging with one or several discrete bandpass filters (1).

The work described here is based on the newly developed SpectraCube™ method of spectral bio-imaging, a combination of Fourier spectroscopy, charge-coupled device (CCD) imaging, light microscopy, and analysis soft-

Fluorescence Imaging Spectroscopy and Microscopy, edited by Xue Feng Wang and Brian Herman. Chemical Analysis Series, Vol. 137.
ISBN 0-471-01527-X © 1996 John Wiley & Sons, Inc.

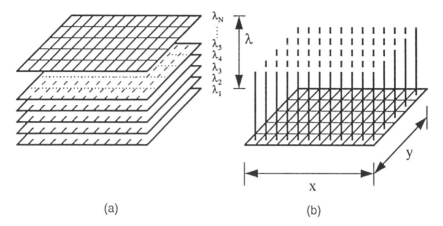

(a)   (b)

Figure 4.1. A spectral image can be viewed as a stack of contiguous images, each measured at a different wavelength, or as a two-dimensional array of vectors, with each vector representing the measured light spectrum at a given pixel.

ware for biological research applications. Spectral bio-imaging is a powerful method for measuring the spectrum of light at every picture element (pixel) of a two-dimensional image. The spectral images acquired by a system based on the SpectraCube™ method constitute a cube of information $I_{xy}(\lambda)$, in which the full spectrum has been measured at each pixel position x, y. Since spectral images bring together the salient features of spectroscopy and imaging, new insights can be gained into the structure and dynamics of life by applying both spectral and image analysis algorithms to the same image data cube. Figure 4.1 is a schematic of a typical spectral image.

Combining spectroscopy with imaging is particularly useful in investigations with fluorescent probes. For example, spectral bio-imaging can be used to identify and map several fluorophores simultaneously in one measurement. In fact, the inherently high spectral resolution of spectral bio-imaging is ideally suited for sorting out fluorescent probes (or other chemical constituents) with overlapping spectra. Similarly, the imaging capabilities of spectral bio-imaging enable the detection, at any location in the image, of subtle spectral shifts (e.g., probes for pH). Spectral bio-imaging is a new modality of bio-imaging, a modality in which the high-resolution spectrum measured at each point in the image adds a new dimension, one that will enable the exploration of cells and tissues in new and exciting ways.

Spectral imaging has been, and continues to be, used successfully in the area of remote sensing to provide important insights into the study of planets, including Earth. By identifying characteristic spectral absorption features, it has been possible to study many aspects of Earth, including environmental changes, important rock-forming minerals, the effects of soil composition on trees (i.e., by measuring shifts in the chlorophyll absorption band), forest fire damage, etc. However, the high cost, large size, and complexity of remote sensing spectral imaging systems (e.g., Landsat, AVIRIS) has limited their use to well-funded agencies (2–6). These systems are not practical for laboratory use with a microscope.

Performance improvements and the cost-effective availability of CCD imaging detectors, low-noise electronics, computers, and image analysis software have enabled the development of a commercially available spectral bio-imaging system for biological microscopy applications. In the following sections, the general structure, elements, and performance considerations of a spectral bio-imaging system based on the SpectraCube™ method will be described. Some potential fluorescence imaging applications, including examples of spectral images acquired with the SpectraCube™ method, will also be presented.

4.2. SPECTRAL BIO-IMAGING SYSTEMS

4.2.1. Overview

Conceptually, a spectral bio-imaging system consists of a measurement system and analysis software. The measurement system includes all of the optics, electronics, and software necessary to acquire a spectral image. The analysis includes all of the software and mathematical algorithms that are necessary to display and analyze spectral images acquired with the measurement system.

4.2.2. System Architecture

The system architecture of a typical spectral bio-imaging system is shown in Figure 4.2. The spectral bio-imaging system attaches, like a conventional CCD camera, to the video port of a microscope. Light is projected from the microscope port into the optical head, which consists of three main elements:

1. An inteferometer (or spectral dispersion element) that enables the separation of light into its spectral components, i.e., different wavelengths
2. A CCD array detector that collects, for all pixels in the array simultaneously, the light intensity required to measure (or calculate) the spectrum at each point in the image
3. Collection and imaging optics that produce a real image on the CCD array while enabling proper operation of the interferometer; in particu-

Optical Head

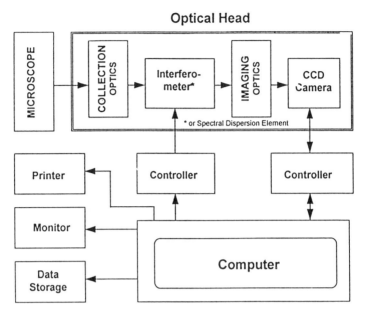

Figure 4.2. System architecture of a typical spectral bio-imaging system. The optical signal (image) originating in the microscope passes through the interferometer (or spectral dispersion element) and is focused on the CCD camera. Both the interferometer and the CCD have their own controllers, which in turn are controlled by the computer. The computer orchestrates the overall acquisition of spectral images and is used to display, analyze, and output spectral images.

lar, the collection optics should be such that all the light projected from the microscope will be collected by the optical head

Both the interferometer and the CCD camera usually require dedicated controllers. The coordination of the interferometer (or spectral dispersion element) with the CCD camera is orchestrated by a dedicated computer equipped with the appropriate image acquisition and processing hardware and software. Because spectral image analysis is one of the most important and challenging aspects of spectral imaging, this important subject will be discussed in Section 4.3.

4.2.3. Fourier Spectroscopy and Spectral Dispersion Methods

The heart of any spectral bio-imaging system is an interferometer (or spectral dispersion element) that enables the separation of the light entering the optical head into its spectral components. Interferometric spectroscopy (better known as Fourier spectroscopy) is one of three methods of separating light into its consituent spectral components. The other two methods, grating and spectral filters, are spectral dispersion methods that might also be considered for a spectral bio-imaging system.

In a grating (monochromator)-based system, only one axis of the CCD array detector (the spatial axis) provides real imagery data; the other (spectral) axis is used to measure the intensity of the light dispersed by the grating. An image can be obtained only after scanning the grating parallel to the spectral axis of the CCD. This inability to visualize the two-dimensional image makes it virtually impossible to choose, prior to making a measurement, a desired region of interest from within the field of view and to optimize the system focus or exposure time. Grating-based spectral imagers are popular for remote sensing applications because an airplane (or satellite) flying over the surface of the Earth provides a natural scanning mechanism. A list of spectral imaging systems based on a monochromatic grating can be found in Ref. 7.

Only the Fourier spectroscopy or filter approaches are able to provide a real two-dimensional image that can be observed, focused, and optimized prior to the measurement. These two methods acquire imagery data in both spatial axes (x and y) simultaneously, thus providing a complete two-dimensional image for every frame time of the CCD camera. A more detailed discussion of these two methods follows.

4.2.3.1. Fourier Spectroscopy

Fourier spectroscopy uses an interferometer and Fourier transform methods to derive the spectrum. The principles of this method are as follows: The interferometer divides the incoming beam (in this case, the light projected from the port of the microscope) into two coherent beams and creates a variable optical path difference (OPD) between them. The beams are then recombined to interfere with each other, and the resulting interference intensity is measured by the detector as a function of the OPD. This intensity vs. OPD function, called an interferogram, is then Fourier transformed to recover the spectrum. Fourier spectroscopy is a well-known technique that is widely used to make high-resolution, high-sensitivity spectral measurements for a variety of applications, mainly in the infrared spectral range (8, 9). When applied to an imaging system, the interferogram is measured individually at each pixel in the CCD array. Fourier transformation thus yields a distinct spectrum at each pixel. A more detailed explanation of Fourier spectroscopy is given in the appendix.

Fourier spectroscopy enjoys several important advantages as a method for measuring spectral data, especially when applied to fluorescence imaging microscopy. These advantages include:

1. High optical throughput, achieved by the efficient collection of the intensities at all wavelengths in the source spectrum during the *entire* acquisition time (unlike a filter-based system, which at any moment only measures the intensity in a single narrow spectral band)
2. High and variable spectral resolution
3. A wide spectral range that can also be extended to the ultraviolet (UV) and infrared (IR) spectral regions.

The high optical throughput advantage of Fourier spectroscopy results in measurements with higher signal-to-noise ratios (SNRs), relative to the SNR measured with a filter-based system, or enables faster measurements. Higher SNR or reduced acquisition times are important when low-light signals are measured in fluorescence microscopy. Furthermore, the advantage of high and variable spectral resolution allows the user to trade off sensitivity, spectral resolution, and acquisition time in a way that is not possible with any other method. Furthermore, interferometers do not polarize the incoming light, thus facilitating the direct measurement of polarization effects.

A disadvantage of interferometric spectroscopy is that the mathematical Fourier transform operation must be performed in order to observe the spectrum. In a filter-based system, on the other hand, the intensity of the spectrum can be measured directly (over a narrow spectral region).

4.2.3.2. Filter-Based Spectral Dispersion Methods

Filter-based dispersion methods can be further categorized into discrete filters and tunable filters. The use of discrete spectral filters typically requires inserting a few filters into a filter wheel, which is rotated, synchronous with the frame rate of the CCD camera. Tunable filters, such as acousto-optic tunable filters (AOTFs) and liquid-crystal tunable filters (LCTFs), have no moving parts and can be randomly tuned to any wavelength in the spectral range of the device. One advantage of using tunable filters as a dispersion method for spectral imaging is their random wavelength access, i.e., the ability to measure the intensity of an image at a number of wavelengths in any desired sequence. However, AOTFs and LCTFs have the disadvantages of (i) limited spectral range (typically $\lambda_{max} = 2\lambda_{min}$), and all radiation that falls outside of this spectral range *must* be blocked; (ii) temperature sensitivity; (iii) poor light throughput; (iv) polarization sensitivity; and (v) in the case of AOTFs, shifting of the image during wavelength scanning. Nevertheless, both LCTFs and AOTFs have been used successfully in spectral imaging systems, primarily for nonbiological applications. The interested reader is referred to Morris et al. (10) and Yu et al. (11) for further details.

4.2.4. Charge-Coupled Device Detectors

The CCD (camera) detector is one of the most critical elements of a spectral bio-imaging system. Although the CCD camera is located at the end of the light path, its performance largely determines the overall spectral imaging performance that can be achieved. For a detailed discussion of CCDs, refer to Chapter 21 in Mason (12). It is assumed that the reader has some familiarity with CCD detectors. A brief qualitative discussion of some of the more important parameters of CCDs and their relevance to spectral bio-imaging follows.

4.2.4.1. Pixel Size

The pixel size of a CCD array detector, typically between 6 and 25 μm along each dimension of the pixel, is important for several reasons. First, and most important, the pixel size defines the best achievable *spatial resolution* of the imaging system. Selecting a CCD array with smaller pixels improves the spatial resolution, allowing higher spatial frequency details to be sampled without aliasing. This reasoning would suggest that it is best to always select the CCD array with the smallest pixel size; however, this is not always the case because smaller pixels also have smaller well capacities for storing photoelectrons. Since the well capacity (and readout noise) define the dynamic range of the CCD detector (i.e., the ratio of the maximum to the minimum signal that can be measured simultaneously), pixel size is proportional to dynamic range. Thus CCD arrays with larger pixels have a proportionately higher dynamic range. In many spectroscopy applications, a large dynamic range is required in order to separate chemical constituents that exhibit subtle spectral differences; hence, spatial resolution and dynamic range must be traded off, depending on the specific requirements of an application.

4.2.4.2. Spectral Response

The spectral response of a CCD camera is primarily a function of the energy band-gap structure of the silicon material on which most CCDs arrays are based. In practice, the spectral range of a typical CCD detector extends from 400 to 1100 nm, with peak quantum efficiencies between 40 and 50%. By applying UV coatings to the surface of the CCD, the spectral response can be extended to the UV region. More expensive back-illuminated CCDs are available with good response in the UV and improved quantum efficiencies (as high as 85%).

4.2.4.3. CCD Camera Noise

There are three primary sources of noise in a typical CCD camera: shot noise, dark noise, and readout noise. Photon shot noise is the result of statistical fluctuations in the signal; it is proportional to the square root of the signal intensity. Dark noise, on the other hand, is caused by electrons that are excited by the CCD's thermal energy. Dark noise is linearly dependent on the square root of the frame integration time of the CCD (typical units are electrons per second), where dark noise is the square root of the dark charge that accumulated in the CCD. Thus, in order to maintain low dark noise when long integration times are required (e.g., to measure low-light signals), it is frequently necessary to cool the CCD (e.g., thermoelectrically or with liquid nitrogen). Readout noise originates primarily in the output electronics of the CCD. The pixel readout noise is proportional to the data readout rate of the CCD camera. Thus, lower readout noise requires low pixel rates, while high pixel rates are usually accompanied by higher noise. Here too a trade-off must be made, depending on the application, between faster readout rate and reduced readout noise.

4.2.4.4.   Chip Architecture and Readout Modes

The chip architecture of the CCD describes the structure of the CCD chip. A typical CCD chip employed in a high-performance camera is either a full-frame CCD or a frame-transfer CCD. In a full-frame CCD the active area encompasses the entire area of the chip. In a frame-transfer CCD, on the other hand, the CCD chip is divided into two areas: a sensing (active) area and a storage area. The photoelectrons that accumulate in the sensing area during a single frame integration time can be transferred to the storage area very quickly, allowing data in the storage area to be read while the sensing area is exposed to acquire the next image. Frame-transfer CCDs can accelerate the acquisition time of multiple images and are thus preferred for spectral bio-imaging. Another advantage of frame-transfer CCDs is that they can often be used effectively without a mechanical shutter, further facilitating higher effective CCD frame rates.

4.2.5.   Performance of a Spectral Bio-Imaging System

Since a spectral bio-imaging system consists of several elements (e.g., an optical head, a CCD camera, and a computer), it is important to characterize its performance as a system. The performance parameters of a spectral bio-imaging system can be categorized into imaging, spectral, and overall performance parameters. A brief discussion of the most important parameters follows.

4.2.5.1.   Imaging Performance

4.2.5.1.1.   Spatial Resolution. The spatial resolution of a spectral bio-imaging system depends on two major parameters: the pixel size and the optical modulation transfer function (MTF) of the system. If the optics are well designed (usually a good assumption), the spatial resolution is essentially limited by the pixel size. In this case, the spatial resolution is approximately equal to $x = 2d/M$, where x is the smallest resolvable dimension in the sample, d is the pixel size, and M is the overall magnification of the optics between the sample and the detector, including the magnification of the microscope and any other magnification effects introduced by the system.

4.2.5.1.2. Field of View. The field of view, X (defined here as the total size of the sample that can be measured), is equal to Nd/M, where N is the number of pixels in a row (or column) of the CCD array and M and d are as defined above.

4.2.5.1.3. Sensitivity. Sensitivity (as defined here) is the lowest light level at which the system achieves reasonable performance. Because the illumination, the spectral response of the CCD, and the transmission of the optics are not uniform in a real system, the sensitivity is actually different for each wavelength. Sensitivity is usually specified in units of lux (lumes/m^2), a photopic unit that represents a "human eye" weighted average of the (spectral) sensitivity. When a high-quality cooled CCD is used, the sensitivity (to first order) improves linearly with the integration time; it is therefore important to specify the sensitivity at a given integration time. A parameter related to the sensitivity is the minimum detectable light level, i.e., the incident light level at which the SNR at the output of the detector is 1. The minimum detectable light level is always lower than the sensitivity.

4.2.5.1.4. Dynamic Range. The dynamic range of a spectral bio-imaging system defines the largest differences in the intensities of incident light that can be observed in one spectral image. The system dynamic range depends largely on the dynamic range of the CCD detector, i.e., the number of meaningful levels into which a full-scale optical signal incident on the CCD can be divided. The CCD dynamic range is typically given by the number of real bits of the CCD, n, and the corresponding number of separable levels is equal to 2^n. There are other factors that can affect the system dynamic range, and in some cases the actual system dynamic range can increase over that of the CCD detector.

4.2.5.2. Spectral Performance

4.2.5.2.1. Spectral Range. The spectral range defines the range of wavelengths over which the system can measure the spectrum. Because the response of a real system is not spectrally uniform, it is important to characterize the spectral response of the system (e.g., via a plot of relative response vs. wavelength). The spectral range is usually given by the minimum and maximum wavelengths at which the spectral response exceeds some threshold (e.g., 5%).

4.2.5.2.2. Spectral Resolution. The spectral resolution defines the ability of a spectral bio-imaging system to distinguish (or spectrally resolve) two signals that are separated by a small spectral difference. Spectral resolution is typically specified by the full width at half-maximum (FWHM) criterion, a parameter that can be checked by measuring the "spread" in the spectrum of a very (spectrally) narrow source (e.g., a laser). Depending on the specific nature of the spectral peaks, the spectral resolution achievable in a real system is often smaller (i.e., better) than the FWHM value.

4.2.5.2.3. Polarization. Here polarization is defined as the degree of polarization of the system, i.e., the transmission of the system as a function of the polarization of the incident light.

4.2.5.3. General Performance

The measurement of a spectral image requires significantly more time than is required to acquire a single frame using a conventional (CCD) imaging system. The reason for this increase in the acquisition time is that the spectral image must be measured at different wavelengths (or different OPDs). The measurement time depends both on the effective frame time (i.e., the sum of integration time, data transfer time, and any additional overhead required to change the wavelength of the OPD) and on the total number of frames that should be measured. The time required to transfer the data from the CCD camera to the computer can impose a significant overhead on the effective frame rate; thus, selection of CCDs with fast readout rates is often desirable. For many high-performance CCD cameras, the readout rate can be increased by selecting a subset (i.e., a region of interest) of the total number of pixels in the CCD array.

4.2.6. The SpectraCube™ Method

The SpectraCube™ method of spectral bio-imaging combines Fourier spectroscopy with a CCD array detector and powerful analysis methods. The interferometer is a special type of common path interferometer known as a Sagnac interferometer. A simplified optical diagram of the SpectraCube™ method is shown in Figure 4.3. In a Sagnac interferometer, the two beams (delayed by a known OPD) travel in the same (common) path, but in opposite directions, and are then recombined, enabling the measurement of interference intensity by the detector. The advantage of a common path interferometer is its intrinsic stability. Disturbances, such as a small shift of one of the optical elements, affect

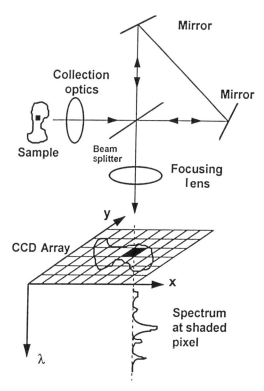

Figure 4.3. Simplified optical diagram of a spectral bio-imaging system based on the SpectraCube™ method. The light emitted by the sample is collected by the fore-optics and enters the Sagnac interferometer. The real image (plus interference fringes) is focused on the CCD array, yielding (on Fourier transformation) the spectrum at each pixel.

both beams the same way and hence have a minor effect on the measurement. In addition to being rugged, this type of interferometer is relatively compact. A more detailed discussion of the Sagnac interferometer is given in the appendix.

The SpectraCube™ interferometer is designed in such a way that a real image (on which the interference fringes generated by the Sagnac interferometer are superimposed) is projected on the CCD detector for any OPD. This image does not move when the OPD is changed. The ability to see a real image of the sample of interest facilitates definition of the region of interest (ROI) over which the spectral image will be measured, as well as optimization of measurement parameters such as focus, integration time, and so on.

During the measurement, the OPD is changed and many frames of data are acquired (and stored) at many different OPDs until the entire cube of interferometric data has been acquired. It is important to realize that each pixel acts very much like an independent Fourier spectrometer, and the acquisition of a spectral image is analogous to the simultaneous measurement of a complete interferogram, in parallel, by tens of thousands of point spectrometers, each focused on one small element of the sample. Fourier transformation of this interferogram data cube yields the complete spectrum at every pixel, i.e., the spectral image.

The spectral bio-imaging system described here, the SD200 (Applied Spectral Imaging Ltd., Migdal Haemek, Israel), is based on the SpectraCube™ method. The CCD camera employed by the SD200 is a $512 \times 512 \times 2$ pixel, thermoelectrically cooled, frame-transfer CCD with 12 bits of dynamic range (Princeton Instruments, Trenton, New Jersey). The pixel size is 15 μm square, and the readout rate is 1 Mixpel per second. This system is based on a 486/66 personal computer. The SD200 achieves a spatial resolution equal to $30/M$ μm and a field of view equal to $8/M$ mm, where M is the effective microscope magnification. The sensitivity of the SD200 is approximately 20 millilux for 100 ms integration times, while the spectral range is 400–1000 nm. Spectral resolutions of 4 nm at 400 nm (14 nm at 800 nm) are achievable. The SD200 system easily attaches to any microscope with a C-mount or F-mount connector and can stand in any orientation during the measurement.

4.3. SPECTRAL IMAGE ANALYSIS

4.3.1. Overview

A spectral image is a three-dimensional array of data, $I(x, y, \lambda)$, that combines precise spectral informaton with spatial correlation. As such, a spectral image is a novel data base that enables the extraction of features and the evaluation of quantities that are difficult, and in some cases impossible, to obtain otherwise. Since both spectroscopy and digital image analysis are well-known fields (13) with an enormous literature, this discussion will focus primarily on the benefits of combining spectroscopic and imaging information in a single (spectral image) data base.

One simple approach to the analysis of a spectral image data base is to perform the analysis of spectral and spatial (image) data separately, i.e., to apply spectral algorithms to the spectral data and (two-dimensional) image processing algorithms to the spatial data. In this case, a two-dimensional monochrome image (or set of images) will be created as a result of spectral analyses that have been performed on each of the image pixels (point operations). These gray scale images can then be further analyzed using image processing and computer vision techniques (e.g., image enhancement, pattern recognition) to extract features and parameters of interest. This handshaking between the spectral- and image-based algorithms can in general be iterated, as shown in the accompanying diagram.

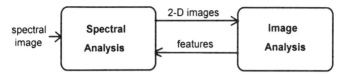

Of course, it is also possible to apply spectral image algorithms based on nonseparable operations, i.e., algorithms that include both local spectral information and spatial correlation between adjacent pixels. The examples given in the following section will be limited to the simpler case where spectral and spatial algorithms are separable.

4.3.2. Displaying Spectral Images

One of the basic needs that arise naturally when dealing with any three-dimensional (3-D) data structure such as a spectral image, is visualizing that data structure in a meaningful way. Unlike other types of 3-D data such as tomographic data, $D(x, y, z)$, where each point represents, in general, a different location (x, y, z) in *space*, a spectral image is a sequence of images representing the intensity of the same two-dimensional *plane* (i.e., the sample) at different wavelengths. For this reason, the most intuitive way to view a spectral image is to view either the image plane (spatial data) or the spectral axis. In general, the image plane can be used to display either the intensity measured at any single wavelength or the gray scale image that results after applying a spectral analysis algorithm, over a desired spectral region, at every image pixel. The spectral axis can, in general, be used to present the resultant spectrum of some spatial operation performed in the vicinity of any desired pixel (e.g., averaging the spectrum from several adjacent pixels).

It is possible, for example, to display the spectral image as a gray scale image similar to the image that might be obtain from a simple monochrome camera. Since such a camera simply integrates the optical signal over the spectral range of the CCD array, the equivalent monochrome CCD camera image can be computed from the 3-D spectral image data base by integrating along the spectral axis as follows:

$$\text{gray scale } (x, y) = \int_{\lambda_1}^{\lambda_2} w(\lambda) \cdot I(x, y, \lambda) \, d\lambda \tag{1}$$

Here $w(\lambda)$ is a general weighting response function that provides maximum flexibility in computing a variety of gray scale images, all based on the integration of an appropriately weighted spectral image over some spectral range. For

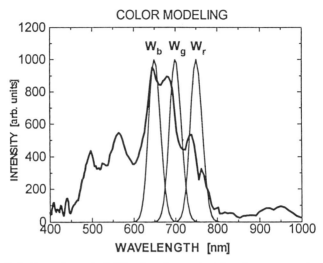

Figure 4.4. Definition of pseudo-RGB colors for emphasizing chosen spectral ranges. The intensity for each pseudocolor is calculated by integrating the area under the curve after multiplication by the appropriate pseudo-RGB spectral filter function.

example, by evaluating Eq. (1) with three different weighting functions, $\{w_r(\lambda), w_g(\lambda), w_b(\lambda)\}$, corresponding to the tri-stimulus response functions for red (R), green (G), and blue (B), respectively, it is possible to display a conventional RGB color image. It is also possible to display meaningful nonconventional (pseudo) color images. As an example of the power of this simple algorithm, consider choosing $\{w_r, w_g, w_b\}$ to be Gaussian functions distributed "inside" a spectrum of interest (Figure 4.4). The resulting pseudocolor image that is displayed in this case emphasizes only data in the spectral regions corresponding to the weighting functions, enabling spectral differences in these three regions to be detected more clearly.

4.3.3. Point Operations

Point operations are mathematical operations that can be performed on the image pixels of a variety of images. For example, in a traditional gray scale image, point operations map the intensity $v_1 \in [0, L]$ at a given image pixel into another intensity $v_2 \in [0, L]$ according to some transformation $v_2 = f(v_1)$. A simple example of a point operation applied to a gray scale image is the multiplication of the intensity at each pixel by some constant. The concept of point operations can also be extended to spectral images, in which each pixel constitutes a spectrum, i.e., an n-dimensional vector $_1(\lambda)$; $\lambda \in [\lambda_1, \lambda_n]$. A point operation applied to a spectral image would thus map the spectrum at any image pixel into a scalar (i.e., an intensity value) according to the transformation

$$v_2 = g(V_1(\lambda)); \qquad \lambda \in [\lambda_1, \lambda_n]$$

The evaluation of the gray scale image in Eq. (1) is an example of this type of spectral image point operation. In the more general case, a point operation maps the spectrum (vector) at any image pixel into another vector in accordance with the transformation

$$_2(l) = g(V_1(\lambda)); \qquad l \in [1, N], \qquad \lambda \in [\lambda_1, \lambda_n]$$

where $N \leq n$.

If one now extends the definition of point operations to include operations between corresponding pixels of *different* spectral images, it is possible to categorize the entire field of spectroscopic calculations as point operations. Furthermore, since the spectrum at every pixel (x_0, y_0) is given by

$$s(x_0, y_0) = I(x_0, y_0, \lambda); \qquad \lambda \in [\lambda_1, \lambda_n]$$

the extraction of spectroscopic information from a spectral image (e.g., determination of absorption wavelengths and ratiometric computations) can be done separately on any pixel of the image. An important example of this type of point spectral arithmetic is the optical density function defined by

$$OD(x, y, \lambda) = -\log \frac{I_s(x, y, \lambda)}{I_b(x, y, \lambda)} \qquad (2)$$

where I_s is the (normal) transmission spectral image through the sample, and I_b is the spectral image measured in the absence of the sample (but under otherwise identical conditions). Note that the optical density is invariant both to the spectral response of the measuring system and to the nonuniformity of the CCD detector.

4.3.4. Spatial-Spectral Operations

In all of the spectral image analysis methods mentioned above, algorithms were applied to the spectral data. The value added by also displaying the spectrally processed data as an image is mostly qualitative, providing the user with the aesthetically pleasing picture. It is also possible, however, to use the available imaging data in much more meaningful ways by applying algorithms that utilize the spatial-spectral correlation inherent in a spectral image. Spatial-spectral operations represent the most powerful types of spectral image analysis algorithms. As an example, consider the following problem:

A tissue contains k cell types stained with k different fluorophores. Each fluorophore has a distinct fluorescence emission spectrum and binds to only one of the k cell types. It is important to find the average fluorescence intensity per cell for each of the k cell types. Solving this task involves the following procedure (subtasks 1–3):

1. Classify each pixel in the image as belonging to one of $k + 1$ classes (k cell types plus a background) due to its spectrum.
2. Segment the image into the various cell types, and count the number of cells of each type.
3. Sum the fluorescence energy contributed by each class and divide it by the total number of cells from the corresponding class.

This procedure makes use of both spectral and spatial data. The relevant spectral data take the form of characteristic cell spectra (i.e., spectral "signatures"), while the spatial data consist of data about various types of cells (i.e., cell blobs), many of which appear similar to the eye. The solution to this problem requires a spectral image (data base). Since the (spectral) λ axis is intrinsically different from the spatial axes (x, y), an intuitively simple analysis approach in which the spectral and spatial features are separable will be considered.

In the above problem, cells can be differentiated by their characteristic spectral signature. Hence, a suitable point operation will be performed to generate a synthetic image in which each pixel is assigned one of $k + 1$ values. Assuming that the fluorescence emission spectra of the different cell types are known to be $s_i(\lambda)$; $i = 1, 2, \ldots, k$, $\lambda \in [\lambda_1, \lambda_n]$ and the measured spectrum at each pixel (x, y) is $s_{x,y}(\lambda)$, $\lambda \in [\lambda_1, \lambda_n]$, the following algorithm is an optional solution to the classification problem (subtask 1):

Let e_i^2 be the error between the measured spectrum and the known spectrum of the fluorophore attached to cell type i. Then, adopting the least-squares criterion, one can write

$$e_i^2 = \sum_{\lambda \in R_\lambda} (s(\lambda) - s_i(\lambda))^2, \tag{3}$$

where R_λ is the spectral region of interest. Each point [pixel (x, y)] in the image can then be classified into one of the $k + 1$ classes using the following criterion:

$$\text{point}(x, y) \in \begin{cases} \text{class } k+1 & \text{if } e_i^2 < \text{threshold, for all } i \in [1, k] \\ \text{class } \rho & \text{otherwise, if } e_\rho^2 < e_i^2, \text{ for all } i \in [1, k] \text{ excluding } \rho \end{cases}$$

$$\tag{4}$$

Subtasks 2 and 3 (image segmentation and calculation of average fluorescence intensity) are now straightforward, using standard computer vision operations on the synthetic image created in accordance with the algorithm described in Eqs. (3) and (4).

Another approach to solving this problem is to express the measured spectrum $s_{x,y}(\lambda)$ at each pixel as a linear combination of the k known fluorescence spectra $s_i(\lambda)$; $i = 1, 2, \ldots, k$. In this case one would find the coefficient vector $C = [c_1, c_2, \ldots, c_k]$ that solves

$$F = \min \sum_{\lambda \in R_\lambda} (s(\lambda) - \hat{s}(\lambda))^2$$

where

$$\hat{s}(\lambda) = \sum_{i=1}^{k} c_i \cdot s_i(\lambda) \tag{5}$$

Solving $\partial F / \partial c_i = 0$; for $i = 1, 2, \ldots, k$ (i.e., find values of c_i that minimize F) yields the matrix equation $C = A^{-1}B$, where A is a square matrix of dimension k with elements

$$a_{m,n} = \left[\sum_{\lambda \in R_\lambda} s_m(\lambda) \cdot s_n(\lambda) \right]$$

and B is a vector defined as

$$b_m = \left[\sum_{\lambda \in R_\lambda} s_m(\lambda) \cdot s(\lambda) \right], \qquad m, n = 1, 2, \ldots, k$$

4.3.5. Similarity Mapping Algorithms

Spectral images allow the comparison of any spectrum selected from within the spectral image to any of the other (tens of thousands of) spectra in the image. The degree of similarity between the spectrum at each pixel and some chosen (reference) spectrum can thus be calculated mathematically (i.e., it is a quantitative result). This procedure, called *similarity mapping*, creates a (two-dimensional) gray level image from the original (three-dimensional) spectral image. In principle, a similarity map algorithm highlights all the pixels in the image that have spectral characteristics similar to those of the selected reference pixel. The similarity map thus enhances the ability of the human eye to differentiate fine details in the stained cell.

There are many algorithms that could be envisioned for generating similarity map images. One form that has been used successfully is given by Eq. (6):

$$G_{x,y} = \frac{1}{\displaystyle\int_{\lambda_1}^{\lambda_2} |I_{x,y}(\lambda) - R(\lambda)| d\lambda} \tag{6}$$

In the equation $I_{x,y}(\lambda)$ is the measured spectrum at pixel x, y and $R(\lambda)$ is the (user-defined) reference spectrum. $G_{x,y}$ is the (scalar) result of the similarity algorithm at pixel (x, y). This algorithm calculates, for each pixel, the absolute difference in the area bounded by the measured spectrum and the reference spectrum over the spectral range $\lambda_1 - \lambda_2$. This "area difference" is computed for every pixel, which is then assigned a gray level that is inversely proportional to the area difference. It should be noted that while the reference spectrum in this example was selected directly from a pixel in the *measured* spectral image, it is also possible to perform similarity mapping using reference spectra that have been stored (e.g., in a library of reference spectra). This approach is useful for automating the analysis of spectral images, in particular, when the location of a chemical constituent with a known spectral signature is of interest.

4.3.6. Practical Considerations

It should be clear that the methods discussed above represent only a small fraction of the possible spectral image analysis methods. Customized algorithms will continue to be developed to support specific spectral bio-imaging methods and applications.

4.4. MEASUREMENT METHODS AND APPLICATIONS

4.4.1. Overview

Spectral bio-imaging systems are potentially useful in all applications in which subtle spectral differences exist between chemical constituents within an image. The measurement can be carried out using virtually any optical system—for example, an upright microscope, an inverted microscope, a dissecting microscope, or even a macro lens. Any standard experimental method can be used, including transmission (bright-field) imaging microscopy and fluorescence imaging microscopy.

Fluorescence measurements can be made with any standard filter cube (consisting of a barrier filter, an excitation filter, and a dichroic mirror) or any customized filter cube for special applications, provided that the emission spectra fall within the spectral range of the system. Spectral bio-imaging can also be used in conjunction with any standard spatial filtering method such as dark field, phase contrast, and even polarized light microscopy. The effects on spectral information when using such methods must, of course, be understood to correctly interpret the measured spectral images. [Refer to Kam (14) for a comprehensive discussion of microscopic techniques.]

There are many experimental methods and specific applications for spectral bio-imaging systems in transmission and fluorescence microscopy. These methods and applications include the following:

Fluorescence Microscopy

- Spectral identification of multiple fluorophores
- Detection of microenvironmental changes in subcellular compartments (e.g., pH) and dyes characterization
- Measurement of fluorescence from natural pigments (e.g., chlorophyll)
- Fluorescence resonance energy transfer (FRET)

Transmission Microscopy

- Measurements of stained histological samples

Other possible applications, which will not be discussed in detail here, include (i) time-resolved spectral imaging (by utilizing the pump and probe techniques with an appropriate external trigger) and (ii) Raman scattering measurements. A more detailed discussion, including examples of spectral images acquired using the SpectraCube™ method and the SD200 spectral bio-imaging system, follows.

4.4.2. Fluorescence Microscopy

4.4.2.1. Advantages of Spectral Bio-Imaging

It is generally agreed that the use of dyes as markers, particularly the use of multiple dyes (15), is one of the most powerful tools for analyzing organic tissues. Fluorescence microscopy is therefore one of the most important experimental methods used in light microscopy (16). The power of fluorescent probes is mainly due to the great variety of biological structures to which specific dyes can be bound (17). For a detailed review of fluorescence probes, see Mason (12) and Ploem (18). The rapid development of new and more sophisticated multicolor fluorescent dye molecules will continue to create the need for more advanced fluorescence imaging techniques that can utilize the full potential of these dyes. For a discussion of the revolutionary impact fluorescent dyes have had, and will continue to have, on the way research is conducted today, see Taylor et al. (19).

Spectral bio-imaging provides several important advantages for fluorescence imaging applications over more traditional filter-based approaches. These advantages include the following:

- Measurement of the complete spectrum, providing much more (quantitative) insight into the actual behavior of dye molecules in the sample of interest
- Simplification of fluorescence image acquisition, enabling, in a single measurement, the separation and mapping of many spectrally overlapping fluorescent probes
- Ability to overcome many of the traditional problems arising from undesirable background luminescence

In fact, by applying sophisticated data analysis algorithms such as multivariate analysis and principal component regression (20), it is possible to analyze many spectrally related parameters simultaneously.

4.4.2.1.1. Measurement of the Complete Spectrum. Measuring the complete spectrum at every image pixel can be an important advantage over filter-based fluorescence techniques. Knowledge of the complete spectrum provides more insight into the bahavior of fluorescent probes (and their binding sites). For example, undesirable or unpredictable spectral shifts that occur in the emission spectrum of a fluorescent probe, due to its microenvironment (e.g., temperature), will not cause errors in determining the concentration of the probe. When the fluorescence intensity is measured with only a bandpass filter, however, such spectral shifts would likely go undetected and might cause significant errors in determining the probe concentration.

4.4.2.1.2. Simplified Image Acquisition. The acquisition of multicolor fluorescence images can be greatly simplified when the power of spectral bio-imaging is combined with the appropriate fluorescent probes. Although a computerized system based on dual filter wheels and dual excitation sources has been shown to be effective in acquiring and registering the emission intensities of five spectrally distinct fluorescent probes (15), a fair degree of sophistication is required to control separately (for each probe) the excitation and emission filters, select the appropriate filter cube, and optimize the focus and exposure. Moreover, the resulting images must usually be registered prior to analysis, which can rarely be done with sufficient satisfaction. Recently, multispectral interference filters have also been used to enable imaging multiple fluorophores (21).

By enabling the simultaneous measurement of the emission spectrum of many fluorescent dyes (including dyes whose emission spectra overlap), spectral bio-imaging overcomes one of the fundamental limitations imposed by filter-based approaches: the need to acquire sequential images of the emissions of multiple fluorescent probes. The advantage of using a spectral bio-imaging system is greatest when all of the fluorescent probes can be excited by a common excitation source. In this case, a single spectral image acquisition can capture the fluorescence emission of an almost unlimited number of dyes, and the need to select nonoverlapping dyes, change filter cubes, change excitation or emission filters, optimize the focus and/or exposure time, or register the images is eliminated.

4.4.2.1.3. Elimination of Undesirable Background Luminescence. Spectral bio-imaging also provides a means for eliminating problems associated with undesirable background luminescence. Fluorescence imaging microscopy is typically performed by using a (fluorescence) filter cube, which ensures that the sample is excited by the desired short wavelengths while allowing only wavelengths in a limited spectral band corresponding to the fluorescence emission of the probe to reach the detector (e.g., the eye or a camera) (12). Since fluorescence intensities are usually several orders of magnitude below the intensity of the excitation source, such background luminescence can never be eliminated perfectly (22). The three primary sources for undesirable background luminescence are as follows:

1. Radiation from the excitation source that is not completely blocked by the dichroic mirror coating and/or the barrier filter
2. Autofluorescence of the sample, and sometimes also from the optical elements, which can contribute significantly to the background fluorescence: the effects of sample autofluorescence can usually be reduced by selecting fluorescent probes whose absorption and emission bands do not overlap with those of the sample being measured; similarly, by choosing optical elements that are appropriately coated to reduce auto-fluorescence, the effects of this latter type of autofluorescence can also be minimized
3. Selection of an inappropriate (or suboptimal) combination of excitation filter, dichroic mirror, and barrier filters

In spite of the best filtering methods available, undesirable background luminescence often makes it difficult, and sometimes impossible, to bring out the relevant fluorescence signal from its background (noise). A spectral bio-imaging system has the advantage of being able to use differences between (i) the shape and spectral range of the fluorescent dye and (ii) the shape and spectral range of the background luminescence (including autofluorescence) to eliminate the effects of undesirable background luminescence. Thus, by applying the appropriate spectral image analysis methods to the emission spectra of fluorescent probes, it is possible to improve the SNR, and hence the accuracy, of fluorescence imaging measurements. For example, linear combinations and principal component algorithms can be used to extract, at every pixel, the exact spectral contribution of background components. The desired dominant background spectrum to be extracted can be determined from a previous calibration measurement or by using statistical methods applied to the entire image. This advantage of a spectral bio-imaging approach is of particular importance for ratio imaging when quantitation of the results is desired. In addition, a spectral bio-imaging system can save time and effort that are otherwise expended in choosing the optimal filters for a specific filter-based measurement.

An example of how a spectral imaging can eliminate undesirable background is shown in Figure 4.5. (see Color Plates). Figure 4.5a shows the fluorescence spectral image of a cell stained with propidium iodide. This image was acquired using the SD200 SpectraCube™-based spectral bio-imaging system attached to an Olympus inverted microscope (IMT2). The sample was illuminated by a mercury source, and the fluorescence intensity was imaged through a DMG filter cube (DM580 dichroic mirror, D590 excitation filter, and BP545 barrier filter). Figure 4.5b is a plot of the fluorescence emission spectrum from three image pixels. Note that each spectrum has two peaks. The peak at 623 nm is due to the actual fluorescence emission spectrum of propidium iodide, while the second peak, at 775 nm, is just a residual of the excitation light that was not completely eliminated by the excitation or barrier filter. This undesirable background luminescence could be eliminated by adding another excitation filter (Olympus PB460) or an appropriate barrier filter. However, if this same measurement had been made with a nonspectral imaging system (without the PB460 filter), the intensity measured at each pixel would be proportional to the integral of the spectra shown in Figure 4.5b, *including* the contribution of the undesirable signal in the 775 nm peak. Correction algorithms exist that could be applied to the traditional CCD image, albeit at the expense of noise, to help this situation (22). However, with a spectral bio-imaging system that measures fluorescence intensity at every pixel as a function of wavelength, it is easy to confine the analysis to those wavelengths that correspond to the emission of the fluorescent dye of interest (i.e., the 623 nm peak).

4.4.2.2. *Spectral Identification of Multiple Fluorophores*

The use of spectral bio-imaging enables the *simultaneous* measurement of *many* dyes in one step. There is no restriction on the type of dye; even dyes that overlap spectrally (e.g., rhodamine and Texas Red) can be identified and their occurrence mapped in an image. If many dyes are to be used simultaneously, careful consideration should be given to their excitation wavelength, fluorescence intensity, and emission spectrum. When this is done properly, the results can be analyzed quantitatively as well. For example, the relative concentration of several proteins can be found in a single measurement. By using standard calibrated dyes, the absolute concentrations can also be determined. One important case where the detection of multiple fluorescent probes can be a significant advantage is fluorescence in situ hybridization (FISH) (23), which is used to analyze gene codes and find possible defects. Malik et al. (24) summarizes experimental results utilizing the SpectraCube™ methods of spectral bio-imaging to measure multicolor fluorescence.

4.4.2.2.1. Detecting Microenvironmental Changes. The use of fluorescent dyes (e.g., when attached to an antibody) is not limited to identifying the existence of certain chemical compounds. Some dyes can also be used to probe actual chemical and physical parameters. For example, a dye whose spectrum changes when it gains or loses a hydrogen atom can serve as a pH indicator. Dyes exhibiting spectral changes due to local electrical potential, pH level, and intracellular ion concentration [e.g., sodium, magnesium, zinc, or free Ca^{2+} (25)] are currently used in a variety of applications.

The use of environmentally sensitive dyes is closely related to ratio imaging, a common analysis method (26). For some of the dyes in use today, environmentally related spectral changes occur in only part of the spectral range. By measuring the ratio of fluorescence emission in two spectral ranges, one can thus study environmental effects, independent of sample thickness, illumination nonuniformity, etc.

The ability of a spectral bio-imaging system to measure the complete spectrum, rather than using current techniques to measure the fluorescence at two (or three) discrete wavelengths, has several advantages for ratio imaging applications. Measurement of the complete spectrum significantly improves the accuracy and sensitivity of ratio imaging. For example, one can use the complete spectrum to measure the ratio of two integrated intensities in two different spectral ranges, and at the same time eliminate all the background intensity by integrating only over the relevant spectral range. It is also possible to use more sophisticated algorithms to analyze the spectral data, such as determining the environmental parameters of interest by fitting the measured spectral data to reference spectra stored in a library. Because spectral analysis is performed at every point in the image, a full morphological analysis of the sample is possible. Spectral bio-imaging thus provides the capability to map (i.e., display as an image) physical and chemical parameters of interest, potentially enabling more innovative and effective research.

By using spectral bio-imaging techniques, it may also be possible to use environmentally sensitive dyes that do not lend themselves to analysis using ratio methods. The detection of dyes sensitive to several environmental parameters (e.g., voltage and pH) could also be achieved by using appropriate analysis algorithms.

4.4.2.2.2.   Measurement of Fluorescence from Natural Pigments. Chlorophyll is a natural pigment exhibiting fluorescence (27). The fluorescence spectrum of chlorophyll has been studied extensively, in part because it is more complicated than the sectrum from a typical fluorescent dye (28). For example, the measurement of chlorophyll fluorescence can be used to probe cell metabolism and to track photosynthesis. Fluorescence also occurs for porphyrins, native cytoplasmic proteins, and other compounds. The ability of a spectral bio-imaging system to measure spectral differences in different parts of a cell provides important insights into the functions of the organelles in a living cell.

An example of a chlorophyll fluorescence spectral imaging is shown in Figure 4.6 (see Color Plates). The algae *Oedogonium* was examined by fluorescence microscopy. Excitation was in the green spectral range (540 nm), and the emission in the red spectral band was measured by the SD100 SpectraCube™-based spectral bio-imaging system. Figure 4.6A reveals the total emitted fluorescence from the specimen. Both the red chlorophyll fluorescence and the reflected green excitation light from cell debris in the specimen can be clearly seen. The individual algal cells exhibiting red fluorescence are only partially visible. The fluorescence spectra from four different pixels of the specimen are shown in Figure 4.7. Specimen A, with an emission peak at 685 nm, was measured from a point in the central algae in which the red fluorescence dominates. Spectrum B, on the other hand, corresponds to a point of cell debris exhibiting green emission, with a peak at 542 nm. Spectra C and D correspond to two other pixels in the algae. By using spectrum A as a reference spectrum for a similarity map, it is possible to precisely locate and determine the relative concentration of chlorophyll in the algae, including portions of the algae obscured by debris (see Figure 4.6B). This (pseudo-colored) figure shows that the chlorophyll, although dispersed, is located in the center of the cell. Figure 4.6C is a similarity map using as a reference spectrum a point in the image at which the red fluorescence is faint. This similarity map demonstrates the existence of different chlorophyll intensities at different cell sites; for example, some of the fluorescence occurs at the periphery of the cell membrane. Thus, by using the similarity mapping function, it is possible to visualize the characteristic fluorescence emitted from specific subcellular points in the cell. Figure 4.6D highlights the spongy structure of the cell debris when a similarity map using the green spectrum from the cell debris (spectrum B) is applied. This last similarity map also reveals the algal cell wall, which reflects the green emitted light. In addition, this similarity shows the location of unicellular algae.

The study of autofluorescence has many other important applications. For example, studies of natural skin fluorescence have identified "spectral finger-prints" corresponding to different tissue constituents and histological organization of the tissue. The use of tissue spectral fingerprints to identify normal

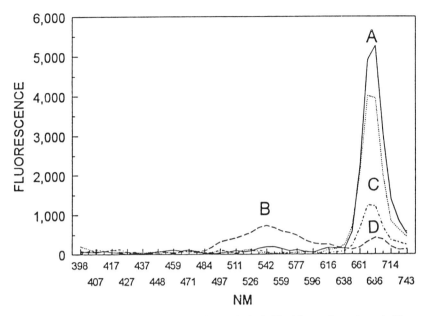

Figure 4.7. Fluorescence spectra from four different pixels (A–D) of the specimen shown in Figure 4.6 (see Color Plates).

from abnormal tissue (29–31) has also been studied extensively. The potential clinical utility of spectral bio-imaging in such medical diagnostic applications appears to be significant.

4.4.2.2.3. Fluorescence Resonance Energy Transfer (FRET). FRET is a fluorescence method that enables the determination of the spatial separation between two fluorophores. In FRET, two different fluorophores, designated the donor and acceptor, are used. This pair of fluorophores is chosen carefully so that when the donor is excited, it can either fluoresce or transfer the energy being absorbed by it to the second fluorophore (the acceptor), causing it to fluoresce. Thus, it is possible to distinguish the donor from the acceptor because of differences in the emission spectrum.

The physical separation between the donor and acceptor is determined from the fact that the efficiency of the energy transfer strongly depends on the distance between the two fluorophores (typically, the efficiency is proportional to the inverse of the sixth power of the separation distance). When these two fluorophores are attached to two different types of molecules or to the same molecule in two different states, measuring FRET by using a spectral bio-imaging system (attached to a microscope) greatly facilitates the ability to distinguish between different molecules (or molecular states) while simultaneously measuring their spatial distribution. A detailed discussion of energy transfer phenomena in microscopy can be found in Herman (32) and Jovin and Arndt-Jovin (33).

FRET is performed by measuring three parameters: (i) the intensity at the donor emission peak when exciting at the donor absorption peak; (ii) the intensity at the acceptor emission peak when exciting at the donor absorption peak; and (iii) the intensity at the acceptor emission peak when exciting at the acceptor absorption peak. These three parameters are then corrected and analyzed to give the separation distances and location of the two fluorophores being studied.

Using a spectral bio-imaging system not only improves the measurement results but also simplifies the measurement. In fact, by using a spectral bio-imaging approach, only two measurements are required, since both the donor and acceptor emission bands can be measured simultaneously when excitation is in the donor absorption band.

By means of spectral bio-imaging, the FRET method might be further developed. For example, one might select for FRET an acceptor fluorophore that is sensitive to changes in the environment and, at the same time that molecular distances are measured, use ratio imaging methods to monitor microenvironmental changes. In a further example, FRET can be used with *two* different acceptors (with different emission spectra), providing information about two separation distances and fluorophore locations. Dual acceptor experiments are very hard to perform with traditional methods. The use of spectral bio-imaging can improve the accuracy of the FRET results, simplify the experimental procedure, and enable more powerful analysis methods.

4.4.3. Transmission Microscopy

Light microscopy is one of the most fundamental techniques for the visualization of cells and tissues in biology and pathology. Transmission microscopy suffers greatly, however, from the inherently low contrast of cell organelles and structural details. Many methods have been developed to improve this contrast, among them staining and spatial filtering techniques [e.g., dark field and polarization (14)]. Spectral bio-imaging is one of the most straightforward methods used to increase the apparent contrast of cells and tissues examined under a transmission microscope, thereby improving dramatically the identification ability and specificity of this microscopic method. The basic approach is to measure a wealth of information (i.e., the spectrum) at every point in an image and to then use this spectral information, in conjunction with spectral analysis methods, as the basis for identifying cellular and subcellular details.

Spectral bio-imaging applied to transmission microscopy can thus improve morphometric analysis (the quantitative measurement of the size, shape, and textural features of cells and tissues) by enabling the evaluation of subtle cytological and histological features, which in turn can yield useful ultrastructural and medical information for diagnostic and prognostic evaluation (34, 35).

To facilitate the histological examination of biological specimens, a variety of staining techniques were developed during the last century using organic stains that specifically bind to different macromolecules in the cells. The molecular basis of these staining techniques continues to be empirical. The most common staining techniques are the Romanowsky–Giemsa stain and hematoxylin–eosin. The Romanowsky–Giemsa staining procedure uses a combination of two dyes: azure B (trimethyl methionine), a thiazine dye, and eosin Y (hydroxyxanthene bromide). The thiazines are cationic dyes and therefore bind to acidic cellular components, whereas eosin is an anionic dye that tends to bind to basic cellular components. It is widely accepted that the use of these two dyes creates the so-called Romanowsky–Giemsa effect, which is expressed as a new dye complex with a specific purple color in some stained sites. The molecular basis of the azure B–eosin complex is not well understood. Some investigators believe that azure B binds to anionic structures such as the phosphate groups of DNA, while eosin simultaneously binds to both an adjacent cationic site on the DNA and azure B. In a more recently proposed model of the azure B–eosin complex, Friedrich et al. (36) have suggested that azure B first binds to phosphodiester residues of the DNA molecule. The authors have hypothesized that the phenyl group of the eosin molecule is the portion that binds to the azure B molecule (which lies in a single plane). The purple color is thus a result of a red shift of the eosin absorption peak, which in turn is caused by the dielectric polarization of bound eosin. The very existence of such an azure B–eosin complex is still questioned by others (36–40).

By using an appropriate method of staining, it is possible to distinguish between subcellular compartments of the cell, and in particular to distinguish the chromatin organization in the nucleus. It is well established, for example, that the ratio between heterochromatin, stained dark blue, and euchromatin, stained pink, is one of the major determinants in the evaluation of cells in tissue sections. Nevertheless, the results obtained from stained specimens remain, to some degree, a matter of experience, art, and subjective interpretation. In order to provide a more objective measure of nuclear chromatins, attempts have been made to study the spectroscopic characteristics of the interaction between organic stains and macromolecules, and thus to evaluate the Romanowsky–Giemsa effect of DNA staining.

Recently, Malik et al. (24) demonstrated the ability to measure quantitatively the ratio of heterochromatin to euchromatin in stained tissue, as well as to identify different cell organelles by using spectral bio-imaging. Figure 4.8 (see Color Plates) is an example of a transmission spectral image of a plasma cell stained by the method of May–Grünwald–Giemsa for blood cells. In the drawing adjacent to Figure 4.8, the different subcellular sites are delineated, and the pixels corresponding to the five spectra are highlighted. Pixel A corresponds to the heterochromatin, and pixel B is selected from within the light blue–stained area of the cytosol. Pixel C is representative of euchromatin, pixel D corresponds to a cytoplasmic vesicle, and pixel E is a point in the cytosol. The complete spectral image consists of 10,000 spectra. Figure 4.9 shows the (uncorrected) transmitted light spectra measured at the five pixels selected from the image of the plasma cell. It is apparent that the transmitted light from the euchromatin (C) and heterochromatin (A) is of low intensity, and exhibits spectral differences. The point spectra taken from the cytosol (B) also exhibit specific spectral changes, in particular differences in the intensity of transmitted light.

The absorption spectra corresponding to these five pixels were also cal-culated by appropriately using the spectrum outside the cell as a reference (see Figure 4.9). From these absorption spectra, it is apparent that each subcellular

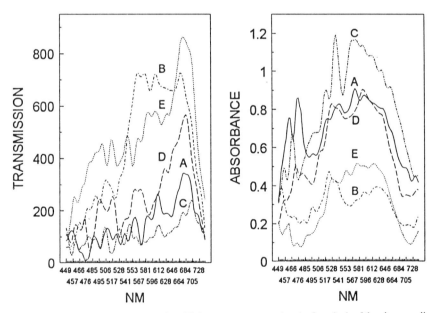

Figure 4.9. The uncorrected transmitted light spectra measured at the five pixels of the plasma cell shown in Figure 4.8 (see Color Plates) and the corresponding absorbance spectra.

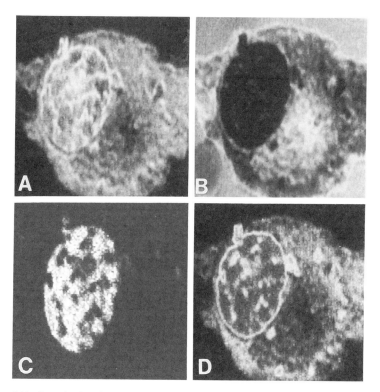

Figure 4.10. Quantitative similarity maps obtained when the spectra shown in Figure 4.9 are used, one at time, as the reference spectrum: (A) the euchromatin network tightly connected across the nucleus (spectrum A); (B) a bright central area in the cytosol corresponding to a large Golgi complex (spectrum B). (C) the nuclear heterochromatin (spectrum C); and (D) peripheral vacuoles surrounding the Golgi area (spectra D and E).

compartment in the image of the stained cell has spectral characteristics consistent with the colors seen by the eye. Because the human eye is limited in its ability to discriminate and separate the different color–macromolecular complexes on the basis of their color, similarity mapping algorithms of the form shown in Eq. (6) can be used to show details not apparent to the eye.

Figure 4.10 shows different gray scale images resulting from applying the similarity mapping algorithm using as a reference the spectra from Figure 4.9. Figures 4.10A and 4.10C demonstrate how the similarity mapping function enables the differentiation between chromatin and heterochromatin. Figure

4.10A shows a similarity map reconstructed by mapping with spectrum A. The fine structure of euchromatin areas in the nucleus is markedly enhanced. One can detect a euchromatin network tightly connected all over the nucleus, even on the upper side of the nuclear envelope. In the center of the nucleus there is a particularly bright site, which may represent the nucleolus; this conjecture is supported by the similarity map in Figure 4.10C using reference spectrum A. Figure 4.10B shows a similarity map built up with reference spectrum B. Two main cell features, both connected to membraneous borders, can be noted in this figure: one in the center of the cell and the other demarcating the outer cell membrane. This new similarity image may support the idea that the central bright array in the cytosol is a large Golgi complex. This interpretation is supported by the similarity map in Figure 4.10D, demonstrating peripheral vacuoles surrounding the Golgi area. The nuclear heterochromatin is shown in Figure 4.10C, after mapping with the reference spectrum C. This image is complementary to Figure 4.10A, but also shows some unexpected connections between the different complexes of the nuclear heterochromatin. Finally, Figure 4.10D, achieved by mapping of the combined spectra of D and E, reveals visicular structures in the cytosol (i.e., cytoplasmic vacuoles). Surprisingly, this similarity map also highlights the nuclear envelope, another membrane surrounding a subcellular compartment.

The spectral cytoplasmic features discussed above, when used for similarity mapping, allow the clear demarcation of components that are believed to represent the nuclear envelope, Golgi cisternae, cytoplasmic vacuoles, and outer cell membrane. Since stained cells dried in air may result in the superposition of cytoplasmic layers, thus reducing the specificity of spectral image analysis methods, it is suggested that aldehyde-fixed cells be used for future investigations with stained cells (thus enabling focusing on specific depths of the cell).

Based on the analysis of spectral images acquired from this May–Grünwald–Giemsa stained cell, there are some indications of the existence of a spectral component that may correlate well with what is called "the purple Romanowsky–Giemsa complex." For example, the absorption spectrum of heterochromatin (Figure 4.10) shows clearly an outstanding absorption peak at 540 nm, which agrees well with what has been described as the "the purple Romanowsky–Giemsa complex" (36–40).

In some cases, spectral images acquired using transmission methods and unstained tissue may also provide useful information similar to the results available from fluorescence microscopy techniques. One of the advantages of combining spectral bio-imaging and transmission microscopy of unstained tissue is the ability to use a "clean" measurement technique (i.e., there is no need to work with potentially toxic dyes or fixation methods) instead of fluorescence methods.

APPENDIX

Overview of Fourier Spectroscopy

Fourier spectroscopy is a well-known and well-established method for measuring the spectrum of light emitted by or transmitted through objects. The spectroscopic technique can be applied to a variety of measurements modes, including transmission, fluorescence, and reflection measurements. Fourier spectroscopy is a powerful analytical technique for materials recognition and characterization; it is used in many fields in science, research, and industry. As detailed in Section 4.2.3, Fourier spectroscopy has some important advantages over other spectroscopic methods.

Fourier spectroscopy was first used by Albert A. Michelson (1852–1931) about a century ago; however, it was not until the 1960s, with the advent of powerful computers, that Fourier spectroscopy emerged as a popular analytical method. In Fourier spectroscopy the spectrum is not measured directly, but must be calculated from the measured data (called the interferogram) using a Fourier transform algorithm that is long and time-consuming to implement without a computer. Today a Fourier transform algorithm can be implemented very quickly on a typical personal computer (PC) using fast Fourier transform (FFT) algorithms. The need to compute the spectrum is thus no longer a practical limitation for Fourier spectroscopy. In fact, the computational power of PCs enables the rapid (i.e., within minutes) calculation of tens of thousands of spectra in a spectral image.

Fourier spectroscopy is based on the principles of optical interferometry. At the heart of every Fourier spectrometer is an interferometer. The most common type of interferometer is the Michelson interferometer, first built by Michelson in 1881. Figure 4.11 is a schematic diagram of a Michelson interferometer. This interferometer serves as a linear OPD generator. The entering light falls on the beam splitter. This beam splitter is usually a mirror that has been silver coated to allow half of the incident light to be transmitted by the beam splitter (shown as a dotted line) and half to be reflected by it (shown as a solid line). These two coherent beams travel to two different flat mirrors (M_1 and M_2), where they are reflected back to the beam splitter. At the beam splitter these two beams are recombined and split again, with half of the interference intensity proceeding to the detector. In a Michelson interferometer, the mirror M_2 can be moved parallel to the incident (solid) beam; therefore, the path traveled by this beam can be different from the path to the fixed mirror (M_1). For example, if the moving mirror (M_2) is positioned at a distance $L + d$ from the beam splitter, and the fixed mirror is positioned at a distance L from the beam splitter, an OPD equal to $2d$ is created between the two beams. The interference intensity of the two beams depends on the OPD. During the measurement the

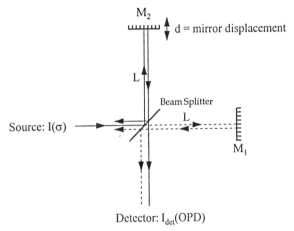

Detector: I_{det}(OPD)

Figure 4.11. Schematic diagram of a Michelson interferometer.

detector records the interference intensity for many OPDs, which results in the interferogram. It is possible to compute the spectrum from the interferogram by Fourier transformation of the interferogram.

In order to show, even intuitively, how the spectrum can be computed from the interferogram, it is necessary to review some basic principles of light. Recall, for example, that it is possible to describe monochromatic light (wavelength = λ_0) as a wave, in terms of its electric field, $E(\sigma_0)$, as follows:

$$E(\sigma_0) = A(\sigma_o)\cdot\cos(2\pi\sigma_0 x - \omega t)$$

Here σ_0 is the wavenumber of the monochromatic light, in units of cm^{-1}, where $\sigma_0 = 1/\lambda_0$ (wavenumber is the preferred unit for Fourier spectroscopy); $A(\sigma_0)$ is the amplitude of the (monochromatic) light of wavenumber σ_0; and ω is the angular frequency (in rad/s) of the traveling wave ($\omega = 2\pi c/\lambda = 2\pi c\sigma$, where c is the speed of light). The intensity of the monochromatic light, $I(\sigma_0)$, is equal to its amplitude squared, $[A(\sigma_0)]^2$, while x is the spatial axis of incidence of the wave and t is time.

Since a general broadband (polychromatic) source can be expressed as the summation of many monochromatic sources, it is possible to describe the general spectrum as an electric field, which is given by an integral of monochromatic electric fields as follows:

$$E = \int A(\sigma)\cdot\cos(2\pi\sigma x - \omega t)\,d\sigma \tag{7}$$

The total intensity of this source is also an integral, given by $I = \int A^2(\sigma)\,d\sigma$.

Using the formulation in Eq. (7) for the (source spectrum) electric field that enters the interferometer, and then splitting and recombining the two beams that have been delayed with respect to one another by an OPD (D), it can be shown that the intensity measured by the detector, $I_{det}(D)$, is

$$I_{det}(D) = \tfrac{1}{2}[\int I(\sigma)\,d\sigma + \int I(\sigma)\cdot\cos(2\pi\sigma D)\,d\sigma] \qquad (8)$$

Here $I_{det}(D)$ is the interferogram, i.e., the interference intensity measured by the detector as a function of OPD. Note that the first term of Eq. (9) is the total intensity of the source,

$$I = \int I(\sigma)\,d\sigma = \int A^2(\sigma)\,d\sigma$$

which does not depend on the OPD. This first term is thus a constant "dc" term in the interferogram. The second "ac" term, however, does depend on the OPD. It is in this second term that the spectral information lies. Readers familiar with Fourier theory will be able to identify this second term as the real part of the Fourier transform of $I(\sigma)$, i.e.,

$$I_{det}(D) = c_1 + c_2\cdot\text{Real}\{FT[I(\sigma)]\}$$

where c_1 and c_2 are constants. To reconstruct the spectrum $I(\sigma)$, one can therefore (inverse) Fourier transform the measured interferogram $I_{det}(D)$.

Figures 4.12a and 4.12b show the interferogram resulting from monochromatic illumination (e.g., from a laser) with $\lambda_0 = 500$ nm ($\sigma_0 = 20{,}000$ cm^{-1}). Note that the interferogram is simply a constant-amplitude cosine wave given by $I_{det}(D) = c_1 + c_2\cdot\cos(2\pi\sigma_o D)$. Figures 4.12c and 4.12d show the interferogram resulting from a narrowband source (also centered at 500 nm) with uniform intensity over a 2000 wavenumber narrowband.

In a real Fourier spectrometer the measured interferogram is actually sampled at discrete OPDs. Thus the Fourier transform performed to compute the spectrum is also a discrete operation, given by

$$I(\sigma) = \Delta x \sum I(D) e^{-i2\pi\sigma D} \qquad (9)$$

where the sum is taken over the total number of points (OPDs) in the interferogram and Δx is the (equally spaced) difference in OPD between contiguous samples of the interferogram.

Three additional operations are usually performed on the interferometric data prior to Fourier transformation to spectral domain: (i) phase correction; (ii) apodization; and (iii) zero filling. The actual algorithms will not be described here in detail; however, a good discussion of these methods can be found in Ref.

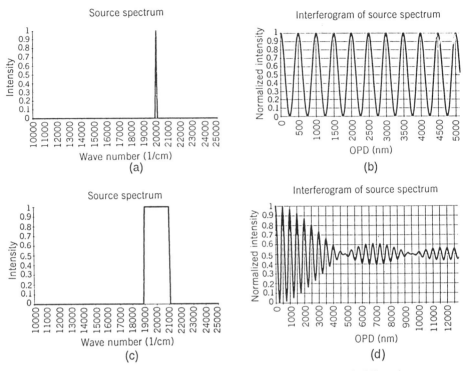

Figure 4.12. Interferogram and spectrum for monochromatic [(a) and (b)] and narrow-band [(c) and (d)] illumination.

8. The phase correction algorithm compensates for effects that introduce imaginary terms into the calculated spectrum. These imaginary terms cannot be totally eliminated in a real interferometric systems. In fact, the reason it is desirable to sample the double-sided interferogram (as in the SpectraCube™ method) is to further reduce phase-related errors. The apodization operation compensates for effects introduced by measuring a finite vs. infinite interferogram. The zero filling operation is an interpolation method that increases the number of data points in the spectrum.

The SpectraCube™ Interferometer

The interferometer used in the SpectraCube™ method is not a Michelson interferometer, but a special type of common path interferometer known as a Sagnac interferometer. An optical diagram of the Sagnac interferometer is

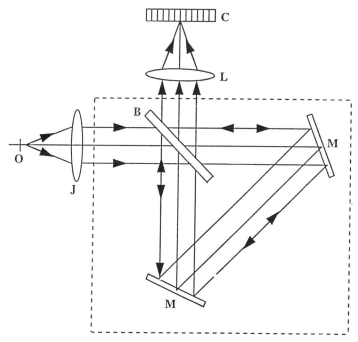

Figure 4.13. Schematic diagram of the SpectraCube™ imaging Sagnac interferometer: O is the object; J, the collection optics; L the imaging optics; and C, the CCD camera. The Sagnac interferometer is shown within the dashed box: B is the beam splitter; M, M are the two mirrors. The optical path difference required for Fourier spectroscopy results from the angular dependence of the Sagnac interferometer.

shown in Figure 4.13. In a common path interferometer, the two beams (delayed by a known OPD) travel in the same (common) path, but in opposite directions, until they are recombined prior to measurement of the interference intensity by the detector. The advantage of a common path interferometer is its intrinsic stability. Disturbances, such as a small shift of one of the optical elements, affect both beams the same way and hence do not contribute to measurement errors. In addition to being rugged, this type of interferometer is relatively compact.

Conceptually, the operation of the Sagnac interferometer is similar to the operation of a Michelson interferometer, the major difference being the manner in which an OPD is created. In a Michelson interferometer an OPD is created by displacing a mirror, while in a Sagnac interferometer the OPD is created by rotation and depends on the angle of incidence of light on the beam splitter.

The interference fringes of the Sagnac interferometer can be superimposed on the real image projected on the CCD array. Thus, because the interferogram is angle dependent, each pixel (with a small instantaneous field of view) measures the intensity at not just one well-defined OPD, but over a small (continuous) part of the interferogram, $I_{det}(D)$. In the SpectraCube™ method, the Sagnac interferometer is set such that each pixel measures an OPD equal to 1/4 of the period of the shortest wavelength in the spectrum to be measured. Thus, if the CCD contains $N \times N$ pixels (and assuming that the CCD array covers the entire field of view of the system), the largest OPD achievable by the system is given by

$$OPD_{max} = \frac{N\lambda}{4} \tag{10}$$

The resolving power, $R = \Delta\lambda/\lambda$ that results is therefore equal to $4/N$, and for a 512×512 CCD array, the spectral resolution at wavelength λ is better than 0.01λ (i.e., better than 4 nm at 400 nm).

REFERENCES

1. Andersson-Engels, S., Johannson, J., and Svanberg, S. (1990). Multicolor fluorescence imaging systems for tissue diagnostics. *Proc. SPIE—Bioimag. Two-Dimens. Spectrosc.* **1205**, 179–189.

2. Maymon, W., and Neeck, S. P. (1988). Optical system design alternatives for the Moderate-Resolution Imaging Spectrometer Tilt (MODIS-T) for the Earth Observing System (EoS). *Proc. SPIE—Recent Adv. Sensors, Radiometry Data Process. Remote Sens.* **924**, 10–22.

3. Dozier, J. (1988). HIRIS—The high resolution imaging spectrometer. *Proc. SPIE—Recent Adv. Sensors, Radiometry Data Process. Remote Sens.*, **924**, 23–30.

4. Vane, G., Chrien, T. G., Reimer, J. H., Green, R. O., and Conel, J. E. (1988). Comparison of laboratory calibrations of the Airborne Visible/Infrared Imaging Spectrometer (AVIRIS) at the beginning and end of the first flight season. *Proc. SPIE—Recent Adv. Sensors, Radiometry Data Process. Remote Sens.*, **924**, 168–178.

5. Schott, J. R. (1989). Remote sensing of the Earth: A synoptic view. *Phys. Today,* September, pp. 72–79.

6. Lewotsky, K. (1994). Hyperspectral Imaging: evolution of imaging spectrometery. *SPIE OE/Rep.,* November, pp. 1–3.

7. ISSIR (International Symposium on Sensing Research), (1992). *Proc. Conf. Maui, Hawaii, 1992,* p. 20.

8. Chamberlain, J. (1979). *The Principles of Interferometric Spectroscopy.* Wiley, New York.

9. Bell, R. J. (1972). *Introductory Fourier Transform Spectroscopy.* Academic Press, London.

10. Morris, H. R., Hoyt, C. C., and Treado, P. J. (1994). Imaging spectrometers for fluorescence and Raman microscopy: Acousto-optic and liquid crystal tunable filters. *Appl. Spectrosc.* **48**, 857–866.

11. Yu, J., Chao, T. H., and Chen, L. J. (1990). Acousto-optic tunable filter (AOTF) imaging spectrometer for NASA applications: System issues. *Proc. SPIE—Opt. Inf.-Process. Syst. Archit. II*, **1347**.

12. Mason, W. T. (1993). *Fluorescence and Luminescent Probes for Biological Activity*, Biol. Tech. Ser. Academic Press, London.

13. Jain, A. K. (1989). *Fundamentals of Digital Image Processing.* Prentice-Hall International, London.

14. Kam, Z. (1987). Microscopic imaging of cells. *Q. Rev. Biophys.* **20**, 201–259.

15. Waggoner, A., DeBiasio, R., Conrad, P., Bright, G. R., Ernst, L., Ryan, K., Nederlof, M., and Taylor, D. (1989). Multiple spectral parameter imaging. *Methods in Cell Biol.* **30**, Part B, 449–478.

16. Lakowicz, J. R. (1983). *Principles of Fluorescence Spectroscopy.* Plenum, New York and London.

17. Waggoner, A. S. (1986). Fluorescence probes for analysis of cell structure, function and health by flow and imaging cytometry. In *Applications of Fluorescence in the Biomedical Sciences* (D. Lansing Taylor, A. S. Waggoner, F. Lanni, R. F. Murphy, R. Birge, Eds.), pp. 3–28. Alan R. Liss, Inc., New York.

18. Ploem, J. S., and Tanke, H. J. (1987). *Introduction to Fluorescence Microscopy.* Oxford University Press, Royal Microscopical Society, London.

19. Taylor, D. L., Nederlof, M., Lanni, F., and Waggoner, A. S. (1992). The new vision of light microscopy. *Am. Sci.* **80**, 322–335.

20. Martens, H., and Naes, T. (1989). *Multivariate Calibration.* Wiley, London.

21. Lengauer, C. et al. (1993). Chromosomal bar codes produced by multicolor fluorescence in situ hybridization with multiple YAC clones and whole chromosome painting probes. *Hum. Mol. Genet.* **2**, 505–512.

22. Benson, D. M., Plant, A. L., Bryan, J., Gotto, A. M., Jr., and Smith, L. C. (1985). *J. Cell Biol.* **100**, 1309–1323.

23. Emanuel. (1993). The use of fluorescence *in situ* hybridization to identify human chromosomal anomalies. *Growth Genet. Horm.* **9**, 6–12.

24. Malik, Z., Cabib, D., Buckwald, R. A., Garini, Y., and Soenksen, D. (1994). A novel spectral imaging system combining spectroscopy with imaging. Applications for biology. *Proc. SPIE—Opt. Imaging Tech. Biomed.* **2319**, 180–184.

25. Tsien, R. Y. (1994). Fluorescence imaging creates a window on the cell. *Chem. Eng. News* **34**, 34–44.

26. Bright, G. R., Rogowska, J., Fisher, G. W., and Taylor, D. L. (1987). Fluorescence ratio imaging microscopy: Temporal and spatial measurements in single living cells. *BioTechniques* **5**, 556–563.

27. Haliwell, B. (1981). *Chloroplast Metabolism: The Structure and Function of Chloroplast in Green Leaf Cells.* Oxford University Press (Clarendon), Oxford.

28. Parson, W. W. (1988). Photosynthesis and other reactions involving light. In *Biochemistry* (G. Zubay, Ed.), 2nd ed., pp. 564–597. Macmillan, New York.

29. Marchesini, R., Brambilla, M., Clemente, C., Maniezzo, M., Sichirollo, A. E., Testori, A., Venturoli, D. R., and Cascinelli, N. (1991). In vivo spectrophotometric evaluation of neoplastic skin pigmented lesions. I. Reflectance measurements. *Photochem. Photobiol.* **53**, 77–84.

30. Bottiroli, G., Croce, A. C., Locatelli, D., Marchesini, R., Pignoli, E., Tomatis, S., Cuzzoni, C., Di Palma, S., Dalfante, M., and Spinelli, P. (1994). Natural fluorescence of normal and neoplastic human colon: A comprehensive 'ex vivo' study. *Lasers Surg. Med.*

31. Profio, E. (1984). Laser excited fluorescence of hematoporphyrin derivative for diagnosis of cancer. *IEEE J. Quantum Electron.* **QE-20**, 1502–1506.

32. Herman, B. (1989). Resonance energy transfer microscopy. In *Fluorescence Microscopy of Living Cells in Culture* (D. Lansing Taylor, and Y.-L. Wang, Eds.), Part B, Chapter 8, pp. 219–243. Academic Press, San Diego, CA.

33. Jovin, T. M., and Arndt-Jovin, D. J. (1989). FRET microscopy: Digital imaging of fluorescence resonance energy transfer: Application in cell biology. In *Cell Structure and Function by Microspectrofluorometry*, (E. Kohen and J. Hirschberg, Eds.), Chapter 5. Academic Press, San Diego, CA.

34. Erler, B. S., Chein, K., and Marchevsky, A. M. (1993). An image analysis workstation for the pathology laboratory. *Mod. Pathol.* **6**, 612–618.

35. Hytiroglou, P., Harpaz, N., Heller, D. S., Liu, Z. Y., Deligdisch, L., and Gil, J. (1993). Differential diagnosis of borderline and invasive serious cystadenocarcinomas of the ovary by computerized interactive morphometric analysis of nuclear features. *Cancer* (Philadelphia) **69**, 88–212.

36. Friedrich, K., Seiffert, W., and Zimmerman, H. W. (1990). Romanowsky dyes and Romanowsky-Giemsa effect 5. Structural investigations of the purpole DNA–AB–EY dye complexes of Romanowsky–Giemsa staining. *Histochemistry* **93**, 247–256.

37. Horobin, R., Curtis, D., and Pindar, L. (1989). Understanding Romanowsky staining 2. The staining mechanism of suspension-fixed cells, including influences of specimen morphology on the Romanowsky–Giemsa effect. *Histochemistry* **91**, 77.

38. Horobin, R., and Walter, K. (1987). Understanding Romanowsky staining 1. The Romanowsky–Giemsa effect in blood smears. *Histochemistry* **86**, 331.

39. Wittekind, D. (1983). On the nature of Romanowsky–Giemsa staining and its significance for cytochemistry and histochemistry: An overall view. *Histochim. J.* **15**, 1029.

40. Marshall, P., and Galbraith, W. (1984). On the nature of the purple coloration of leucocyte nuclei stained with azure B–eosin Y. *Histochim. J.* **16**, 793.

CHAPTER

5

ACOUSTO-OPTIC TUNABLE FILTERS AND THEIR APPLICATION IN SPECTROSCOPIC IMAGING AND MICROSCOPY

XIAOLU WANG

Brimrose Corporation of America
Baltimore, Maryland 21236

E. NEIL LEWIS

Laboratory of Chemical Physics
National Institute of Diabetes, Digestive and Kidney Diseases
National Institutes of Health
Bethesda, Maryland 20892

5.1. INTRODUCTION

Microscopic or macroscopic spectral imaging is the determination of spatially distributed and chemically/biologically distinct elements in heterogeneous material. It is a powerful tool for studying a wide range of materials, including biological materials, polymers, and semiconductors. For example, progress in determining the specificity of new chemical dyes as molecular markers has enabled scientists to observe subcellular processes in living cells, in real time, using spectroscopic techniques such as fluorescence imaging spectroscopy. Recent advances in the field of spectral imaging have been made through the development of a number of different types of technology, including digital imaging processing hardware and software, as well as focal-plane array detectors and continuously tunable, image-quality spectral filters. These advances have enabled the rapid collection of chemically specific images that have high contrast and high spatial resolution. This, in turn, has made possible the observation of structures in heterogeneous materials that were not visible using traditional methods.

Fluorescence Imaging Spectroscopy and Microscopy, edited by Xue Feng Wang and Brian Herman. Chemical Analysis Series, Vol. 137.
ISBN 0-471-01527-X © 1996 John Wiley & Sons, Inc.

The construction of a fully functional spectroscopic imaging microscope requires the integration of traditional microscope optics with a focal-plane array detector, an image processing system, and tunable optical filters. A system configured for both absorption and emission measurements may have a tunable optical filter placed in front of the optical source as well as in front of the focal-plane array detector. Traditional mechanical interference filter wheels are the most widely used monochromator for multispectral imaging. They are easily attached to excitation light sources or inserted between the microscope and the camera. They have from 3 to 10 slots for standard 25 mm diameter interference filter wheels. The out-of-band rejection, image resolution, and peak transmittance of such filters are high. The disadvantages of filter wheels are limited wavelength selection, image shift, relatively low switching speed (0.1–0.5 s) and vibration, which can be a major problem in many kinds of experimental configurations. Interference filters also tend to deteriorate when exposed to heat and intense light.

Diffraction grating monochromators are convenient because they are tunable over a wide wavelength range and have variable bandwdth. However, these systems are difficult to integrate into a microscope and, once installed, are expensive, bulky, and relatively slow in acquiring complete spectral scans. This is especially true when uniform illumination is important.

Two new approaches to spectroscopic imaging have been introduced in the past few years. These approaches use continuously tunable image-quality spectroscopic filters such as the liquid-crystal tunable filter (LCTF) or the acousto-optic tunable filter (AOTF) as both the source and image filtering devices.

LCTFs are birefringent filters that use the phase retardation between the ordinary and extraordinary rays passing through a liquid crystal to create constructive and destructive interference (Hoyt and Benson, 1992). By stacking a number of stages in series, a single passband is obtained in a manner similar to that of a multicavity interference filter. However, because each stage is electronically tunable, the passband can be varied continuously throughout the visible and near-IR spectral regions. LCTFs are made of planar optical elements, which give them high image transmission fidelity and a straight-through optical path. This design also makes them compact (6–30 mm thick) and easy to integrate into a microscope. In addition, there is no measurable image shift as a function of wavelength tuning when LCTFs are placed between the sample and the camera. This feature is particularly important when images recorded at different wavelengths are ratioed and in situations where accurate spatial information is critical. Throughput appears to be sufficient for most applications in visible absorption and fluorescence imaging applications. After taking into consideration transmission, acceptance angle, and aperture size, LCTFs provide one-quarter to one-third of the light flux that an interference filter with comparable bandwidth can provide.

AOTFs are also electronically addressable optical filters that provide rapid, random wavelength access, broad spectral coverage, and moderately high spectral resolution (Wang, 1992). AOTFs function by the interaction of light with a traveling acoustic wave through an anisotropic crystal. This chapter reviews the operating principle of AOTFs and their recent applications in spectroscopic imaging.

5.2. ACOUSTO-OPTIC TUNABLE FILTERS

5.2.1. Operating Principles and Device Configurations

AOTF operation is based on the acoustic diffraction of light in an anisotropic medium. The device consists of a piezoelectric transducer bonded to a birefringent crystal (Chang, 1981; Wang, 1992). When the transducer is excited by an applied radio frequency (rf) signal, acoustic waves are generated in the medium, which in turn produce a periodic modulation of the index of refraction of the crystal. This provides a moving phase grating that, under appropriate conditions, will diffract portions of an incident light beam. For a fixed acoustic frequency, only a limited band of optical frequencies can satisfy the phase-matching condition and be cumulatively diffracted. However, by tuning the rf frequency, the center optical passband is changed accordingly, and the phase-matching condition is maintained.

In principle, both the isotropic and anisotropic Bragg diffractions can be utilized for the spectral filtering mechanism. However, a filter based on isotropic diffraction is impractical, since its optical passband depends on the angular aperture of an incident beam and is usable only with well-collimated light (on the order of milliradians). The angular aperture limitation results from the fact that a change in the angle of the incident light will introduce a momentum mismatch. For an incident light beam of finite divergence, the width of the passband is greatly increased. Furthermore, the diffracted beam is deflected to a different angle for each wavelength.

An anisotropic acousto-optic filter, however, can maintain a narrow passband over a large angular change in the incident beam. Anisotropic acousto-optic diffraction involves a 90° rotation of the polarization plane of the diffracted wave. Since the refractive indices for ordinary and extraordinary light in a birefringent crystal are not the same, it is possible to choose the direction of acoustic wave propagation so that the group velocity for both the incident and diffracted light is collinear. This process is referred to as noncritical phase matching (NPM). Under the NPM condition, the maximum compensation for the momentum mismatch due to angular deviation of the incident light beam is achieved by the angular change in birefringence. Hence,

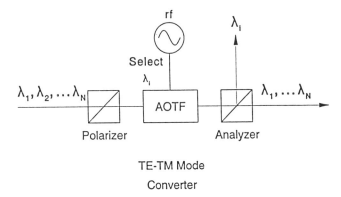

TE-TM Mode

Converter

Figure 5.1. Schematic illustration of an AOTF. It operates like a polarization mode converter sandwiched between a pair of crossed polarizers. Wavelength tuning is accomplished by changing the rf frequency.

the NPM is maintained to the first order over a large angular change in the incident light beam. The angular field of view for a noncollinear AOTF can be as high as $\pm 20°$ in the near-IR region (Dwelle and Katzka, 1987). If an AOTF is sandwiched between crossed polarizers, only those spectral wavelengths within the AOTF passband are polarization converted (Figure 5.1). Wavelengths outside the passband are not rotated and are therefore blocked.

AOTFs generally fall into two categories, depending on the configurations of the acousto-optic interaction. Figure 5.2 is a so-called collinear AOTF

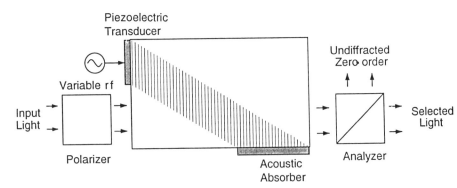

Figure 5.2. Schematic illustration of a crystalline quartz collinear AOTF that produces monochromatic output that must be separated from the input with a polarizer.

based on a quartz crystal in which the incident light, the diffracted light, and the acoustic wave all interact collinearly in a birefringent crystal (Chang, 1974a). In this configuration, even though the acoustic phase velocity is collinear with the direction of light propagation, acoustic group velocity makes an angle with respect to the wave normal. This is called "acoustic walk-off" and is common among different acousto-optic devices. As the result of the acousto-optic interaction, some of the incident light within the filter spectral passband is diffracted. The polarization of this diffracted beam is orthogonal to that of the incident beam. Polarizers can then be used to separate the collinear, yet orthogonally polarized, zero- and first-order beams.

Figure 5.3A illustrates a noncollinear AOTF based on tellurium dioxide (TeO_2), in which the acoustic and optical waves propagate at different angles through the crystal (Chang, 1974b). In this configuration, the zero- and first-order beams are physically separated, so that the filter can be operated without polarizers. As shown in the figure, for a randomly polarized input (k_i), the diffracted light intensity is directed into two physically separated first-order beams, k_d^o and k_d^e, which are orthogonally polarized. For linear polarized input (p or s polarization), there is only one diffracted beam with its polarization $90°$ rotated from the input state. The two orthogonally polarized first-order beams do not separate until they exit the crystal, where they diverge at a fixed angle. The angle of the deflected beam—and therefore the spatial location of the image—does not change with changing wavelength. To use a noncollinear AOTF as a tunable filter, the zero-order beam is blocked, allowing only the diffracted monochromatic component to interact with the sample. The angle between the zero- and first-order components that exit the filter is a function of device design and is typically a few degrees.

Various AOTF devices have been designed and demonstrated in the last two decades, based on both collinear and noncollinear modes of operation. TeO_2 is the preferred AOTF material due to its high acousto-optic figure of merit. This crystal can be used in the visible and infrared (IR) regions up to $5.5\,\mu m$ but cannot be utilized for ultraviolet (UV) applications due to its transmission cutoff near 380 nm. For the UV region, crystalline quartz is the preferred material, although its acousto-optic figure of merit is only $\sim 1/500$ that of TeO_2. This translates into lower diffraction efficiency and higher rf powers required to drive the crystal. For example, a quartz device may need up to 10–20 W of rf power for efficient operation in comparison to a TeO_2 device, which needs only 1–2 W. For reference, typical performance parameters of commercially available AOTFs are listed in Table 5.1. A detailed discussion of the theory of operation, as well as the design criteria and performance of these devices, appears in the following section.

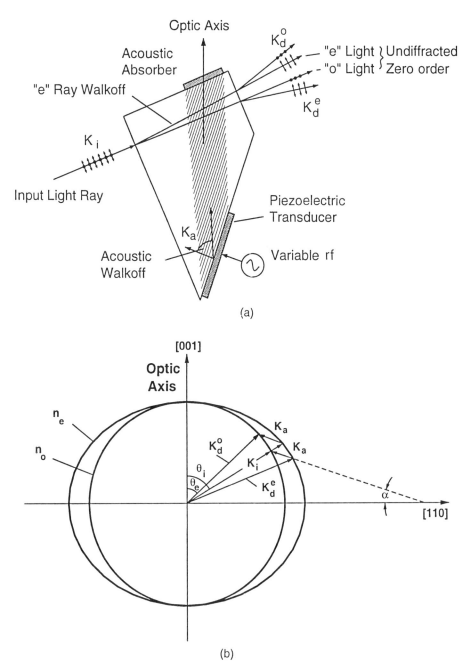

Figure 5.3. Schematic illustration of the acousto-optic diffraction process in a TeO$_2$ noncollinear AOTF: (A) the paths of acousto-optic interaction and the device geometry; (B) **k**-vector diagram superimposed on the crystal's refractive index ellipsoid.

Table 5.1. Typical Specifications for Commercially Available AOTFs

	TeO_2 AOTFs	Quartz AOTFs
Wavelength coverage	0.38–5.5 μm	0.2–1 μm
Tuning range (single transducer)	1 octave	1 octave
Spectral resolution power ($\lambda/\delta\lambda$)	10–1000	100–2000
Optical aperature	0.1–1.5 cm^2	0.1–5 cm^2
Field of view angle	5°–15°	2°–5°
Deflection angle ($\theta_d - \theta_i$)	3°–9°	0°
Tuning spped	4–20 μsec	14–35 μsec
Optical efficiency (single polarization input)	10–90%	20–90%
Extinction ratio	> 10^3	> 10^3
Input rf power	0.5–3 W	5–30 W
Input rf value	20–200 MHz	50–220 MHz
Piezoelectric transducer material	LiNbO$_3$	LiNbO$_3$

Source: Courtesy of Brimrose Corporation of America, Baltimore, Maryland.

5.2.2. AOTF Characteristics

The majority of successfully demonstrated AOTF spectral imaging systems are based on TeO_2 noncollinear AOTFs. For the purpose of general discussion, several important parameters will be derived for the noncollinear AOTF based on a TeO_2 crystal. The collinear AOTF is considered a simplified case of the noncollinear AOTF with an incident angle, θ_i, of 90°.

Figure 5.3 shows the crystal geometry, and the passage of light and acoustic waves, through the filter (Glenar, et al., 1994). Figure 5.3B is a **k**-vector diagram superimposed on the crystal's index ellipsoid. A randomly polarized input ray (k_i) containing multiple wavelengths enters the crystal at a fixed angle θ_i from the optic axis. It consists of an extraordinary component k_i^e of magnitude $2\pi n_e/\lambda$ (λ is the vacuum wavelength and n_e is the extraordinary refractive index) and an ordinary component k_i^o of magnitude $2\pi n_o/\lambda$ (n_o is the ordinary refractive index). The acoustic wavevector $\mathbf{k}_a$ has magnitude $2\pi f_a/V_a$, with f_a the rf frequency and V_a the acoustic velocity. Diffraction occurs only for the wavelength that satisfies the specific momentum matching condition $k_d = k_i \pm k_a$.

5.2.2.1. Tuning Relationship

According to the momentum matching condition, the wavelength of diffracted light can be varied simply by changing the applied rf frequency. The tuning relationships (selected wavelength vs. rf frequency) are generally different for the two orthogonally polarized output beams. The tuning relation for ordinary output light is (Gass and Sambles, 1991; Glenar et al., 1994; Yano and Watanabe, 1976)

$$f_{E \to O} = \frac{V_a}{\lambda} n_e(\theta_i) \{ \sin(\theta_i - \alpha) - (\sin^2(\theta_i - \alpha) + [(n_o/n_e)^2 - 1]\sin^2\theta_i)^{1/2} \} \quad (1)$$

The velocity is calculated from the characteristic sound velocity modes (Ohmachi and Uchida, 1970) in TeO_2

$$V_a^2 = V_{[110]}^2 \cos^2\alpha + V_{[001]}^2 \sin^2\alpha \quad (2)$$

with $V_{[110]} = 616 \, \text{m/s}$ and $V_{[001]} = 2104 \, \text{m/s}$; α is the angle off $[110]$ axis. In Eq. (1), the wavelength dependence of the refractive indices n_o, n_e has been taken into account by measuring and fitting to a Sellmeier dispersion model (Uchida, 1971):

$$n_o^2 - 1 = \frac{1.4351 \times 10^{-4}}{(1/134.2)^2 - (1/\lambda)^2} + \frac{1.6621 \times 10^{-5}}{(1/263.8)^2 - (1/\lambda)^2} \quad (3)$$

$$n_e^2 - 1 = \frac{1.5678 \times 10^{-4}}{(1/134.2)^2 - (1/\lambda)^2} + \frac{2.2248 \times 10^{-5}}{(1/263.1)^2 - (1/\lambda)^2} \quad (4)$$

where λ is in nanometers.

For an extraordinary incident ray in the crystal, the extraordinary refractive index n_e is also a function of the angle off the optical axis (θ):

$$n_e(\theta) = \left(\frac{\cos^2\theta}{n_o^2} + \frac{\sin^2\theta}{n_e^2} \right)^{-1/2} \quad (5)$$

The tuning expression for extraordinary output light has a similar form:

$$f_{O \to E} = \frac{V_a}{\lambda} n_e(\alpha - \pi/2) \{ C(\theta_i) + (C(\theta_i)^2 - [(n_0/n_e)^2 - 1]\sin^2\theta_i)^{1/2} \} \quad (6)$$

with

$$C(\theta_i) = \frac{n_e(\alpha - \pi/2)}{n_o} \left[\sin\alpha \cos\theta_i - \left(\frac{n_o}{n_e} \right) \cos\alpha \sin\theta_i \right] \quad (7)$$

As shown in Eqs. (1) and (6), the tuned wavelength changes approximately as

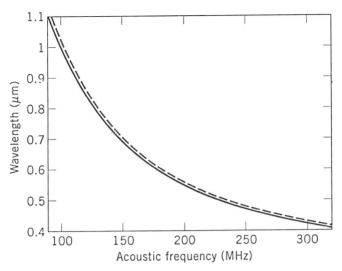

Figure 5.4. The tuning curve for a TeO_2 noncollinear AOTF device with the design parameters $\alpha = 13.5°$ and $\theta_i = 32.5°$. The solid line corresponds to the diffracted extraordinary light, and the dashed line corresponds to the diffracted ordinary light.

$1/\lambda$, but this relation is not exact, since n_o/n_e is also wavelength dependent. Based on Eqs. (1) and (6), Figure 5.4 shows a tuning curve of a TeO_2 noncollinear AOTF device. The center transmission wavelength increases with decreasing rf frequency. The wavelength differences between two outputs increase with the wavelength and are between 2.5 and 7 nm. Depending on the device configuration, the difference can range from several nanometers to tens of nanometers. In a device in which the design parameters are $\alpha = 13.5°$ and $\theta_i = 32.5°$, only the extraordinary input satisfies the NPM condition. The tuning relation for the ordinary input is highly dependent on the input ray direction because the NPM condition is not satisfied.

5.2.2.1.1. Wavelength Tuning Range. The effective wavelength tuning range is generally much smaller than the optical transmission range of the acousto-optic medium. The limiting factor is the electro-acoustic bandwidth of the piezoelectric transducer, which is typically less than an octave. Broader wavelength coverage is possible using multiple transducers on the crystal (Chang, 1974a). For example, some comercially available AOTFs have tunable ranges of 400–1000, 1200–2400, and 1250–4500 nm using a double transducer design.

5.2.2.1.2.   Tuning Speed. The tuned wavelength can be varied very rapidly, since the filter requires only the acoustic transit time, $t = K/V_a$, to diffract a new wavelength once the rf input frequency has been changed; K is the acoustic transit distance and is approximately equal to the optical aperture in a noncollinear device. For example, in a TeO_2 crystal, if the acoustic wave propagates $10°$ off the [110] axis, from Eq. (2), $V_a = 708$ m/s, and if the optical aperture is 7 mm, then the random tuning speed is about 10 μs. Typically, the tuning speed is from several to tens of microseconds, depending on the device configuration and aperture size.

5.2.2.1.3.   Sequential, Random Access, and Simultaneous Multiple Wavelength Modes. When switching between two randomly selected wavelengths, an AOTF, unlike a grating or prism, does not have to sweep over the wavelengths in between. It can just move from one to another by switching between the corresponding rf frequencies. Furthermore, an AOTF can be programmed to simultaneously pass multiple wavelengths because the grating is formed by an acoustic wave, which can be varied, rather than by fabrication. This is accomplished by injecting more than one excitation rf frequency into the device. This property can be exploited to simultaneously measure multiple chemical components or to implement powerful matched filtering algorithms. Different mathematical transformations, such as Hadamard transforms, can be used to analyze the data and greatly increase the sensitivity to important spectral components while suppressing background contributions (Schaeberle et al., 1994). The maximum number of rf frequencies allowed is limited by the rf power-handling capability of the piezoelectric transducer and the minimum rf power required for efficient single-channel operation. It is typically limited to fewer than 10 channels with a commercially available visible (near-IR noncollinear TeO_2 AOTF. There are low-power AOTFs under development that may be able to accommodate more than 100 channels simultaneously.

5.2.2.1.4.   Utilization of Both Diffracted Outputs. As shown in Figure 5.3A, the diffracted light intensity is directed into two physically separated first-order beams that are orthogonally polarized. It would be efficient to combine them to increase the optical throughput or use them for spectropolarimetry. However, these applications require both orders to have the same output wavelength and to satisfy the NPM condition. As indicated by Eqs. (1) and (6), the tuning relationships are different for the two orthogonally polarized outputs. In addition, only one output typically satisfies the NPM condition. Specially designed TeO_2 noncollinear AOTFs have been manufactured with design parameters ($\alpha = 18.9°$ and $\theta_i = 55.6°$) chosen such that both outputs have the same tuning relationship and simultaneously satisfy the NPM in the near-IR wavelength region. Even with a standard AOTF, for a given α it is possible to choose an input angle that allows the orthogonally polarized diffracted beams to have the same tuning relation. However, this does affect spectral resolution.

5.2.2.2. Transmission Efficiency and Spectral Resolution

The peak transmission T_0 is given by the ratio of diffracted light to that of incident light (Chang, 1981).

$$T_0 = \sin^2\left(\frac{\pi^2 M_2}{2\lambda^2} P_d L^2\right)^{1/2} \tag{8}$$

where L is the acousto-optic interaction length; P_d is the acoustic power density; and M_2 is a normalized acousto-optic figure of merit that is completely determined by the media propeties, given here:

$$M_2 = \frac{n_o^3 n_e^3 p^3}{\rho V_a^3} \tag{9}$$

Here ρ is the mass density of the medium and p is the elasto-optic coefficiency. For infinitely long interaction length, only the light wave with the wavelength that exactly satisfies the phase-matching condition will be transmitted through the AOTF. As a result of the finite interaction length, the width of the filter bandpass is broadened. The passband shape for a uniform transducer and collimated input light is (Glenar et al., 1994)

$$T_\lambda = T_o \, \text{sinc}^2\left(0.886 \frac{\lambda - \lambda_c}{\Delta\lambda_{FW}}\right) \tag{10}$$

where λ_c is the wavelength of peak transmission. It is a sinc^2 function and has multiple sidelobes; $\Delta\lambda_{FW}$ is the spectral resolution of the tunable filter, defined as full width at half-maximum (FWHM) of the main lobe, as shown in Figure 5.5. In the small α approximation, it is given as (Chang, 1981)

$$\Delta\lambda_{FW} = \frac{1.8\pi\lambda^2}{bL\sin^2\theta_i} \tag{11}$$

where b is the dispersion constant given by

$$b = 2\pi\left(\Delta n - \lambda \frac{\partial \Delta n}{\partial \lambda}\right) \approx 2\pi \, \Delta n \tag{12}$$

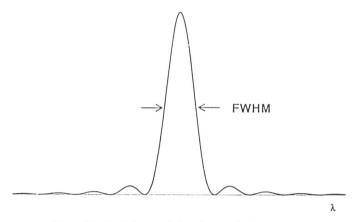

Figure 5.5. Optical transmission characteristics of AOTFs.

and where $\Delta n = n_e - n_o$. The spectral resolution of the selected light depends on the device configuration and the wavelength of operation, and is typically from several nanometers to tens of nanometers in the visible and near-IR regions. Peak transmission can be as high as 98%, with the intensity divided between the two orthogonally polarized outputs for randomly polarized input.

5.2.2.2.1. Transmission Control and Intensity Modulation. One of the useful and unique features of the AOTF, as indicated in Eq. (8), is its ability to precisely and rapidly adjust the intensity of the diffracted light by varying the acoustic power. It can be used to provide a flat response over the entire wavelength range of the instrument and can also be used with lock-in (phase- and frequency-sensitive) detection schemes for low-light signal detection.

5.2.2.2.2. Variable Bandpass. It is noted from Eq. (11) that unlike a diffraction grating, the FWHM of the AOTF is not constant. Instead, it changes according to the square of the wavelength. For example, the bandpass of a typical visible AOTF, operating between 400 and 700 nm, changes between 2 and 6 nm when scanned over the whole wavelength region.

5.2.2.2.3. Sidelobes and Apodization. The sidelobe structure is caused by the abrupt onset and cutoff of the acousto-optic interaction, which resembles a rectangular waveform. The Fourier transform of such an envelope is a sinc^2 function that falls off as λ^2. For optimum conditions (weak acousto-optic interaction), the first sidelobe has a level of $-13\,\mathrm{dB}$ (4.7%). Theoretically, the sidelobe structure should be symmetrical about the central peak. However, in practice, the sidelobe intensities relative to the main peak are often asymmetrical, with individual intensities varying with the acousto-optic interaction. This imperfection in sidelobe structure may be caused by excessive drive power or phase nonuniformity during the acousto-optic interaction. Reduction of sidelobes in the AOTF bandpass can be achieved by amplitude apodization of the acoustic wave. For example, if the acoustic field distribution is tailored using a Hamming window function, the highest sidelobe is -40 dB below the main peak (Harris, 1978). Various other window functions can also be used for apodization. In practice, two types of amplitude apodizations have been used to suppress the sidelobes. In the noncollinear AOTF, spatial apodization is achieved using a weighted excitation of the transducer, whereas in the collinear AOTF, an acoustic pulse, apodized in time, is launched into the acousto-optic medium (Chang and Katzka, 1978; Weverka et al., 1983). A reduction of more than -30 dB has been achieved experimentally, a level at which the transmission profile of the AOTF will be satisfactory for most spectroscopic applications.

5.2.2.2.4. Out-of-Band Rejection Ratio. The out-of-band rejection ratio is mainly determined by the optical quality of the crystals. Optical nonuniformities, defects and strains can cause background scattering (Chang, 1974b). This is especially true for high refractive index materials such as TeO_2 ($n = 2.2$). In practice, a 33 dB (2×10^3) rejection ratio was reported without using polarizers (Glenar et al., 1994). The use of a pair of polarizers can help to further increase the rejection ratio, especially in the case of a large angular aperture device, but the effectiveness is limited by the depolarization due to scattering.

5.2.2.3. Optical Aperture and Angular Field of View

In practice, the linear aperture of the AOTF is limited by the acoustic beam height in one dimension and by acoustic attenuation across the optical aperture in the other dimension. The height is limited by the acoustic power density requirement, diffraction efficiency, and maximum power-handling capability of the transducer. It is also limited by the electrode size. As acoustic attenuation in a crystal medium is proportional to the square of the acoustic frequency, it is not usually a severe constraint in the near-IR and IR regions that correspond to lower rf frequencies. In the visible spectral region, due to the higher rf frequencies, the aperture size is limited to less than 15 mm in TeO_2. For the slow shear wave, the acoustic attenuation is about $18 \, dB/\mu s \cdot GHz^2$ (Uchida, 1972). Therefore, for a device with an optical aperture of 12 cm ($\sim 17 \, \mu s$) operating at 100 MHz, the loss due to attenuation is about 3 dB (50%). For quartz AOTFs, the acoustic attenuation is much lower. There are collinear devices in which the acoustic propagation distance can be as long as 250–300 mm.

The effective aperture is also limited by the deflection angle (the angle of separation betwen the undiffracted and diffracted light). The deflection angle outside of the medium is (Xu and Stroud, 1992)

$$\theta_i - \theta_d = \Delta n \sin 2\theta_i \qquad (13)$$

As indicated, the crystal with a larger birefringence (Δn) is desirable for achieving a better separation between the undiffracted an diffracted light without using polarizers. Also, as shown in both Eq. (13) and Figure 5.3B, the diffracted beam appears at one characteristic deflection angle that is smallest when k_a lies along the [110] axis ($\theta_i = 90°$; $\alpha = 0°$) and increases with α up to a practical limit of 8° or 9°. Diffracted and undiffracted images exiting the crystal are spatially separated after propagating a certain distance. This distance is related to the input aperture size as well as to the separation angle. The larger the image, the longer the distance that is required to separate them. In designing a compact instrument, a size limitation imposed on the entire imaging system therefore limits the aperture size. A pair of polarizers may be used to reduce this constraint, but this is not feasible in the case where both diffracted beams are utilized.

An additional consideration is the angular field of view (FOV) of the AOTF. This parameter is proportional to the total amount of light that can be collected by the device. The angular aperture is defined as the angle for which the intensity at the exit is equal to 50% of the incident monochromatic light intensity. To a first-order approximation, the polar angular FOV outside of the medium is approximately given by

$$\Delta\theta_i^o \approx n\left(\frac{\lambda}{\Delta nL|F_\theta|}\right)^{1/2} \qquad (14)$$

$n = (n_e + n_o)/2$, an averaged refractive index, where

$$F_\theta = 2\cos^2\theta_i - \sin^2\theta_i \qquad (15)$$

In the visible region, $\Delta\theta_i^o$ is typically several degrees. The azimuthal angular FOV outside of the medium is approximately given by

$$\Delta\phi_i^o \approx n\left(\frac{\lambda}{\Delta nLF_\phi}\right)^{1/2} \qquad (16)$$

where

$$F_\phi = 2\cos^2\theta_i + \sin^2\theta_i \tag{17}$$

For a collinear AOTF, $\theta_i = 90°$, so that $\Delta\theta_i^o = \Delta\phi_i^o$, the polar angular FOV, equals the azimuthal angular FOV.

5.2.2.4. Spatial Resolution

An AOTF, like an optical grating, is not an imaging system, and therefore no definition of spatial resolution exists. However, the AOTF may impose limitations on the overall spatial resolution of an imaging system due to its limited linear aperture size and acceptance angle. Currently, the largest optical aperture size for commercially available visible wavelength AOTFs is limited to $12 \times 12\,\text{mm}$, and the acceptance angle is typically less than $10°$. The maximum number of image elements resolvable according to the Rayleigh criterion is determined b the angular and linear apertures of the filter and by the wavelength of the light. The maximum number of resolvable image elements, N_p (in the polar plane) and N_a (in the azimuthal plane) is given by

$$N_p < \frac{\Delta\theta_i^o A}{\lambda}, \qquad N_a < \frac{\Delta\phi_i^o A}{\lambda} \tag{18}$$

where A is the optical aperture. The spatial resolution is

$$R_s = \frac{\Delta\theta^o}{\lambda}, \qquad \frac{\Delta\phi^o}{\lambda} \tag{19}$$

For example, if the optical aperture is $12 \times 12\,\text{mm}$ and the acceptance angle is $\pm 7°$, the calculated maximum number of resolvable image elements is 2933 and the spatial resolution is 244 lines/mm, at a wavelength of 500 nm. Diffraction limited resolution has been achieved in the azimuthal plane, but spatial resolution can be substantially lower in the polar plane (acousto-optic diffraction plane) due to nonselective dispersion in the AOTF. By properly suppressing the dispersion, near-diffraction limited spatial resolution can be achieved with only a modest reduction of image contrast.

To a first-order approximation, the changes of the deflection angle are negligible when an AOTF sweeps around a center wavelength with a limited bandwidth, such as tens of nanometers. However, there is a dispersion that results in the wavelength dependence of the deflection angle. When the AOTF operates over a broad spectral range, such as several hundreds of nanometers, this dependence can cause the diffracted image to shift across the detector with tuning. The shifting magnitude can be estimated by using Eqs. (13) and (3)–(4). For example, if an AOTF is operating between 400 and 700 nm with an incident angle $\theta_i = 32.5°$, the estimated maximum angular shift is $\sim 1.8°$. This shift is certainly undesirable for many applications, such as image ratioing at different wavelengths. By using post-digital-imaging processing, the shift can be corrected because the image shift is a fixed function of the wavelength in an imaging system. It is also shown that the dispersion has a substantial influence on the angular resolution of the elements of the filtered image and causes image blurring. For an input image, its output spatial element with different wavelengths will be angularly broadened. The lower the spectral resolution of the AOTF, the more severe the blurring effect.

These unwanted effects can be well suppressed by proper selection of the angle of the filter output face (Epikhin and Kallinikov, 1989). In this method, the spectral drift of the diffraction angle in the filter is compensated for by the spectral drift of the refraction angle at the inclined output face. Another method of suppression is to use successive anisotropic diffraction in two noncollinear filters (Kalinnikov and Statsenko, 1990). In this scheme, the beam that has been diffracted in the first AOTF is separated out and diffracted into the second AOTF, acquiring its own polarization and initial direction of propagation. Here the second device compensates for the dispersion of the diffraction angle, and also for the "drift" caused by the spectral dependence of the refractive index. The additional advantage of the dual AOTF scheme is the possibility of achieving a high extinction ratio. One modification of this method is a scheme that passes light through the same AOTF twice by reflection from a mirror.

5.3. APPLICATIONS IN SPECTRAL IMAGING

The AOTF can be employed in a spectral imaging system either as a source filter, an image filter, or a combination of both. When the source filtering approach is used, the AOTF is placed directly in front of a broadband soure or multiline laser. Only the wavelength selected by the filter is transmitted, creating a tunable monochromatic source. This filtered light can be passed through illumination optics such as a microscope condenser before being transmitted or reflected by the sample and imaged onto a two-dimensional detector array. This method can be applied to fluorescence excitation ratioing and absorption spectroscopy.

A more difficult approach is image filtering, which may be more generally applied to absorption as well as emission spectroscopies such as fluorescence and Raman. In this approach, a broadband source, filtered source, or laser may be allowed to impinge on the sample, while an AOTF is placed in front of a two-dimensional detector array for multispectral image tuning. Kurtz et al. (1987) first described the application of an AOTF in microscopy. In the following section, more recent examples of AOTF-based spectral imaging systems will be reviewed.

5.3.1.   Light Source Filtering

5.3.1.1.   Excitation Ratio Imaging in Confocal Microscopy

AOTFs can be combined with an incoherent light source or a multiline laser to produce a fast tunable light source. One such application is in fluorescence spectroscopy. Fluorescent probes are capable of indicating such diverse properties as ion concentration, pH, and electric potential in live cells and tissues. In acquiring kinetic data from fluorescent probes, it is often necessary to monitor two or more excitation or emission wavelengths, using their ratio to cancel out the intensity of excitation and dye concentration and to get an accurate estimation of the target ion concentration. Rapid switching between two or more wavelengths is also highly advantageous, since capturing many types of fluorescent images requires both high spatial and high temporal resolution. Images must be recorded before samples are damaged by bleaching or photodynamic effects. In addition, many important biochemical processes occur in the 10–100 ms time frame. The conventional strategy is to peform such measurements using a mechanical rotating filter wheel, which is slow, difficult to synchronize with the image acquisition hardware, and causes mechanical vibration, which is particularly undesirable in microscopy. The wavelength switching times for such devices may be less than 10 Hz. Many of these problems can be entirely avoided by using an AOTF as the wavelength selector. The AOTF has no moving parts, and the wavelengths are selected by a high-speed frequency synthesizer. Additionally, by coupling the AOTF to a broadband light source or a multiline laser and driving the AOTF with two rf frequencies, two excitation wavelengths can be generated simultaneously. Furthermore, each wavelength can be modulated electronically at different frequencies, and lock-in amplifiers can be used to demodulate the fluorescent emission into its two components.

Kenneth Spring, at the National Institutes of Health's Laboratory of Kidney and Electrolyte Metabolism, has developed a confocal laser scanning microscope (CLSM) that captures multiwavelength fluorescent images of single living cells at video frame rates (Nitschke and Spring, 1995; Wang, 1994a). A multiwavelength argon–ion laser source allows quantitative measurements to be made from fluorescent indicators by ratioing images at two different wavelengths. This requires rapid wavelength and power switching among the various laser lines. Since electromechanical movement of a filter assembly is too slow, an AOTF has been incorporated into the system. Thus, both laser power and wavelength can be controlled in the 0.1–1.0 ms time frame (Chatton and Spring, 1993). The addition of an LCTF to one of the two photomultiplier tube (PMT) detectors also permits rapid spectral scanning of the fluorescence emission. An illumination intensity of only 40 µW/cm^2 at the back focal plane of the microscope objective allows high-quality fluorescence and transmitted light images of living cells to be recorded with minimal bleaching and photodynamic damage.

Figure 5.6 is a schematic diagram of the optical path of the CLSM. It is designed around a commercially available Odyssey CLSM (Noran Institute, Middelton, Wisconsin). The CLSM is connected to an inverted microscope (Diaphot, Nikon, Inc., Melville, New York). Multiple wavelength laser light is first made monochromatic by an AOTF (Brimrose, Baltimore, Maryland). The beam is then expanded and shaped for the acousto-optic deflector (AOD). The resultant first-order diffracted beam, constituting the X scan, passes

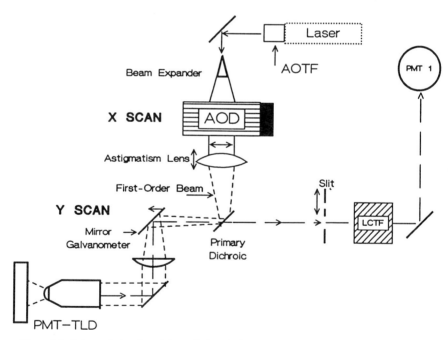

Figure 5.6. Schematic drawing of the optical path of the CLSM showing an AOTF being used for fast wavelength switching in fluorescence excitation ratio imaging. Reprinted with permission from the *Journal of the Microscopy Society of America* (Spring 1995), **1**(1), 1–110.

through a movable lens and shaping optics to the primary dichroic mirror. The reflected beam falls on the mirror galvanometer, where it is scanned in the *Y* direction and focused to fill the back focal plane of the objective lens. Excitation light that passes through the specimen falls on the diode detector to form the transmitted light image. Fluorescent signals arising from the preparation are descanned in the *Y* direction by the mirror galvanometer and passed through the primary dichroic, a slit, and an LCTF (Cambridge Research & Instrumentation, Inc., Cambridge, Massachusetts). The light is then detected by a side-on PMT detector. Figure 5.7 (see Color Plates) shows confocal images of living renal epithelial cells with their lateral intercellular spaces stained with BCECF, a pH indicator. The figure shows a transmission light image and a color-coded fluorescent ratio image recorded at excitation wavelengths of 457 and 488 nm.

5.3.1.2. Near-Infrared Spectroscopic Microscopy

Figure 5.8 is a schematic of an AOTF spectroscopic microscope (Treado et al., 1992a). The system integrates a visible/near-IR TeO_2 AOTF (Brimrose, TEAF—0.6–1.2L) with a metallurgical microscope (Olympus BH-2). Image detection is provided by a low-noise, 16 bit CCD detector (Spex Industries, New Jersey). The detector is liquid nitrogen cooled ($-140\,°C$) and contains 576×348 pixels, each $20\,\mu m^2$. In operation, the CCD is used as a focal-plane array, while wavelength selectivity is provided by the AOTF and a quartz tungsten halogen lamp. Data collection is initiated by tuning the AOTF through hundreds of nanometers at predetermined wavelength intervals, recording one frame with the focal-plane array detector at each wavelength. One data set may contain several hundred individual image frames and may be many megabytes in size. The data can be considered either as separate sample images at discrete wavelengths or as a matrix of individual absorption spectra for each pixel element in the image. An alternative strategy is to tune the AOTF to one or more wavelengths corresponding to known absorption or emission maxima, where numerous spectral frames can be collected and averaged to provide maximum sensitivity and image contrast. The frames may be collected rapidly over a period of several seconds or more slowly over time frames extending to hours or days for kinetic studies. From these data, images may be constructed that yield information on the chemical, spatial, and temporal properties of a sample with respect to highly selective spectroscopic markers.

The instrument design is entirely solid state and contains no moving parts. It can be configured for both absorption and reflectance spectroscopy. Images can be simultaneously collected at high spatial resolution (1 μm) and moderate spectral resolution (2 nm). Figure 5.9 is an image of water microdroplets collected with the imaging system tuned to 960 nm (A) and 850 nm (B) (Treado

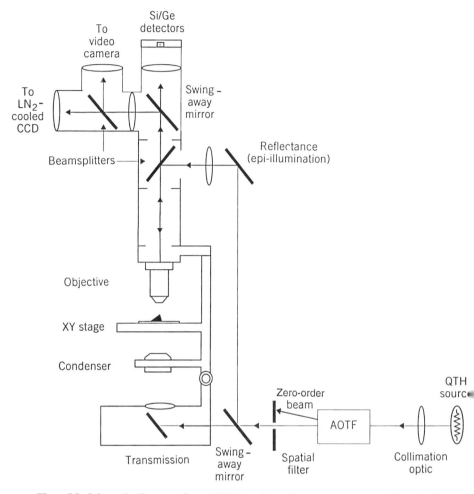

Figure 5.8. Schematic diagram of an AOTF-based imaging spectrometer optical train. The figure shows the arrangement of the modified microscope with respect to the AOTF illumination scheme for both transmission and reflectance measurements, as well as the relative positions of the CCD, video camera, and point detectors. Reprinted with permission from *Applied Spectroscopy* (1992), **46**(4), 553–559.

et al., 1992a). The darkening of the droplets in (A) relative to (B) corresponds to the vibrational absorption of water, specifically the second overtone OH stretch at 960 nm. For presentation, the images are corrected for instrument response and background contributions. The reference bar in (B) corresponds to 7 µm.

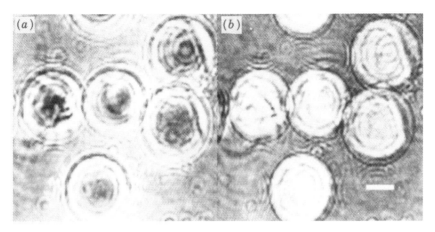

Figure 5.9. Microscope image of water microdroplets collected at 960 nm (A) and 850 nm (B). Reprinted with permission from *Applied Spectroscopy* (1992), **46**(4), 553–559.

5.3.2.   Image Filtering

Image filtering can be more generally applied to vibrational as well as fluorescent imaging. Image filtering is more difficult than source filtering because an AOTF is not a classic optical component and cannot be easily placed in front of a two-dimensional detector. Device design alterations and optical system compensations must be considered in order to achieve near-diffraction limited spatial resolution. An understanding of the mode of operation of the AOTF and in-depth optical design skills are required to produce an AOTF-based imaging system with the required spectral resolution and image fidelity. While many design parameters and trade-offs must be taken into consideration when designing such a system, there are now commercially available AOTF imaging subsystems that can be readily interfaced with microscopes and telescopic systems.

5.3.2.1.   *Raman Spectroscopic Microscopy*

Figure 5.10 diagrams an AOTF spectral imaging microscope operating in an image filtering mode suitable for Raman, fluorescence or absorption spectral imaging (Lewis et al., 1994; Soos, 1992; Treado et al., 1992b). The AOTF (Brimrose, Baltimore, Maryland) is integrated within an infinity-corrected refractive microscope (BHS-2, Olympus), which performs magnification and

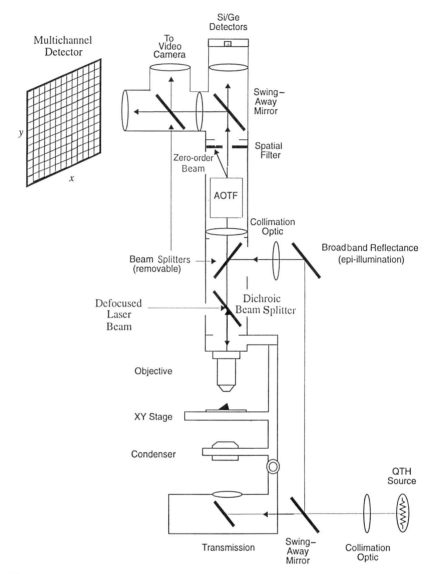

Figure 5.10. Schematic illustration of the acousto-optic filtered Raman/near-IR imaging microscope. The optical train shows the path of the incident laser beam and Raman scattered light. Samples are mounted on standard microscope slides, and 180° Raman scattering is collected and spectrally filtered with the AOTF. Holographic Raman filters are placed after the AOTF to eliminate intense Rayleigh scattering before the image is focussed onto a liquid nitrogen–cooled CCD. To perform NIR absorption imaging, an Si CCD was substituted for an InSb NIR focal-plane array detector, an NIR AOTF was substituted for the visible model. Reprinted from *American Laboratory*, **26** (9), 16 (1994). Copyright 1994 by International Scientific Communications, Inc.

image collection and finally presents the image to the detector. To perform near-IR absorption imaging, the Si CCD (Spectrum-1, Spex Industries, Inc., Edison, New Jersey) is substituted for an InSb near-IR focal-plane array detector (ImagIR, Santa Barbara Focalplane, Goleta, California), as well as a near-IR AOTF and appropriate optics.

Figure 5.11 (see Color Plates) shows three images of a lipid–peptide mixture of L-asparagine and dipalmitoylphosphatidylcholine (DPPC). Figure 5.11a is a bright-field image, and Raman images are shown in Figures 5.11b and 5.11c. The methylene symmetric stretching mode image (Figure 5.11b) indicates the spatial distribution primarily of DPPC, while the NH_2 stretching mode image (Figure 5.11c) indicates only the position of L-asparagine. The image is presented in pseudocolor, with the highest Raman intensity corresponding to yellow and the lowest to blue. Figure 5.12 shows two spectra obtained from different pixels of the data set, which correspond to regions of pure DPPC (curve A) and pure L-asparagine (curve B).

Figure 5.13 shows composite AOTF images of OH absorption in onion epidermis (Treado et al., 1994). The left panel shows a nonabsorbing image collected at 1340 nm, where sample opacity is due to scattering and refractive index contributions occurring primarily at the epidermal cell membrane interfaces. The right panel shows an image collected at 1440 nm and exhibits much higher relative opacity due to OH absorption. The size bar equals 15 μm.

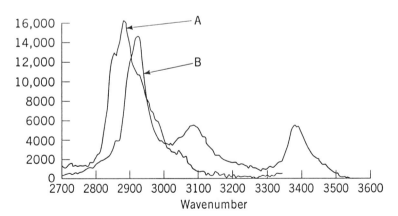

Figure 5.12. Raman spectra constructed from a single image pixel of DPPC (A) and L-asparagine (B) collected between 2700 and 3500 cm^{-1} at 5 cm^{-1} increments. Reprinted from *American Laboratory*, **26**(9), 16(1944). Copyright 1994 by International Scientific Communications, Inc.

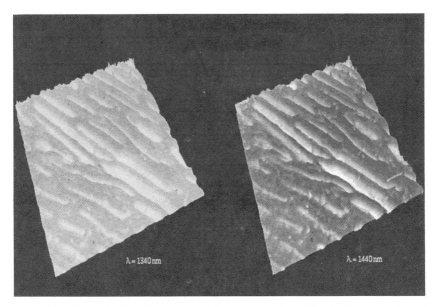

Figure 5.13. Composite AOTF images of OH absorption in onion epidermis with the use of a 10X objective and averaging of 10 frames per wavelength increment. (Left) Nonabsorbing image collected at 1340 nm. Sample opacity is due mainly to scattering and refractive index contributions that occur primarily at the epidermal cell membrane interfaces. (Right) Vibrational spectral image collected at 1440 nm containing higher relative opacity than the nonabsorbing image due to OH absorption. Size bar equals 15 μm. Reprinted with permission from *Applied Spectroscopy* (1994), **48**(5), 607–615.

5.3.2.2.   Fluorescence Microscopy

Patrick Treado, at the University of Pittsburgh, is performing multispectral fluorescence microscopy using an AOTF. Figure 5.14 shows images of two 15 μm diameter polystyrene microspheres that have been tagged with two different fluorescent dyes (Morris et al., 1994). Figure 5.14A shows the microspheres in bright-field microscopy, while the corresponding fluorescence images are shown in Figures 5.14B and 5.14C. Excitation at 560 nm was accomplished with an interference filter placed in front of an Hg lamp. Figure 5.14B is collected by tuning the AOTF to 602 nm, a wavelength approximately corresponding to the emission maximum of the orange dye–labeled sphere. This sphere appears at the top left of the image. In Figure 5.14C the AOTF is tuned to 640 nm, and while emission from the orange dye–labeled sphere is still significant at this wavelength, the red dye–labeled sphere is also emitting. This sphere appears at the bottom right of the image frame.

A Brightfield Microscopy

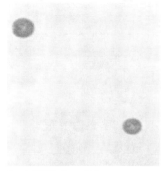

B AOTF λ = 602 nm

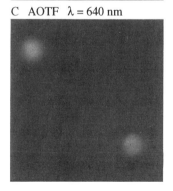

C AOTF λ = 640 nm

Figure 5.14. (A) Bright-field image of two 15 μm diameter microspheres collected with a 20X objective (NA 0.46). The microspheres were tagged with different fluorophores. (B) and (C) are fluorescence images collected through the AOTF tuned to the characteristic fluorescence maxima of 602 and 640 nm, respectively. Reprinted with permission from *Applied Spectroscopy* (1994), **48**(7), 857–865.

The two fluorescence emission images presented in Figure 5.14 were removed from a single data set comprising a series of images collected by sequentially tuning the AOTF from 560 to 730 nm at 2 nm intervals. Spectral vectors extracted from x, y coordinates corresponding to the centers of the two differently tagged microspheres are presented in Figure 5.15. Curve A corre-

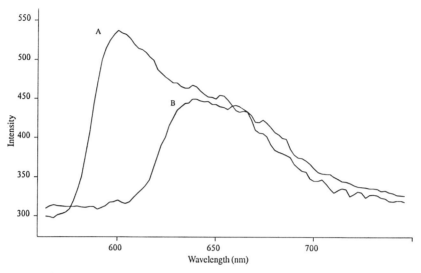

Figure 5.15. Fluorescence spectra of two different fluorophores. Reprinted with permission from *Applied Spectroscopy* (1994), **48**(7), 857–865.

sponds to the emission spectra of the orange sphere, and curve B corresponds to the red sphere.

5.3.2.3. Photoluminescent Volumetric Imaging

The study of the physical interaction between immiscible liquids, such as oil and water, and solid surfaces, such as rock or soil particles, is useful in fields as diverse as environmental cleanup, enhanced oil recovery, particle dialysis, and coating of ceramics. Montemagno and Gray are using an AOTF in an imaging system designed to study microscopic and macroscopic effects in a fluid/fluid/solid model system. The imaging system can view the interfaces in three dimensions under static conditions and in a variety of perturbations (Montemagno and Gray, 1995; Wang, 1994b). A typical model system consists of one or two immiscible fluids and crushed fused silica glass with particle sizes ranging between 70 and 700 μm. To allow undistorted visible imaging of the system, the fluids are index matched to the silica, and the interfaces are preferentially excited with a combination of polar and nonpolar fluorophores. The fluorophore in the polar fluid is usually a biotype surfactant—a long organic chain with polar end groups. Such molecules are naturally attracted to the fluid/fluid and fluid/solid interfaces. Three-dimensional images are constructed from a series of two-dimensional images as follows.

The 514.5 nm output of an argon–ion laser is expanded to a 25 mm beam and then condensed in one dimension using a long-focal-length cylindrical lens. This shaped beam is then directed into the multiphase sample under study, producing a sheet of excited sample, typically only 100 μm deep. The fluorescence from a finite area is then imaged onto a cooled CCD. An AOTF is positioned in front of the CCD camera to allow spectral filtering of the fluorescence. After the two-dimensional image is recorded at a number of discrete wavelengths, the sample is moved perpendicular to the FOV, where the image collection process is repeated. In this manner, a volume of up to 1.5 cm³ can be recorded as a series of image slices. A computer and image reconstruction software convert the data into a false-color three-dimensional image in which the effective resolution is better than 106 line pairs/mm. The ability of the AOTF to rapidly record images at several specific wavelengths is critical, because it distinguishes between the two fluorophores and, hence, the distribution of the polar and nonpolar liquids. In addition, the fluorescence emission of a fluorophore can be subtly shifted, depending on the nature of the interface to which it is aggregated. Such spectral shifts can be used to study the system in greater detail, and continuously tunable imaging filters such as AOTFs are able to measure these shifts. Figure 5.16 (see Color Plates) shows a false-color image of a two-phase-mixture model system (aqueous fluid and quartz sand) using the method described. Quartz grains range in size from 300 to 520 μm, and the entire image cube is approximately 7 × 7 × 7 mm.

5.3.3.   Time-Resolved Fluorescence Microscopy

Rapid, continuous wavelength scanning for both the excitation and emission components of a fluorescence imaging system would be extremely valuable for many types of applicatios. For example, a time-resolved fluorescence microscope (TRFM) may be constructed by utilizing two synchronized AOTFs. Figure 5.17 is a schematic illustration of a proposed TRFM for implementation of fluorescence *in situ* hybridization (FISH) and fluorescence *in situ* polymerase chain reaction (FIS-PCR) (X. Wang, X. F. Wang, and B. Herman, NIH grant application, private communications, 1993). These techniques will open up new possibilities for the location and analysis of protein, DNA sequences, and RNA in tissues, cells, and chromosomes. Based on the lifetime difference between a fluorescence signal and background contributions, a TRFM can provide the ability to discriminate between highly specific fluorescence signals and improve the detecting sensitivity of the system so that it performs comparably to or even more sensitively than radioisotopic assays.

Several TRFM systems have been developed to allow the discrimination and detection of fluorescence signals with long lifetimes (Marriott et al., 1991;

Seveus et al., 1992). These spectrometer systems use a rotating disc or chopper to regulate the opening and closing of the excitation and emission light paths in the microscope. They are complex, bulky, expensive, and difficult to synchronize; in addition, they have mechanical vibration problems. As a result, they can only be used in research laboratories and are very difficult to integrate into commercial products.

In the system proposed in Figure 5.17, one AOTF acts as the wavelength tunable monochromatic excitation light source, as well as a very fast optical switch to generate an excitation pulse on the order of a few microseconds. A second AOTF is synchronized with the first one and turned on after a time delay. Unwanted autofluorescence is a rapidly decaying process with lifetimes in the range of 1–100 ns, whereas the signals of interest can have lifetimes in the range of 1 μs to 10 ms. Therefore, if the detection is delayed and synchronized within this time frame, no autofluorescence will be observed in the image. Consequently, the emitted intensity from fluorophores that have long lifetimes can be observed without interference from autofluorescence

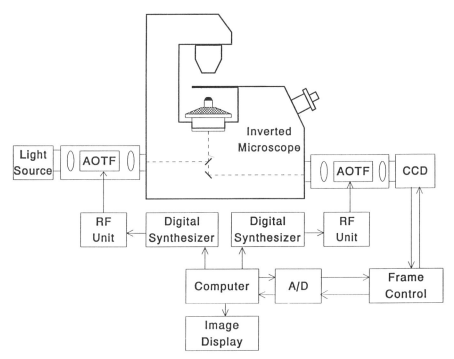

Figure 5.17. Schematic illustration of a proposed TRFM that uses two synchronized AOTFs for both source filtering/modulation and gating/image filtering.

and scattered light, leading to a substantial increase in detection sensitivity. The recently developed long lifetime probes, such as trivalent lanthanide ion chelates (Beverloo and Tanker, 1992; Sevens et al., 1992), with a decay time on the order of several hundred microseconds, are a perfect match for the speed of the AOTF. This AOTF-TRFM is advantageous in that it is compact and offers an all-solid-state solution with no moving parts or mechanical vibration.

5.4. CONCLUSIONS AND FUTURE TRENDS

The development of AOTF technology in the last two decades has resulted in a practical, general-purpose spectroscopic device. While its application in spectroscopic imaging is still relatively new, developments in AOTF design and fabrication continue to enhance its appeal to developers of spectral imaging systems. In the visible, near-IR and mid-IR spectral regions, TeO_2 noncollinear AOTFs will remain the preferred device for many applications. They are easy to fabricate and are generally of good quality. Commensurate with improvements in AOTF design will be more elaborate imaging system implementations. Such systems will utilize diffracted outputs to achieve polarization-insensitive operation and polarimetry, and will exploit the property of simultaneous multiple-wavelength selection. This can be utilized to implement powerful matched filtering algorithms tailored to maximize the performance of the imaging spectrometer in numerous specialized applications.

In the UV region, few birefringent crystals transmit. Those that do have a typically low AO figure of merit due to low indices of refraction, relatively high acoustic velocities, and small photoelastic tensor components. In addition, a small angle of deflection, due to a small birefringence, requires the use of polarizers for most applications of even noncollinear AOTFs. New materials in this region will have to be discovered if AOTFs are to achieve the required performance for many practical purposes. Improvements based on existing crystals will consist of operating these AOTFs using a power-saving acoustic resonance technique or at higher power levels with the improvement of piezoelectric bonding and thermal cooling techniques. In the longer wavelength infrared region, there are several materials that have the required high figure of merit. Chief among them are Tl_3AsSe_3, Hg_2Cl_2, $PbBr_2$, and HgS (Gottlieb et al., 1987, 1992; Singh et al., 1989). These crystals transmit up to 40 μm but are rather difficult to grow or fabricate. In addition, most of them have poor thermal characteristics, which could be a problem in the far-IR region. However, improvements in the growth of these crystals are expected, leading to improved optical quality and increased physical size, making them suitable for device manufacturing.

Another area of active development is integrated-optic acousto-optic tunable filters (IAOTFs). All the AOTF devices discussed so far are based on acousto-optic interaction in bulk crystal media. Acousto-optic interaction also exists in surface waves in which both optical and acoustic waves are confined within a thin surface layer of several microns (Binh and Livingstone, 1980; Ohmachi and Noda, 1977). An IAOTF based on surface wave interaction has several advantages over traditional AOTFs including lower rf driving power, more compact size, less weight, and higher spectral resolution. An IAOTF has been successfully used as the monochromatic wavelength selective element in a spectrometer (Wang et al., 1994). The performance of the system was demonstrated by measuring the absorption spectra of two samples in the near-IR region. The device required less than 70 mW of rf drive power and exhibited an FWHM resolution of 1.2 nm. In addition, the device was exceptionally compact, measuring $25 \times 3 \times 1$ mm and weighing less than 1 g. This IAOTF offers several attractive design features that have utility in a variety of applications, particularly for miniature sensor instruments.

Technological developments in both AOTFs and focal-plane array detectors will soon enable higher spectral and spatial resolution imaging to be performed from the UV through the IR spectral region. The potential impact of such technology is far-reaching, and the areas of application overlap those where conventional optical spectroscopic approaches have proven valuable. These areas include biological materials, polymers, semiconductors, and even remote sensing and real-time industrial process imaging.

ACKNOWLEDGMENTS

The authors gratefully acknowledge P. J. Treado, K. R. Spring, and C. D. Montemagno for providing figures for the manuscript as well as useful discussions. The authors thank C. C. Hoyt for helpful comments and information about LCTFs. The authors would also like to thank V. Pelekhaty, Q.X. Li, and D. C. Fedele for their help in preparing this manuscript.

REFERENCES

Beverloo, H. B., and Tanke, H. J. (1992). Preparation and microscopic visualization of multicolor luminescent immunophores. *Cytometry* **13**, 561.

Binh, L. N., and Livingstone, J. (1980). A wide-band acousto-optic TE–TM mode converter using a doubly-confined structure. *IEEE J. Quantum Electron.* **QE-16**(8), 964–971.

Chang, I. C. (1974a). Tunable acousto-optic filter utilizing acoustic beam walkoff in crystal quartz. *Appl. Phys. Lett.* **25**(6), 323–324.

Chang, I. C. (1974b). Noncollinear acousto-optic filter with large angular aperture. *Appl. Phys. Lett.* **25**(7), 370–372.

Chang, I. C. (1981). Acousto-optic tunable filters. *Opt. Eng.* **20**(6), 824–829.

Chang, I. C., and Katzka, K. (1978). Tunable acousto-optic filters with apodized acoustic excitation. *J. Opt. Soc. Am.* **68**, 1449.

Chatton, J.-Y., and Spring, K. R. (1993). Light source and wavelength selection for widefield fluorescence microscopy. *MSA Bull.* **23**, 324–333.

Dwelle, R., and Katzka, P. (1987). Large field of view AOTFs. *Proc. SPIE—Int. Soc. Opt. Eng.* **753**, 18–21.

Epikhin, V. M., and Kalinnikov, Y. K. (1989). Compensation of spectral drift of the diffraction angle in a noncollinear acousto-optic filter. *Sov. Phys.—Tech. Lett. (Engl. Transl.)* **34**(2), 227–228.

Gass, P. A., and Sambles, J. R. (1991). Accurate design of a noncollinear acousto-optic tunable filter. *Opt. Lett.* **16**, 429–431.

Glenar, D. A., Hillman, J. J., Saif, B., and Bergstralh, J. (1994). Acousto-optic imaging spectropolarimetry for remote sensing. *Appl. Opt.* **33**(31), 7412–7424.

Gottlieb, M., Goutzoulis, A. P., and Singh, N. B. (1987). Fabrication and characterization of mercurous chloride acousto-optic devices. *Appl. Opt.* **26**(21), 4681–4687.

Gottlieb, M., Goutzoulis, A. P., and Singh, N. B. (1992). High-performance acousto-optic materials: Hg_2Cl_2 and $PbBr_2$. *Opt. Eng.* **31**(10), 2110–2117.

Harris, F. J. (1978). On the use of window functions for harmonic analysis with the discrete Fourier transform. *Proc. IEEE* **66**(1), 51–83.

Hoyt, C. C., and Benson, D. M. (1992). Merging spectroscopy and digital imaging enhances cell research. *Photonics Spectra* **26**(11), 92–97.

Kalinnikov, Y. K., and Statsenko, L. Y. (1990). Use of acousto-optical filters for imaging filtration. *Sov. Phys.—Tech. Lett. (Engl. Transl.)* **34**(9), 1050–1051.

Kurtz, I., Dwelle, R., and Katzka, R. (1987). Rapid scanning fluorescence spectroscopy using an acousto-optic tunable filter. *Rev. Sci. Instrum.* **58**(11), 1996–2003.

Lewis, E. N., Treado, P. J., and Levin, I. W. (1994). Near-infrared and Raman spectroscopic imaging. *Am. Lab.*, June, pp. 16–21.

Marriott, G., Clegg, R. M., and Jovin, T. M. (1991). Time resolved imaging microscopy. *Biophys. J.* **60**, 1374.

Montemagno, C. D., and Gray, W. G. (1995). Photoluminescent volumetric imaging: A technique for the exploration of multiphase flow and transport in porous media. *Geophys. Res. Lett.* **22**, 425–428.

Morris, H. R., Hoyt, C. C., and Treado, P. J. (1994). Imaging spectrometers for fluorescence and Raman microscopy: Acousto-optics and liquid crystal tunable filters. *Appl. Spectrosc.* **48**(7), 856–857.

Nitschke, R., and Spring, K. R. (1995). Electro-optic wavelength selection enables confocal ratio imaging at low light levels. *J. Microsc. Soc. Am.* **1**(1), 1–11.

Ohmachi, Y., and Noda, J. (1977). LiNbO$_3$ TE–TM mode converter using collinear acousto-optic interaction. *IEEE J. Quantum Electron.* **QE-13**(2), 43–46.

Ohmachi, Y., and Uchida, U. (1970). Temperature dependence of elastic, dielectric and piezoelectric constants in TeO$_2$ single crystals. *J. Appl. Phys.* **41**, 2307–2311.

Schaeberle, M., Turner, J., and Treado, P. J. (1994). Multiplexed acousto-optic tunable filter (AOTF) spectral imaging microscopy. *Proc. SPIE—Int. Soc. Opt. Eng.* **2173**, 11–20.

Seveus, L., Kojola, H., and Soini, E. (1992). Time-resolved fluorescence imaging of europium chelate label in immunohistochemistry and *in situ* hybridization. *Cytometry* **13**, 329.

Singh, N. B., Denes, L. J., Gottlieb, M., and Mazelsky, R. (1989). Advances in the growth of Tl$_3$AsSe$_3$ crystal: A high efficiency acousto-optic and opto-electronic material. *Proc. SPIE—Int. Soc. Opt. Eng.* **1118**; 35–41.

Soos, Y. (1992). NIH team builds Raman imaging microscope. *Photonics Spectra* **26**(8), 91.

Treado, P. J., Levin, I. W., and Lewis, E. N. (1992a). Near infrared acousto-optic filtered spectroscopic microscopy: A solid-state approach to chemical imaging. *Appl. Spectrosc.* **46**(4), 553–559.

Treado, P. J., Levin, I. W., and Lewis, E. N. (1992b). High-fidelity Raman imaging spectrometry: A rapid method using an acousto-optic tunable filter. *Appl. Spectrosc.* **46**, 1211–1216.

Treado, P. J., Levin, I. W., and Lewis, E. N. (1994). Indium antimonide (InSb) focal-plane array (FPA) detection for near-infrared imaging microscopy. *Appl. Spectrosc.* **48**(5), 607–615.

Uchida, N. (1971). Optical properties of single-crystal paratellurite (TeO$_2$). *Phys. Rev. B: Solid State* [3] **4**, 3736–3745.

Uchida, N. (1972). Acoustic attenuation in TeO$_2$. *J. Appl. Phys.* **43**(6), 2916.

Wang, S. (1994a). AOTF allows imaging of cell process. *Photonics Spectra* **28**(5), 22–23.

Wang, S. (1994b). AOTF probes surface interfaces. *Laser Focus World*, February, p. 125.

Wang, X. (1992). Acousto-optic tunable filters spectrally modulate light. *Laser Focus World*, May, pp. 173–180.

Wang, X., Vaughan, D. E., and Pelekhaty, V. (1994). A novel miniature spectrometer using an integrated acousto-optic tunable filter. *Rev. Sci. Instrum.* **65**(12), 3653–3658.

Weverka, R., Katzka, P., and Chang, I. C. (1983). Bandpass apodization techniques for acousto-optical tunable filters. *IEEE Sonic Ultrasonic Symp.*, RR-6.

Xu, J., and Stroud, R. (1992). *Acousto-Optic Devices: Principles, Design and Applications*, p. 416. Wiley, New York.

Yano, T., and Watanabe, A. (1976). Acousto-optic TeO$_2$ tunable filter using far-off-axis anisotropic Bragg diffraction. *Appl. Opt.* **15**, 2250–2258.

CHAPTER

6

CONFOCAL MICROSCOPY OF SINGLE LIVING CELLS

JOHN J. LEMASTERS

Laboratories for Cell Biology
Department of Cell Biology and Anatomy
University of North Carolina at Chapel Hill
Chapel Hill, North Carolina 27599-7090

6.1. INTRODUCTION

Conventional wide-field light microscopes create images whose effective depth of field at high power is 2–3 μm. Since the resolving power of optical microscopy is about 0.2 μm, superimposition of detail within this plane of focus obscures structural detail that would otherwise be resolved. In addition, for specimens thicker than this depth of field, light from out-of-focus planes creates diffuse halos around objects under study. These halos are especially prominent in fluorescence microscopy. Confocal microscopy eliminates these undesirable artifacts by generating thin submicron optical slices through thick specimens (Minsky, 1961; Wilson, 1990). Confocal sections minimize superimposition of detail and exclude light from out-of-focus planes. As a consequence, images of remarkable detail and resolution are generated. Recently, UV–visible laser scanning confocal microscopes have become commercially available that expand the range of confocal applications to include UV-excited fluorophores. Increasingly, confocal microscopy has become an essential analytical tool to study the structure and physiology of living cells.

6.2. OPTICAL PRINCIPLE OF CONFOCAL MICROSCOPY

A confocal microscope scans a focused spot of light across the specimen (Figure 6.1). Spot diameter is diffraction limited, or about 0.2 μm for a high

Fluorescence Imaging Spectroscopy and Microscopy, edited by Xue Feng Wang and Brian Herman. Chemical Analysis Series, Vol. 137.
ISBN 0-471-01527-X © 1996 John Wiley & Sons, Inc.

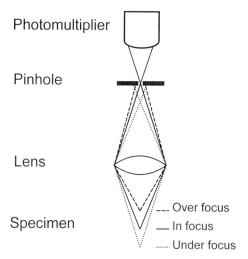

Figure 6.1. Principle of confocal microscopy. Light from the specimen is focused by the objective lens through a pinhole aperture to a photomultiplier. Light from out-of-focus planes above and below the specimen plane is not focused on the pinhole and is not transmitted to the photomulti-plier. Thus, only light arising from the in-focus specimen plane is detected.

numerical aperture (NA) objective lens. Light fluoresced or reflected from the specimen is separated from the illuminating beam of light by a mirror or dichroic reflector and is focused by the objective lens onto a pinhole aperture. Light from above and below the focal plane misses the pinhole opening and strikes the wall of the aperture instead (Figure 6.1). Thus, only light from a narrow in-focus plane passes through the pinhole to strike a photodetector beyond. In this way, the photodetector "sees" light from only a very narrow plane of focus. Two-dimensional images are generated as the illuminating spot of light moves across the specimen. Such scanning is achieved using vibrating mirrors or a rotating disk containing multiple pinholes in a spiral arrangement (Nipkow disk). Reflected and fluoresced light passes back through the scan generator, a process that "descans" the returning light so that it can be focused on a pinhole. When a Nipkow disk in what is called tandem-scanning confocal microscopy is used, con-focal images are viewed directly and recorded by photographic film (Petran et al., 1968). In laser scanning confocal microscopy, images are stored in computer memory and displayed on a monitor (White et al., 1987).

6.3. PRACTICAL CONSIDERATIONS

Because confocal microscopy collects only a fraction of the total fluorescence emitted by a sample, greater excitation energies must be used to image this fluorescence compared to conventional epi-illumination. As a result, photo-damage and photobleaching become greater problems in confocal micros-copy, especially for studying living cells when repeated measurements over time may be required. To minimize photobleaching and photodamage, stable fluorophores should be used while operating the confocal microscope at low laser power, high detector sensitivity, maximum objective NA, and the least magnification possible. Photomultiplier circuits should be set to their highest gain for maximal sensitivity.

Objective lenses vary considerably in light transmission. In general, light transmission varies with the square of the NA. Thus, the higher the NA, the greater the throughput of light. Because light transmission also depends on the number of lens elements and type of glass, empirical measurement of light transmission by various lenses may be necessary for some applications. Unlike conventional light microscopes, confocal micro-scopes have "zoom" features that permit magnification to be varied over a fivefold or greater range. Increasing magnification concentrates laser energy into smaller volumes of sample, thereby increasing the severity of photodamage in the imaged region. Thus, zooming should not exceed that required to address the questions under study. In particular, empty magni-fication exceeding the resolving power of the microscope system (maximally about 0.2 µm) only serves to accelerate photodynamic effects. Desirable in this regard are high-NA objective lenses that produce an initial magnification of 40X to 60X. At the same zoom settings, such lenses produce as much as six times less photodamage than 100X lenses. Recently, high-NA water immersion objective lenses have become available for use in confocal micros-copy. Such lenses not only have high light transmittance, but also eliminate aberrations and distortions associated with imaging deep into aqueous samples.

In commercial confocal microscopes, laser power exceeds by orders of magnitude that required for imaging. For almost all applications, laser intensity can be attenuated by 100- to 3000-fold without loss of image quality by using neutral density filters and by reducing the power setting of the laser. Even at this level of attenuation, some fluorophores, for example fluorescein and acridine orange, may still be prone to photobleaching and/or phototoxic-ity. Such fluorophores should be avoided, since other fluorophores, such as Texas red and rhodamine dyes, are quite stable under the same conditions. Indeed, by observing the precautions listed above, hundreds of images can be obtained with negligible photobleaching and photodamage.

Confocal microscopes produce optical slices of defined thickness through thick specimens. For a high-NA lens, thickness of the confocal sections can reach a theoretical limit of about 0.5 μm. The thickness of confocal sections decreases as the detector pinhole is made smaller (Figure 6.2). Since not all applications require the thinnest possible confocal section, sensitivity can be increased by opening the pinhole aperture. Doubling the diameter of the pinhole quadruples sensitivity but only about doubles the thickness of the optical slice (Figure 6.2). For this reason, most laser scanning confocal microscopes are equipped with variable pinhole apertures. For light-sensitive specimens, a larger pinhole setting may be desirable so that laser power can be attenuated to an acceptable level. Conversely, pinhole diameter can be decreased to reduce slice thickness and increase resolution in the z-axis. However, below a minimum pinhole size, confocal slice thickness no longer decreases as pinhole diameter decreases, although image intensity continues to

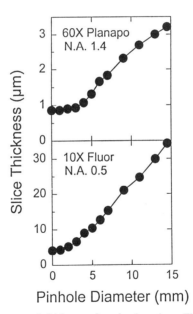

Pinhole Diameter (mm)

Figure 6.2. Pinhole diameter and thickness of confocal sections. The full-width half-maximal thickness of confocal sections was determined for a $60 \times$ NA 1.4 Planapochromat objective lens and a $10 \times$ NA 0.5 Fluor lens as the pinhole micrometer setting was varied. A Biorad MRC-600 laser scanning confocal microscope (Hercules, California) operating in the reflectance mode was used to image a front-reflecting mirror tilted at a slight angle. Full-width half-maximal depth of field using 488 nm illumination was determined from the width of the reflected band of light in the confocal image and the angle of the mirror.

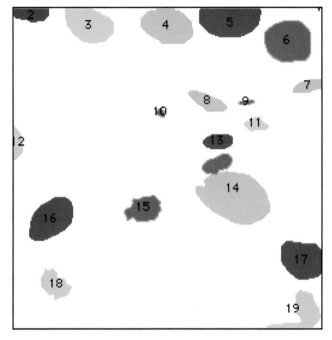

Figure 2.6. Image of cell masks.

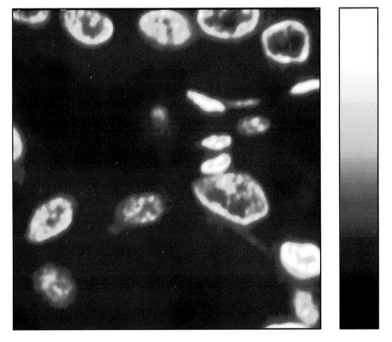

Figure 2.12. Pseudocolored image and look-up table.

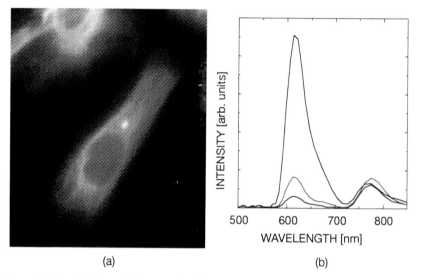

(a)   (b)

Figure 4.5. The color image of a human cell stained with propidium iodide (reproduced from the full spectral image). The spectra of three points from the image are shown in (b). The peaks at 623 nm correspond to the dye. The other peaks (at 775 nm) are due to excitation light that was not blocked by the fluorescence excitation cube. Correction for the effects of the residual excitation light is straightforward when the full spectral image has been measured, but can be a problem if only the intensity through a bandpass filter is measured. The sample was prepared with the courtesy of Anett Howard, Sigma Immuno Chemical, Rehovot, Israel.

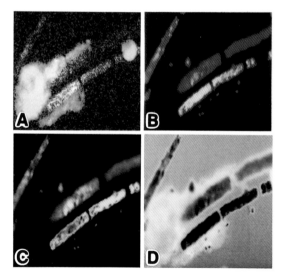

Figure 4.6. The fluorescence from an *Oedogonium* sp. alga: (A) the total emitted fluorescence from the specimen; (B) similarity mapping with spectrum A as a reference, showing localized chlorophyll in the algae; (C) similarity mapping using as references the faint red fluorescence spectrum; (D) the spongy structure of the debris reconstructed by similarity mapping with the green spectrum from the cell debris (spectrum B).

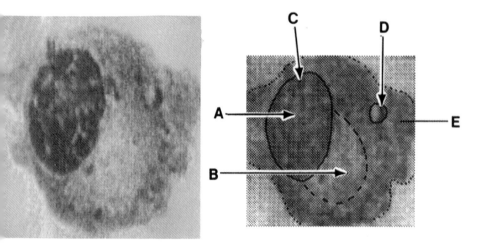

Figure 4.8. A plasma cell (left) stained by the routine method of May–Grünwald–Giemsa for blood cells. In the adjacent drawing (right) the different subcellular sites are demarcated, and the five pixels (A–E) corresponding to the five spectra are highlighted: (A) heterochromatin; (B) a light blue-stained area of the cytosol; (C) euchromatin; (D) a cytoplasmic vesicle; and (E) a point in the cytosol.

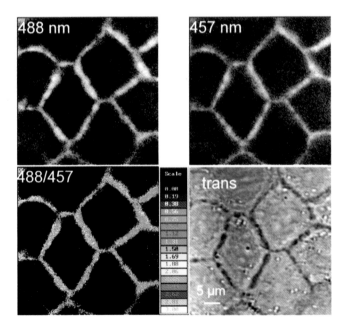

Figure 5.7. Confocal images of living renal epithelial cells (MDCK) with their lateral intercellular spaces stained with BCECF (a pH indicator) at excitation wavelengths of 457 nm and 488 nm; a transmission light image; and a color-coded ratio of the 457/488 image. Reprinted with permission from the *Journal of the Microscopy Society of America* (Spring 1995), **1** (1), 1–11.

A   B   C

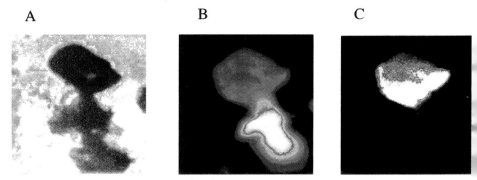

Figure 5.11. Bright-field (A) and Raman images (B,C) of a mixture of DPPC (lipid model) and L-asparagine (protein model). (B) The lipid portion of the sample is shown by recording the asymmetric CH$_2$ stretching mode at 2880 cm^{-1}. (C) L-Asparagine is shown by recording the Fermi resonance NH$_2$ stretching mode at 3390 cm^{-1}. The size bar corresponds to 100 µm. Reprinted from *American Laboratory* **26** (9), 16 (1994). Copyright 1994 by International Scientific Communications Inc.

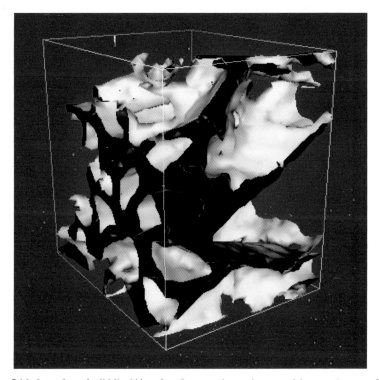

Figure 5.16. Isosurface of solid-liquid interface. In a two-phase-mixture model system (aqueous fluid and quartz sand), the false-color image produced by the AOTF and CCD camera reveals the interfaces. Quartz grains range in size from 300 to 520 µm, and the cube in this image is about 7 x 7 x 7 mm. Reprinted with permission from *Laser Focus World* (February 1994).

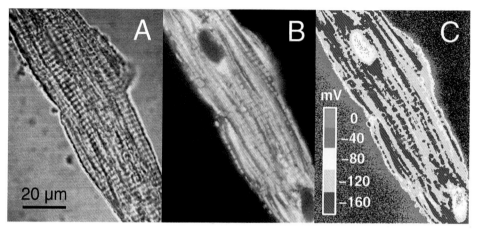

Figure 6.7. Distribution of electrical potential in an adult cardiac myocyte. Panel A is a transmitted light image of a cultured adult rabbit cardiac myocyte loaded with TMRM. Panel B is the image obtained using a gamma setting of +4 and 568 nm illumination from an argon–krypton laser. In Panel C, the image is pseudocolored to show the intracellular distribution electrical potential. After Chacon et al. (1994).

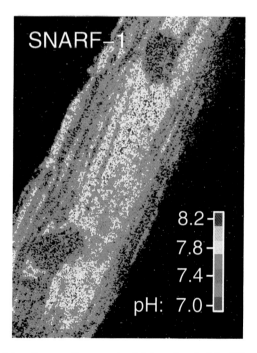

Figure 6.9. Intracellular pH in a cardiac myocyte. A 1 day cultured adult rabbit cardiac myocyte was loaded with SNARF-1. After excitation with 568 nm, the emitted fluorescence was imaged simultaneously at >584 ± 5 nm and >620 nm. The images were ratioed and pseudocolored to represent distribution of pH using an *in situ* calibration. After Chacon et al. (1994).

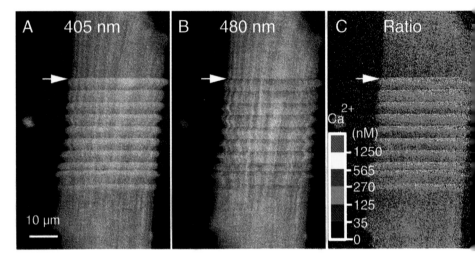

Figure 6.11. Ratio imaging of cytosolic free Ca^{2+} with indo-1 during excitation–contraction coupling in a cardiac myocyte. Adult rabbit cardiac myocytes were loaded with 5 μM indo-1 acetoxymethyl ester for 1 h at 37° C. With an excitation of 351 nm, fluorescence images of indo-1 were collected at emission wavelengths of 395–415 nm (A) and 470–490 nm (B) as the cell was electrically stimulated at 0.5 Hz for 10 s during the 40 s scan. Panel C is the ratio image scaled to Ca^{2+} concentration obtained by dividing panel A by panel B after background subtraction. Ca^{2+} transients occurred in both mitochondrial and nonmitochondrial regions identified by rhodamine 123 labeling (not shown). After Ohata et al. (1994).

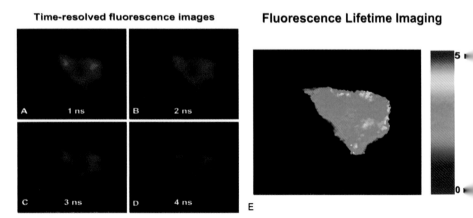

Figure 10.8. (Left) Time-resolved fluorescence images and (right) lifetime imaging of Ca^{2+} in BALB fibroblasts. A BALB/C-3T3 cell is labeled with 10 μM Calcium Crimson-AM, and time-resolved fluorescence intensity images are observed after pulse excitation in (A) 1 ns, (B) 2 ns, (C) 3 ns, and (D) 4 ns. A fluorescence lifetime image calculated from (A) and (D) is shown in (E).

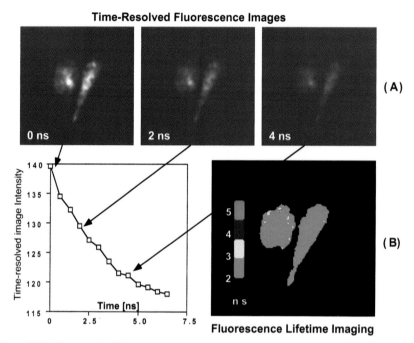

Figure 10.9. Time-resolved fluorescence images and lifetime imaging for two BALB/C-3T3 fibro-blasts microinjected with 20 mM Calcium Crimson–salt and imaged using time-resolved FLIM.

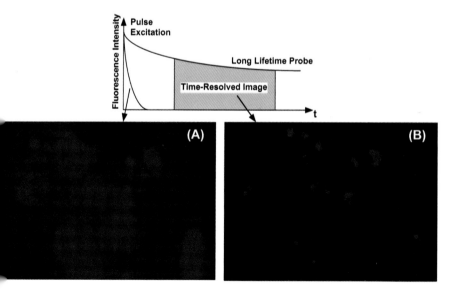

Figure 10.10. FISH images for HPV DNA 16. (A) Fluorescence image detected by the conventional fluorescence detection method. The image includes the signal from HPV 16 DNA and, at the same time, other autofluorescence and nonspecific signals. (B) Fluorescence image detected by time-resolved FLIM using long-lifetime probe (Eu^{3+} chelate). Detection sensitivity has been significantly improved by removing autofluorescence and other background.

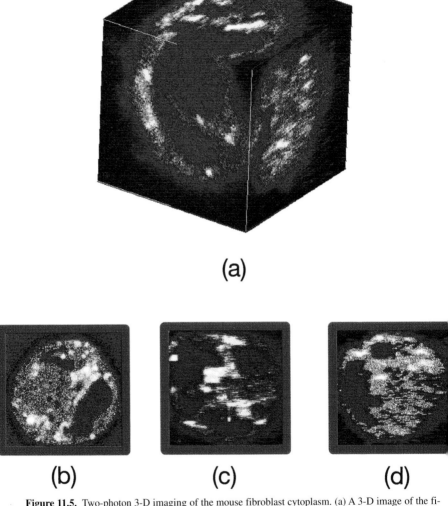

Figure 11.5. Two-photon 3-D imaging of the mouse fibroblast cytoplasm. (a) A 3-D image of the fibroblast cell reconstructed from 35 z-sections; (b) a typical x-y section; (c) a typical y-z section; (d) a typical z-x section.

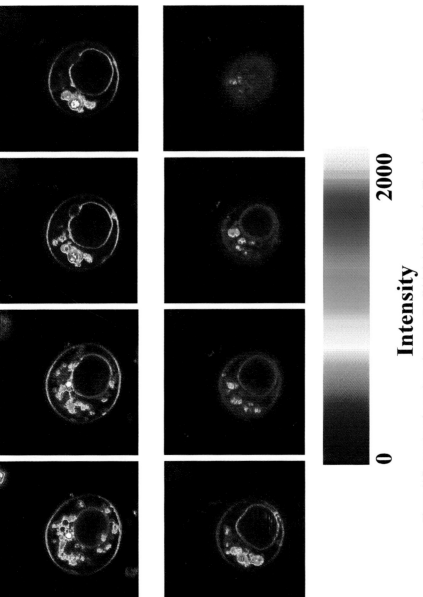

Figure 11.7. A series of z-sections of a mouse T cell labeled with Laurdan. The planes are 2.5 μm apart. The x-y pixel size is 0.14 μm. The plasma, nuclear, and vacuole membranes are clearly visible.

Intensity

0   2000

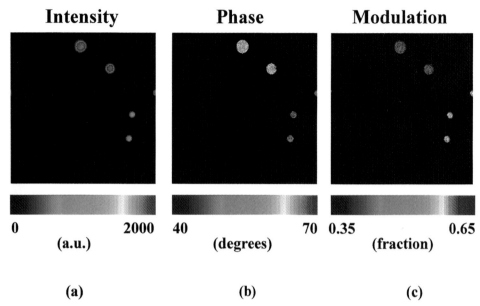

Figure 11.8. Time-resolved images of orange and red fluorescent latex spheres. (a) The intensity image; (b) the phase image; (c) the modulation image.

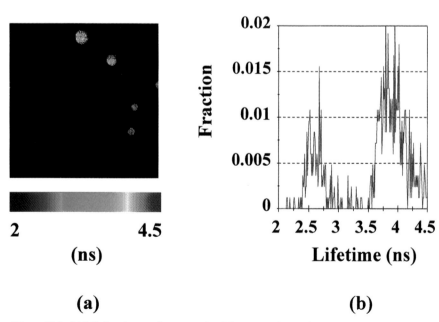

Figure 11.9. (a) Lifetime image of orange and red fluorescent latex spheres; (b) a histogram of the number of pixels in the image with a particular lifetime. The lifetimes of the orange and red fluorescent spheres are 4.3 and 2.9 ns, respectively, which closely match the values obtained with a conventional fluorometer.

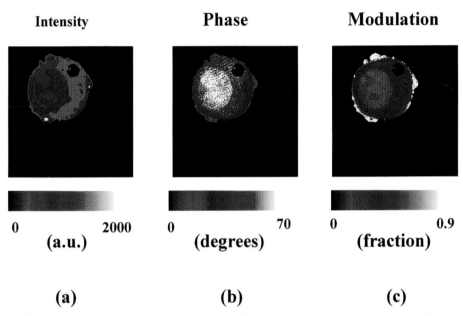

Intensity	Phase	Modulation
0 2000	0 70	0 0.9
(a.u.)	**(degrees)**	**(fraction)**
(a)	**(b)**	**(c)**

Figure 11.10. Time-resolved images of a mouse fibroblast cell that has been double labeled with EtBr, a nucleic probe, and DiO, a membrane probe. (a) The intensity image; (b) the phase image; (c) the modulation image.

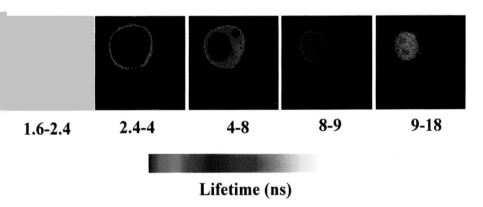

1.6-2.4   **2.4-4**   **4-8**   **8-9**   **9-18**

Lifetime (ns)

Figure 11.11. Lifetime sections of the mouse fibroblast cell. Five distinct lifetime regions can be visualized: the nucleus, $18 > \tau > 9$ ns; the nucleoli, $9 > \tau > 8$ ns; the cytoplasm, $8 > \tau > 4$ ns; the plasma membrane, $4 > \tau > 2.4$ ns; and localized lipid structures on the plasma membrane, $2.4 > \tau > 1.6$ ns.

Intensity

Phase

Modulation

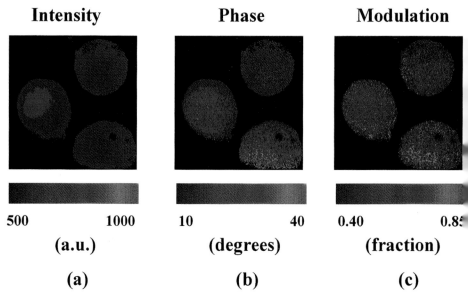

500   1000   10   40   0.40   0.8!
(a.u.)   (degrees)   (fraction)
(a)   (b)   (c)

Figure 11.12. Time-resolved images of live mouse fibroblast cells loaded with the calcium indicator Calcium Green. (a) The intensity image; (b) the phase image; (c) the modulation image.

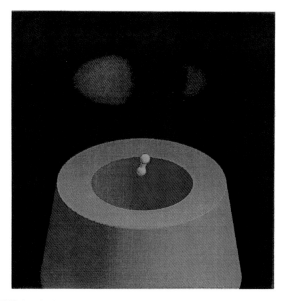

Figure 13.11. NFO for single-molecule localization and detection: portrait of a single molecule. The subwavelength optical source is shown beneath a molecule, represented by the green dumbell (not to scale). Optical excitation results in the red fluorescence emission pattern shown at the top. By comparison of the emission structures seen in the near-field scans, the orientation of individual molecules can be determined. Reprinted with permission from E. Betzig and R.J. Chichester, Single molecules observed by near field scanning optical microscopy. *Science* **262**, 1422–1425 (1993). Copyright 1993 American Association for the Advancement of Science.

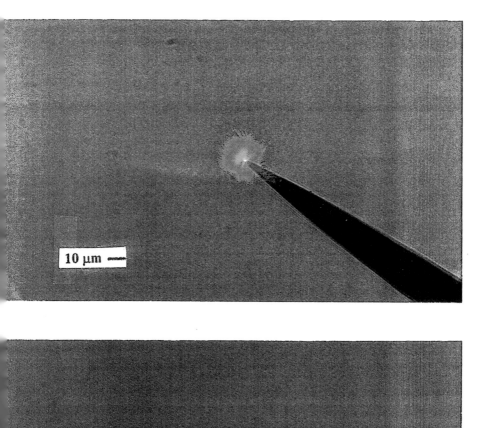

Figure 13.14. Photograph of a subwavelength optical fiber probe and sensor. Top: microphotograph of a nanofabricated optical fiber tip; bottom: microphotograph of an NFO chemical sensor.

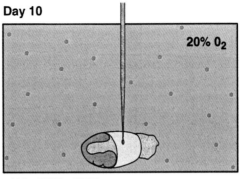

Day 10

20% O$_2$

pH in EEF = 7.53

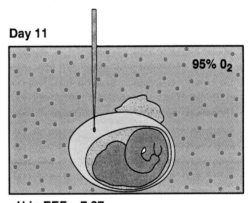

Day 11

95% O$_2$

pH in EEF = 7.37

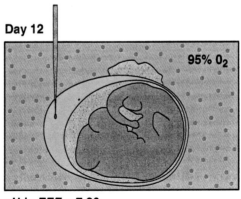

Day 12

95% O$_2$

pH in EEF = 7.26

Figure 13.21. Rat embryo insertion. Acidification of EEF occurs as a function of advancing gestational age. Measurements of pH are made in single viable rat conceptuses *in vitro* using ultramicrofiber-optic pH sensors.

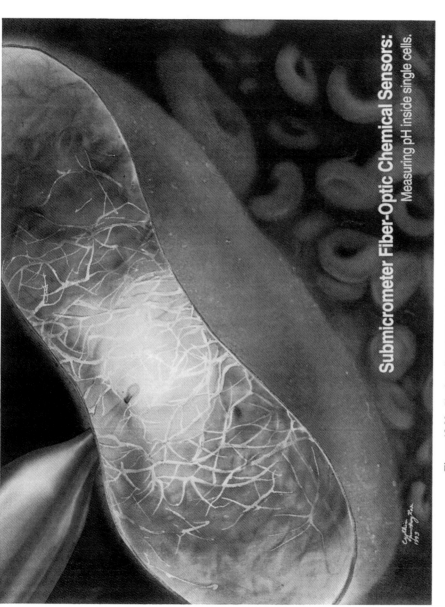

Figure 13.22. Single-cell analysis by a subwavelength optical biochemical sensor.

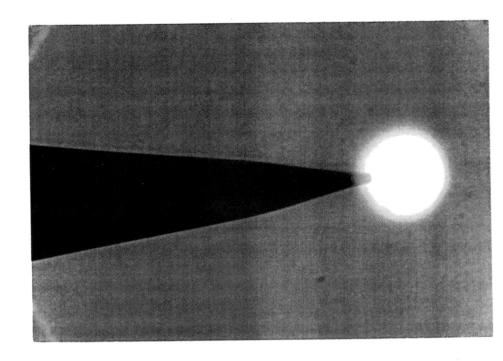

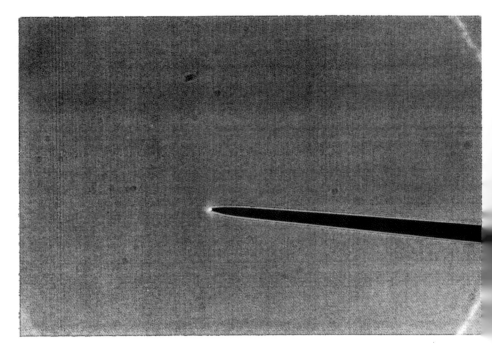

Figure 13.25. Exciton light sources. Top: perylene inside a micropipette; bottom: perylene on an optical fiber probe.

decline. Thus, overly small pinhole diameters should be avoided, especially with light-sensitive specimens.

6.4. MULTICOLOR FLUORESCENCE

For laser scanning confocal microscopy, the wavelength of excitation light—and hence the selection of fluorophores—is determined by the type of laser used. Argon lasers emit 488 and 514 nm; helium–neon lasers, 543 nm; argon–krypton lasers, 488, 568, and 647 nm; and UV–argon lasers, 531 and 488 nm. The argon–krypton laser, with three well-separated lines, is perhaps the most useful of these lasers for biological applications. The 488 and 568 nm lines can simultaneously excite green-fluorescing dyes like fluo-3 and rhodamine 123 and red-fluorescing dyes like tetramethylrhodamine methylester (TMRM). Red and green fluorescence can then be detected simultaneously using different photomultipliers. Simultaneous detection of different fluorophores is a particular advantage when high time resolution is needed.

6.5. PREPARATION OF CELLS

For viewing by confocal microscopy, cells must be attached to No. $1\frac{1}{2}$ glass coverslips. Cells usually adhere poorly to glass. To improve adherence, coverslips may be treated with an extracellular matrix material, such as type I collagen or laminin. Briefly, coverslips are rinsed in ethanol, dried, and placed in plastic Petri dishes. From a long-tip Pasteur pipette, 1 or 2 drops (about 100 µL) of a solution of type I collagen (1 mg/mL in 0.1% acetic acid) or laminin (0.1 mg/mL in Tris-buffered saline) are spread out across the coverslips. After air-drying overnight, the culture dishes are rinsed with washes of Hank's solution. Aliquots of freshly isolated or trypsinized cells are then added in the usual fashion. Cells become firmly adherent in 1 h to a few hours. To view the plated cells, the overslips are mounted in a specimen chamber and placed on the stage of the confocal microscope. Usually an inverted microscope is used. The cells are then viewed from underneath, permitting easy access from above to the medium bathing the cells.

6.6. IMAGING CELL VOLUMES AND SURFACES

Size, shape, and surface topography are basic aspects of cell structure. Confocal microscopy combined with three-dimensional image reconstruction techniques can provide this basic structural information for individual living cells. The experimental strategy is to label the cytoplasm with a fluorescent dye and then collect confocal images as the focal plane is moved through the entire thickness of the cell. Subsequently, the serial confocal sections are reconstructed to reveal three-dimensional structure. Calcein, whose fluorescence is unaffected by physiological changes in ion concentration and pH, is a useful fluorophore for this purpose (Chacon et al., 1994; Zahrebelski et al., 1995). Calcein is loaded by incubating cells with its acetoxymethyl ester. Intracellular esterases cleave the esters, trapping the free acid form of calcein in the cytoplasm. Cultured hepatocytes and myocytes load well when incubated with 3–5 μM calcein–AM for 15–60 min.

In calcein-loaded cells, confocal images are collected as serial sections. Many confocal microscope systems do this automatically, using stepper motors attached to the focus knobs of the microscope to change focus as images are collected and stored. Subsequently, the serial images are transferred to a high-speed imaging workstation for reconstruction or "rendering" in three dimensions. Presently, we use a Silicon Graphics computer and software developed by Vital Images for this purpose.

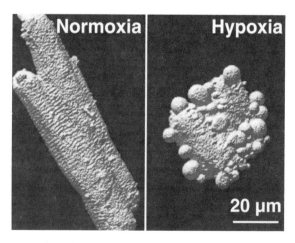

Figure 6.3. Reconstruction of a calcein-labeled myocyte. A cultured adult rabbit cardiac myocyte was loaded with calcein. Before and 40 min after exposure to 2.5 mM NaCN and 20 mM 2-deoxyglucose (chemical hypoxia), confocal images were collected in 1 μm increments with a Biorad MRC-600 laser scanning confocal microscope. Three-dimensional distribution of fluorescence was reconstructed using a Silicon Graphics imaging workstation (Mountain View, California) operating VoxelView software (Vital Images, Fairfield, Iowa). Contour-dependent shading enhances the perception of depth and surface detail. After Chacon et al. (1994).

Figure 6.3 illustrates the results of a volume rendering of a cultured adult cardiac myocyte loaded with calcein. The reconstruction reveals a wealth of surface detail. In particular, regularly spaced corrugations are the dominant surface feature, representing impressions of underlying mitochondria aligned along the sarcomeres of the contractile apparatus. Interestingly, these structures are not recognizable in single confocal images, but only become evident in the three-dimensional reconstruction. Volume renderings obtained this way rival scanning electron micrographs in detail and clarity, but unlike scanning electron micrographs, they can be collected repeatedly over time from the same living cell. Indeed, the structure of the myocyte in Figure 6.3 changed markedly during observation as a consequence of hypoxia.

Retention of calcein fluorescence during hypoxic stress indicates continued cell viability. Loss of cell viability is signaled by abrupt loss of the fluorophore. Also, nuclear uptake of dyes like trypan blue and propidium iodide occurs at the onset of cell death (Herman et al., 1988; Lemasters et al., 1987). In particular, the red nuclear fluorescence of propidium iodide ($1-4\,\mu M$) is easily imaged by confocal microscopy as an indicator of cell killing.

6.7. VISUALIZATION OF ORGANELLES

Several fluorophores label specific organelles of living cells. By exciting with multiline lasers, two or more of these probes may be imaged simulta-neously. Rhodamine 123 and rhodamine–dextran are convenient labels of mitochondria and lysosomes, respectively, that can be used together (Gores et al., 1989; Johnson et al., 1981). Simple incubation of cells with $0.1-1\,\mu M$ rhodamine 123 for 15–30 min at 37 °C provides excellent mitochondrial loading. Lysosomal loading requires overnight incubation with $1\,\mu g/mL$ rhodamine–dextran. Hepatocytes can also be preloaded by an intraper-itoneal injection of 20 mg rhodamine–dextran/100 g body weight into rats 1 day prior to the isolation procedure. In cells coloaded with rhodamine 123 and rhodamine–dextran, mitochondria are identified by green fluo-rescence and lysosomes by red fluorescence (Lemasters et al., 1993). In addition, many other organelle-specific fluorophores have been described that can be adapted to confocal microscopy—for example, BODIPY (4,4-difluoro-4-bova-3a,4a-diaza-5-indacene)–ceramide derivatives for Golgi (Pagano et al., 1991) and carbocyanine dyes for endoplasmic reticulum (Lee and Chen, 1988). The fluorescence of these organelle-specific probes can be monitored for hours, if needed, inside living cells. By means of a transmitted light detector, confocal fluorescence can be simultaneously compared with conventional nonconfocal brightfield images to show, for example, lysosomal

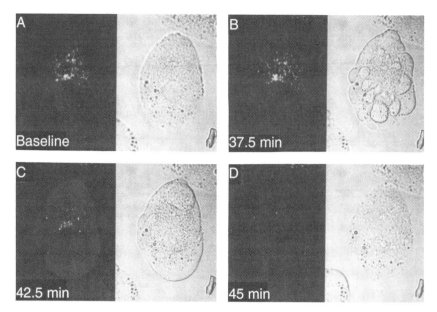

Figure 6.4. Lysosome disintegration during chemical hypoxia. Hepatocytes were loaded with rhodamine–dextran to label lysosomes. Transmitted light images (right) and confocal fluorescence images (left) were recorded before (A) and various times after (B–D) chemical hypoxia with 2.5 mM KCN and 0.5 mM iodoacetate. (C) Lysosomal breakdown, causing release of rhodamine–dextran diffusely into the cytoplasm. (D) Rupture of a bleb and loss of viability. After Zahrebelski et al. (1995).

disintegration accompanying cell surface changes just prior to loss of viability of hypoxic hepatocytes (Figure 6.4).

Although not formally an organelle, the cytosol may be specifically labeled by incubating cells with 3–5 μM calcein–AM for 15–45 min at 37 °C. Under these conditions, calcein loads virtually exclusively into the cytosol. The nucleus also loads because of the free exchange of small solutes between the cytosol and the nucleus. Membranous organelles, especially mitochondria, exclude calcein and appear as dark, round voids in the green cytosolic calcein fluorescence (Figure 6.5) (Nieminen et al., 1995). That these voids are indeed mitochondria can be confirmed by coloading cells with TMRM, which is taken up by mitochondria like rhodamine 123. When simultaneous images of green-fluorescing calcein and red-fluorescing TMRM are collected, it is seen that each TMRM-labeled mitochondrion corresponds to a dark void in the calcein image (Figure 6.5). Occasional voids in calcein fluores-

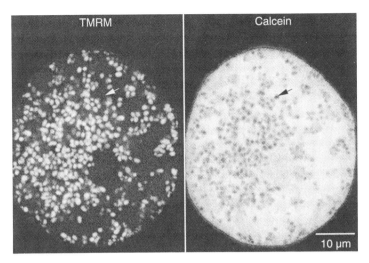

Figure 6.5. Resolution of cytosolic and mitochondrial fluorescence in a cultured hepatocyte. A cultured rat hepatocyte was coloaded with TMRM and calcein acetoxymethyl ester. In the calcein image, dark holes are seen that correspond mostly to mitochondria in the TMRM image. Occasional voids in the calcein image (arrow) do not correspond to TMRM-labeled mitochondria and are probably lysosomes. After Nieminen et al. (1995).

cence do not correspond to TMRM-labeled mitochondria. Presumably, such structures represent some other membranous organelles, most likely lysosomes.

In cells coloaded with calcein and TMRM, it becomes possible to document the onset of the mitochondrial permeability transition after hypoxic and oxidative stress. The mitochondrial permeability transition is caused by opening of a very high conductance pore, or "megachannel," in the mitochondrial inner membrane that has a molecular weight cutoff of about 1200 Da (Gunter and Pfeiffer, 1990). This pore opening produces mitochondrial depolarization and uncoupling of oxidative phosphorylation. In intact cells, onset of the mitochondrial permeability transition causes loss of mitochondrial TMRM fluorescence, which indicates mitochondrial depolarization (Figure 6.6). Simultaneously, calcein fluorescence enters mitochondria, which indicates the development of nonspecific permeability of the mitochondria to small solutes. Onset of the mitochondrial permeability transition may be an important event causing cell death in hypoxic, toxic, and reperfusion injury (Nieminen et al., 1995; Zahrebelski et al., 1995).

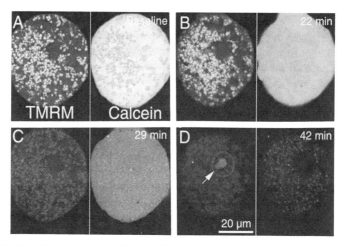

Figure 6.6. Mitochondrial permeability transition in a cultured hepatocyte induced by *t*-butyl-hydroperoxide. A cultured hepatocyte was co-loaded with TMRM and acetoxymethyl ester. After a baseline image was collected (A), 100 μM *t*-butylhydroperoxide (*t*-BuOOH) was added (B–D). *t*-BuOOH caused redistribution of calcein from cytosol to mitochondria and loss of TMRM fluorescence (B,C), events signifying the onset of the mitochondrial permeability transition. These changes preceded loss of cell viability, as documented by leakage of cytosolic calcein and nuclear labeling with propidium iodide (arrow in panel D). After Nieminen et al. (1995).

6.8. DISTRIBUTION OF ELECTRICAL POTENTIAL

6.8.1. Theory

Monovalent cationic fluorophores with delocalized charge, such as TMRM and rhodamine 123, distribute electrophoretically into charged compartments in accordance with the Nernst equation:

$$\Delta\Psi = -60 \log F_{in}/F_{out} \tag{1}$$

where F_{in} and F_{out} are fluorophore concentrations inside and outside the compartment of interest and $\Delta\Psi$ is the electrical potential difference. Thus, a 10:1 uptake ratio of fluorophore signifies a $-60\,mV$ gradient, a 100:1 uptake ratio signifies a $-120\,mV$ gradient, and so forth. For excitable cells like cardiac myocytes or neurons, a plasma membrane $\Delta\Psi$ of $-90\,mV$ and a mitochondrial $\Delta\Psi$ of up to $-150\,mV$ are typical values. Since these $\Delta\Psi$ values are additive, mitochondria inside a myocyte are $240\,mV$ more negative than those in the extracellular space. Such a large electrical gradient drives fluorophore uptake to a concentration ratio of 10,000:1 in mitochondria relative to the outside of the cell. With 8-bit frame memories storing only 256 levels of intensity, measurement of large gradients on a linear scale is impossible. So-called gamma circuits, long used in scanning electron microscopy, must be used instead (Chacon et al., 1994). Gamma circuits apply a logarithmic function to condense a very large signal range into the available 256 gray levels of video memory.

To measure the distribution of intracellular electrical potential, several basic steps are involved:

1. Cells must be loaded with a monovalent cationic fluorophore under conditions permitting equilibrium distribution in response to intracellular gradients of Ψ.
2. By means of a gamma circuit, confocal images of the full range of fluorescence intensities (including fluorescence in the extracellular space) are collected.
3. Gamma circuit intensity values for each pixel are converted to a linear scale of fluorescence intensity.
4. Extracellular fluorescence intensity is averaged and divided into intracellular fluorescence on a pixel-by-pixel basis.
5. The Nernst equation (Eq. 1) is applied to each pixel to calculate $\Delta\Psi$ between each point within the cell and the extracellular space.
6. Pseudocolor maps of Ψ are made to display the intracellular distribution of Ψ, as shown in Figure 6.7 (see Color Plates).

6.8.2. Loading of Cells

Several cationic fluorophores distribute inside cells in response to electrical gradients. The validity of measurement of intracellular electrical potential depends on ideal behavior by these fluorophores. Some fluorophores do not behave ideally. For example, rhodamine 123 binds nonspecifically to the mitochondrial matrix in a fashion unrelated to electrical potential, and its fluorescence is quenched as it is accumulated (Emaus et al., 1986). High matrix concentrations of rhodamine 123 also inhibit the oligomycin-sensitive mitochondrial ATP synthase. Methyl and ethyl esters of tetramethylrhodamine seem to lack these undesirable attributes and are better suited for estimating electrical potential by confocal imaging (Ehrenberg et al., 1988; Farkas et al., 1989).

To load TMRM, cultured cells are incubated with 0.5–1 μM TMRM for 15–30 min at 37 °C. They are then placed on the microscope stage in culture medium or a HEPES-buffered Krebs–Ringer's solution. A small amount of

TMRM (50–150 nM) is retained in the incubation buffer throughout the experiments to keep the extracellular dye concentration constant and to maintain the equilibrium distribution of the fluorophore inside the cells. By means of an air curtain incubator or stage heater to maintain the temperature, experiments can commence after a few minutes of incubation and equilibration.

6.8.3. Conversion of Fluorescence to Ψ

To establish appropriate imaging conditions for Ψ quantitation, true zero for fluorescence intensity must be established. To achieve this, the black level (dark current) is set to zero while focusing within the coverslip below the cell. Through the use of an image histogram, the black level is adjusted so that 50% of the pixels exhibit zero intensity. Since negative voltages from the photomultiplier circuits are recorded as pixel intensities of zero, when half of the pixels have a positive value (pixel intensity > 0) and half have a negative value or zero (pixel intensity $= 0$), average pixel intensity will be very close to zero. After adjustment of the black level, the focal plane is moved back to the specimen and images are collected.

With a Bio-Rad MRC-600 laser scanning confocal microscope, TMRM images are collected at a pinhole setting of 3 using an enhanced (gamma) circuit setting of $+4$. Output of the circuit is given by

$$V_c = 1 + A\ln(V_{PMT} + e^{-1/A}) \qquad (2)$$

where V_c is output voltage; V_{PMT} is input voltage from the photomultiplier; and A is 0.206. Accordingly, when V_{PMT} is 0 V, V_c is 0 V (0 pixel intensity), and when V_{PMT} is 0.99 V, V_C is 1 V (255 pixel intensity). At V_{PMT} between 0 and 0.99 V, V_c is an exponential function of V_{PMT}. Equation (2) can be solved for V_{PMT}:

$$V_{PMT} = e^{(V_c - 1)/A} - e^{-1/A} \qquad (3)$$

Using an area histogram for intensities in the extracellular space and applying Eq. (3), we can calculate the average V_{PMT} of the extracellular space. Since extracellular Ψ is zero (ground) and assuming that intracellular fluorophore concentration is proportional to fluorescence intensity, we calculate Ψ anywhere inside the cell relative to outside as

$$\Psi = -60\log(V_{PMT(in)}/V_{PMT(out)}) \qquad (4)$$

where $V_{PMT(in)}$ is V_{PMT} for any pixel inside the cell, as calculated from Eq. (3).

To display the distribution of Ψ inside cells, pseudocolor look-up tables are generated to convert pixel values of 0 to 255 (corresponding to V_c of 0 to 1 V) to Ψ. A different look-up table is needed for each different value of $V_{PMT(out)}$. Our program in BASIC to construct these pseudocolor look-up tables in a format compatible with Bio-Rad's SOM and COMOS confocal software is available on request.

Figure 6.7 (see Color Plates) illustrates Ψ determined in a cultured adult rabbit cardiac myocyte. Panel A is a brightfield image of the myocyte. This image was collected simultaneously with TMRM fluorescence using a scanning transmission attachment. Because the bright-field image is not confocal, mitochondria, although abundant, are not easily discerned. Panel B shows the confocal TMRM fluorescence image collected using the enhance circuit. Mitochondria are fluorescent spheres and rods. In collecting the fluorescence image, the black level was first set to zero while focusing within the coverslip, as described above. In panel C, individual pixel values were converted to a linear fluorescence scale using Eq. (3). After dividing by average extracellular fluorescence, intracellular Ψ was calculated by applying the Nernst equation to each pixel (Eq. 4) and displayed using different colors to represent different electrical potentials. In areas under the plasma membrane, in the nucleus, and in the open spaces between mitochondria, pseudocoloring shows an electrical potential of about $-80\,mV$. Since Ψ of the extracellular medium is zero, plasma membrane $\Delta\Psi$ is $-80\,mV$, as expected for myocytes. Mitochondria show heterogeneity of Ψ, ranging between -120 and $-160\,mV$. Heterogeneity is a consequence of the small size of mitochondria relative to the thickness of the confocal section. Hence, not all mitochondria extend completely through the confocal slice. Consequently, fluorophore uptake for many mitochondria is underestimated, since observed pixel intensity represents an average of mitochondrial and cytosolic fluorophore concentration. In this experiment, apparent mitochondrial Ψ was as great as $-160\,mV$. Since cytosolic Ψ was $-80\,mV$, the difference, $-80\,mV$, represents a minimum estimate of mitochondrial $\Delta\Psi$.

6.9. pH IMAGING

6.9.1. Loading of Ion-Indicating Probes

The development by Tsien and colleagues of the Ca^{2+}-indicating fluorescent probes fura-2, indo-1, and fluo-3 has revolutionized the study of ion homeostasis in single living cells and prompted the creation of new probes for other ions (Grynkiewicz et al., 1985; Minta et al., 1989). To measure ions inside cells, ion-reporting fluorophores are loaded as their lipid-soluble acetoxymethyl or acetate ester derivatives. Endogeneous esterases hydrolyze the esters to release free acid forms of the ion-indicating fluorophores, which remain trapped in the cytoplasm. The temperature of loading affects the intracellular distribution of many probes. Loading at 37 °C favors cytosolic loading, whereas loading at 4 °C promotes both mitochondrial and cytosolic loading (Nieminen et al., 1995). Apparently, cytosolic esterases are so active at higher temperatures that fluorophore esters cannot cross the cytosol to enter mitochondria without being hydrolyzed first. Lower temperatures reduce esterase activity, allowing esters to diffuse into mitochondria and be trapped there by mitochondrial esterases. Probe loading varies from one type of cell to another, and conditions for optimal loading of fluorophores must always be determined empirically.

6.9.2.   Principles of Ratio Imaging

Inside cells, the intensity of probe fluorescence depends on the free concentration of the ion it measures. Fluorescence also depends on the amount of fluorophore in the microscopic light path and will vary due to differences in cell thickness, presence of organelles excluding the probe, variable probe loading, probe leakage, and probe photobleaching. To correct for differences in probe concentration and effective path length, a ratioing procedure can be used that takes advantage of ion-induced shifts of the fluorescence spectrum (Figure 6.8). For example, the fluorescence of SNARF-1 (seminaphthorhoda-fluor) excited at 568 nm increases with increasing pH at an emission wavelength of 640 nm but remains constant at 585 nm. Thus, the 640:585 nm fluorescence ratio is proportional to pH but is independent of SNARF-1 concentration. By dividing fluorescence at one wavelength by that at the other (ratioing), the effects of variable probe concentration and path length on signal strength are canceled out. Because SNARF-1 has a pK_a of about 7.5, it is ideal for monitoring cytosolic and mitochondrial pH (Chacon et al., 1994).

The steps involved in ratio imaging are illustrated in Figure 6.8.

1. Cells are loaded with an ion-indicating fluorophore, such as SNARF-1 for pH.
2. Images of the loaded cells are collected at two emission wavelengths. At one wavelength, fluorescence intensity increases with increasing ion concentration, whereas at the other wavelength, fluorescence decreases or stays the same.
3. Background images are collected from adjacent cell-free areas of the coverslip or while focusing within the coverslip.
4. Background images at each wavelength are subtracted from the corresponding cell images on a pixel-by-pixel basis.

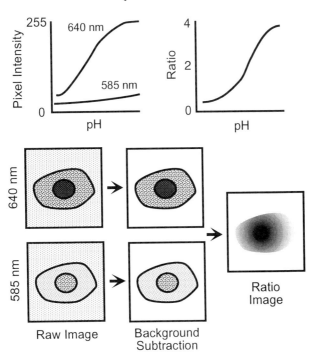

Figure 6.8. Principle of ratio imaging. See text for details.

5. Images at each wavelength after background subtraction are divided into each other on a pixel-by-pixel basis to generate a ratio image.
6. Ratios images are converted to ion maps of using a standard curve.

6.9.3. Loading Cells with SNARF-1 and Collecting Images

SNARF-1 is efficiently excited by the 568 nm line of an argon–krypton laser. For confocal microscopy, fluorescence is imaged at emission wavelengths of > 620 nm (pH sensitive) and 585 ± 5 nm (pH insensitive). The ratio of fluorescence at these two different wavelengths is proportional to pH. Cells are loaded by incubation with 3–5 μM SNARF-1 acetoxymethyl ester for 15–45 min. To improve mitochondrial loading, cells can be incubated at 4 °C for up to 2 h. Cells are then washed and placed on the microscope stage. Using two detectors, fluorescence emitted at 585 nm and > 620 nm is imaged simultaneously. After subtraction of background, the images are divided into each other on a pixel-by-pixel basis, and the resulting ratios are converted to pH using a standard curve generated in situ by incubation of SNARF-1- loaded cells with 10 μM nigericin and 5 μM valinomycin in modified Krebs–Ringer–HEPES buffer where chloride ion is replaced by gluconate to minimize swelling.

Figure 6.9 (see Color Plates) illustrates the distribution of intracellular pH in a cultured adult rabbit cardiac myocyte, as determined by ratio imaging of SNARF-1. Intracellular pH shows marked heterogeneity. Areas under the plasma membrane and in the nucleus have a pH of about 7.2, but mitochondrial regions have a pH of 7.8–8.2. Thus, mitochondrial ΔpH is close to 1 unit.

6.9.4. Mitochondrial Protonmotive Force

The protonmotive force, Δp, defined by Mitchell (1966), is the driving force for ATP formation by oxidative phosphorylation, where

$$\Delta p = \Delta \Psi - 60 \Delta pH \tag{5}$$

By using TMRM to measure $\Delta \Psi$ and SNARF-1 to measure ΔpH, we can evaluate the components of Δp for mitochondria in single living cells. In the examples of Figures 6.7 and 6.9, mitochondrial Δp is at least $-140\,mV$ in resting adult cardiac myocytes.

6.10. Ca^{2+} IMAGING

6.10.1. Nonratiometric Imaging

Fluo-3 is a useful visible wavelength fluorophore for imaging Ca^{2+} by confocal microscopy. Ca^{2+} binding to fluo-3 has a K_d of about 400 nM and produces a 30–80-fold increase in fluorescence intensity (Minta et al., 1989). However, Ca^{2+} binding does not cause a shift in the fluorescence spectrum of fluo-3. Hence, ratioing cannot be used to calibrate fluo-3 fluorescence to free Ca^{2+} concentration. Nonetheless, fluo-3 is a useful probe for determining relative changes in free Ca^{2+}. Green-fluorescing fluo-3 can also be used in combination with red-fluorescing dyes such as TMRM. Depending on cell type and loading temperature, fluo-3 loaded as its acetoxymethyl ester can enter mitochondria, permitting measurement of Ca^{2+} in both mitochondrial and cytosolic compartments.

Cultured cells are loaded with 10–20 μM fluo-3 acetoxymethyl ester for 15–45 min at 37 °C for 2 h at 4 °C. Cells can also be loaded with TMRM, as described above. To permit simultaneous imaging of fluo-3 and TMRM

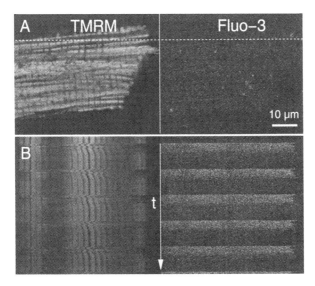

Figure 6.10. Mitochondrial and cytosolic free Ca^{2+} during the contraction of a cardiac myocyte. (A) TMRM (left) and fluo-3 (right) fluorescence was imaged from an unstimulated adult rabbit cardiac myocyte. TMRM fluorescence identified the distribution of mitochondria. Fluo-3 fluorescence was weak and diffuse in the unstimulated cell. The dashed line in panel A identifies the y-axis position where an x vs. time line scan was subsequently collected in panel B. (B), An x vs. time line scan was collected as the myocyte was electrically stimulated at 0.75 Hz. Note the rise and fall of fluo-3 fluorescence after each stimulation. Stimulation also initiated cell shortening, as seen by lateral movement in the TMRM image. After Chacon et al. (1993).

fluorescence, the 488 nm and 568 nm lines of an argon–krypton laser are directed to the specimen by a double dichroic mirror. Green fluo-3 fluorescence is reflected by a 560 nm dichroic mirror through a 522 nm barrier filter to one photodetector. Red TMRM fluorescence is transmitted by the dichroic mirror through a 595 nm long pass filter to a second photomultiplier. By comparing the fluo-3 and TMRM images, the relative distribution of Ca^{2+} in mitochondrial and cytosolic compartments can be determined (Figure 6.10A) (Chacon et al., 1993; Ohata et al., 1994).

6.10.2. Line Scanning

Ordinarily, acquisition of a confocal image requires several seconds of scanning. To improve time resolution, line scanning can be performed. Line scanning creates x vs. time images, as illustrated in Figure 6.10B. Here a cultured myocyte was loaded with fluo-3 and TMRM, and simultaneous x vs. time images were collected of green fluo-3 fluorescence and red TMRM fluorescence. The TMRM scan shows vertical striations representing mitochondria. The darks spaces between stripes are the cytosol. In contrast to TMRM fluorescence, fluo-3 fluorescence is diffuse and is equal in the cytosolic and mitochondrial compartments. The myocyte was electrically stimulated to induce contractions as line scans were collected. With each contraction, fluo-3 fluorescence increased in both the cytosol and mitochondria. TMRM fluorescence was unaffected except for movement artifacts producing waves in the vertical stripes of TMRM fluorescence.

Image processing steps can quantify cytosolic and mitochondrial changes of fluorescence as follows:

1. In the TMRM image, threshold intensity levels are picked to identify pixels of high and low intensity. High pixels represent mitochondria, whereas low pixels represent cytosol or nucleus. Intermediate pixels are ignored since they represent overlap of cytosol and mitochondria.
2. Cytosolic regions are separated from nuclear areas by outlining the latter with a mouse.
3. In the fluo-3 images, intensity values for pixels corresponding to the cytosol, mitochondria, and nucleus are averaged to yield relative Ca^{2+} concentrations in each. Such image processing shows for the first time that mitochondrial free Ca^{2+} changes rapidly and cyclically during excitation–contraction coupling (Chacon et al., 1993; Ohata et al., 1994).

6.10.3. Ratiometric Imaging of Ca^{2+}

UV laser scanning confocal microscopes now permit use of indo-1 for ratiometric measurement of ca^{2+}. When indo-1 is excited with 351 nm light from a UV–argon laser, fluorescence emitted at 450 nm increases as free Ca^{2+} increases, whereas fluorescence at 480 nm decreases. Thus, the 405:480 nm fluorescence ratio can be used to quantify free Ca^{2+}. Myocytes are loaded with 2–5 µM indo-1–acetoxymethyl ester for 1 h at 37 °C or for 2 h at 4 °C. Loading at 37 °C produces cytosolic labeling, whereas loading at 4 °C produces both cytosolic and mitochondrial uptake of the fluorophore. After background subtraction, 480 nm images are divided into 405 nm images on a pixel-by-pixel basis. The resulting ratios (R) are converted to Ca^{2+} as follows:

$$[Ca^{2+}] = K_d((R - R_{min})/(R_{max} - R))(S_f/S_b) \qquad (6)$$

where K_d is the dissociation constant for indo-1 (250 nM); R_{min} and R_{max} are ratio values obtained through the microscope optics in indo-1 containing buffer (100 µM indo-1 pentapotassium salt, 100 mM KCl, 1 mM $MgCl_2$, 1 mM EGTA, and 10 mM MOPS, pH 7.0) with 0 and 5 mM $CaCl_2$, respectively; and S_f and S_b are fluorescence intensities at 480 nm in 0 and 5 mM $CaCl_2$ buffer, respectively. Since the UV laser also excites endogenous NADH and NADPH, control experiments with unloaded cells should be performed to determine that autofluorescence is negligible compared to the signal from indo-1.

When adult cardiac myocytes are loaded with indo-1 and electrically stimulated during the collection of scans, 405 nm fluorescence increases and 480 nm fluorescence decreases after each simulation [Figure 6.11 (see Color Plates)]. Significantly, there is little difference between cytosolic and mitochondrial Ca^{2+}, confirming that both mitochondrial and cytosolic Ca^{2+} increase after each electrical stimulation (Ohata et al., 1994).

6.11. CONCLUSION

Confocal microscopy is a powerful technique used to image the shape, topography, and intracellular distribution of ions and electrical potential of single living cells. Because of the three-dimensional resolving power of confocal microscopy, ion content and electrical potential can be determined for various subcellular compartments, particularly the cytosol and mitochondria. By measuring mitochondrial pH and electrical gradients, total mitochondrial protonmotive force can also be determined. By using stable fluorophores and low levels of illumination, we can collect hundreds of images of living cells with little phototoxicity and photobleaching. In the future, confocal microscopy promises to provide unique new insights into the structure and function of living cells.

ACKNOWLEDGMENTS

This work was supported in part by Grants HL48769, AG07218, and DK37034 from the National Institutes of Health.

The author is indebted to numerous former and present colleagues and collaborators for their contributions to the development of the confocal microscopic work described here: Ms. kristin Al-Ghoul, Dr. Enrique Chacon, Dr. Ian S. Harper, Dr. Brian Herman, Dr. Anna-Liisa Nieminen, Dr. Hisayuki Ohata, Mr. Ting Qian, Mr. Jeffrey M. Reece, Mr. Samuel A. Tesfai, and Dr. George Zahrebelski.

REFERENCES

Chacon, E., Harper, I. S., Reece, J. M., Herman, B., and Lemasters, J. J. (1993). Mitochondrial and cytosolic Ca^{2+} transients during the contractile cycle of cultured cardiac myocytes: A laser scanning confocal microscopic study. *Biophys. J.* **64**, A106.

Chacon, E., Reece, J. M., Nieminen, A.-L., Zahrebelski, G., Herman, B., and Lemasters, J. J. (1994). Distribution of electrical potential, pH, free Ca^{2+}, and cell volume inside cultured adult rabit cardiac myocytes during chemical hypoxia: A multiparameter digitized confocal microscopic study. *Biophys. J.* **66**, 942–952.

Ehrenberg, B., Montana, V., Wei, M.-D., Wuskell, J. P., and Loew, L. M. (1988). Membrane potential can be determined in individual cells from the Nernstian distribution of cationic dyes. *Biophys. J.* **53**, 785–794.

Emaus, R. K., Grundwald, R., and Lemasters, J. J. (1986). Rhodamine 123 as a probe of transmembrane potential in isolated rat liver mitochondria: Spectral and metabolic properties. *Biochim. Biophys. Acta* **850**, 436–448.

Farkas, D. L., Wei, M.-D., Febbroriello, P., Carson, J. H., and Loew, L. M. (1989). Simultaneous imaging of cell and mitochondrial membrane potential. *Biophys. J.* **56**, 1053–1069.

Gores, G. J., Nieminen, A.-L., Wray, B. E., Herman, B., and Lemasters, J. J. (1989). Intracellular pH during 'chemical hypoxia' in cultured rat hepatocytes: Protection by intracellular acidosis against the onset of cell death. *J. Clin. Invest.* **83**, 386–396.

Grynkiewicz, G., Poenie, M., and Tsien, R. Y. (1985). A new generation of Ca^{2+} indicators with greatly improved fluorescence propeties. *J. Biol. Chem.* **260**, 3440–3450.

Gunter, T. E., and Pfeiffer, D. R. (1990). Mechanisms by which mitochondria transport calcium. *Am. J. Physiol.* **258**, C755–C786.

Herman, B., Nieminen, A.-L., Gores, G. J., and Lemasters, J. J. (1988). Irreversible injury in anoxic hepatocytes precipitated by an abrupt increase in plasma membrane permeability. *FASEB J.* **2**, 146–151.

Johnson, L. V., Walsh, M. L., Bockus, B. J., and Chen, L. B. (1981). Monitoring of relative mitochondrial membrane potential in living cells by fluorescence microscopy. *J. Cell. Biol.* **88**, 526–535.

Lee, C., and Chen, L. B. (1988). Dynamic behavior of endoplasmic reticulum in living cells. *Cell (Cambridge, Mass.)* **54**, 37–46.

Lemasters, J. J., DiGuiseppi, J., Nieminen, A.-L., and Herman, B. (1987). Blebbing, free Ca^{2+} and mitochondrial membrane potential preceding cell death in hepatocytes. *Nature (London)* **325**, 78–81.

Lemasters, J. J., Chacon, E., Zahrebelski, G., Reece, J. M., and Nieminen, A.-L. (1993). Laser scanning confocal microscopy of living cells. In *Optical Microscopy: Emerging Methods and Applications* (B. Herman and J. J. Lemasters, Eds.), pp. 339–354. Academic Press, New York.

Minsky, M. (1961). Microscopy apparatus. U.S. Patent 3,013,467.

Minta, A., Kao, J. P. Y., and Tsien, R. Y. (1989). Fluorescent indicators for cytosolic calcium based on rhodamine and fluorescein chromophores. *J. Biol. Chem.* **264**, 8171–8178.

Mitchell, P. (1966). Chemiosmotic conpling in oxidative and photosynthetic phosphorylation. *Biol. Rev. Cambridge Philos. Soc.* **41**, 445–502.

Nieminen, A.-L., Saylor, A. K., Tesfai, S. A., Herman, B., and Lemasters, J. J. (1995). Contribution of the mitochondrial permeability transition to lethal injury after exposure of hepatocytes to t-butylhydroperoxide. *Biochem. J.* **307**, 99–106.

Ohata, H., Tesfai, S. A., Chacon, E., Herman, B., and Lemasters, J.J. (1994). Mitochondrial Ca^{2+} transients in adult rabbit cardiac myocytes during excitation-contraction coupling. *Circulation* **90** (Part 2), I–632.

Pagano, R. E., Martin, O. D., Kang, H. C., and Haugland, R. P. (1991). A novel fluorescent ceramide analogue for studying membrane traffic in animal cells: Accumulation at the Golgi apparatus results in altered spectral properties of the sphingolipid precursor. *J. Cell. Biol.* **113**, 1267–1279.

Petran, M., Hadravsky, M., Egger, M. D., and Galambos, R. (1968). Tandem-scanning reflected-light microscopy. *J. Opt. Soc. Am.* **58**, 661–664.

White, J. G., Amos, W. B., and Fordham, M. (1987). An evaluation of confocal versus conventional imaging of biological structures by fluorescence light microscopy. *J. Cell. Biol.* **105**, 41.

Wilson, T. (1990). *Confocal Microscopy.* Academic Press, London.

Zahrebelski, G., Nieminen, A.-L., Al-Ghoul, K., Qian, T., Herman, B., and Lemasters, J. J. (1995). Progression of subcellular changes during chemical hypoxia to cultured rat hepatocytes: A laser scanning confocal microscopic study. *Hepatology.* **21**, 1361.

CHAPTER

7

FLUORESCENCE RESONANCE ENERGY TRANSFER

ROBERT M. CLEGG

Max Planck Institute for Biophysical Chemistry
Molecular Biology Department
D-37077 Göttingen, Germany

7.1. INTRODUCTION

This chapter is an account of fluorescence resonance energy transfer (FRET), with an emphasis on those topics that are of interest to FRET microscopy. The recent literature on FRET microscopy is reviewed, with some illustrations of certain practical and analytical aspects; however, this chapter is not a review of the methods of FRET microscopy or of FRET measurements in general. This can be found in the reviews and original papers already available (see references below). Several topics are discussed that are more concerned with the fundamental principles of FRET. FRET microscopy has matured to the point where quantitative imaging measurements are being attempted that were not possible earlier, and this places more demand on the analysis of FRET data; therefore, it is opportune to discuss the general FRET literature together with the topic of FRET microscopy. Because the environment in a biological sample (e.g., a cell in a microscope) is inherently more complex than in a cuvette experiment, it is expedient to be aware of some of the more detailed aspects of FRET; this knowledge can be gained by examining the rate equations in more detail. This is especially important when interpreting data from highly organized structures (e.g., membranes and multisubunit supramolecular structures) and very heterogeneously organized structures.

In this chapter, two different "classical" derivations of FRET are summarized and compared briefly with a quantum mechanical derivation; both classical derivations reveal distinguishing characteristics of the dipole–dipole FRET mechanism that may not be so apparent in a quantum mechanical derivation. The orientational aspects of FRET are also discussed, with

Fluorescence Imaging Spectroscopy and Microscopy, edited by Xue Feng Wang and Brian Herman. Chemical Analysis Series, Vol. 137.
ISBN 0-471-01527-X © 1996 John Wiley & Sons, Inc.

examples, using a simple (but useful) quantitative model that conveys clearly some aspects of κ^2 with important consequences for quantitative FRET interpretations. An attempt is made to present these theoretical topics on a useful and practical level but to retain the quantitative characteristics. These more theoretical discussions are considered appropriate for this chapter because the future of FRET microscopy is definitely progressing toward more quantitative analyses. Many future practitioners of FRET microscopy may not have time to familiarize themselves with some of the more fundamental aspects of FRET, and an understanding of these features is helpful when interpreting data from complex systems. The milieus of many biological systems observed in microscopy are heterogeneous and far from the standards of relatively clean systems that are studied in cuvettes. Spectroscopic experiments in such complex milieus can only be interpreted within certain limits, although these limits are being continually extended. To utilize the quantitative possibilities of new methods, and to avoid the frustations of artifacts and of overinterpreting data, familiarity with derivations of the fundamental equations is helpful.

The topics and publications discussed in this chapter undoubtedly reflect partially the interests of the author, but an attempt has been made to review the relevant literature. The chapter is organized as follows: (i) the general aspects of FRET are introduced; (ii) the importance of FRET in biology and microscopy is discussed; (iii) the photophysical fundamentals are considered in more detail; (iv) the interdependence among the first-order deactivation rates of decay from the excited state and their effect on FRET measurements are considered; (v) orientation effects are discussed in detail; (vi) the effects of distance and orientation distributions are reviewed and a short discussion of FRET measurements on membrane systems is given as an example; (vii) some new techniques are reviewed; and finally, (viii) some of the outstanding contributions to the FRET literature in the last few years, in particular those related to FRET microscopy, are reviewed. The papers reviewed were selected because they lead to a better understanding of important biological structures, or introduce new or improved methods of FRET measurements and analyses.

7.1.1. A Selection of General FRET References

An excellent review of FRET (not microscopy), with emphasis on methods and applications (different from those selected in this review), has recently been published (Wu and Brand, 1994). A comprehensive and useful list of R_0 values for over 70 donor–acceptor pairs is given, and an extensive selected bibliography, also of earlier literature, is provided. Another recent review, supplementing the topics in this chapter and in the review by Wu et al., has also just appeared (Selvin, 1995). There have been several excellent general reviews of

FRET in the last 30 years, with an emphasis on biological and biochemical applications (Cheung, 1991; Clegg, 1992; Dewey, 1991; Fairclough and Cantor, 1978; Jovin, 1979; Kleinfeld, 1988; Schiller, 1975; Steinberg, 1971; Steinberg et al., 1983; Stryer, 1960, 1978; Szöllösi et al., 1987a; Wilkinson, 1965) and other fields (Bennett and Kellogg, 1967; Bojarski and Sienicki, 1990; Förster, 1959, 1960; Ganguly and Chaudhury, 1959; Ghiggino and Smith, 1993; Lamola, 1969; Valeur, 1989; Wolf, 1967). Some general reviews of fluorescence also have excellent discussions about FRET (Brand and Witholt, 1967; Cantor and Tao, 1971; Eyring et al., 1980; Förster, 1951; Lakowicz, 1983, 1991; Laurence, 1957; Stryer, 1968). Some papers that are helpful to consult are Beddard (1983), Bennett (1964), Bennett et al. (1964), Chang and Filipescu (1972), Dewey (1992), Dexter (1953), Dow (1968), Förster (1946, 1947, 1948, 1949a,b), Gösele et al. (1976), Hassoon et al. (1984), Haugland et al. (1969), Hauser et al. (1976), Kellogg (1970), Klafter and Blumen (1984), Kuhn (1970), Latt et al. (1965), Lin (1973), Lin et al. (1993), Mugnier et al. (1985), Perrin (1927), Robinson and Frosch (1962, 1963), Scholes and Ghiggino (1994), Speiser (1983), Stryer and Haugland (1967), and Stryer et al. (1982). The titles of the references are self-explanatory. Many aspects of energy transfer are involved in photosynthetic systems, including FRET, and this field of research is very active; however, because this field is rather specialized, the reader is referred to a recent review (Dau, 1994) for references to part of this extensive literature.

7.2. FRET IN THE MICROSCOPE

In principle, the measurement of FRET in a microscope can provide the same information that is available from the more common macroscopic solution measurements of FRET; however, FRET microscopy has the additional advantage that the spatial distribution of FRET efficiency can be visualized throughout the image, rather than registering only an average over the entire object of interest. Because energy transfer occurs over distances of 10–100 Å, a FRET signal corresponding to a particular location within a microscope image provides an additional magnification surpassing the optical resolution ($\sim 0.3\,\mu m$) of the light microscope (Kauzmann, 1957). Thus, within a voxel of microscopic resolution, FRET resolves average donor–acceptor (D–A) distances beyond the microscopic limit down to the molecular scale. This is one of the principal and unique benefits of FRET for microscopic imaging: not only colocalization of the D- and A-labeled probes within $\sim 0.09\,\mu m^2$ can be seen, but intimate interactions of molecules labeled with D and A can be demonstrated. There have been several general reviews of FRET microscopy: Bottiroli et al. (1992), Cardullo et al. (1991), Dunn et al. (1994), Herman (1989),

Jovin and Arndt-Jovin (1989a, b), Jovin et al. (1989), Mátyus (1992), Uster (1993), and Uster and Pagano (1989).

Spectroscopic measurements in the microscope can furnish a wealth of information regarding structures and molecular interactions. Fluorescence measurements are very sensitive and often relatively easy to perform. However, because the sample environment is often complex, highly structured, and heterogeneous, the results can be very difficult to quantify. It cannot be overemphasized that a successful quantitative analysis of FRET data depends on the availability of correct controls and on the ability to choose the best method available for carrying out the FRET measurements on particular samples.

7.3. QUALITATIVE AND QUANTITATIVE APPLICATIONS OF FRET

FRET can provide qualitative evidence for determining simply whether two molecular entities are close together—intra- or intermolecular—on a molecular scale ($< \sim 100$ Å); it can also be used quantitatively for measuring distances between molecular entities and for providing structural information. Examples of qualitative applications are immunoassays, discrimination of intact multimolecular complexes from dissociated ones, presence of associated heterogeneous molecular complexes, and detection of the hybridization state of DNA, RNA, or DNA–RNA hybrids. The quantitative measurements are employed to determine actual intramolecular distances and orientations between different parts of a macromolecule, intermolecular distances between macromolecules in supramolecular complexes, and structural and geometrical aspects of single macromolecules or of multisubunit molecular assemblies. Although the majority of FRET applications in microscopy have been more qualitative, many present studies are designed to obtain quantitative information.

7.4. WHAT IS FRET?

Fluorescence resonance energy transfer (FRET) is the physical process by which energy is transferred nonradiatively from an excited molecular chromophore (the donor D) to another chromophore (the acceptor A) by means of intermolecular long-range dipole—dipole coupling (Förster, 1946, 1951). The essential requirements for effective transfer over distances from 10 to 100 Å are that the fluorescence spectrum of D and the absorbances spectrum of A overlap adequately, and that the quantum yield of D (ϕ_D) and the absorption coefficient of A (ε_A) be sufficient (e.g., usually $\phi_D \geq .01$ and $\varepsilon_A \geq 1000$). In addition, for the dipole–dipole vectorial interaction to occur, the transition dipoles of D and A must either be oriented favorably relative to each other or one (or both) must have a certain degree of rapid rotational freedom; this latter condition is often (but not always) satisfied for chromophores attached through molecular linkers to biomolecules in solution. The transfer is nonradiative; that is, D does not actually emit a photon and A does not absorb a photon.

7.5. THE RATE OF THE TRANSFER OF ENERGY IN FRET

The process of energy transfer can be depicted mechanistically as a transition between two states[1]

$$(D^*, A) \xrightarrow{k_T} (D, A^*)$$

where k_T is the rate constant of FRET between a particular D–A pair; A is said to be sensitized by D. This transfer is designated as long-range because D and A do not touch and the transfer can take place over several molecular diameters. The mutual dipole–dipole interaction energy provoking the transfer of energy between D and A is very weak, often only ~ 2–$4\,cm^{-1}$, compared to the spectroscopic energies that are interchanged, $\sim 15,000$–$40,000\,cm^{-1}$. The long-range dipole nature of this type of energy transfer was first correctly conjectured by Perrin (1927), although his quantitative theory falsely concluded that energy transfer could take place over distances much longer ($\sim 1000\,Å$) than were observed (see below). A physical theory predicting the correct distance dependence was later proposed by Förster, first classically (see below) (Förster, 1946, 1951) and subsequently quantum mechanically (Förster, 1948, 1960, 1965). Förster's theory predicted that energy could be transferred by a resonance dipole–dipole mechanism over distances ~ 10 and $\sim 100\,Å$, depending on the spectroscopic parameters of D and A; usually D and A should not be closer than $\sim 10\,Å$ to avoid strong ground-state D–A interactions or transfer by exchange interactions (Dexter, 1953), although see Ernsting et al. (1988). Förster's theory correctly described the rate of energy transfer to be

$$k_T = \frac{1}{\tau_D}\left(\frac{R_0}{R}\right)^6 \tag{1}$$

---

[1] The asterisk will be used to denote excited states; (D^*, A) refers to a state in which D is in the excited state and A is in the ground state.

where τ_D is the fluorescence lifetime of D (i.e., the measured lifetime in the absence of A); R is the distance between D and A; and R_0 is a distance parameter calculated from the spectroscopic and mutual dipole orientational parameters of D and A (see Eq. 12) when $R = R_0, k_T = 1/\tau_D$; that is, the rates of emission and of FRET are equal at $R = R_0$. This equation applies to a single configuration of D and A; distributions of D and A must be averaged appropriately over distances and orientations (see below). Förster showed that his theory correctly explained many data involving sensitized emissions that had been puzzling spectroscopists for some time.

7.6. THE EFFICIENCY, E, OF ENERGY TRANSFER

The "efficiency of energy transfer," E, is a quantitative measure of the number of quanta that are transferred from D to A. The "quantum yield" of energy transfer, E, is defined as

$$E = \frac{\text{number of quanta transferred from D to A}}{\text{number of quanta absorbed by D}} \tag{2}$$

From a dynamic point of view, E is the ratio of k_T to the total sum of rates for all processes by which an excited D molecule can return to its ground state, including k_T (see Section 7.11).

$$E = \frac{k_T}{k_T + \sum_{i \neq T} k_i} = \frac{k_T}{1/\tau_M} \tag{2'}$$

where the subscript i refers to the different pathways of deactivation from the D* state; τ_M is the measured fluorescence lifetime of D in the presence of A; E can be measured in numerous ways (see the general references in Section 7.1.1 for discussions of the different experimental and analytical methods used to determine E); steady-state and time-resolved (nanosecond) measurements are employed. The choice of the measurement method depends on the application and the information desired. In the most popular method of determining E, the steady-state donor fluorescence intensity, F_D, is measured from a sample containing only D and from a corresponding sample containing both D and A, F_{DA}. If the values of F_D and F_{DA} are normalized to their respective concentrations of D, the efficiency can be calculated according to the following equation:

$$E = 1 - \frac{F_{DA}}{F_D} = \frac{1}{1 + (R/R_0)^6} \tag{3}$$

This equation is for a single D–A pair. The efficiency of transfer decreases as the distance between D and A increases; the change in E is dramatic in the range $0.5R_0 < R < 1.5R_0$. Similar relations defining E apply to measurements of the acceptor emission (i.e., if A is fluorescent; A must not fluoresce in order for FRET to take place) and direct time-resolved measurements (see Eq. 1).

7.7.   USE OF FRET IN MANY RESEARCH FIELDS

Subsequent to Förster's pioneering series of studies, others promptly realized the unique opportunity provided by FRET for observing hitherto inaccessible physical phenomena related to the relative dispositions and separations of molecules in solutions and solids on a molecular scale. Especially attractive was (and still is) the novel opportunity to measure molecular separations between two molecules, D and A, directly. The utility of FRET is derived from the strong dependence of the rate and efficiency of energy transfer on the sixth power of the distance R between D and A in Eqs. (1) and (3). This strong dependence of E on R has led to the often quoted reference (Stryer, 1978) to FRET as a "spectroscopic ruler." Since about 1950, the scientific literature on FRET has grown continually, reflecting the many innovative applications of FRET encompassing multifarious scientific disciplines, such as physics, chemistry, biology, and polymer science (see the references listed in Section 7.1.1).

7.8.   WHY USE FRET IN BIOLOGY?

Perhaps the greatest impact of FRET has resulted from its application to biological macromolecules. Biological systems are complex molecular organizations involving macromolecules (e.g., proteins, nucleic acid polymers, fatty acids, lipids, and polycarbohydrates) that associate to form supramolecular structures (e.g., multiprotein assemblies, membranes, chromosomes) that, in turn, comprise the different compartments of living cells and their respective parts. An outstanding feature of all biological systems is the intricate and interdependent organization of their various structures. In order to understand and manipulate the functions of biological systems, it is imperative to study the structures of their components and to determine how these components are associated into larger molecular organizations. FRET is unique in its capacity to supply accurate spatial information about molecular structures at distances ranging from ~ 10 to ~ 100 Å. In many cases, geometrical information can also be determined. FRET is observed only if D and A are in proximity (approximately $R < 2R_0$). This possibility of detecting and quantifying molecular proximities has led to numerous applications of FRET in biology over the past 30 years and continues to generate much of the present interest in the subject.

7.9. THE PHOTOPHYSICAL MECHANISM OF DIPOLE–DIPOLE ENERGY TRANSFER

7.9.1.   Some Derivations of the Rate k_T of Energy Transfer in FRET

In the previous sections we have discussed physical mechanisms of FRET rather superficially, without describing the physics involved. To grasp the physical mechanism of FRET at the most fundamental level, a quantum mechanical (QM)[2] derivation is required; this has been presented often in the literature (Bennett, 1964; Dexter, 1953; Dow, 1968; Förster, 1948; Robinson and Frosch, 1962), and new, more expanded accounts still appear (Klafter and Blumen, 1984; Lin, 1973; Lin et al., 1993; Scholes and Ghiggino, 1994). There are certain aspects that can be derived only from a QM viewpoint—for instance, exchange processes ("Dexter" transfer) (Dexter, 1953) and symmetry and spin rules. However, Förster's equations describing dipole–dipole long-range FRET can be derived on classical grounds, and the classical derivations are particularly edifying.

7.9.2.   Förster's Original Classical Derivation

To the author's knowledge, Förster's original classical derivation (Förster, 1946) has never been presented in English for a wider audience and is interesting historically. By taking into account the Stokes shift of the fluorescence spectrum relative to the absorption spectrum and by considering broad spectra instead of assuming line spectra, Förster was able to correct an earlier oversight of Perrin's (1927) and to predict, also using classical arguments, the correct distance dependence of the transfer of energy.

Förster considered two neighboring, electrically charged classical mechanical oscillators (Hertzian oscillators) coupled through coulomb electrostatic interactions. If the two oscillators have the same resonance frequency, they will exchange energy, similar to two resonating pendulums. Förster assumed that only one of the oscillators, the "donor," is initially vibrating. Classically, the "donor" oscillator will either tend to radiate energy into its surroundings or exchange its energy (in both directions) with the "acceptor" oscillator if the two

---

[2] Here QM will refer to "quantum mechanical" and will be used to indicate the places where the classical terminology actually refers to a quantum mechanical effect.

oscillators are close enough. The two oscillators[3] interact over a distance of R as dipoles, with dipole moments of magnitude μ and an interaction energy of $E_{ia} \approx \kappa \mu^2/n^2 R^3$. Here, for completeness, we have included the factor κ, which is the orientation factor between the two dipoles (see Eq. 6 and Figure 7.1), and divided by n^2, where n is the index of refraction of the solution (both factors were not included in Förster's original calculation because it was, in his terminology, an *Überschlagsrechnung*, i.e., an approximate calculation). Förster calculated the distance between the two oscillators where the time for radiative emission of a classical oscillating dipole[4] (Greiner, 1986; Jackson, 1962; Kauzmann, 1957) ($\tau_{rad} \approx 3hc^3/\mu^2\omega^3$) and the time for exchange of energy between the two oscillators[5] ($\tau_{int} \approx h/E_{ia} = hn^2R^3/\kappa\mu^2$) are equal, i.e., $R \equiv R_0$ when $\tau_{rad} = \tau_{int}$; at this distance between D and A, half of the energy that would have been radiated by D will be transferred nonradiatively to A. If the two oscillators are always in exact resonance (i.e., Perrin's case), one can easily calculate that $R_0 = \lambda/2\pi$, where λ is the wavelength of the oscillation $[\lambda = (c/n)/(\omega/2\pi)]$, c is the velocity of light, and ω is the radial frequency; this would result in effective transfer over ~ 1000 Å, which was not observed.

Förster points out that this excessively long estimate for R_0 results from the assumption of exact resonance of line spectra. In actuality, the spectra are broadened rapidly by intramolecular and solvent interactions; as a result of this spectral broadening, the condition of exact resonance between the two oscillators will be met only at short intervals during the long time required for energy transfer (if the interaction energy E_{ia} of two classical oscillators is very small compared to their vibrational energy, they will exchange energy only over many periods; in this case, the number of periods is many orders of magnitude). This small probability of simultaneous exact resonance is due to the very rapid dephasing of the molecular energies. This dephasing of the resonance between the two "narrowband" oscillators considerably lengthens the time that was estimated for the energy exchange between the two oscillators, and this leads to a much smaller R_0. Förster then proceeded to calculate the time required for energy to be exchanged between two oscillators with

---

[3] For this approximate calculation, Förster assumed that the two oscillators have identical amplitudes.

[4] The mean rate of energy radiated from a oscillating dipole is

$$\dot{E} = (16/3)(\pi^4 c/\lambda^4)\mu_0^2 = (1/3)(\omega^4/c^3)\mu_0^2 = (1/3)(\omega^3 h\omega/hc^3)\mu_0^2$$

(Greiner, 1986; Jackson, 1962; Kauzmann, 1957); therefore, the time of radiation is $\tau_{rad} \approx h\omega/\dot{E} = 3hc^3/(\mu_0^2\omega^3)$.

[5] The energy of interaction is seen as an oscillating vibration: $E_{int} = h\omega_{int} \approx h/\tau_{int}$; therefore, $\tau_{int} \approx h/E_{int}$.

broadened spectra. He calculated the probability that the spectral regions of the two oscillators overlap, and that simultaneously the energies of both oscillators are concurrently nearly the same within the small interaction energy E_{ia}. Defining (retaining his notation) the width of the spectral frequencies to be Ω (assuming both spectral widths to be the same) and the overlap region to be Ω', this probability is $w = (\Omega'/\Omega) \cdot (E_{ia}/\hbar\Omega)$. The first term is the probability of spectral overlap of D and A, and the second term is the probability that the frequency within the broadband of frequencies, Ω, will fall within the relatively narrow bandwidth of $E_{int}/\hbar$ (see footnotes 4 and 5). The rate of transfer calculated without this "dephasing" correction ($1/\tau_{int} \approx \kappa\mu^2/\hbar n^2 R^3$; see above) is then multiplied by w to give the rate of transfer with dephasing:

$$k_T \approx 1/\tau'_{int} = w/\tau_{int} = \mu^4\kappa^2\Omega'/(\hbar^2 n^4 R^6 \Omega^2)$$

We now see, in contrast to the earlier derivation of Perrin (1927), the dependence of the transition rate on $1/R^6$ that is the trademark of FRET. By solving for the distance R_0 where $\tau_{rad} = \tau'_{int}$ (i.e., setting the radiative and nonradiative transfer rates equal), Förster calculated

$$R_0^6 = 3(c^3/\omega^3)\mu^2\kappa^2(\Omega'/\Omega^2)(1/\hbar n^4) = 9(\lambda^6/(2\pi)^6)(\kappa^2/n^4)(\Omega'/\Omega^2)(1/\tau_{rad})$$

where in the last equality the used $\tau_{rad} = 3\hbar \cdot c^3/\mu^2\omega^3$ to replace μ^2 and $c = \lambda\omega/2\pi$.

By reviewing the above derivation, one sees that the $1/R^6$ dependence of the transfer rate results from the doubly multiplicative appearance of E_{ia}, once in the calculation of τ_{int} between two classical dipoles and once from the correction to τ_{int} due to the dephasing of the two spectra (finite spectral widths). This insistence on a resonance condition[6] within the broad spectrum of frequencies for successful energy transfer (essentially a condition for energy exchange between two classical oscillators) is the reason that Förster explicitly emphasized the term resonance in the name FRET. According to this derivation, if the two dipoles exchanged energy only at a single frequency (line spectra), they could exchange energy over longer distances. The resonance requirement together with the dephasing of the oscillators within the broadened spectral overlap shortens the distance for effective interaction between D and A considerably. Förster emphasizes that although FRET

---

[6] In a broad sense, resonance simply means that the frequencies of the two separate sepactral transitions in D and A must be equal; that is, for any D–A pair, the energy lost by D is gained by A. This is true for any one-step transfer of energy, including the absorption or emission of a photon.

involves an exchange of quanta, and that therefore the most fundamental analysis of the physical mechanism must be based on a QM derivation, FRET is essentially a classical resonance effect. We also see that the dependence of the transfer rate (and of R_0^6) on the spectral overlap, on the donor fluorescence lifetime in the absence of acceptor (τ_{rad}), on the κ^2 orientation term, and on the index of refraction are correctly predicted (see Eq. 12). In spite of the approximations, one sees that for reasonable values of $(\Omega'/\Omega^2)(1/\tau_{rad})$ the distance of FRET interaction has been decreased from approximately $\lambda/2\pi$ to $< 0.1\lambda/2\pi$. Förster later published his exact QM derivation of k_T and R_0 (Förster, 1948), and still later published another classical derivation more similar to the method discussed in the next section (Förster, 1951), but many essentials of the QM derivation are already apparent, and clearly presented, in his original classical approximate derivation.

7.9.3. A Second Classical Derivation

Different classical approaches were considered later by Ketskemety (1962), in more detail by Kuhn (1970, 1977), and later by Steinberg et al. (1983). The classical derivations can be carried out quantitatively, and the result agrees exactly with that of the QM derivation. We sketch the approach of Kuhn. The electric field of an oscillating dipole (the donor) can be expressed in R and θ component form as follows (Jackson, 1962; Stratton, 1941):

$$
\begin{aligned}
E_\theta &= (n^{-2})(1/R^3 - ik/R^2 - k^2/R)\sin\theta \cdot \mu_0 \cdot \exp(i\omega(t - Rn/c)) \\
E_R &= 2(n^{-2})(1/R^3 - ik/R^2)\cos\theta \cdot \mu_0 \cdot \exp(i\omega(t - Rn/c))
\end{aligned}
\tag{4}
$$

where $k = n(2\pi/\lambda)$, n is the index of refraction, and θ is the angle between the dipole axis and the direction vector connecting the donor and acceptor (see Figures 7.1 and 7.2). The $1/R$ term dominates far from the dipole ($R > \lambda$), and only this field contributes to the radiation of the energy and is therefore called the "radiation" or far field [energy flux $= (1/2)\cdot \mathbf{E} \times \mathbf{H}$]; the terms proportional to $1/R^2$ and $1/R^3$ in Eq. (4) do not contribute to the radiation.[7] The $1/R^3$ term dominates close to the oscillator ($R \ll \lambda$); this term has the same form as the static dipole field and is appropriately called the "static" or near field. The near-field terms oscillate, but the oscillatory near-field energy components do not radiate. Due to the far-field terms, the D dipole will radiate energy at a rate

---

[7] The coupling of the oscillator to the radiation field is very weak. The frequency of the molecular electric oscillators is $\sim 10^{15}\,s^{-1}$, and spontaneous emission occurs at a frequency of $\sim 10^8\,s^{-1}$; if the coupling to the radiation field was not so weak, we would not observe FRET, which is also a result of very weak coupling between the dipoles.

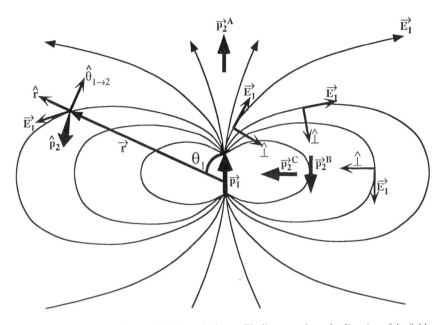

Figure 7.1. The field $\mathbf{E}_1$ of a static dipole $\mathbf{p}_1$ is shown. The lines run along the direction of the field. The carets over the symbols stand for unit vectors. The unit vectors $\hat{\perp}$ are perpendicular to $\mathbf{E}_1$ and emphasize those vectors that are perpendicular to the field, where, if $\mathbf{p}_2$ is parallel to a $\hat{\perp}$ vector, then $\mathbf{E}_1 \cdot \mathbf{p}_2 = 0$. Of course, all $\hat{\perp}$ vectors that rotate around the $\mathbf{E}_1$ vector are perpendicular to $\mathbf{E}_1$; $\mathbf{r}$ is the distance vector from $\mathbf{p}_1$ to $\mathbf{p}_2$. The angle θ_1 and the unit vectors $\hat{\theta}_{1 \to 2}$ and $\hat{r}$ are defined in the text (see Section 7.14). See text for further explanations.

of $w_{rad}^D = \mu_0^2 \omega^4 n/(3c^3)$(Jackson, 1962), which agrees with the section above (but with n explicitly expressed) and is derived from the $1/R$ terms of Eq. (4) by the expression $I = E_0^2\,(cn/8\pi)$, where I is the intensity of light with an electric field of field strength E_0 (Jackson, 1962); only the E_θ term contributes to the radiation (only its electric field component is perpendicular to the direction in which the radiation is traveling). The interaction between two oscillating dipoles that leads to FRET takes place in the near-field zone. Here the dominant $1/R^3$ term of the electric field has a radial as well as a transverse component (see Eq. 4), and we must consider both vector components. In the near-field zone, the electric field of the donor $\mathbf{E}(t)_{nf,D}$ can be written as

$$\mathbf{E}(t)_{nf,D} \approx \mathbf{E}_{nf,0,D}\exp(i\omega(t - Rn/c)) = \{E_{nf,R}\cdot\hat{R} + E_{nf,\theta}\cdot\hat{\theta}\}\exp(i\omega(t - Rn/c))$$
$$= \{2\cos\theta_D\cdot\hat{R} + \sin\theta_D\cdot\hat{\theta}\}(\mu_0/n^2)(1/R^3)\exp(i\omega(t - Rn/c)) \qquad (5)$$

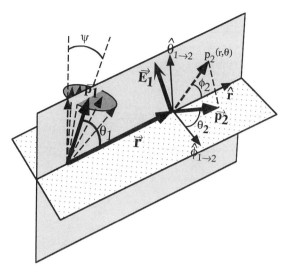

Figure 7.2. This figure describes the coordinate systems used in the text, especially for discussing the orientation factor κ. Here $\mathbf{E}_1$ is the field of dipole $\mathbf{p}_1$ at the location of dipole $\mathbf{p}_2$. The right-handed Cartesian coordinate system $(\hat{r}, \hat{\theta}_{1\rightarrow 2}, \hat{\phi}_{1\rightarrow 2})$ is explained in the text: $\mathbf{r}$ is the distance vector from $\mathbf{p}_1$ to $\mathbf{p}_2$; the angle θ_1 is the angle between $\mathbf{p}_1$ and $\mathbf{r}$. For the case of a cone, the central dotted line is in the direction of $\mathbf{p}_0$, which is the central vector of the cone. The darker shaded plane is the plane formed by the two vectors $\mathbf{p}_1$ and $\mathbf{r}$: $\mathbf{E}_1$ lies in this plane, as does the $\hat{\theta}_{1\rightarrow 2}$ axis vector; $\hat{\phi}_{1\rightarrow 2}$ lies in the lightly shaded plane that is perpendicular to the other plane. In general, $\mathbf{p}_2$ is not in either of these two planes. The angle ψ is the angle between $\mathbf{p}_0$ and $\mathbf{p}_1$. See text (Section 7.14.3.) for further explanations.

If a second dipole $\boldsymbol{\mu}_A$ (the acceptor) is in the near-field zone of the donor, the field strength of the vectorial component of the electric field of the donor $\mathbf{E}(t)_{nf,D}$ acting on the acceptor dipole is the dot product $|\mathbf{E}_{0,D}|_{D\cdot A} = \hat{\boldsymbol{\mu}}_A \cdot \mathbf{E}_{0,nf,D}$, where the cap on $\hat{\boldsymbol{\mu}}_A$ implies a normalized vector. That is, the effective field of the donor acting on the acceptor depends on the relative orientation of A and D. This can be expressed as $|\mathbf{E}_{0,D}|_{D\cdot A} = (\mu_0/n^2)(1/R^3)\kappa$, where κ is the well-known orientation factor of two dipoles interacting; κ is

$$\kappa = \{2\cos\theta_D \hat{r} \cdot \hat{\boldsymbol{\mu}}_A + \sin\theta_D \hat{\theta}_D \cdot \hat{\boldsymbol{\mu}}_A\} = \{2\cos\theta_D \sin\theta_A \cos\phi_A + \sin\theta_D \sin\theta_A \sin\phi_A\}$$

$$(6)$$

For the definition of the angles, see Section 7.14.3 and the corresponding

Figure 7.2. At this point Kuhn (1965, 1970, 1977) considers the rate at which an A molecule absorbs energy w_{abs}^A from the oscillating near-field of the donor $E(t)_{nf,D}$ as though the oscillating field were that of a light wave [the same procedure was suggested and carried out earlier by Ketskeméty (1962)]. The rate of absorption of energy from an oscillating electric field by an absorbing molecule is proportional to the square of the effective field strength of D at the location of A:

$$w_{abs}^A = \alpha(|\mathbf{E}_{0,D}|_{D \cdot A})^2 \tag{7}$$

Kuhn (1970) shows with a simple argument that

$$\alpha = -(dI/dl)/C_A N_{Av} = 3(\ln 10/N_{Av})(c \cdot n/8\pi)\varepsilon_A$$

is the correct expression for relating α to the extinction coefficient of A, ε_A; N_{Av} is Avogadro's number; and C_A is the effective concentration of the acceptor in a shell of thickness l close to D; (dI/dl) is the change in the intensity of the "nonradiative!" near field (treated similarly to an absorber in a radiative field), where

$$I = (cn/8\pi)(|\mathbf{E}_{0,D}|_{D \cdot A})^2 \tag{8}$$

By combining the expressions for α and $|\mathbf{E}_{0,D}|_{D \cdot A}$, the rate w_{abs}^A can be calculated. Then, by setting $w_{abs}^A = w_{rad}^D$, the distance between D and A where the rate of radiation of D and the rate of absorbance by A are equal can be calculated. This distance R_0 is then found to be

$$R_0 = [9(\ln 10)\kappa^2 c^4 \varepsilon_A/(128\pi^5 N_{Av} n^4 v^4)]^{1/6} \tag{9}$$

This expression for R_0 is the same as that calculated by Förster with his QM calculation (Förster, 1948; Ketskeméty, 1962), except that the thermal and solvent broadening of the spectral bandwidths of the emission and absorption spectra of the D and A molecules and the quantum yield of D have not been taken into account. This is done simply by replacing ε_A/v^4 by the "overlap integral":

$$J(v) \equiv \int_0^\infty [\varepsilon_A(v)/v^4] f_D(v) dv \tag{10}$$

where $f_D(v)$ is the normalized quantum emission distribution (fluorescence spectrum) of D, i.e., where

$$\int_0^\infty f_D(v) dv = 1 \tag{11}$$

and $\varepsilon_A(v)$ is the spectrum of the absorption coefficient. The expression must also be multiplied by the quantum yield of the donor in the absence of the acceptor ϕ_D (to take into account all other deactivation pathways that do not lead to emission). After multiplying by ϕ_D and substituting $J(v)$, Eq. (9) becomes

$$R_0 = [9(\ln 10)\kappa^2 c^4 \phi_D \cdot J(v)/(128\pi^5 N_{Av} n^4)]^{1/6} \tag{12}$$

Equation (12) is exactly the expression for R_0 derived by Förster, quantum mechanically (Förster, 1948) and classically (Förster 1951).

In Eq. (12) the spectrum variable v is given in s^{-1} units. If instead we use $\bar{v} = v/c$ (in units of cm^{-1}), then

$$R_0 = [9(\ln 10)\kappa^2 \phi_D \cdot J(\bar{v})/(128\pi^5 N_{Av} n^4)]^{1/6} \tag{12'}$$

In Eq. (12') R, R_0, and $\bar{v}$ are in cm units, $\varepsilon_A(\bar{v})$ is in cm^2/mol, and $J(\bar{v})$ is in cm^6/mol units. If these are the units, then $R_0^6 = 8.79 \times 10^{-28}(n^{-4}\phi_D\kappa^2 \cdot J(\bar{v}))$; if $\varepsilon_A(\bar{v})$ is in $cm^{-1} mol^{-1}$ units, then $R_0^6 = 8.79 \times 10^{-25}(n^{-4}\phi_D\kappa^2 \cdot J(\bar{v}))$. Often the spectra are taken in wavelength units; then the overlap integral is

$$J(\bar{v}) = J(\lambda) \equiv \int_0^\infty [\varepsilon_A(\lambda) \cdot \lambda^4] f_D(\lambda) d\lambda \tag{10'}$$

where again

$$\int_0^\infty f_D(\lambda) d\lambda = 1 \tag{11'}$$

This classical derivation is interesting from several points of view. It shows quite clearly that the $1/R^6$ dependence of the rate of FRET arises because the energy transfer process occurs in the nonradiative near-field zone of the oscillating dipole of D (the same applies to Förster's derivation above). There is no effect of a radiated photon from D (of the energy $hv = hc/\lambda$, where $\lambda = 3000$–7000 Å) on an A molecule in the near-field zone of D (e.g., with $R \leq \lambda/100$); this is not compatible with the uncertainty principle. The origin of the constants in the expression for R_0 can also be clearly followed, as well as the origin of κ^2.

Förster later published a second different classical derivation (Förster, 1951) that is somewhat similar to the second derivation above (Eqs. 4–12). He calculated the rate of energy uptake by the classical acceptor dipole (bathed in the oscillating near field of the donor) in terms of the radiation rate of the donor, first for the case where the two classical oscillators are exactly in resonance. Then, by introducing the spectral dispersion of the oscillator strengths of D and A, and correcting for the time that the D oscillator is in the proper resonance condition with the A oscillator, he arrived (after transforming to the measurable spectroscopic variables) at Eq. (12). This is similar to the correction for the finite bandwidth to the time-dependent QM transition rate constant between two solitary states (Atkins, 1983) that leads to Fermi's "golden rule" (see below).

7.10. COMPARING THE TWO CLASSICAL DERIVATIONS WITH EACH OTHER AND WITH THE QM DERIVATION

Many aspects of the two classical derivations are similar, e.g., the use of the near field of the dipoles; the use of oscillating classical dipoles rather than the coulomb interaction between the D and A molecules that leads quantum mechanically, in the dipole approximation, to the presence of the absorption and emission transition moments in the expression for k_T. The equations for the electric field components of an oscillating dipole radiation are used to derive the perturbation interaction. However, some aspects are quite different. Förster emphasizes that the transfer is a classical resonance phenomenon; without the resonance condition, he could not arrive at the $1/R^6$ dependence of k_T. Kuhn maintains that this is a misnomer because there is no explicit need to invoke resonance in his derivation in order to arrive at the $1/R^6$ dependence (see, however, footnote 6 concerning the term *resonance*). The resonance that Förster discusses results from the requirement that the two broad-spectrum oscillators have simultaneously the same energies (frequencies) in order to exchange energy effectively by resonance. In the derivations, the overlap integral is introduced only to account for the requirement that the frequency ranges of the two spectra overlap, so that it is possible for the two oscillators to exchange energy. In Förster's original derivation, forced energy uptake by the acceptor through the interaction with the donor is separated into two parts, leading eventually to the $1/R^6$ dependence—first, in the calculation of the time for resonance energy exchange, $\tau_{int} \approx h/E_{int}$ and, second, in calculating the probability of the acceptor's having the same narrow band of frequencies in resonance with the donor oscillation (see above).

So that the interested reader can compare the classical and QM derivations, we briefly summarize the major points of the QM derivation. In the QM derivation of FRET, it is convenient to express the rate of the transition between the two states

$$(D^*, A) \xrightarrow{k_T} (D, A^*)$$

according to Fermi's "golden rule" (Atkins, 1983; Schiff, 1968), which is $k_T = (2\pi/\hbar)\beta^2 \cdot \rho(E)$, where β is the interaction matrix (in the case of FRET $\beta \propto 1/R^3$, a dipole–dipole interaction). It is assumed that the transfer is slow compared to the relaxation of the vibrational excitations, so that the energy is not transferred from the initially vibrationally excited states of D^*; therefore, a rate equation can be used. It is also assumed that the interaction between D and A is very weak, so that their individual molecular wavefunctions are not perturbed. The spectra are those of the individual D and A species. The interaction matrix $\beta = \langle \Psi_{D^*A} | H' | \Psi_{DA^*} \rangle$ expresses the coupling of the two states with wavefunctions representing Ψ_{D^*A} and Ψ_{DA^*} by the dipole operator $H' = \kappa |\mu_D||\mu_A|/R^3$, where $|\mu_D|$ and $|\mu_A|$ are the magnitudes of the transition dipoles; κ is the orientation factor between the two dipoles and is given in Eq. (6). The electronic and vibrational parts of the wavefunctions can be separated, leading to the expression

$$k_T = C(2\pi/\hbar) \cdot \rho(E) \cdot \kappa^2 (|\mu_D||\mu_A|/R^3)^2 (\langle \chi_{D^*} | \chi_D \rangle \langle \chi_{A^*} | \chi_A \rangle)^2$$

The last term of the sum in the equation is the product of the Frank Condon factors. This expression for k_T is for a pair of narrowband states. By appropriately transforming the variables and summing over all the resonance states, this expression for k_T can be shown to be equal to Eq. (1) with R_0 as given in Eq. (12) (Dexter, 1953; Dow, 1968; Förster, 1948, 1951, 1965). Also, in the QM derivation, the overlap integral is calculated from the transformation of the transition dipole–dipole interaction term from a "state representation" to an "energy representation" that is then transformed into the experimental variables of an absorption coefficient and a quantum yield with overlapping spectral distributions. The second classical derivation above is more closely related to the QM derivation than the original Förster classical derivation (Förster, 1946).

There is an important point that can be seen from the QM derivation: k_T depends directly on the magnitudes of the transition dipoles of D and A, $|\mu_D|$ and $|\mu_A|$; therefore, it is dependent on the selection rules of these spectroscopic transitions. Looking at Eqs. (1) and (12), we see that if either of the transitions is not fully allowed, the rate of the transfer will be smaller than if both of the transitions are fully allowed. However, the relative transfer efficiency (i.e., the number of transfers per excitation event during the excited lifetime of D^*) depends on the product $k_T \cdot \tau_D$; thus, the donor transition can even be spin forbidden, and the transfer efficiency can still be close to 1 because τ_D is then very long. If, on the other hand, the acceptor is either symmetry or spin forbidden, both the rate and the transfer efficiency will be too small to measure (Lamola, 1969). Thus

$$D^*(triplet) + A(singlet) \rightarrow D(singlet) + A^*(singlet)$$

and

$$D^*(triplet) + A(triplet) \rightarrow D(singlet) + A^*(triplet)$$

are just as *efficient* as

$$D^*(singlet) + A(singlet) \rightarrow D(singlet) + A^*(singlet)$$

and

$$D^*(singlet) + A(triplet) \rightarrow D(singlet) + A^*(triplet)$$

even though the rates of transfer for the first two cases are very small. This can be seen using very long-lived lanthanides ($\tau_D \sim 1$ ms) as donors (Selvin and Hearst, 1994; Selvin et al., 1994). Lanthanides (Kwiatkowski et al., 1994; Saha et al., 1993; Selvin, 1995; Selvin and Hearst, 1994; Selvin et al., 1994) will most certainly play important roles as donors in future FRET microscopy experiments, as will other long-lived donors, because the more rapidly decaying background fluorescence can be essentially completely suppressed.

7.11. COMPETITION BETWEEN DIFFERENT DEACTIVATION PATHWAYS FROM THE EXCITED D* STATE

Due to the heterogeneous environment within many biological samples, it is quite possible that rates of deactivation corresponding to the different physical and chemical pathways of deactivation from the D* state are differentially affected at different locations of a sample. This will complicate the interpretation of a quantitative FRET image. The following is a discussion of a simple formalism used widely in spectroscopy to analyze such situations. We will see that one can sometimes take advantage of these "extra" pathways of deactivation to determine the efficiency of FRET.

An excited D* molecule can become deactivated in parallel via several pathways; all pathways of deactivation can influence the efficiency of FRET. The common pathways of deactivation are represented in Eq. (13) by their corresponding rate constants:

$$D^* \xrightarrow{\ k_F, k_{ic}, k_{isc}, k_Q, k_{pb}, k_T\ } D \tag{13}$$

where k_F = fundamental radiative emission rate; k_{ic} = rate of internal conversion; k_{isc} = rate of intersystem crossing (e.g., from the singlet to the triplet or vice versa); k_Q = quenching rate (often a pseudo-monomolecular rate constant; k_{pb} = rate of photodestruction (photobleaching); and k_T = rate of FRET. The total rate constant for a molecule to leave an excited state is equal to the sum of the individual rate constants of all possible pathways for leaving the excited state (i.e., $\sum_i k_i$ from Eq. 13). The individual rate constants of all the deactivation pathways are independent of each other (under any particular condition). The statistical processes applying to single molecules are interpreted in terms of macroscopic rate constants of deactivation on the measured macroscopic ensemble. The rate of deactivation via any particular pathway is

(number of excited molecules per second following a particular deactivation pathway i)

= (the separate individual rate constant of this particular pathway, k_i)

$$\times \left(\text{the number of excited molecules, i.e., } \int_{\substack{\text{illuminated}\\ \text{volume}}} [D^*] \, dV \right) \qquad (14)$$

Equation (14) is valid for any individual deactivation pathway.

The rate constants of deactivation in Eq. (14) can be measured directly in fast dynamic (nanosecond) experiments or the efficiency (i.e., quantum yield; see Eq. 2) of one of the processes can be measured in the presence and absence of another process that competes with it (from the D* state). In the latter method, we can measure steady-state signals using continuous illumination, or we can monitor the steady-state concentration of [D] or the concentration of some product of a photoreactive pathway away from the excited state D*. The quantum yield of any particular process is expressed identically to Eqs. (2) and (2′), where we simply replace the energy transfer term by the equivalent terms for the process in question (E is the quantum yield of FRET). One simply forms the ratio either of the quantum yield of a monitored process (or some quantity proportional to it) or of the rate of change of a steady-state concentration, in the presence of another competing process to that in its absence. Then, by writing the ratio in terms of the rates, an unknown rate constant can be determined in terms of the other rate constants.

As an example, assume that we have N processes, each with a rate constant of k_i, and we can measure the quantum yield of process m:

$$\phi_m = k_m \left/ \sum_1^N k_i \right. \qquad (15)$$

We want to know the rate constant of a process p, k_p. We can measure the quantum yield (or a quantity proportional to it) of process m in the presence

$$\phi_{m,+p} = k_m \left/ \left(k_p + \sum_{i \neq p} k_i \right) \right. \qquad (16)$$

and in the absence

$$\phi_{m,-p} = k_m \left/ \sum_{i \neq p} k_i \right. \qquad (16')$$

of process p. We then form the following ratio:

$$\frac{\phi_{m,-p}}{\phi_{m,+p}} = \frac{k_m/\sum_{i \neq p} k_i}{k_m/(k_p + \sum_{i \neq p} k_i)} = \frac{(k_p + \sum_{i \neq p} k_i)}{\sum_{i \neq p} k_i} = 1 + \frac{k_p}{\sum_{i \neq p} k_i} \qquad (17)$$

If we know $\sum_{i \neq p} k_i$ (e.g., a lifetime measurement in the absence of process p; see below), we can determine k_p using Eq. (17). This method is very common in photochemistry and fluorescence (Lamola, 1969); examples are the Stern–Volmer plot of determine quenching constants (Lakowicz, 1983) (where $k_p \equiv k_q = k'_q[Q]$ and $\tau_{-p} = 1/\sum_{i \neq p} k_i$ = fluorescence lifetime without quencher), triplet sensitization of photochemical reactions (Lamola, 1969), and sensitized phosphorescence.

We also easily see that

$$\frac{\phi_{m,+p}}{\phi_{m,-p}} = \frac{\sum_{i \neq p} k_i}{(k_p + \sum_{i \neq p} k_i)} = \frac{(k_p + \sum_{i \neq p} k_i) - k_p}{(k_p + \sum_{i \neq p} k_i)} = 1 - \phi_p \qquad (18)$$

Equation (18) says that we can measure the quantum yield of process p by measuring the quantum yield of process m with and without p; we can also use the rates (or, rather, lifetimes) of process m in Eq. (18) because $\phi_m = k_m \tau_m$, and $1/\tau_m$ is the rate of process m that we measure. Equations (17) and (18) are very useful for FRET (especially in the microscope), because they provide a way to measure the efficiency of FRET indirectly by following quantitatively another process that may be easier to measure than measuring FRET directly. It should be realized that the process m that we measure does not have to involve fluorescence in order to measure FRET efficiencies.

We illustrate the use of Eq. (14) in more detail for the case of FRET to show how the steady-state fluorescence intensity and the fluorescence lifetime of a donor decrease in the presence of an acceptor. This is the most common way to measure E of FRET (Eq. 3). We assume normal illumination conditions such that $[D^*] \ll [D]$, where $[D^*]$ = number of excited molecules per unit volume.

In the presence of an acceptor, the total rate of deactivation of the donor (i.e., the number of D* molecules disappearing per second per unit volume) is

$$d[D^*_{+A}]/dt = -(k'_F + k_T)[D^*_{+A}]$$

where $k'_F = \sum_{i \neq T} k_i = (1/\tau_D)$; $(1/\tau_D)$ is the rate without FRET, and k_T is given by Eq. (1). In the absence of energy transfer, the rate of deactivation of the donor is $d[D^*_{-A}]/dt = -k'_F[D^*_{-A}]$. Assuming a short pulse of light, we can integrate these simple rate equations to calculate the time dependence of [D*]:

$$[D^*_{+A}] = [D^{*0}_{+A}]\exp(-(k'_F + k_T)t) \quad \text{and} \quad [D^*_{-A}] = [D^{*0}_{-A}]\exp(-k'_F t)$$

the superscript 0 refers to the zero time quantity. If in both cases the total concentrations of $[D_{+A}]$ and $[D_{-A}]$ are equal, and if the excitation intensities are the same, then $[D^{*0}_{+A}] = [D^{*0}_{-A}]$ because the absorption coefficients of $[D_{+A}]$ and $[D_{-A}]$ are equal (if they are not, the theory of Förster transfer cannot be applied). The steady-state concentration of D* for a constant light source (equivalent to many continuously applied, closely spaced, short light pulses) is proportional to the time integral of a one-pulse experiment, i.e., $[D^*]_{ss} \propto \int [D^*] dt$; from the integrals, we can easily derive $[D^*_{+A}]_{ss}/[D^*_{-A}]_{ss} = k'_F/(k'_F + k_T)$. According to Eq. (14),

$$[d[(D^*)]/dt]_F = -k_F[D^*] \Rightarrow [d[D^*]_{ss}/dt]_F = -k_F[D^*]_{ss}$$
$$\propto (\text{the fluorescence intensity, } F_{ss})$$

Here the subscript F in the term $[d[D^*]/dt]_F$ (also for k_F without the prime) refers to the deactivation process of radiative emission alone; $[d[D^*]_{ss}/dt]_F$ is to be understood as a flux, i.e., as the number of D* molecules undergoing deactivation by emitting photons per unit time at steady state—this is the intensity of the steady-state fluorescence, F_{ss}; the steady-state concentration of D*, $[D^*]_{ss}$, does not actually change. Therefore, using the above relations, the ratio of the steady-state fluorescence intensities with and without acceptor is (see Eq. 18)

$$\frac{F_{ss, +A}}{F_{ss, -A}} = \frac{k'_F}{(k'_F + k_T)} = \frac{\tau_{D, +A}}{\tau_{D, -A}} = 1 - E \tag{19}$$

where we see the connection between the static and the dynamic experiment evaluations of E; $\tau_{D, +A}$ and $\tau_{D, -A}$ are the fluorescence lifetimes of D with and without A; $k'_F/(k'_F + k_T) \leq 1$, so we see that the steady-state fluorescence of D and the lifetime of fluorescence (the inverse of the rate constants) decrease in the presence of FRET. Equation (19) demonstrates two of the common ways to measure E: by measuring either the steady-state fluorescence intensities of the donor, $\pm A$ (and correcting for differences in [D] in the two experiments), or by measuring the fluorescence decay of D, $\pm A$ (actually, in case of multiple lifetimes, the mean fluorescence decay times are usually used). Of course, the extent of D and A labeling or their relative concentrations must be taken into account.

All deactivation processes (quenching, FRET, photochemical destruction), in addition to the fundamental molecular radiative process, lead to a decrease in the fluorescence quantum yield because fewer absorbed quanta will be emitted as photons. The steady-state rate of photon emission (intensity of fluorescence) decreases in the presence of additional deactivation processes because [D*] is less; on the other hand, the total rate constant for leaving the excited state (this is the measured rate of fluorescence decay) is equal to the sum of all deactivation rate constants and increases for every additional deactivation process. This increase in the rate constant of deactivation corresponds to a decrease in the measured fluorescence lifetime (see Eq. 19).

7.12. THE INTERACTION BETWEEN FRET AND DYNAMIC QUENCHING

Förster (1949a,b) showed, by steady-state dynamic quenching experiments (before he could measure lifetimes directly), that FRET was a dynamic process rather than being due to the formation of a static complex between D and A. The decrease in the steady-state fluorescence of a donor by a dynamic quencher—measured by increasing the concentration of a collisional quencher—is diminished in the presence of an acceptor. This is usually done by making a Stern–Volmer plot to determine the quenching rate k_Q, see Eq. (7). k_Q is usually equal to a real rate constant k'_Q times the quencher concentration, i.e., $k_Q = k'_Q[Q]$; but $[Q] \gg [D^*]$, so k_Q is essentially constant. Using the same reasoning for Eqs. (13–19) and assuming only dynamic processes of deactivation, the ratio of the steady-state fluorescence intensities of a donor with and without an acceptor, both in the presence of different concentrations of the quencher, is [see Eqs. (18) and (19)]

$$\frac{F_{ss,+A}}{F_{ss,-A}} = \frac{(k'_F + k_Q)}{(k'_F + k_Q + k_T)} = 1 - E \qquad (20)$$

Förster measured quenching curves of D by increasing the concentration of the quencher [Q], thereby increasing k_Q, and compared the fluorescence intensity curves with and without the acceptor. Since he knew from viscosity studies that quenching was truly a dynamic process, he could conclude from the [Q] dependence of Eq. (20) that FRET reduced the fluorescence of D by a competitive dynamic rate effect. A static complex between D and A would not have changed the measured quenching curve for D. Förster did not take into account that k_T (in the denominator of Eq. 20) is also reduced by decreasing the quantum yield of the donor due to quenching, but this was a secondary effect in his measurements.

Quenching can be used to measure the efficiency of FRET (by varying the concentration of a quencher) if allowance is made for the effect of donor quenching on R_0 (Gryczynski et al., 1988). The method is based on the modulation of R_0 via the quantum yield ϕ_D (see Eq. 12) by varying the concentration of a quencher. This method is especially useful for investigating distributions in R; the distance distribution can be determined by fitting the dependence of the FRET efficiency on a variable R_0 [see Eqs. (43) and (44) and the references in the corresponding paragraph]. This quenching method has been employed recently to detect the flexible motions and conformations of IgE and IgG1 bound and unbound to their receptors (Zheng et al., 1991, 1992). The measurement of FRET can be modulated significantly by the presence of quenchers, and this could have measurable effects on the estimated values of E, as shown in Eq. (20). In a heterogeneous environment such as a biological cell, quenchers can have disparate concentrations, and there are a variety of potential quenchers at different locations. This is important to keep in mind when interpreting FRET microscopy images in complex biological objects.

7.13. THE EFFECT OF FRET ON THE RATE OF PHOTODESTRUCTION (PHOTOBLEACHING)—pbFRET

In the above examples, we have assumed in the integration of the kinetic equations that all the excited molecules eventually return to the ground state. If there are deactivation processes leading to the irreversible destruction of the D species, the number of D molecules will decrease. The change in the total concentration of [D] due to irreversible destruction of the D* molecules proceeds orders of magnitude more slowly than the nanosecond processes that compete with fluorescence deactivation because the photodestruction of D proceeds only from the minuscule population of molecules in the D* state. This slow rate of photobleaching is very useful for measuring FRET in biological samples, as has been effectively demonstrated by Jovin and Arndt-Jovin (1989a,b). $[D^*] \ll [D]$ for normal illumination conditions; for a 1 ns lifetime and an absorption constant of $10^5 \, \text{cm}^{-1} \cdot \text{mol}^{-1}$, $[D^*]/[D] \approx 10^{-10}$. On the other hand, the rate of irreversible photodestruction of the D* molecules competes quite effectively on the nanosecond time scale with the

other deactivation processes such as fluorescence and quenching. Each D*
molecule of the excited population follows a certain deactivation pathway,
producing a photon, a ground state D molecule, or a photodestructed
D species—all on the nanosecond time scale. However, due to the enormous
pool of ground state D molecules, the population of D* molecules that are
deactivated is continuously replenished, also on the nanosecond time scale,
and the fluorescence remains approximately constant over many cycles of
excitation. The $[D^*]_{ss}$ concentration may be reduced significantly from its
level in the absence of the photodestructive pathway, just as quenching or
FRET deactivation pathways also decrease $[D^*]_{ss}$. But the large change in the
total population of D molecules due to photodestruction will still be slow. This
slow rate of photodestruction of D will be reduced in the presence of any
additional path of deactivation from D*, just as the kinetically based concentration quenching curve (Stern–Volmer plot; Eq. 17) of the previous paragraph shows a decreased rate of quenching in the presence of FRET. In
particular, the effect of FRET on the slow photodestruction rate can be
described by the ratio of the rates of photodestruction of D (the slow process)
with and without FRET with an expression similar to Eqs. (18) and (20) (Jovin
and Arndt-Jovin, 1989a; Kuhn et al., 1972); assuming that the concentration of
$[D]$ is the same $\pm A$:

$$\frac{[d[D]/dt]_{pb,+A}}{[d[D]/dt]_{pb,-A}} = \frac{(k'_F + k_{pb})}{(k'_F + k_{pb} + k_T)} = 1 - E \qquad (21)$$

where the subscript pb refers to the term *photobleaching*. It should be
remembered that k_{pb} is on the order of ns^{-1} and $[d[D]/dt]_{pb}/[D]$ is on the
order of s^{-1}. Equation (21) is valid only if $[D]$ is the same with and without A.
If not, the ratio must be appropriately corrected (see Eq. 22). This method is
often referred to as *donor*-pbFRET and is a convenient method for imaging
distributions of FRET in the microscope (Jovin and Arndt-Jovin, 1989a;
Kubitscheck et al., 1991, 1993; Szabò et al., 1992; Young et al., 1994). It is
a kinetic measurement and therefore is insensitive to absolute signals. Only
relatively simple digital imaging instrumentation is required, and the time
constants to be compared are in the second range and easy to measure
accurately. The disadvantage is that two photobleaching imaging measurements must be compared, usually on different samples, over a longer time, and
the positions within a complex object must be in register; this may be difficult,
especially for nonfixed samples.

It has been shown that the total integrated number of photons emitted from
a dye molecule until photodestruction ensues, $[N_{pb,tot}]_{t \to \infty}$, is independent of
the quantum yield, the length of time that the sample is illuminated, and the
absorption coefficient of the dye (Hirschfeld, 1976). In this paper, the rate of

photodestruction $[d[D]/dt]_{pb}$ was shown to be equal to the ratio of the rate of photons emitted by D (proportional to the fluorescence intensity) at the initial time, $[dN_{pb}/dt]_{t \to 0}$ to $[N_{pb,tot}]_{t \to \infty}$. Measuring these quantities in the presence and absence of FRET, and taking their ratios, we see from Eq. (21) that another expression for determining E from photodestruction data can be derived:

$$\alpha_{\pm A}\{[d[D]/dt]_{pb+A}/[d[D]/dt]_{pb-A}\}_{t\to 0} = F_{ss,+A}/F_{ss,-A} = \tau_{D,+A}/\tau_{D,-A} = 1 - E$$
$$= \{[dN_{pb}/dt]_{t\to 0}/[N_{pb,tot}]_{t\to\infty}\}_{+A}/\{[dN_{pb}/dt]_{t\to 0}/[N_{pb,tot}]_{t\to\infty}\}_{-A} \quad (22)$$

Here $\alpha_{\pm A}$ is the correction factor of this ratio if $[D]$ is not equal for the two measurements with and without A; $[N_{pb,tot}]_{t \to \infty}$ is a direct measure of the number of fluorescence molecules and is not influenced by changes in quantum yield or light intensity; therefore, $[N_{pb,tot}]_{t \to \infty}$ is an excellent measurable normalization constant. This avoids the necessity of an independent determination of $[D]$ for the two measurements in order to determine $\alpha_{\pm A}$. Equations (21) and (22) have both been used to calculate E from a microscopic sample using photodestruction (Jovin and Arndt-Jovin, 1989a; Kubitscheck et al., 1991; Szabò et al., 1992).

A recent publication (Young et al., 1994) has employed a different method of analyzing *donor*-pbFRET data; the authors note that distributions of photobleaching rates are usually present, rather than a single component, and conclude that this should be taken into account in the analysis. This issue has also been addressed by an earlier study (Szabò et al., 1992). Young et al. (1994) present a method that essentially weights the analysis in the earlier portion of the bleaching curve (faster components) with multiple exponentials rather than integrating the entire curve (Jovin and Arndt-Jovin, 1989a) or accounting for the pbFRET decay with a single exponential (Kubitscheck et al., 1991, 1993) (although the pbFRET data are well represented by a single exponential in the last two studies after correcting for the photodestruction of the autofluorescence). A similar treatment, fitting the early part of the quenching curve, was also used earlier; see the next paragraph (Barth et al., 1966; Kuhn et al., 1972). With present instrumentation, the entire bleaching process can easily be recorded and analyzed with multiexponentials. The recent reports (Jovin and Arndt-Jovin, 1989a,b; Kubitscheck et al., 1993; Szabò et al., 1992; Young et al., 1994) should be consulted for discussions of data analysis, experimental methods, and other references on pbFRET.

In a series of very early studies (Barth et al., 1966; Drexhage et al., 1963; Kuhn, 1965), the possibility of using photodestruction rates to measure FRET was realized and measured. These authors demonstrated the diminishing rate of photodestruction in the presence of FRET of systems on monolayers with

D and A dyes separated by well-defined distances. They analyzed their data with the expression

$$\alpha_{\pm A}\{[dl[D]/dt]_{\text{pb},+A}/[d[D]/dt]_{\text{pb},-A}\}_{t\to 0}$$
$$= F_{\text{ss},+A}/F_{\text{ss},-A} = \tau_{\text{D},+A}/\tau_{\text{D},-A} = 1 - E$$

[see Eqs. (18–22)] and compared their measured photobleaching data with other determinations of FRET. It is interesting that they measured the amount of photodestruction by making absorption measurements on the monolayers, not by measuring fluorescence; this underscores the point that fluorescence does not have to be measured in order to quantify FRET. The work by this group is innovative and interesting, and many of their methods and ideas could be adapted to biological samples in the microscope; two of their reviews should be consulted for details (Kuhn and Möbius, 1993; Kuhn et al., 1972).

These authors (Barth et al., 1966; Kuhn, 1965) also showed that FRET can be detected and quantified by measuring the photodestruction of the acceptor that has been sensitized by a donor, i.e., *acceptor*-pbFRET as opposed to the *donor*-pbFRET discussed above. The acceptor will also undergo photodestruction because, on accepting the energy from the donor, the acceptor has the same propensity to undergo photodestruction as it would if it were directly excited. An independent account of *acceptor*-pbFRET has recently been published (Mekler, 1994). Mekler discusses the analysis and assumptions, and the technique is demonstrated on FRET measurements using 2,5-bis(5-*tert*-butyl-2-benzoxasolyl)thiophen as a donor and acridine orange as an acceptor. It is claimed that this method allows the measurement of exceptionally accurate efficiencies of 10^{-2} to 10^{-3}!

The effect that the photodestruction of sensitized acceptors has on the *donor*-pbFRET has not yet been considered or taken into account in the measurements of *donor*-pbFRET. If sensitized acceptors can readily be photobleached, the donors closest to acceptors (i.e., donors that have not yet become photolyzed) will bleach more rapidly after their closely neighboring acceptors have been removed by photodestruction. Unfortunately, this effect will be most pronounced for those D–A pairs that have the highest efficiency of FRET. This would be expected to affect especially the analysis of microscope pbFRET measurements if the rate constants for acceptor photodestruction are the same or higher than those for the donor. Effectively, this would lead to an underestimation of the efficiency of FRET and could also lead to artifacts concerning the distribution of FRET efficiency. It is interesting to note, however, that bleaching of the sensitized acceptors would not affect the total number of photons emitted by the donors $[N_{\text{pb,tot}}]_{t\to\infty}$ discussed above (Hirschfeld, 1976) (see Eq. 22).

In an earlier study of photobleaching in digital imaging microscopy (without FRET)(Benson et al., 1985), it was pointed out that the photobleaching rate in individual cells is often spatially very heterogeneous, as it also is for extrinsic probes. These authors pointed out that the appropriate corrections of photobleaching rates must be taken into account to obtain the "zero-time" image correctly, especially the rapidly photobleaching components. This heterogeneity of the rates of photodestruction may be difficult to correct if different cells must be used to obtain the photobleaching data with and without acceptor for pbFRET microscopy, especially if high bleaching rates (high intensities) are employed. At any rate, care must be taken to avoid such heterogeneous bleaching rates, or these rates must be taken into account in the measurement process or the data analysis; a single exponential decay (after correcting for autofluorescence photodestruction) is evidence that the photobleaching rate is homogeneous (Kubitscheck et al., 1991).

7.14. ORIENTATIONAL EFFECTS OF BIOLOGICAL SAMPLES

Biological specimens encountered in fluorescence microscopy are usually complex objects consisting of supramolecular structures and large organizations of associated and aggregated molecular components. Spectroscopic probes that interact with biological material are often oriented and undergo less rotational motion than they do in less organized surroundings. The proximity of immobilized molecular species leads naturally to directional interactions between neighboring molecular species, and the mutual orientation of closely spaced D and A molecules affects the efficiency of FRET through κ^2 variations. κ^2 can also vary considerably throughout the sample. Therefore, one must be especially careful in quantitatively interpreting FRET measurements made on highly structured biological samples, and it is expedient to inquire more carefully into the nature of the κ^2 effects on a FRET measurement.

7.14.1. The Mutual Orientation of D and A Molecules Is an Important Factor in FRET Measurements

The orientation factor κ^2 is often considered the limiting factor for estimating accurate D–A distances from FRET data; κ^2 can vary from 0 to 4, although the actual values of κ^2 rarely, if ever, extend over this total range. In general, we do not have sufficient information (often none at all) regarding D–A orientations, and approximations must be made. This problem has been discussed in detail in the literature. Dale and Eisinger (1974; Dale et al., 1979; see below) have made a detailed analysis enabling the min–max limits of κ^2 to be determined experimentally. They have presented a method to estimate the maximum error in κ^2 values representing our level of ignorance concerning the relative orientations of the D–A transition moments; see the discussion below and Eq. (36). The estimates of possible orientations of D and A are determined from fluorescence anisotropy measurements. The D and A fluorescence anisotropy values, and consequently the range of κ^2 values, depend on the extent to which the D and A molecules rotate while D is in its excited state. A good example of this type of analysis can be found in the recent work of Censullo and coworkers (1992); they have applied it to check the accuracy of determining the geometry of actin filaments from FRET data. There is often disagreement concerning the applicability of certain approximations in specific circumstances. Potential FRET users must acquire sufficient familiarity with the problem to be able to estimate the probable errors and understand how certain approximations affect their interpretations of FRET data.

Often κ^2 is treated superficially in the FRET literature, although excellent general theoretical treatments exist. In the following discussion, the example of a completely randomly oriented acceptor (or donor) molecule is used to demonstrate the magnitude of the effects that κ^2 can have on FRET measurements and to help provide a basic understanding of the physical origins of these effects. No claim is made for new theoretical results that are not already available in the more thorough accounts of κ^2. Rather, the calculations are presented such that the arguments and interpretations can easily be followed by most readers. Although the model treated below is of limited generality, it leads to results that are applicable to many important experimental situations.

7.14.2. The Interaction Between Two Electric Dipoles

It has been shown above that the magnitude of the time-dependent field strength of the near-field zone of an oscillating dipole ($1/R^3$ term, Eqs. 27 and 28) is identical to that of a static dipole. The electric interaction between two electric dipoles, either static or time varying, is a vector dot product between the dipole moment of one dipole with the electric field of the other, and κ is the factor describing their relative orientation. Therefore, no generality is lost by discussing κ in terms of classical static dipoles. The rate of the energy transfer k_T is proportional to κ^2 because k_T is proportional to the square of the interaction energy between D and A.

7.14.3. The Classical Description of Two Interacting Dipoles p_1 and p_2

The interaction energy $U_{p_1 \to p_2}$ between two dipoles, p_1 and p_2, can be expressed as the vector dot product between the electrical field E_1 of p_1 and the moment of the second dipole $\mathbf{p}_2$, i.e., $U_{p_1 \to p_2} = \mathbf{E}_1 \cdot \mathbf{p}_2$; $\mathbf{E}_1$ of the dipole $\mathbf{p}_1$ is depicted in Figure 7.1 ($\mathbf{E}_1$ follows the designated lines of force; the vectors labeled $\hat{\perp}$ are unit vectors in a direction perpendicular to $\mathbf{E}_1$ (shown in the plane of the paper)]. The electrostatic field of a permanent dipole is the gradient of the equipotential surface at any point and can be expressed as follows (Jackson, 1962; see Figures 7.1 and 7.2):

$$\mathbf{E}_1 = \frac{|p_1|}{r^3} \{2\cos\theta_1 \hat{r} + \sin\theta_1 \hat{\theta}_{1 \to 2}\} \tag{23}$$

where the capped symbols are unit vectors defining the coordinate system at the position under consideration (the position of $\mathbf{p}_2$); see Figures 7.1 and 7.2. Note that $\mathbf{r}$ is the distance vector from $\mathbf{p}_1$ to the point of interaction, r is the magnitude of this vector, and $\hat{r}$ is the unit vector of $\mathbf{r}$; $|p_1|$ is the magnitude of $\mathbf{p}_1$; θ_1 is the magnitude of the polar angle between the vectors $\mathbf{p}_1$ and $\mathbf{r}$. The axes of the right-hand Cartesian coordinate system, with its origin at $\mathbf{p}_2$, are defined with reference to the displacement vector $\mathbf{r}$ between the dipoles and the relative orientation of $\mathbf{p}_1$ and $\mathbf{r}$; $(\hat{r}, \hat{\theta}_{1 \to 2}, \hat{\phi}_{1 \to 2})$ are the unit vectors of this axis system. Here $\hat{\theta}_{1 \to 2}$ is perpendicular to $\hat{r}$ and lies in the plane of $\mathbf{p}_1$ and $\hat{r}$; $\hat{\phi}_{1 \to 2}$ is directed out of the page at the location of $\mathbf{p}_2$. In general, $\mathbf{p}_2$ is not in either of the planes in Figure 7.2.

7.14.4.   Some General Conclusions Regarding κ^2

By examining Figures 7.1 and 7.2 together with Eq. (23), many features concerning the dipole orientations, in particular κ^2 of the FRET experiment, become evident. Recalling that the rate of energy transfer is proportional to the square of the interaction energy between the transition dipoles, we have

$$k_T \propto (U_{p_1 \to p_2})^2 = (\mathbf{E}_1 \cdot \mathbf{p}_2)^2 = \frac{|p_1|^2 |p_2|^2}{r^6} \{2\cos\theta_1 \hat{r} \cdot \hat{p}_2 + \sin\theta_1 \hat{\theta}_{1 \to 2} \cdot \hat{p}_2\}^2$$

$$= \frac{|p_1|^2 |p_2|^2}{r^6} \kappa^2 \tag{24}$$

Here κ^2 is the square of the term in the $\{\cdots\}$ braces (see also Eq. 28).
    First, before considering κ^2, we examine the coefficients of κ^2. Just as in the case of Förster energy transfer, the $1/r^6$ dependence arises from the dependence of k_T on the square of the dipole–dipole interaction energy $(U_{p_1 \to p_2})^2$. Quantum mechanically, $(U_{p_1 \to p_2})^2$ corresponds to the square of the "perturba-

tion matrix"[8] defining the rate of energy transfer between two interacting "transition" dipoles of D and A according to Fermi's "golden rule" (Kauzmann, 1957; Schiff, 1968). The rate of energy transfer is proportional to the emission quantum yield of the donor (the QM equivalent of $\propto |p_1|^2$) and the absorption coefficient of the acceptor (the QM equivalent of $\propto |p_2|^2$). The expression relating the dipoles (squared) to the overlap integral is discussed above. The dependence of κ^2 on the relative orientation of the two dipoles (QM—transition dipoles) is clearly demonstrated in Figure 7.1. Wherever $\mathbf{p}_2$ is perpendicular to $\mathbf{E}_1$ (i.e., $\mathbf{E}_1 \cdot \mathbf{p}_2 = 0$), there is no energy transfer, regardless of how close the two dipoles are. The selected examples given in many textbooks and general articles on FRET can be deceptive (although correct); for instance, it is usually pointed out that a dipole with the orientation of the dipole $\mathbf{p}_2^C$ in Figure 7.1 cannot participate in energy transfer; this is obvious from Figure 7.1 because $\mathbf{E}_1 \cdot \mathbf{p}_2 = 0$ (see Eq. 24). Because of cylindrical symmetry, this is of course also true for all $\mathbf{p}_2$ vectors that are perpendicular to $\mathbf{p}_1$ and also lie in the singular plane that includes and is perpendicular to the $\mathbf{p}_1$ vector (see Figure 7.1). In the literature it is often stated that the two transition dipoles must be perpendicular to each other for FRET to vanish. Figure 7.1 shows the fallacy of this statement; the situation is much more general. For every position in space there is an orientation of $\mathbf{p}_2$ where $\mathbf{E}_1 \cdot \mathbf{p}_2 = 0$, so that $\kappa^2 = 0$; however, in most cases, $\mathbf{p}_1$ must not be perpendicular to $\mathbf{p}_2$. This is a trivial observation according to Eq. (24) (Dale and Eisinger, 1974; Förster, 1951), but may not be initially obvious. A similar situation pertains to the orientations of the two dipoles corresponding to maximum FRET efficiencies (maximum κ^2) at any selected separation of $\mathbf{p}_2$ and $\mathbf{p}_1$. The two "classical examples (Förster, 1951) are $\mathbf{p}_2^A$ and $\mathbf{p}_2^B$ in Figure 7.1, where $\kappa^2 = 4$ and 1. But obviously, at every position in space there will be one orientation of $\mathbf{p}_2$ relative to $\mathbf{p}_1$ that will exhibit maximum energy transfer, even though $\mathbf{p}_2$ and $\mathbf{p}_1$ are not parallel. The simple picture in Figure 7.1 of one dipole placed in the electric field of another dipole allows us to envisage easily the relative angular dependence of the dipole–dipole interaction between $\mathbf{p}_1$ and $\mathbf{p}_2$ of transition dipoles at any relative orientation and at any distance between them (see also Steinberg et al., 1983).

7.14.5. Rapid and Complete Random Dynamic Orientation of Either A or D with a Unique Static Orientation of the Complementary Chromophore D or A

We consider the simple example of a dynamically and completely randomly oriented acceptor molecule to demonstrate some general properties of κ^2. For

---

[8]See Sections 7.9.2 and 7.9.3 for a short discussion of the QM treatment of the transition rates between two electronic states of interacting chromophores.

this case we take advantage of the $(\hat{r}, \hat{\theta}_{1\rightarrow2}, \hat{\phi}_{1\rightarrow2})$ coordinate system at $\mathbf{p}_2$ shown in Figure 7.2. We can expand the expression in Eq. (24) to give

$$
\begin{aligned}
\kappa^2 &= \{2\cos\theta_1 \hat{r}\cdot\hat{p}_2 + \sin\theta_1\hat{\theta}_{1\rightarrow2}\cdot\hat{p}_2\}^2 \\
&= \{2\cos\theta_1 \sin\theta_2 \cos\phi_2 + \sin\theta_1 \sin\theta_2 \sin\phi_2\}^2 \qquad (25) \\
&= \{4\cos^2\theta_1 \sin^2\theta_2 \cos^2\phi_2 + 4\cos\theta_1 \sin\theta_1 \sin^2\theta_2 \cos\phi_2 \sin\phi_2 \\
&\quad + \sin^2\theta_1 \sin^2\theta_2 \sin^2\phi_2\}
\end{aligned}
$$

where the spherical angles θ_2 and ϕ_2 are defined within the coordinate system at the position of $\mathbf{p}_2$ shown in Figure 7.2. θ_2 is the angle between $\mathbf{p}_2$ and the $\hat{\phi}_{1\rightarrow2}$ axis, and ϕ_2 is the angle between the projection of $\mathbf{p}_2$ onto the $(\hat{\theta}_{1\rightarrow2}, \hat{r})$ plane; see Figure 7.2.

The assumption for our simple model (and this is also often assumed in the literature) is that $\mathbf{p}_2$ has a spherically uniform angular distribution *at all times* during the process of energy transfer. This means that the orientation of this dipole is rapidly randomized within the time scale of FRET; therefore, in the calculation of κ^2, the orientation of this dipole can be "dynamically averaged" at all times. We choose the acceptor in our example, but it could just as well be the donor; the dipole–dipole interaction is symmetric in the sense of interchanging D and A. This approximation is especially appropriate for the "rapid diffusion" limit of FRET (Mersol et al., 1992; Stryer et al., 1982; Thomas et al., 1978) or the use of degenerate or almost spherically symmetric metal ions (Kirk et al., 1993). κ^2 is averaged over the rotational space of the acceptor, assuming equal probability of every orientation:

$$
\langle \kappa^2 \rangle_d = \frac{1}{4\pi} \int_0^{2\pi} \int_0^{\pi} \{2\cos\theta_1 \hat{r}\cdot\hat{p}_2 + \sin\theta_1\hat{\theta}_{1\rightarrow2}\cdot\hat{p}_2\}^2 \sin\theta_2 \, d\theta_2 \, d\phi_2 \qquad (26)
$$

where $\sin\theta_2 \, d\theta_2 \, d\phi_2$ is the surface element on a sphere of radius 1 appropriate for integrating in spherical coordinates, and the 4π in the denominator results from the normalization $\int_0^{2\pi}\int_0^{\pi} \sin\theta_2 \, d\theta_2 \, d\phi_2 = 4\pi$. Since in our example the distribution over the angles θ_2 and ϕ_2 is assumed to be uniform at all times, the integration over θ_2 and ϕ_2 is independent of the values of θ_1 (and r). Substituting Eq. (25) in Eq. (26) and carrying out the integrations results in

$$
\langle \kappa^2 \rangle_d = \{\tfrac{4}{3}\cos^2\theta_1 + \tfrac{1}{3}\sin^2\theta_1\} = \{\cos^2\theta_1 + \tfrac{1}{3}\} \qquad (27)
$$

Equation (27) is the correct expression for calculating the effective κ^2 for the case of a dynamically averaged, randomly oriented $\mathbf{p}_2$ and a fixed $\mathbf{p}_1$. The value of $\langle \kappa^2 \rangle_d$ depends in a very simple way on θ_1. Even though $\mathbf{p}_2$ is completely randomly oriented, $\langle \kappa^2 \rangle_d$ in Eq. (27) has a maximum of 4/3 at $\theta_1 = 0$ and a minimum of 1/3 at $\theta_1 = \pi/2$ (Dale and Eisinger, 1974; Stryer, 1978). The completely random dynamic orientation of $\mathbf{p}_2$ has reduced the difference between the κ^2 min–max value pertaining to two fixed dipoles, Eq. (24), from 0 (at $\theta_1 = \pi/2$) and 4 (at $\theta_1 = 0$) to 1/3 (at $\theta_1 = \pi/2$) and 4/3 (at $\theta_1 = 0$); however, the averaged value $\langle \kappa^2 \rangle_d$ still depends strongly on θ_1 (but it is independent of R). Note that $\langle \kappa^2 \rangle_d = 2/3$ when $\theta_1 = 54.74°$; this angle is called the magic angle in fluorescence anisotropy (see below); $\langle \kappa^2 \rangle_d = 2/3$ also for the case where both D and A are dynamically randomly oriented (see the next section).

7.14.6. The Effect of a Distribution of Orientations of the Complementary FRET Chromophore (D in Figure 7.2)

We can easily average the value of $\langle \kappa^2 \rangle_d$ over any specific distribution of angles $\theta_1, \phi_1, f(\theta_1, \phi_1)$. A general formula for this is

$$\langle \kappa^2 \rangle_d^{\langle \theta_1, \phi_1 \rangle} = \frac{\displaystyle\int_{\phi_{1,\min}}^{\phi_{1,\max}} \int_{\theta_{1,\min}}^{\theta_{1,\max}} f(\theta_1, \phi_1) \left\{ \cos^2\theta_1 + \frac{1}{3} \right\} \sin\theta_1 \, d\theta_1 \, d\phi_1}{\displaystyle\int_{\phi_{1,\min}}^{\phi_{1,\max}} \int_{\theta_{1,\min}}^{\theta_{1,\max}} f(\theta_1, \phi_1) \sin\theta_1 \, d\theta_1 \, d\phi_1} \tag{28}$$

If there is a uniformly random angular distribution of $\mathbf{p}_1$, then $f(\theta_1, \phi_1) = 1$ for all θ_1 and ϕ_1. In this case

$$\int_0^{2\pi} \int_0^{\pi} f(\theta_1, \phi_1) \sin\theta_1 \, d\theta_1 \, d\phi_1 = 4\pi \tag{29}$$

and the integration over $d\theta_1 \, d\phi_1$ in Eq. (28) results in $\langle \kappa^2 \rangle_d^{\langle \theta_1, \phi_1 \rangle} = 2/3$. This case of "dynamically rapid averaging" of both the donor and the acceptor is the customary assumption in the FRET literature.

This assumption that $\langle \kappa^2 \rangle_d^{\langle \theta_1, \phi_1 \rangle} = 2/3$ is often made for two reasons: (i) sufficient orientation information is lacking, and one hopes that the relative orientations of the transition dipoles of D and A do not have a significant effect (however, the possible error in this assumption can be partially quantified; see below); and (ii) subsequent FRET efficiency calculations and distance estimates are usually greatly simplified if κ^2 is assumed to be constant for all D–A pairs. When dealing with distributions of distances between D and A, this assumption is almost always made, usually because the calculations would otherwise by very complex (Eis and Millar, 1993; Lakowicz et al., 1991a,b; Wu and Brand, 1992) (but see van der Meer et al., 1993). The term "dynamically rapid" means that the angular distributions of both D and A are implicitly assumed to be uniformly distributed at all times; this is only possible if both the D and A molecules achieve isotropic rotational equilibrium in a time much shorter than the emission lifetime of the donor. The important point is that both D and A must rotate rapidly for the rigorous limit of $\langle \kappa^2 \rangle_d^{\langle \theta_1, \phi_1 \rangle} = 2/3$ to hold. This dynamic averaging is not, however, required in order that $\kappa^2 = 2/3$, as can be easily seen from Eqs. (25), (27), and (33) and in the discussion below. It has been emphasized (Haas and Katchalski-Katzir, 1978; Stryer, 1978) that $\langle \kappa^2 \rangle_d^{\langle \theta_1, \phi_1 \rangle} = 2/3$ is often a reasonable approximation even if the D and A molecules do not have total orientational randomization, due to mixed spectroscopic transitions.

The error in calculating an R_0 value can be quite small, even though κ^2 varies strongly throughout the sample. For instance, in Eq. (27) the min–max values for $\langle \kappa^2 \rangle_d$ are $1/3 < \langle \kappa^2 \rangle_d < 4/3$; the corresponding min–max errors in the R_0 values (remembering that $R_0 \propto (\kappa^2)^{1/6}$) are $0.88 R_0^{\langle 2/3 \rangle} < R_0^{\langle 2/3 \rangle} < 1.12 R_0^{\langle 2/3 \rangle}$—where $R_0^{\langle 2/3 \rangle}$ is the value of R_0 calculated assuming that $\langle \kappa^2 \rangle_d = \langle \kappa^2 \rangle_d^{\langle \theta_1, \phi_1 \rangle} = 2/3$. The 12% error for this case is a *maximum* possible error in R_0 (Dale and Eisinger, 1974; Dale et al., 1979; Stryer, 1978), and this error is often well within the accuracy that one can achieve in determining R_0. We see that a $\pm 100\%$ error in the estimated value of κ^2 leads to only a 12% maximum error in the value of R_0. This does not mean that such a small variation in R_0 cannot be observed in certain cases (E is still a strong function of κ^2). But unless we have very reliable information concerning the angular distributions of D and A, it is often impossible to make more reliable, accurate estimates of distances.

The min–max values of Eq. (27) refer only to the cases $\theta_1 = \pi/2$ and $\theta_1 = 0$; however, very seldom is θ_1 exactly either of these extreme values. Usually θ_1 is distributed over a range of values. The question arises concerning the effect of variable extents of disorientation of $\mathbf{p}_1$ on the values of $\langle \kappa^2 \rangle_d$. The effect of a distribution of θ_1 values on the extreme values and averages of κ^2 can be nicely demonstrated with our simple model.

7.14.7. Angular Variation in $\mathbf{p}_1$

We assume that $\mathbf{p}_1$ can rotate rapidly and freely within the constraints of a cone, and we calculate the average $\langle \kappa^2 \rangle_d^{\langle \theta_1, \phi_1 \rangle}$ (Figure 7.2) (Dale et al., 1979). Equation (28) can be easily integrated over a nonuniform distribution of the donor by using the coordinate system in Figure 7.2: $\mathbf{p}_1^0$ is the singular $\mathbf{p}_1$ at the center of the cone (the direction of $\mathbf{p}_1^0$ is shown as a dotted line); it has a polar angle θ_1^0 relative to the $\mathbf{r}$ vector. The polar orientation of any $\mathbf{p}_1$ relative to $\mathbf{p}_1^0$ *inside* the cone can be described by the polar angle ψ. The maximum angular deviation of $\mathbf{p}_1$ from $\mathbf{p}_1^0$ in the cone is ψ_M. We define a coordinate system such that the z-axis is the $\mathbf{p}_1^0$ vector, the z- and y-axes define a plane containing both

p_1^0 and r; χ is the azimuthal angle of $\mathbf{p}_1$ about $\mathbf{p}_1^0$. With these definitions, the orientations of $\mathbf{r}$ and any $\mathbf{p}_1$ can be defined Eq. (30):

$$\hat{r} = \cos\theta_1^0 \hat{z} + \sin\theta_1^0 \hat{y}$$
$$\hat{p}_1 = \cos\psi\hat{z} + \sin\psi\sin\chi\hat{y} + \sin\psi\cos\chi\hat{x} \tag{30}$$

In terms of these angles (see Figure 7.2) we can write

$$\langle\kappa^2\rangle_d = \{\cos^2\theta_1 + \tfrac{1}{3}\} = \{(\hat{r}\cdot\hat{p}_1)^2 + \tfrac{1}{3}\} \tag{31}$$

where

$$(\hat{r}\cdot\hat{p}_1)^2 = \cos^2\theta_1^0\cos^2\psi + 2\cos\theta_1^0\sin\theta_1^0\cos\psi\sin\psi\sin\chi + \sin^2\theta_1^0\sin^2\psi\sin^2\chi \tag{31'}$$

This expression must be integrated and normalized according to Eq. (28).We assume that $\hat{p}_1$ is *uniformly* distributed in the cone. The integration in Eq. (28) can be performed easily in terms of the coordinate variables (θ_1^0, ψ, χ):

$$\langle\kappa^2\rangle_d^{\langle\theta_1,\phi_1\rangle} = \langle\kappa^2\rangle_d^{\langle\psi_M,\chi\rangle} = \frac{\displaystyle\int_0^{2\pi}\int_0^{\psi_M}\left\{(\hat{r}\cdot\hat{p}_1)^2 + \frac{1}{3}\right\}\sin\psi\,d\psi\,d\chi}{\displaystyle\int_0^{2\pi}\int_0^{\psi_M}\sin\psi\,d\psi\,d\chi} \tag{32}$$

$\sin\psi\,d\psi\,d\chi$ is again the surface element on a sphere of radius 1 needed for the integration using spherical variables. Substituting Eq. (31') into Eq. (32) and carrying out the integration over χ leads to Eq. (33):

$$\langle\kappa^2\rangle_d^{\langle\psi_M,\chi\rangle} = \frac{\cos^2\theta_1^0\displaystyle\int_0^{\psi_M}\cos^2\psi\sin\psi\,d\psi}{\displaystyle\int_0^{\psi_M}\sin\psi\,d\psi} + \frac{\sin^2\theta_1^0\displaystyle\int_0^{\psi_M}\sin^3\psi\,d\psi}{2\displaystyle\int_0^{\psi_M}\sin\psi\,d\psi} + \frac{1}{3} \tag{33}$$

Equation (33) is a simple and very useful expression.[9] It is the appropriate value of κ^2 to use when the acceptor is completely dynamically rotationally

---

[9] The analytical solutions of the integral can be written as

$$\int_0^{\psi_M}\sin\psi\,d\psi = 1 - \cos(\psi_M)$$

$$\int_0^{\psi_M}\cos^2\psi\sin\psi\,d\psi = \{1 - (\cos^3(\psi_M))\}/3$$

$$\int_0^{\psi_M}\sin^3\psi\,d\psi = \{2 + (\cos^2(\psi_M) - 3)\cos(\psi_M)\}/3$$

disoriented, and the donor is rapidly and uniformly distributed in a cone with a maximum half-cone angle of ψ_M, where the center of the donor cone of rotational freedom is oriented at an angle of θ_1^0 to $\mathbf{r}$. Of course, the same equations apply if the rotational characteristics of the acceptor and the donor are exchanged, that is, complete disorientation of the donor and partial (in a cone) disorientation of the acceptor.

7.14.8. Numerical Evaluation of $\langle \kappa^2 \rangle_d^{\langle \psi_M, \chi \rangle}$

Figure 7.3 represents $\langle \kappa^2 \rangle_d^{\langle \psi_M, \chi \rangle}$ as a function of θ_1^0 (abscissa) for different values of half angles of the cone, $\psi_M = 0°$ to $\psi_M = 90°$, in 5° steps. Also shown is the θ_1^0 dependence of the ratio of the calculated R_0 value to the R_0 value that would pertain if $\kappa^2 = 2/3$. This is proportional to $(\kappa^2/(2/3))^{1/6}$. The angle at which all curves intersect is $\theta_1^0 = 54.74°$. We have already seen (see Eq. 27) that even with no rotational freedom of the donor, $\langle \kappa^2 \rangle_d = 2/3$, when $\theta_1 = 54.74°$. Now we see that also for the case where $\mathbf{p}_1$ rotates freely in a cone, $\langle \kappa^2 \rangle_d^{\langle \theta_1, \phi_1 \rangle} = 2/3$, when $\theta_1^0 = 54.74°$, regardless of the maximum extent of wobble of $\mathbf{p}_1$ (i.e., the value of ψ_M) within the cone. This latter fact results from the symmetry of the distribution within the cone. The angle of 54.74° is the magic angle encountered in fluorescence polarization.[10] Figure 7.3 should be compared to Figure 20 (right side) of Dale and Eisinger (1974).

As expected, the variation (i.e., the min–max values) in $\langle \kappa^2 \rangle_d^{\langle \psi_M, \chi \rangle}$ decreases as the cone angle ψ_M increases. At $\psi_M = 90°$ the dipole is completely randomly oriented, and $\langle \kappa^2 \rangle_d^{\langle \psi_M, \chi \rangle} = 2/3$ for all θ_1^0. This limit is, of course, the same as we calculated previously in the text following Eqs. (28) and (29) using the variables θ_1, ϕ_1 for integrating Eq. (28) with complete dynamic random orientation of *both* D and A. For $\theta_1^0 < 54.74°$ the limits are $4/3 > \langle \kappa^2 \rangle_d^{\langle \psi_M, \chi \rangle} > 2/3$ and $1.12 > R_0^{\langle \psi_M, \chi \rangle}/R_0^{\langle 2/3 \rangle} > 1$; for $\theta_1^0 > 54.74°$ the limits are $1/3 < \langle \kappa^2 \rangle_d^{\langle \psi_M, \chi \rangle} < 2/3$ and $0.88 < R_0^{\langle \psi_M, \chi \rangle}/R_0^{\langle 2/3 \rangle} < 1$. This graph in Figure 7.3 can be used to estimate the uncertainty in FRET measurements if reliable estimates of θ_1^0 and ψ_M are available and if the model of Figure 7.2 applies. This is a simple case of the graphs presented by Dale et al. (1979). Estimates of ψ_M values can be made from fluorescence polarization measurements (Dale et al., 1979, and see the

---

[10] In a fluorescence instrument, if the angle between the linear polarizers of the excitation and emission beams is 54.74°, the fluorescence measurement will be insensitive to the effects of linear polarization. There is an analogous insensitivity of the κ^2 values (i.e., $\langle \kappa^2 \rangle = 2/3$) under certain circumstances in FRET. In our case there is random dynamic and isotropic rotational averaging of the acceptor (or donor). Even with no rotational averaging of the other dipole, Eq. (27), the case of $\langle \kappa^2 \rangle = 2/3$ occurs when the angular orientation of the isotropic dipole relative to the fixed dipole is the "magic angle" $\theta_1^0 = 54.74°$.

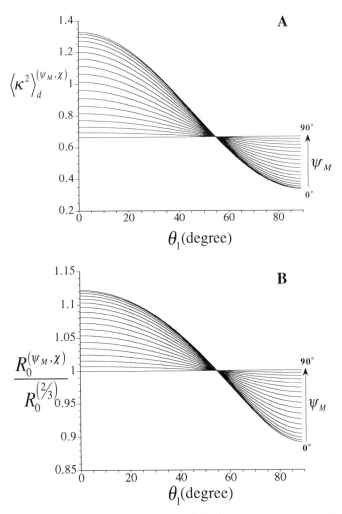

Figure 7.3. These are simulations of Eq. (33) for $\langle \kappa^2 \rangle_d^{(\psi_M, \chi)}$, referring to a random distribution inside of the cone of maximum half-angle ψ_M. See text for details.

discussion below). Provided that other parameters are well determined, the FRET measurements can be used to estimate θ_1^0 and ψ_M.

7.14.9. Distribution Restricted to the Surface of a Cone

Equation (33) has been derived for a uniform distribution of $\mathbf{p}_1$ throughout the volume of the cone. If a probe is in a lipid membrane or attached to a protein that is inserted in the membrane, it may be more likely in some cases that the

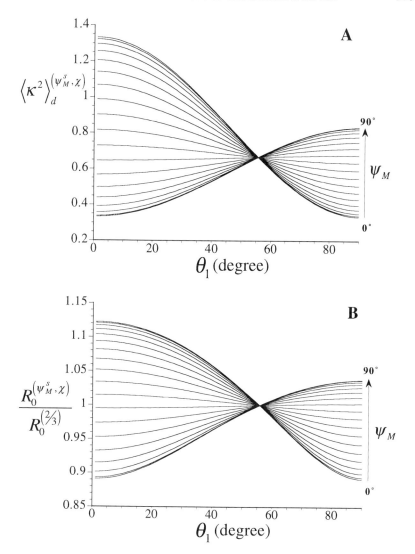

Figure 7.4. These are simulations of Eq. (34) for $\langle \kappa^2 \rangle_d^{(\psi_M^s, \chi)}$, referring to a random distribution on the surface of a cone of half angle ψ_M. See text for details.

transition dipole is more restricted and rotates rapidly only on the surface of a cone (i.e., $\psi_M \equiv \psi_M^s$). In this case the expression for the orientation factor becomes

$$\langle \kappa^2 \rangle_d^{\langle \psi_M^s, \chi \rangle} = \cos^2\theta_1^0 \cos^2\psi_M^s + \frac{\sin^2\theta_1^0 \sin^2\psi_M^s}{2} + \frac{1}{3} \tag{34}$$

If there is no rotational freedom of $\mathbf{p}_1$, then $\psi_M^s = 0$ and we have the same situation as for Eq. (27). Figure 7.4 shows a plot of $\langle \kappa^2 \rangle_d^{\langle \psi_{M} \cdot \chi \rangle}$ vs. θ_1^0 for different values of ψ_M^s. If $\psi_M^s = \pi/2$, then the cone becomes a circular plane and $\mathbf{p}_1$ rotates in a plane rather than on the surface of cone; in this case, $\langle \kappa^2 \rangle_{d,\text{max}}^{\langle \psi_{M} \cdot \chi \rangle} = 5/6$ and $\langle \kappa^2 \rangle_{d,\text{min}}^{\langle \psi_{M} \cdot \chi \rangle} = 1/3$. If $\psi_M^s = 54.74^\circ$ in Eq. (34), then $\cos^2 \psi_M^s = 1/3$ and $\sin^2 \psi_M^s = 2/3$; then, because $\cos^2 \theta_1^0 + \sin^2 \theta_1^0 = 1$, $\langle \kappa^2 \rangle_d^{\langle \psi_{M} \cdot \chi \rangle} = 2/3$ for all values of θ_1^0. This is a good example of the fact that the freedom of dipole orientations must extend in a three-dimensional space (i.e., not only in a plane) in order for κ^2 to tend to $2/3$ for all values of θ_1^0. Figure 7.4 should be compared to Fig. 20 (left side) of Dale and Eisinger (1974).

7.14.10.   When Can We Assume That $\langle \kappa^2 \rangle_d = 2/3$?

One motive for this model exercise has been to emphasize that even with considerable dynamic rotational freedom of the D and A transition moments—in our simple model, complete isotropic rotational freedom in A (or D) and partial rotational freedom within a restricted cone for the conjugate D (or A)—there is still considerable variation in κ^2. That the assumption of $\langle \kappa^2 \rangle = 2/3$ may not be appropriate in many circumstances has been stressed, especially by Dale and Eisinger (1974); they have given an extensive, detailed account of the possible min–max values for κ^2 that apply in many different circumstances. The question of whether it is warranted to assume that $\langle \kappa^2 \rangle = 2/3$ is peculiar to each experimental situation and depends on the type of information one is trying to obtain from the data. For instance, for the conditions of the model in Figure 7.3, $R_0^{\langle \psi_{M} \cdot \chi \rangle} / R_0^{\langle 2/3 \rangle}$ is relatively insensitive to the choice of κ^2 and the assumption that $\langle \kappa^2 \rangle = 2/3$ will yield estimates of the distance between D and A within $\pm 12\%$ maximum error. This is within the experimental error of the measurement in many experiments (including error in sample labeling, preparation, etc.). However, if it is desired to extract more structural information directly from the FRET efficiency curves (e.g., by varying R), the variation in κ^2 enters into the expressions directly and the effect can be considerable. The model presented here with the approximations applying to Figures 7.3 and 7.4 is useful for many situations using FRET—for instance, when using certain metals as donors (Mersol et al., 1992; Selvin and Hearst, 1994; Selvin et al., 1994; Stryer et al., 1982) or acceptors (Cronce and Horrocks, 1992; Kirk et al., 1993). It is easy to extend this simple model to encompass variations in θ_1^0, including simultaneous variations in $\mathbf{r}$. The experimenter is advised to become familiar with the more detailed treatments of this topic (Dale and Eisinger, 1974; Dale et al., 1979; Haas and Katchalski-Katzir, 1978; Steinberg et al., 1983). The discussion here has provided only an introduction to the complex problem of κ^2.

7.14.11. The Use of Fluorescence Polarization of D and A to Estimate the Maximum Error in κ^2

It is possible to make reasonable estimates of the maximum and minimum values of κ^2 of a general nature, provided that both D and A fluoresce (Dale et al., 1979). This is done by measuring the axial depolarization factors (see Eq. 35) for D and A. The depolarization factor is a measure of the average depolarization that the molecule has undergone before emitting a photon. In practice, what is interesting for FRET is the extent of molecular rotation of both D and A molecules during the excited lifetime of D; these fast rotations will partially randomize the angular correlations between D and A. If the rotations within the constraints of maximum angular deviation are rapid compared with the fluorescence lifetime D, and if the constraints on the probe's angular freedom are not too stringent, this can lead to considerable reduction in the range of possible κ^2 values, as we have seen. An account of a donor chromophore with restricted rotation embedded in a membrane, and acceptors freely diffusing (translationally and rotationally) outside the membrane (Mersol et al., 1992), illustrate one practical use of the equations developed above; here it is shown that even within the rapid diffusion limit of the acceptor, if the geometry is not spherically symmetric, the energy transfer rate is strongly dependent on the chromophore orientation in the membrane.

To follow up the discussion above, we will sketch the analysis by Dale et al. (1979) for the case where both D and A rotate rapidly in a cone. This is the most frequently employed estimate of the min–max values of $\langle \kappa^2 \rangle$ in the literature, and it is useful to see how the general analysis agrees with our Eqs. (27–34). We will use the notation of that paper, and we limit our attention here to the free rapid rotation of the D and A transition dipoles within their respective cones. The maximum angular deviation of a transition dipole in a cone, defined by the polar cone angle ψ_M [see Eqs. (32) and (33)], can be related to the ratio of two limiting anisotropies:

$$r'_0/r_0 = \{ \langle d^x \rangle \}^2 = \{ \cos \psi_M (1 + \cos \psi_M)/2 \}^2 \tag{35}$$

Here $\langle d^x \rangle$ is the depolarization factor and r_0 is the anisotropy of the randomly oriented fluorophore where absolutely no rotation takes place; r_0 values are usually taken from the literature, where they have been determined in the solid, or glassy, state or under conditions of very high viscosity. r'_0 is the limiting anisotropy following the fast rotational motion in the cone, but where no further rotation follows within the lifetime of D. r'_0 is usually the measured limiting anisotropy at short times in the solution conditions where the FRET is measured. r'_0 is determined either directly from time-resolved fluorescence anisotropy measurement or from the intercept of a Perrin plot (Lakowicz,

1983). The cone itself is usually considered to rotate more slowly on a longer time scale (e.g., the rotation of a protein molecule).

Once we have experimentally determined the depolarization factors for D and A, a dynamically averaged value of the orientation factor can be written as follows (Dale et al., 1979);

$$\langle \kappa^2 \rangle = \kappa^{x2} \langle d_D^x \rangle \langle d_A^x \rangle + \tfrac{1}{3}(1 - \langle d_D^x \rangle) + \tfrac{1}{3}(1 - \langle d_A^x \rangle) + \cos^2 \Theta_D \langle d_D^x \rangle (1 - \langle d_A^x \rangle)$$

$$+ \cos^2 \Theta_A \langle d_A^x \rangle (1 - \langle d_D^x \rangle), \tag{36}$$

where $\langle d_D^x \rangle$ and $\langle d_A^x \rangle$ are the axial depolarization factors of D and A; $\kappa^{x2} = (\sin \Theta_D \sin \Theta_A \cos \Phi - 2 \cos \Theta_D \cos \Theta_A)^2$, Θ_D and Θ_A are the angles between the center cone vectors for the D and A transition dipoles ($\mathbf{p}_1^0$ of Figure 7.2) and the $\mathbf{r}$ vector between the dipoles; and Φ is the angle between the two planes that are defined by the D or A dipoles with $\mathbf{r}$ vector. This equation is valid for any generalized depolarization factors, and it is not restricted to the case of Eq. (35). We notice in Eq. (36) that if $\langle d_A^x \rangle = 0$ and $\langle d_D^x \rangle = 1$, we have $\langle \kappa^2 \rangle = \cos^2 \Theta_D + 1/3$, which is exactly Eq. (27)—the case for isotropic random orientations for A (i.e., $\langle d_A^x \rangle = 0$) and a selected singular orientation for D (i.e., $\langle d_D^x \rangle = 1$). If we allow the D transition dipole to rotate in a cone, we can substitute the analytical expressions for $\langle d_D^x \rangle$ from Eq. (35) into Eq. (36), setting $\langle d_A^x \rangle = 0$, and arrive at the same expression for $\langle \kappa^2 \rangle$ as that given in Eq. (33). It should be clear now, keeping in mind our earlier discussion of the derivation of Eqs. (27–33), how to use Eq. (36) to estimate the minimum and maximum values for different orientations and degrees of rotational freedom of D and A. The reader should consult the original reference (Dale et al., 1979) for an excellent discussion and many graphic examples, and should also study other approaches (Steinberg et al., 1983). In addition a recent paper (Censullo et al., 1992) discusses in detail the application of these formulas to multiple acceptors and presents many contour curves of some useful and generally applicable examples.

7.15. DISTRIBUTIONS IN *R*

This topic is included because many of the FRET measurements in the microscope most surely involve distributions of distances (and angles) between D and A; this section describes how this situation is handled in the more sophisticated cuvette FRET literature. With the evolution of more quantitative fluorescence microscopy, this will be more important for future FRET microscopy experiments; as discussed below, the consideration of distributions is important for interpreting FRET measurements in biological membranes. Until now we have assumed that the distances between D and A are the same for all D–A pairs. This is not usually the case unless D and A are attached to the same molecule, and even then there may be enough translational freedom of the dyes so that a distribution of distances $\rho_{DA}(R)\,dR$ must be presumed. There are three major "time" categories of R distributions: (i) D and A undergo very rapid translational diffusion (within their constraints), so that $\rho_{DA}(R)$ remains constant during the transfer process—this is normally not applicable unless the fluorescence lifetimes of D are very long (microseconds to seconds), for instance, in "diffusion-enhanced FRET" (Stryer et al., 1982); (ii) the molecules do not undergo significant translational Brownian motion during the energy transfer process; and (iii) energy transfer and translational diffusion occur on similar time scales. In addition, the orientational distributions (i.e., κ^2) may vary, and the two motions may even be coupled; the analysis is quite formidable if D and A are translationally and orientationally redistributed during the energy transfer process (Haas et al., 1978; Steinberg and Katchalski, 1968). The general case can be extremely complex, and simplifications must be introduced to commence with a reasonable analysis.

A static distance distribution is often a reasonable approximation. The effects of the R and κ^2 distributions can easily be accounted for in terms of the time-resolved fluorescence signals. The decay of the donor fluorescence $F(t)$ following a short pulse of light $\delta(t)$ (delta function response) can be expressed as

$$F(t) = \sum_i \int_{R_{min}}^{R_{max}} \alpha_i \exp\{-(t/\tau_i)[1 + (R_{0,i}/R)^6]\}\rho_{DA}(R)\,dR \qquad (37)$$

Equation (37) expresses the time-resolved fluorescence decay of D (with A at a single distance R from D) with multiple fluorescence components with decay times $\tau_i/(1 + (R_{0,i}/R)^6)$ (and corresponding amplitudes α_i), where τ_i is the ith decay time in the absence of FRET (see Eq. 1). It is assumed in Eq. (37) that κ^2 is the same for all R values (usually it is assumed that $\kappa^2 = 2/3$); this permits integration only over R. By writing the subscript i in $R_{0,i}$, the possibility has been included that $R_{0,i}$ can be different for every fluorescence decay component i. If the τ_i values are influenced only by dynamic interactions in the excited state of D*, then it is reasonable that $R_{0,i}$ varies as $R_{0,i} = \bar{R}_0(\tau_i/\bar{\tau})^{1/6}$ (Albaugh and Steiner, 1989), where $\bar{\tau}$ is the average fluorescence decay time [see Eq. (41)]; $R_{0,i}$ could also depend on R (see, e.g., the discussion of static orientations below). In addition, there may be large populations of donors in heterogeneous samples that are not involved in FRET; they will decay with the intrinsic relaxation times τ_i, complicating the analysis. Often Gaussian distributions in R are assumed.

$$\rho_{DA}(R) = \frac{1}{(2\sigma\pi)^{1/2}} \exp\left(-\left(\frac{R - \bar{R}}{2^{1/2}\sigma}\right)^2\right) \qquad (38)$$

or a Lorentian distribution. Even if κ^2 can vary, it may be possible to work with an ensemble average of $\langle R_0 \rangle$, as discussed in Sections 7.14.8, 7.14.10, and 7.14.11. However, in general, a distribution in κ^2 may have to be included in the analysis, as in the following equation:

$$F(t) = \sum_i \iiint_{\text{all angles}} \int_{R_{\min}}^{R_{\max}} \alpha_i \exp\{-(t/\tau_i)[1 + \kappa_i^2(R'_{0,i}/R)^6]\}\rho_{DA}(R)\sin\theta_D \sin\theta_A$$

$$\cdot dR\, d\theta_D\, d\phi_D\, d\theta_A\, d\phi_A \tag{39}$$

where $\kappa_i^{1/3} R'_{0,i} = R_{0,i}$ (i.e., $R'_{0,i}$ is the value of $R_{0,i}$ without κ_i^2 factor), and κ_i^2 [see Eq. (6)] can also be a function of R. Similar integrations over distributions can be written for the steady-state parameters and for the FRET efficiency. A variable κ_i^2 makes the situation even more complex. In general, the fluorescence decay curves must be analyzed by fitting them directly to the integral equations. There have been many publications dealing with this problem (Eis and Millar, 1993; Hochstrasser et al., 1992; Lakowicz et al., 1991b; Wiczk et al., 1988). In a recent study (Steiner et al., 1991), FRET was used to investigate separations between different positions in engineered calmodulin with multiple donor lifetimes; the data were fitted to $\rho(R)_{DA}$ distributions, to different R_0 values for the separate lifetimes, and to distributions in $\langle \kappa^2 \rangle$. Incomplete labeling of macromolecules (for intramolecular FRET) with acceptor can lead to serious errors in time-resolved distance distribution FRET analyses; a method for correcting for this has been presented (Lakowicz et al., 1991a). The effects of diffusion during FRET were minimized by using highly viscous solutions. Analysis of fluorescence time decay in the presence of different spatial distributions of probes in micelles has been considered (Barzykin, 1991).

The relation between the measured efficiency, E, of FRET and Eqs. (37) and (39) is unambiguous if D and A are separated by a single distance $\bar{R}$, i.e., $\rho_{DA}(R) = \delta(R - \bar{R})$, and if all $R_{0,i}$ values are equal to a single R_0. The ratio of the time integrals of Eq. (37)—i.e., the steady-state fluorescence—with and without acceptors present (assuming that [D*] is the same in both cases) is directly related to E.

$$\frac{F_{DA,ss}}{F_{D,ss}} = \frac{\int_0^\infty \sum_i \alpha_i \exp\{-(t/\tau_i)[1 + (R_0/R)^6]\}\, dt}{\int_0^\infty \sum_i \alpha_i \exp\{-(t/\tau_i)\}\, dt} = \frac{1}{[1 + (R_0/R)^6]} = 1 - E \tag{40}$$

The average decay time is defined as

$$\langle \tau \rangle = \sum_i \alpha_i \tau_i^2 / \sum_i \alpha_i \tau_i \tag{41}$$

and the ratio of the $\langle \tau \rangle$ values with and without acceptors corresponding to Eq. (40) is also simply related to E:

$$\frac{\langle \tau_{DA} \rangle}{\langle \tau_D \rangle} = \frac{1}{[1 + (R_0/R)^6]} = 1 - E \qquad (42)$$

Therefore, even if D* decays with multiple fluorescence components, the steady-state and time-resolved measurements, according to Eqs. (40) and (42), will agree, provided that D and A are separated by a single R value and provided that the $R_{0,i}$ values are all equal. Equations (40) and (42) should be compared to Eqs. (2) and (3), where it was implicitly assumed that only one donor lifetime exists.

If there is a distribution $\rho_{DA}(R)$ of distances, then the signal ratios can be shown to be (Albaugh et al., 1989; Cantor and Pechukas, 1971)

$$\frac{F_{DA,ss}}{F_{D,ss}} = \int_{R_{min}}^{R_{max}} \frac{\rho_{DA}(R)}{[1 + (R_0/R)^6]} dR = 1 - E \qquad (43)$$

Here *a single* R_0 has been assumed, i.e., D and A rotate rapidly compared to the fluorescence lifetimes, so that the orientations can be dynamically averaged within their constraints (Dale et al., 1979). This makes $\langle \kappa^2 \rangle$ independent of R; usually it is assumed that $\kappa^2 = 2/3$. If the distribution is approximated by Eq. (37), the plot of E vs. R/R_0 becomes less steep as σ increases (keeping $\bar{R}$ constant) (Cantor and Pechukas, 1971). This is due to the tails of the distribution. The effect on the corresponding expression for the average lifetimes is

$$\frac{\langle \tau_{DA} \rangle}{\langle \tau_D \rangle} = \frac{\displaystyle\int_{R_{min}}^{R_{max}} \rho_{DA}(R)[1 + (R_0/R)^6]^{-2} dR}{\displaystyle\int_{R_{min}}^{R_{max}} \rho_{DA}(R)[1 + (R_0/R)^6]^{-1} dR} = 1 - E \qquad (44)$$

Equations (43) and (44) show that if there is a distribution in R, then $F_{DA,ss}/F_{D,ss}$, and $\langle \tau_{DA} \rangle/\langle \tau_D \rangle$ are generally not equal. This fact can be used as an indication of a distribution in R.

7.15.1. Additional Orientation Distributions

In addition to an R distribution, if the $R_{0,i}$ values are different (as mentioned above, it is reasonable to assume that $R_{0,i} = \bar{R}_0(\tau_i/\bar{\tau})^{1/6}$ (Albaugh and Steiner, 1989)), a detailed analysis becomes much more difficult, even if the $R_{0,i}$ values are still considered to be independent of R and have global average values of $\langle \kappa_i^2 \rangle$. The literature should be consulted for discussions (see, e.g., Albaugh and Steiner, 1989, and references above). If there is also a distribution of κ^2 values, and if these distributions do not relax faster than the fluorescence lifetimes (e.g., within a cone of rotational freedom), a detailed analysis becomes very difficult; this is definitely beyond the present capabilities of FRET microscopy. The simultaneous temporal relaxation of fluorescence and D–A orientations is much more difficult to analyze in order to extract the FRET parameters (Beecham and Haas, 1989; Haas and Steinberg, 1984; Haas et al., 1978, 1988; Klein et al., 1976; Lakowicz et al., 1990a, b, 1991b, 1994; Maliwal et al., 1993; Steinberg and Katchalski, 1968; van der Meer et al., 1993).

An extreme case is presented by a completely static distribution of mutual D–A orientations; that is, the D and A molecules do not undergo Brownian motion (translation or rotation) at all in the time scale of fluorescence decay. This model is probably applicable if dyes are bound tightly to macromolecules and if their anisotropies are high. The simplest case of *static* but *random* D–A orientations and distances has been dealt with extensively (Dale and Eisinger, 1974, 1976; Dale et al., 1979; Steinberg, 1968; Steinberg and Katchalski, 1968; Steinberg et al., 1983). In this case, even though the distributions are random, no average $\langle \kappa^2 \rangle$ exists independent of R (Dale et al., 1979; Haas and Katchalski-Katzir, 1978; Steinberg et al., 1983). The $\langle \kappa^2 \rangle$ values for the D–A pairs that are separated by distances $R \approx \leq R_0$ tend to zero as R tends to approach zero. This is a consequence of the shape of the electric field surrounding an electric dipole; there is a decrease in the relative number of effective static mutual orientations of D–A pairs as R decreases. As R increases with $R > R_0$, $\langle \kappa^2 \rangle$ approaches a value of 2/3 (Steinberg et al., 1983).

A recent publication (Wu and Brand, 1992) has considered the effect of a "static" distribution of D and A orientations on a recovered distance distribution (of form Eq. 37) from a fit to a model of Eq. (37) (where it is assumed that $\kappa^2 = 2/3$—dynamic averaging), where the "data" were simulated according to Eq. (39), including the orientation of static random dipoles. This is an excellent example of the complexities that can arise in interpreting FRET data and demonstrates several important facets of interpreting the fitted parameters. The fluorescence decay was simulated with a random static orientation distribution with either a single distance R or a distribution of R values; the authors assumed only one fluorescence decay component (i.e., $i = 1$ in Eqs. 37 and 39). The reader is referred to this paper for details and references; here we wish only to point out some of its major conclusions that are of general interest for understanding the effect that static orientational distributions can have on fitting data with the assumption that only an R distribution is present. Wu and Brand concluded that if $R > R_0$, then κ^2 approaches 2/3 and the error in the estimate of the mean distance (approximate maximum of the fitted distribution or the distance determined from steady-state measurements) is small. However, if $R < R_0$, the error made in determining $\rho_{DA}(R)$ by assuming only a distribution in R (Eq. 37) becomes substantial; the mutual orientations are much more influential at smaller distances (see also the above paragraph). Thus, it may sometimes be advisable to use dyes with shorter R_0 values or to employ quenchers to reduce R_0, keeping in mind, of course, that the error of the measurement increases for lower efficiencies. Also, when $R > R_0$, the distributions could be fitted by symmetrical distributions (Eq. 37); otherwise, skewed distributions were required. In general, the recovered fitted R distributions are skewed toward longer distances because at shorter distances there are many more orientations that produce little FRET (Dale et al., 1979). Dynamic averaging (e.g., when $\kappa^2 = 2/3$) increased the FRET efficiency compared to the static case, as expected. The major sensitivity of FRET on D–A orientations is found when $R < R_0$. The width of the recovered fitted $\rho_{DA}(R)$ distributions is often considerable even if only a single R is present in the simulations of the data. It is important to keep in mind that the average $\langle \kappa^2 \rangle$ for any particular distribution, and the average distances, are determined by real molecular distributions, and that these true distributions are not taken into account in the min–max error estimations (Dale et al., 1979) discussed above.

7.16. HOW DO DISTRIBUTIONS IN R AND R_0 AFFECT FRET MICROSCOPY?

There is no doubt that extreme or complex distributions in R or κ^2, and the presence of multiple lifetime components, will impede the determination of accurate estimates of R, and that problems encountered in "clean" cuvette experiments will be compounded in FRET microscopy. Equations (37) through (44) have been discussed briefly in order to indicate general tendencies and to point out that quantitative analyses of FRET data can become very elaborate. Distributions in R are to be expected in many biological samples investigated in the microscope, and owing to the organized structures and the mixed environments in a biological cell, variations in κ^2 and consequently variable R_0 values may be present. Differential quenching environments and other solvent effects add to the troubles. One should be aware of these issues and be prepared to be confronted with seemingly incompatible results that could arise due to these distributions (see also Section 7.17).

For many FRET microscopy measurements it is more important to detect the differences between FRET efficiencies at different locations in a sample, or to determine simply the presence or absence of FRET, than to measure accurate values of E; this has been the objective of most FRET microscopy measurements to date. However, it is now possible to quantify microscopic images reliably and accurately with both steady-state and time-resolved measurements. New technologies are becoming available in imaging FRET microscopy (see Section 7.20). For instance, it is now possible to extend nanosecond time-resolved measurements to the microscope; mean fluorescence lifetimes can be measured simultaneously at thousands of locations in a microscope image (Gadella et al., 1994—see also Chapter 10 in this book). Improvements in the methods of quantifying fluorescence detection in the microscope and the availability of time- or phase-resolved fluorescence microscopy provide more accurate and higher contrast, spatially resolved FRET measurements. This will lead to a better understanding of the underlying dynamics and organization of intact biological structures on a molecular scale.

After discussing various seemingly insurmountable uncertainties that seem to bar a reliable quantitative interpretation of FRET data, it is comforting to know that despite these impediments the average distances determined by steady-state FRET measurements, assuming that $\kappa^2 = 2/3$, are frequently in reasonable agreement with values determined by other methods (Stryer, 1978). This is often the case despite a wide R distribution indicated by time-resolved studies or in spite of a considerable range of κ^2 values indicated by anisotropy measurements. A spectacular example of this has been demonstrated in the actin/myosin field. There have been over 60 FRET studies in this field, and 26 distances have been determined. It has been determined that many of these distances are in excellent agreement with X-ray data; the results from many investigations using different D–A pairs agree, and the global results lead to an internally consistent model (Censullo et al., 1992; dos Remedios et al., 1987; Miki et al., 1992; O'Donoghue et al., 1992). Similar conclusions for different systems have been made by others (Amir et al., 1992; Haas et al., 1988; Wu and Brand, 1992). In many cases, the appropriateness of making the simplifying assumption $\langle \kappa^2 \rangle \approx 2/3$ is supported by the fact that many dyes have multiple transition moments that participate conjointly in the spectroscopic transitions; these transition moments are often not parallel to each other, and this can lead to effective κ^2 values very close to 2/3 (Haas and Katchalski-Katzir, 1978).

In his early works Förster analyzed a uniform static distribution in three-dimensional space of many acceptors surrounding excited donors[11] (Förster, 1949a, b), assuming that $\kappa^2 = 2/3$ and one fluorescence D component. The decay of the fluorescence signal, calculated from Eq. (37), is an average over the

---

[11] Because the concentration of excited donors [D*] is usually very small (i.e., [D*] ≪ [D]), it is almost always true that each D* is surrounded by many acceptors, rather than acceptors being surrounded by many Đ* molecules.

distribution $\rho_{DA}(R)$; the distribution of A molecules surrounding a D* molecule depends on the [A] concentration. The result of summing the randomly situated, individual exponentially decaying components is a nonexponential fluorescence decay (with an exponential decay with $\sqrt{t}$ dependence).

7.17. INTERPRETATION OF FRET MEASUREMENTS MADE ON BIOLOGICAL MEMBRANES

An important and biologically relevant example of D–A distributions exists when D and A molecules are dispersed in pseudo-two-dimensional systems. Lipid–lipid, protein–lipid, and protein–protein proximity relations in biological membranes play an important role in the functioning of cells. For instance, the organization of lipids and proteins within biological membranes is especially important for extra- and intracellular communication. Such membrane organizations correspond to the experimental systems in many FRET microscope studies, and the analyses of these systems (so far usually not measured in an imaging situation) can be carried over to FRET microscopy. The bilayer membrane is a closely packed, crowded environment, and donor and acceptor molecules with good spectral overlap can communicate by FRET over lateral distances of several lipid molecular diameters.

7.17.1. Recent FRET Studies in Membrane Systems of Interest to FRET Microscopy

Many FRET studies dealing with the distributions of components in biological membranes and cells (not in the microscope) have been published in the last several years; a few titles of selected topics with references are listed: membrane fusion (Arts et al., 1993; Martin et al., 1994), intermembrane exchange and mixing of lipid components (Mancheño et al., 1994; Silvius and Zuckermann, 1993; Sunderland and Storch, 1993; Walter and Siegel, 1993), lipid flip-flop (Fattal et al., 1994), vesicle aggregation (ter Beest and Hoekstra, 1993), flow cytometry (Chan et al., 1979; Jovin, 1979; Szöllösi et al., 1984, 1987a,b; Trón et al., 1984), protein–lipid interaction in membranes (Cortese and Hackenbrock, 1993; Gawrisch et al., 1993; Luan and Glaser, 1994; McLean and Hagaman, 1993; Narayanaswami and McNamee, 1993; Narayanaswami et al., 1993; Pap et al., 1993; Ward et al., 1994), protein–protein interactions in membranes (Adair and Engelman, 1994; Baker et al., 1994; Ben-Efraim et al., 1993; Corbalan-Garcia et al., 1993; Gazit and Shai, 1993; Hanicak et al., 1994; Horowitz et al., 1993; Kubitscheck et al., 1993; Mata et al., 1993; Remmers and Neubig, 1993; Ringsdorf et al., 1993; Stefanova et al., 1993), DNA–lipid interactions (Kinnunen et al., 1993; Mustonen et al., 1993), and the interaction of ions with lipid membranes (Holmes et al., 1993; Sanderson et al., 1993). FRET is an excellent technique for following these processes.

7.17.2. Theoretical Descriptions of FRET Distributions in Membranes

Several theories describing the FRET between a donor and a distribution of randomly placed acceptors in biological membranes have been published in the last several years. Shaklai et al. (1977) showed that if a donor is placed at a height R_M above the membrane (e.g., in a membrane protein) and acceptors are randomly distributed in the surface of the membrane, that the ratio of the donor steady-state intensity without (F_D) and with (F_{DA}) an acceptor (if $R_M > 1.7R_0$ and if κ^2 is independent of the distance between any D–A pair) is proportional to the surface concentration of the acceptor σ_A, i.e., $F_D/F_{DA} = 1 + K_q\sigma_A$, where K_q is a constant, $K_q = \pi/2R_0^6/R_M^4$. The rate of FRET, k_T', is $k_T' = K_q\sigma_A/\tau_D$. The dependence on the fourth power of R_M results from the integration of the $1/R^6$-dependent FRET signal over the two-dimensional surface; R_M is the closest distance at which an acceptor can approach a donor. This same dependence of the rate of FRET on the shortest distance between two planar molecular layers, each layer containing either D or A, was found much earlier by Drexhage et al. (1963) (see also Kuhn, 1968); these authors expressed the rate of energy transfer as $k_T' = \gamma\sigma_A(R_0')^6/(R_M)^4$, where R_0' is the normal R_0 without the κ^2 term, and the constant γ corresponds to different orientations of D and A molecules in their respective membranes (including random orientations of the transition dipoles in the plane of the membranes). This analysis included the dependence of κ^2 on the distance between the D–A pairs. Several values of γ were given for different orientations of D and A in their respective bilayers (values of γ between $3\pi/4$ and $3\pi/16$). By defining $d_0^4 = (3/2)\gamma\sigma_A(R_0'(2/3))^6$, the authors derived the simple expression for the measured ratio of spectroscopic intensities $F_D/F_{DA} = (1 + (d_0/R_M)^4)$, similar to the findings of Shaklai et al. The important relations derived in both of these (and the following) studies are the $\sim 1/R_M^4$ dependence and the direct dependence on σ, allowing molecular parameters to be determined. Following the work of Shaklai et al. and Kuhn et al., (the work of the Kuhn group—Barth et al., 1966; Inacker and Kuhn, 1974; Inacker et al., 1976; Kuhn and Möbius, 1993; Kuhn et al., 1972—on this topic has not been noticed by many in the biophysical literature) a number of important theoretical treatments of similar problems of D and A distributions in membrane systems have appeared, together with experimental analyses (Baird et al., 1979; Baker et al., 1994; Dale et al., 1981; Davenport et al., 1985; Dewey, 1991; Dewey and Hammes, 1980; Eisinger and Flores, 1982; Estep and Thompson, 1979; Fleming et al., 1979; Fung and Stryer, 1978; Gutierrez-Merino, 1981; Gutierrez-Merino et al., 1987; Haigh et al., 1979; Hasselbacher et al., 1984; Holowka and Baird, 1983; Koppel et al., 1979; Kubitscheck et al., 1993; Matkó et al., 1988, 1993; Sklar et al., 1980; Szöllösi et al., 1987a; Valenzuela et al., 1994; Vanderkooi et al., 1977; Wolber and Hudson, 1979; Yguerabide, 1994). The reader should consult these references for details concerning the relevant situations for the different practical employments. These methods of analysis can be used under appropriate circumstances for quantifying FRET measurements in the microscope on biological systems containing membrane surfaces.

7.18. FRET IN SYSTEMS WITH SPATIAL RESTRICTIONS

Polymer surfaces, organized molecular self-assemblies, glasses, lipid vesicles, protein aggregates, and solid solutions are examples of molecular systems that are confined in their movements, configurations, and rotations. Such confinement influences the thermodynamic and dynamic features of the subcomponents, and it becomes important to characterize the geometry of such a system. As we have seen, the efficiency of FRET is sensitive to distribution functions, and these distributions are in turn controlled by the constraints that act on the local environment of the "matrix." An approach has been developed to categorize the geometrical nature of locally heterogeneous systems, as opposed to homogeneous systems, using time-resolved FRET. The geometries are described by either simple geometrical forms or fractal models. The distribution of the D and A molecules is described in a generalized way, and the dependence of the time decay of the donor fluorescence is fitted to models to obtain coefficients that can then be used to describe the nature of the surrounding geometry (e.g., the dimensionality of the space over which the interactions between D and A chromophores take place). This information can give clues to the underlying molecular geometry of pores and extended networks. Space limitations do not allow us to review this interesting field, but these ideas can certainly be applied in FRET microscopy, and the interested reader should consult the original literature on FRET in extended organized molecular structures and clusters (Baumann and Fayer, 1986; Blumen and Klafter, 1986; Dewey, 1992; Drake et al., 1991; Klafter and Drake, 1989; Kost and Breuer, 1991; Liu et al., 1993; Schleicher et al., 1993; Zumofen et al., 1990). Some related works on the development of expressions for FRET in different molecular dimensions are Gösele et al. (1975, 1976) and Hauser et al. (1976).

7.19. USE OF FRET FOR STUDYING PROTEINS AND NUCLEIC ACIDS AND IN BIOTECHNOLOGY

FRET has been applied to a wide variety of systems in molecular biology and biochemistry. Although the vast majority of these experiments deal with homogeneous solutions containing only a few purified components, many of these systems could eventually be applied to experimental situations in the microscope. Traditionally, FRET has been used mostly to measure protein structures, interactions, and dynamics; some of the reviews referenced in Section 7.1 contain an introduction to this vast literature. However, the applications of FRET have expanded to many other categories of biological macromolecules (as shown above). For a brief overview of FRET in 1993/94 the reader is referred to a recent review by Clegg (1995). An abridged review of a few topics follows.

Recently, there has been a substantial increase in interest in applying FRET to detect and to study structural and dynamic aspects of DNA and RNA structures; these studies are important for applications of FRET microscopy. The hybridization of simple duplexes of DNA and RNA can easily be detected by observing FRET between conjugated labels on complementary single strands of DNA (Cardullo et al., 1988; Clegg et al., 1993; Morrison et al., 1989; Selvin et al., 1994), and their three-dimensional structures can be mapped (Clegg et al., 1993). The stereochemical structures (Clegg et al., 1992, 1994; Cooper and Hagerman, 1990; Lilley and Clegg, 1993; Murchie et al., 1989) and dynamics (Eis and Millar, 1993; Hochstrasser et al., 1992) of the more complex branched DNA four-way junctions can be determined. The extent of bending of DNA and RNA duplexes with nucleotide bulges of different lengths (Gohlke et al., 1994) has been observed. Unwinding of double-stranded DNA by DNA-unwinding enzymes can be followed (Houston and Kodadek, 1994), providing a convenient FRET microscopic assay for helicases. The thermodynamic properties of oligonucleotide stability can be determined (Clegg et al., 1993; Morrison and Stols, 1993; Morrison et al., 1989). In another recent publication (Mergny et al., 1994), several D–A pairs were compared and it was demonstrated that FRET is an excellent technique for detecting hairpins and observing translocation products of genes between two chromosomes. New lanthanide chelates conjugated to DNA open up new possibilities for sensitive time-resolved detection (Kwiatkowski et al., 1994; Saha et al., 1993; Selvin, 1995; Selvin and Hearst, 1994; Selvin et al., 1994); these probes will be especially important for hybridization detection in FRET microscopy. New heterodimers of "FRET dyes" are being developed that can lead to sensitive detection in the FRET microscope by producing long Stokes shifts when bound to DNA (Benson et al., 1993a, b; Rye et al., 1993). Energy transfer between DNA-binding dyes has been used in solution for a long time (Genest and Wahl, 1983; LePecq and Paoletti, 1967); this technique was employed relatively early in the microscope (Langlois and Jensen, 1979; Latt et al., 1979). The two-dye FRET technique used in flow cytometry ("Dutch" Boltz et al., 1994; Vinogradov, 1994) will be useful for FRET microscopy.

FRET is being used more often for biotechnological applications, and many new assays are being developed that will be useful in the FRETmicroscope. The literature in this field is very important for future applications in the FRET microscope. We only have room to list a few references on this interesting subject (Matayoshi et al., 1990; Morrison, 1988; Ullman and Khanna, 1981; Ullman et al., 1980) and a few recent papers (Amin et al., 1993; Ozinskas et al., 1993; Rice et al., 1993).

7.20. SOME USEFUL NEW DEVELOPMENTS FOR FRET MICROSCOPIC MEASUREMENTS

7.20.1. pbFRET

This relatively new method in FRET microscopy has already been discussed. See below for a summary of recent publications.

7.20.2. FLIM–FRET

Fluorescence lifetime imaging microscopy (FLIM) (measurements of fluorescence nanosecond lifetimes at every pixel of a digital electronic image) in the time and frequency domains has recently been demonstrated by several groups (see Gadella et al., 1994, for recent references, and Chapter 10 in this volume). FLIM can be employed to measure time-resolved FRET directly at every point of an image simultaneously. FLIM–FRET has been used to image endosome fusion in single cells (Oida et al., 1993); four to seven images were recorded with a CCD camera at different time delays (smallest time delay = 5 ns). FLIM–FRET can significantly improve selectivity and differentiability in microscopic images and will be employed more often in the near future.

7.20.3. Polarization Measurements

It has been suggested that sensitive FRET measurements on cells (in the microscope and flow cytometer) could be made by analyzing the polarization of the donor or acceptor emission (Matkó et al., 1993). This would avoid certain calibration procedures, and the method could be extended to a full imaging situation.

7.20.4. FRET Photochemical Labeling

The energy transferred from a donor brings an acceptor into the excited state, and if the acceptor is an efficient photochemical agent, the excited acceptor can specifically photolabel neighboring sites that are within $\sim R_0$ of the donor. A short review of this technique has appeared (Peng et al., 1994) that deals especially with membrane-associated and channel-forming domains of proteins. This novel application of FRET is seldom used (essentially only by this one research group), but the improved coupling yields with the acceptors now available, and the specificity of labeling achievable by FRET-directed photochemistry, make it a very attractive technique for measuring molecular proximities between multiple sites, with the added advantage of being able to separate subsequently the residues that have photoreacted with the acceptor for identification.

7.21.   A SELECTION OF FRET MICROSCOPY PUBLICATIONS

Almost 20 years ago, the first measurement of FRET from a biochemical sample in a fluorescence microscope was made (Fernandez and Berlin, 1976). In this early study, the distribution of concanavalin A (con A) on cell surfaces was measured by labeling con A with DANS (dimethylaminonaphthalene sulfonate) and RITC (rhodamine isothiocyanate); the distribution was determined to be nonrandom. Fluorescence from a single cell was selected by focusing the excitation light on part of the specimen, and the fluorescence from the top optical port of the microscope was passed through a monochrometer and focused onto a photoncounting photomultiplier. Changes in the efficiency of FRET from larger areas of the microscopic object (e.g., the cell surface) on which the excitation light was focused were determined in several ways: (i) by measuring the steady-state ratio of the acceptor fluorescence, to the donor fluorescence, I_A/I_D; (ii) by measuring directly the lifetime of the donor (see Eq. 19); and (iii) by acquiring time-resolved spectra (i.e., emission spectra extending over the D and A emission wavelengths were taken at different delay times, ranging from a few nanoseconds to ~ 10 ns) following a short excitation pulse. The last method combines spectral and temporal information; because the distances between D and A were statistically distributed on the cellular surface, those donors emitting at longer times tend to be statistically distributed at greater distances from acceptors—and therefore they contribute more to the total emission spectra at longer times—than those emitting at shorter times. The sensitivities of all three measurements on this system depend on the D/A ratio, and this ratio can be optimized to manifest the largest change for the case at hand. Fernandez and Berlin anticipated the possibility of making imaging FRET measurements down to the limits of the resolution of the microscope.

Probably the first published example of a FRET image in a microscope was used to measure the thickness of a stepladder of overlapping monolayers resting on a gold surface (as acceptor), with a layer of donor (CY9) on top (Kuhn and Möbius, 1971; Kuhn et al., 1972; Peters, 1971); these layers were then used to measure the thickness of an erythrocyte ghost embedded in arachidate monolayer assemblies, also laying on gold with a layer of donor spread on the top. Following the pioneering study of Fernandez and Berlin, FRET microscopy was intermittently applied until recently, when interest in this technique increased rapidly. This increased interest is partially due to the spectacular advances in electric digital image acquisition and processing; modern digital imaging cameras and associated computer hardware and image analysis software are now commercially available, making it possible to measure FRET easily and rapidly at every point of an image simultaneously. Methods of measurement that were not possible earlier can be carried out conveniently, and the existing procedures for measuring FRET in microscopes can benefit from many of the new imaging technologies. FRET microscopy is rapidly becoming a familiar supplement to the repertoire of the fluorescence microscopist, as shown by the numerous interesting papers reviewed here.

The earlier FRET microscopy experiments did not record images but measured FRET from an extended illuminated object. The ratio and the above-mentioned time-resolved methods (Fernandez and Berlin, 1976) were used to follow the redistribution of con A receptors on the cell surface (of whole cells) during the period of myoblast fusion (Herman and Fernandez, 1982). The association of peptide antigen and the major histocompatibility complex protein, in the presence of helper T cells, was studied in a unique FRET microscope (Watts et al., 1986). In this study, evanescent wave excitation FRET microscopy was used to excite selectively only the donor-labeled peptide molecules that are within 800 Å of a planar membrane surface (this reduces the donor background by almost 3 orders of magnitude and also reduces autofluorescence); the spectrum of the emission from the D–A system was collected through a monochrometer, and the efficiencies of FRET were calculated from the spectral data. No FRET images were made in the above studies. Perhaps the first report of FRET imaging microscopy was by Uster and Pagano (1986), who described the construction of a FRET microscope in which images were recorded on film. The authors measured donor quenching and acceptor sensitization in their FRET microscope to study model membranes, to identify intercellular organelles labeled with lipid probes, and to follow the internalization of lipid and protein probes.

A sensitive, rapid-repetition, time-resolved streak-camera microscope has been used to measure FRET signals (Kusumi et al., 1991); FRET measurements were used to measure the fusion of endosomes in single cells by exciting only selected parts of the cells. This group has also used FRET efficiencies determined from spectral data (selected with filters and images recorded on film) in an optical microscope to study the lysosomal disintegration of low-density lipoprotein (LDL) particles; D- and A-labeled lipids were incorporated into LDL particles, and the disruption of these particles was evident from the decrease in FRET efficiency. The pbFRET technique was used to measure the affinity of haptens to monoclonal IgE bound to its type I Fc_ε receptor; measurements performed in solution and in the microscope were compared, and the affinity constants determined by fluorescence titrations agreed well with those of FRET titrations. Calibration procedures for FRET microscopy, involving the absorption, emission, and fluorescence polarization properties of the D–A pair and the optical properties of the microscope, have been discussed (Ludwig et al., 1992) and verified by comparing microscope FRET efficiencies (between two DNA-binding dyes—Hoechst 33342 and acridine orange) to those calculated from data taken with a fluorometer. Possible measurement artifacts in FRET microscopic measurements of biological material are discussed.

7.22. SOME RECENT APPLICATIONS

The $Fc_\varepsilon RI$ (cell surface type I receptor for Fc_ε) aggregation state on membranes of mast cells has been studied on single cells by *donor*-pbFRET by labeling specific receptor ligands (IgEs or Fab fragments of Fc_ε monoclonal antibodies) with fluorescein and rhodamine (Kubitscheck et al., 1993); see also Section 7.20.1. Because the FRET components are densely distributed, it was necessary to analyze the FRET data with a numerical calculation using cylindrical models of the receptor–ligand complexes in order to distinguish aggregated from randomly distributed proteins. From their data, the authors conclude that $Fc_\varepsilon RI$ is monovalent and randomly distributed in the plasma membrane. The general method of analysis presented in their paper (i.e., comparing the ratio of FRET efficiencies of two labeled receptor ligands of different sizes) is especially useful for analyzing FRET in membranes when the D and A components are present in high density.

The analysis of pbFRET bleaching data has been considered in detail by Young et al. (1994) (see also above). These authors note that distributions of photobleaching rates are usually present, rather than a single component, and conclude that this should be taken into account in the analysis. With present instrumentation, the entire bleaching process can be recorded and analyzed.

In a recent application of pbFRET in a microscope, it has been shown that glycosyl–phosphatidylinositol (GPI)-anchored proteins are clustered in Golgi membranes and transport vesicles before they are delivered to the cell surface (Hannan et al., 1993); the donor and acceptor for the FRET measurements were FITC-Fab- and Texas Red-Fab-labeled fragments made from antibodies against the GPI-anchored protein that is sorted in normal cells. The rate of photobleaching on delivery of the GPI-anchored protein to the surface of wild-type cells was slower when D/A-labeled proteins were present than when only D-labeled proteins were present, indicating that the proteins are clustered (i.e., display FRET) before delivery to the surface. However, this was also true for mutant cells that are known not to sort the protein to the apical surface, showing that clustering is not sufficient for apical sorting.

An interesting variation of *donor*-pbFRET has recently been published (Mekler, 1994). The efficiency of FRET is determined from the kinetics of photobleaching of the acceptor that has been sensitized by a donor (i.e., *acceptor*-pbFRET). It is claimed that *acceptor*-pbFRET allows exceptionally accurate efficiencies of 10^{-2} to 10^{-3} to be measured!

The pbFRET method was also used to investigate the overlapping binding of two strongly competing antibodies; the binding of one of the antibodies is blocked by a polyanionic compound [aurintricarboxylic acid (ATA)] that blocks the human immunodeficiency virus (HIV)-binding epitope of CD4 (Szabò et al., 1993). The efficiencies were used to estimate the range of possible distances between the binding sites, and the FRET measurements suggested that the competing antibodies bind to overlapping but nonidentical sites.

The hypothesis that both superoxide anions (O_2^-) and lactoferrin are simultaneously delivered in granules to erythrocyte target phagosomes during phagocytosis has been tested by a combination of Søret band transmission light microscopy (to register O_2^- entry into the erythrocyte) and FRET microscopy (Maher et al., 1993). Energy transfer between FITC-labeled anti-lactoferrin IgG (lactoferrin is a specific granule marker) and anti-sheep red blood cells (anti-SRBC) was observed following antibody-dependent phago-cytosis into the erythrocytes, showing the parallel delivery of lactoferrin with superoxide.

Steady-state FRET measurements in the confocal microscope have been used to study hybrid formation following microinjection of separately labeled (with fluorescein and rhodamine) oligonucleotides into living cells (Sixou et al., 1994). The role of ssDNA-binding proteins and nucleases in a cellular environment is discussed.

The increase in the proximity of LFA-1 and microfilaments induced by ligation of the T cell receptor CD3 complex has been demonstrated by a microscope FRET study (Poo et al., 1994) in which both imaging and photon-counting techniques were used.

The spectral and target specificity characteristics of N-(7-dimethylamino-4-methylcoumarinyl) maleimide (a thiol-labeling fluorophore, donor) and propidium iodide (a DNA-binding fluorophore, acceptor) have been used in a microscope imaging FRET study employing a steady-state ratio technique, where chromatin of spermatozoa during sperm maturation was analyzed at different levels of condensation (Bottiroli et al., 1994). The availability of the cysteine thiol groups on the protamines (the basic proteins used to pack DNA tightly into a sperm head) was probed by the fluorescent thiol reagent, and the efficiency of energy transfer to the DNA-bound dye was found to increase as the extent of cysteine reaction increased.

In a remarkable application of FRET confocal microscopy (Bacskai et al., 1993) (see also the earlier paper by Adams et al., 1991), the cAMP-dependent dissociation of protein kinase A into catalytic (C) and regulatory (R) subunits (when C was labeled with fluorescein and R with rhodamine; the labeled associated subunits display significant energy transfer) was used as a dynamic FRET assay to determine the level of free cAMP in sensory neurons; a steep gradient of the FRET signal in response to a uniform application of seratonin (induces cAMP release) and other cAMP activators indicated a correspondingly steep spatial gradient of cAMP. These FRET results (Bacskai et al., 1993) are interpreted as indicating compartmentation or sequestering of cAMP within the cell. In addition, the dissociated C subunit of the kinase translocates slowly into the nucleus if [cAMP] is persistently high; thus, strong [cAMP] could lead to phosphorylation and therefore to stimulation of transcription factors.

FRET imaging microscopy has been used to show that the tumor suppressor protein p53 and the human papillomavirus E6 protein exist as a complex in the cytoplasm of cervical carcinoma cells (Liang et al., 1993). An indirect immunoassy using fluorescein and rhodamine secondary antibodies for specific antibodies against the p53 and E6 proteins showed significant fluorescence increases attributable (see their methods) to FRET in the cytoplasm. This indicates not only colocalization of the p53–E6 proteins in the cytoplasm but also their close association; this may account for the E6 inactivation of p53 tumor suppressor function in the nucleus. A recent short survey of FRET microscopy has been included in a general article on ratio fluorescence microscopy (Dunn et al., 1994).

7.23. CONCLUSION

In the last few years, the number of applications of FRET has grown considerably in many disparate research fields. There are many reasons. A large variety of new dye-conjugated molecules have been synthesized that couple specific biological functions with convenient spectroscopic properties for FRET, and many of these are commercially available. The improvements in and sophistication of FRET data analysis, especially of temporally resolved fluorescence data, and the development of highly sensitive and reliable steady-state and time-resolved fluorescence instrumentation have increased considerably the possibilities for FRET measurements. Accurate data sets can be acquired in a variety of diverse experimental conditions, and molecular models can be fitted to the total data in global data analyses. Attempts are now being made to use FRET for detecting and quantifying complex molecular properties and interactions of macromolecules, such as their intramolecular dynamics and conformational and spatial distribution functions. This progress has been particularly dramatic in biochemistry and molecular biology, where many new and reliable techniques of synthesis and purification have become available to the researcher, making it possible to construct macromolecules according to design with the desired functional properties; these macromolecules can often be conjugated with extrinsic chromophores at defined sites.

The recent progress in the performance and availability of sophisticated image data acquisition and image analysis has also made it possible to contemplate more quantitative FRET experiments in optical microscopes. We are already seeing progressive incorporation of the rich data acquisition and analysis techniques available in fluorescence cuvette experiments into micro-scale imaging environments; by making spatially resolved FRET measure-ments in an image, the distribution of molecular-scale interactions over distances of microns provides important information for understanding the intricate interactions in functioning biological systems. Capitalizing on the enormous potential of FRET to gather structural and dynamic molecular information from *in situ* functioning biological systems with complex, interac-tive, and highly variable environments represents a challenge to the physical and biological researcher. This endeavor will surely profit from the integrated effort of multidisciplinary research and the interactive cooperation between several different scientific disciplines. It was this noticeable trend by various research laboratories toward more quantitative applications of FRET in microscopy, with all of its advantages, possibilities, and complexities, that provided the incentive to elaborate on some of the more quantitative funda-mental aspects of FRET measurements in this chapter on FRET microscopy.

ACKNOWLEDGMENTS

It has been my pleasure to work and collaborate with the following people on the applications of FRET to study nucleic acid structures, and I thank them also for many discussions: D. M. J. Lilley, C. Gohlke, A. I. H. Murchie, F. Stühmeier, F. Walter, G.Vámosi, and A. Zechel. For collaborations and for many interesting discussions on various aspects of microscopy and spectroscopy, I thank T. M. Jovin, D. Arndt-Jovin, G. Marriott, and D. Gadella. I am also grateful to many of these people, and to P. Schneider and O. Holub, for reading and criticizing the manuscript. I also thank M. Clegg for assistance in proofreading and B. F. Clegg for assistance in the literature search.

REFERENCES

Adair, B. D., and Engelman, D. M. (1994). Glycophorin a helical transmembrane domains dimerize in phospholipid bilayers: A resonance energy transfer study. *Biochemistry* **33**, 5539–5544.

Adams, S. R., Harootunian, A. T., Buechler, Y. J., Taylor, S. S., and Tsien, R. Y. (1991). Fluorescence ratio imaging of cyclic AMP in single cells. *Nature (London)* **349**, 694–697.

Albaugh, S., and Steiner, R. F. (1989). Determination of distance distribution from time domain fluorometry. *J. Phys. Chem.* **93**, 8013–8016.

Albaugh, S., Lan, J., and Steiner, R. F. (1989). The effect of distribution of separations upon intramolecular distances in biopolymers, as determined by radiationless energy transfer. *Biophys. Chem.* **33**, 71–76.

Amin, M., Harrington, K., and von Wandruszka, R. (1993). Determination of steroids in urine by micellar HPLC with detection by sensitized terbium fluorescence. *Anal. Chem.* **65**, 2346–2351.

Amir, D., Krausz, S., and Haas, E. (1992). Detection of local structures in reduced unfolded bovine pancreatic trypsin inhibitor. *Proteins: Struct., Funct., Genet.* **13**, 162–173.

Arts, E. G. J. M., Kuiken, J., Jager, S., and Hoekstra, D. (1993). Fusion of artificial membranes with mammalian spermatozoa. Specific involvement of the equatorial segment after acrosome reaction. *Eur. J. Biochem.* **217**, 1001–1009.

Atkins, P. W. (1983). *Molecular Quantum Mechanics.* Oxford University Press, Oxford.

Bacskail, B. J., Hochner, B., Mahaut-Smith, M., Adams, S. R., Kaang, B.-K., Kandel, E. R., and Tsien, R. Y. (1993). Spatially resolved dynamics of cAMP and protein kinase A subunits in *Aplysia* sensory neurons. *Science* **260**, 222–226.

Baird, B. A., Pick, U., and Hammes, G. G. (1979). Structural investigation of reconstituted chloroplast ATPase with fluorescence measurements. *J. Biol. Chem.* **254**, 3818–3825.

Baker, K. J., East, J. M., and Lee, A. G. (1994). Localization of the hinge region of the Ca^{2+}-ATPase of sarcoplasmic reticulum using resonance energy transfer. *Biochim. Biophys. Acta* **1192**, 53–60.

Barth, P., Beck, K. H., Drexhage, K. H., Kuhn, H., Möbius, D., Molzahn, D., Röllig, K., Schäfer, F. P., Sperling, W., and Zwick, M. M. (1966). *Optische und elektrische Phänomene an monomolekularen Farbstoffschichten.* Verlag Chemie, Weinheim.

Barzykin, A. V. (1991). Spatial distribution of probes in a micelle: Effects on the energy transfer time-resolved measurables. *Chem. Phys.* **155**, 221–231.

Baumann, J., and Fayer, M. D. (1986). Excitation transfer in disordered two-dimensional and anisotropic three-dimensional systems: Effects of spatial geometry on time-resolved observables. *J. Chem. Phys.* **85**, 4087–4107.

Beddard, G. S. (1983). Energy migration in disordered systems. In *Time-Resolved Fluorescence Spectroscopy in Biochemistry and Biology* (R. B. Cundall and R. E. Dale, Eds.), Vol. 69, pp. 451–461. Plenum, New York.

Beecham, J. M., and Haas, E. (1989). Simultaneous determination of intramolecular distance distributions and conformational dynamics by global analysis of energy transfer measurements. *Biophys. J.* **49**, 1225–1236.

Ben-Efraim, I., Bach, D., and Shai, Y. (1993). Spectroscopic and functional characterization of the putative transmembrane segment of the minK potassium channel. *Biochemistry* **32**, 2371–2377.

Bennett, R. G. (1964). Radiationless intermolecular energy transfer. I. Singlet → singlet transfer. *J. Chem. Phys.* **41**, 3037–3040.

Bennett, R. G., and Kellogg, R. E. (1967). Mechanisms and rates of radiationless energy transfer. *Prog. React. Kinet.* **4**, 215–238.

Bennett, R. G., Schwenker, R. P., and Kellogg, R. E. (1964). Radiationless intermolecular energy transfer. II. Triplet → singlet transfer. *J. Chem. Phys.* **41**, 3040–3041.

Benson, D. M., Bryan, J., Plant, A. L., Gotto, A. M., Jr., and Smith, L. C. (1985). Digital imaging fluorescence microscopy: Spatial heterogeneity of photobleaching rate constants in individual cells. *J. Cell. Biol.* **100**, 1309–1323.

Benson, S. C., Mathies, R. A., and Glazer, A. N. (1993a). Heterodimeric DNA-binding dyes designed for energy transfer: Stability and applications of the DNA complexes. *Nucleic Acids Res.* **21**, 5720–5726.

Benson, S. C., Singh, P., and Glazer, N. (1993b). Heterodimeric DNA-binding dyes designed for energy transfer: synthesis and spectroscopic properties. *Nucleic Acids Res.* **21**, 5727–5735.

Blumen, A., and Klafter, J. (1986). Influence of restricted geometries on the direct energy transfer. *J. Chem. Phys.* **84**, 1397–1401.

Bojarski, C., and Sienicki, K. (1990). Energy transfer and migration in fluorescent solutions. In *Photochemistry and Photophysics* (J. F. Rabek, Ed.), Vol. 1, pp. 1–57. CRC Press, Boca Raton, FL.

Bottiroli, G., Croce, A. C., and Ramponi, R. (1992). Fluorescence resonance energy transfer imaging as a tool for *in situ* evaluation of cell morphofunctional characteristics. *J. Photochem. Photobiol. B:Biol.* **12**, 413–416.

Bottiroli, G., Croce, A. C., Pellicciari, C., and Ramponi, R. (1994). Propidium iodide and the thiol-specific reagent DACM as a dye pair for fluorescence resonance energy transfer analysis: An application to mouse sperm chromatin. *Cytometry* **15**, 106–116.

Brand, L., and Witholt, B. (1967). Fluorescence measurements. *Methods Enzymol.* **11**, 776–856.

Cantor, C. R., and Pechukas, P. (1971). Determination of distance distribution functions by singlet–singlet energy transfer. *Proc. Natl. Acad. Sci. U.S.A.* **68**, 2099–2101.

Cantor, C. R., and Tao, T. (1971). Application of fluorescence techniques to the study of nucleic acids. *Procedures Nucleic Acid Res.* **2**, 31–93.

Cardullo, R. A., Agrawal, S., Flores, C., Zamecnik, P. C., and Wolf, D. E. (1988). Detection of nucleic acid hybridization by nonradiative fluorescence resonance energy transfer. *Proc. Natl. Acad. Sci. U.S.A.* **85**, 8790–8794.

Cardullo, R. A., Mungavon, R. M., and Wolf, D. E. (1991). Imaging membrane organization and dynamics. In *Biophysical and Biochemical Aspects of Fluorescence
Spectroscopy* (T. G. Dewey, Ed.), pp. 231–260. Plenum, New York.

Censullo, R., Martin, J. C., and Cheung, H. C. (1992). The use of the isotropic
orientation factor in fluorescence resonance energy transfer (FRET) studies of the
actin filament. *J. fluoresc.* **2**, 141–155.

Chan, S. S., Arndt-Jovin, D. J., and Jovin, T. M. (1979). Proximity of lectin receptors on
the cell surface measured by fluorescence energy transfer in a flow system. *J.
Histochem. Cytochem.* **27**, 56–64.

Chang, D. S. C., and Filipescu, N. (1972). Unusually weak electronic interaction between two aromatic chromophores less than 10 Å apart in a rigid model molecule. *J.
Am. Chem. Soc.* **94**, 4170–4175.

Cheung, H. C. (1991). Resonance energy transfer. In *Topics in Fluorescence Spectroscopy* (J. R. Lakowicz, Ed.), Vol. 3, pp. 127–176. Plenum, New York.

Clegg, R. M. (1992). Fluorescence resonance energy transfer and nucleic acids. *Methods
Enzymol.* **211**, 353–388.

Clegg, R. M. (1995). Fluorescence resonance energy transfer. *Curr. Opin. Biotechnol.* **6**,
103–110.

Clegg, R. M., Murchie, A. I. H., Zechel, A., Carlberg, C., Diekmann, S., and Lilley,
D. M. J. (1992). Fluorescence resonance energy transfer analysis of the structure of
the four-way junction. *Biochemistry* **31**, 4846–4856.

Clegg, R. M., Murchie, A. I. H., Zechel, A., and Lilley, D. M. J. (1993). Observing the
helical geometry of double-stranded DNA in solution by fluorescence resonance
energy transfer. *Proc. Natl. Acad. Sci. U.S.A.* **90**, 2994–2998.

Clegg, R. M., Murchie, A. H., and Lilley, D. M. J. (1994). The solution structure of the
four-way DNA junction at low-salt conditions: A fluorescence resonance energy
transfer analysis. *Biophys. J.* **66**, 99–109.

Cooper, J. P., and Hagerman, P. J. (1990). Analysis of fluorescence energy transfer in
duplex and branched DNA molecules. *Biochemistry* **29**, 9261–9268.

Corbalan-Garcia, S., Teruel, J. A., and Gomez-Fernandez, J. C. (1993). Intramolecular
distances within the Ca^{2+}-ATPase from sarcoplasmic reticulum as estimated
through fluorescence energy transfer between probes. *Eur. J. Biochem.* **217**, 737–744.

Cortese, J. D., and Hackenbrock, C. R. (1993). Motional dynamics of functional
cytochrome *c* delivered by low pH fusion into the intermembrane space of intact
mitochondria. *Biochim. Biophys. Acta* **1142**, 194–202.

Cronce, D. T., and Horrocks, W. D. Jr. (1992). Probing the metal-binding sites of cod
parvalbumin using europium(III) ion luminescence and diffusion-enhanced energy
transfer. *Biochemistry* **31**, 7963–7969.

Dale, R. E., and Eisinger, J. (1974). Intramolecular distances determined by energy
transfer. Dependence on orientational freedom of donor and acceptor. *Biopolymers*
13, 1573–1605.

Dale, R. E., and Eisinger, J. (1976). Intramolecular energy transfer and molecular
conformation. *Proc. Natl. Acad. Sci. U.S.A.* **73**, 271–273.

Dale, R. E., Eisinger, J., and Blumberg, W. E. (1979). The orientational freedom of molecular probes. The orientation factor in intramolecular energy transfer. *Biophys. J.* **26**, 161–194.

Dale, R. E., Novros, J., Roth, S., Edidin, M., and Brand, L. (1981). Application of Forster long-range excitation energy transfer to the determination of distributions of fluorescently-labelled Concanavalin A-receptor complexes at the surfaces of yeast and of normal and malignant fibroblasts. In *Fluorescent Probes* (G. S. Beddard, and M. A. West, Eds.), pp. 159–181. Academic Press, New York.

Dau, H. (1994). Molecular mechanisms and quantitative models of variable photosystem II fluorescence. *Photochem. Photobiol.* **60**, 1–23.

Davenport, L., Dale, R. E., Bisby, R. H., and Cundall, R. B. (1985). Transverse location of the fluorescent probe 1,6-diphenyl-1,3,5-hexatriene in model lipid bilayer membrane systems by resonance excitation energy transfer. *Biochemistry* **24**, 4097–4108.

Dewey, T. G. (1991). Fluorescence energy transfer in membrane biochemistry. In *Biophysical and Biochemical Aspects of Fluorescence Spectroscopy* (T. G. Dewey, Ed.), pp. 197–230. Plenum, New York.

Dewey, T. G. (1992). Fluorescence resonance energy transfer on fractals. *Acc. Chem. Res.* **25**, 195–200.

Dewey, T. G., and Hammes, G. G. (1980). Calculation of fluorescence resonance energy transfer on surfaces. *Biophys. J.* **32**, 1023–1036.

Dexter, D. L. (1953). A theory of sensitized luminescence in solids. *J. Chem. Phys.* **21**, 836–850.

dos Remedios, C. G., Miki, M., and Barden, J. A. (1987). Fluorescence resonance energy transfer measurements of distances in actin and myosin. A critical evaluation. *J. Muscle Res. Cell Motil.* **8**, 97–117.

Dow, J. D. (1968). Resonance energy transfer in condensed media from a many-particle viewpoint. *Phys. Rev.* **174**, 962–976.

Drake, J. M., Klafter, J., and Levitz, P. (1991). Chemical and biological microstructures as probed by dynamic processes. *Science* **251**, 1574–1579.

Drexhage, K. H., Zwick, M. M., and Kuhn, H. (1963). Sensibilisierte Fluoreszenz nach strahlungslosem Energieübergang durch dünne Schichten. *Ber. Bunsenges. Phys. Chem.* **67**, 62–67.

Dunn, K. W., Mayor, S., Myers, J. N., and Maxfield, F. R. (1994). Applications of ratio fluorescence microscopy in the study of cell physiology. *FASEB J.* **8**, 573–582.

"Dutch" Boltz, R. C., Fischer, P. A., Wicker, L. S., and Peterson, L. B. (1994). Single UV excitation of Hoechst 33342 and ethidium bromide for simultaneous cell cycle analysis and viability determinations on in vitro cultures of murine B lymphocytes. *Cytometry* **15**, 28–34.

Eis, P. S., and Millar, D. P. (1993). Conformational distributions of a four-way DNA junction revealed by time-resolved fluorescence resonance energy transfer. *Biochemistry* **32**, 13852–13860.

Eisinger, J., and Flores, J. (1982). The relative locations of intramembrane fluorescent probes and of the cytosol hemoglobin in erythrocytes, studied by transverse resonance energy transfer. *Biophys. J.* **37**, 6–7.

Ernsting, N. P., Kaschke, M., Kleinschmidt, J., Drexhage, K. H., and Huth, V. (1988). Sub-picosecond time-resolved intramolecular electronic energy transfer in bichromophoric rhodamine dyes in solution. *Chem. Phys.* **122**, 431–442.

Estep, T. N., and Thompson, T. E. (1979). Energy transfer in lipid bilayers. *Biophys. J.* **26**, 195–208.

Eyring, H., Lin, S. H., and Lin, S. M. (1980). *Basic Chemical Kinetics.* Wiley, New York.

Fairclough, R. H., and Cantor, C. R. (1978). The use of singlet–singlet energy transfer to study macromolecular assemblies. *Methods Enzymol.* **48**, 347–379.

Fattal, E., Nir, S., Parente, R. A., and Szoka, F. C., Jr. (1994). Pore-forming peptides induce rapid phospholipid flip-flop in membranes. *Biochemistry* **33**, 6721–6731.

Fernandez, S. M., and Berlin, R. D. (1976). Cell surface distribution of lectin receptors determined by resonance energy transfer. *Nature (London)* **264**, 411–415.

Fleming, P. J., Koppel, D. E., Lau, A. L. Y., and Strittmatter, P. (1979). Intramembrane position of the fluorescent tryptophanyl residue in membrane-bound cytochrome b_5. *Biochemistry* **18**, 5458–5464.

Förster, T. (1946). Energiewanderung und Fluoreszenz. *Naturwissenschaften* **6**, 166–175.

Förster, T. (1947). Fluoreszenzversuche an Farbstoffmischungen. *Angew. Chem. A.* **59**, 181.

Förster, T. (1948). Zwischenmolekulare Energiewanderung und Fluoreszenz. *Ann. Phys. (Leipzig)* **2**, 55–75.

Förster, T. (1949a). Experimentelle und theoretische Untersuchung des zwischenmolekularen Übergangs von Elektronenanregungsenergie. *Z. Naturforsch.* **4A**, 321–327.

Förster, T. (1949b). Versuche zum zwischenmolekularen Übergang von Elektronenanregungsenergie. *Z. Elektrochem.* **53**, 93–100.

Förster, T. (1951). *Fluoreszenz Organischer Verbindungen.* Vandenhoeck & Ruprecht, Göttingen.

Förster, T. (1959). Transfer mechanisms of electronic excitation. *Discuss. Faraday Soc.* **27**, 7–17.

Förster, T. (1960). Transfer mechanisms of electronic excitation energy. *Radiat. Res., Suppl.* **2**, 326–339.

Förster, T. (1965). Delocalized excitation and excitation transfer. In *Modern Quantum Chemistry* (O. Sinanoglu, Ed.), Vol. 3, pp. 93–137. Academic Press, New York.

Fung, B. K.-K., and Stryer, L. (1978). Surface density determination in membranes by fluorescence energy transfer. *Biochemistry* **17**, 5241–5248.

Gadella, T. W. J. J., Jovin, T. M., and Clegg, R. M. (1994). Fluorescence lifetime imaging microscopy (FLIM): Spatial resolution of microstructures on the nanosecond time scale. *Biophys. Chem.* **48**, 221–239.

Ganguly, S. C., and Chaudhury, N. K. (1959). Energy transport in organic phosphors. *Rev. Mod. Phys.* **31**, 990–1017.

Gawrisch, K., Han, K.-H., Yang, J.-S., Bergelson, L. D., and Ferretti, J. A. (1993). Interaction of peptide fragment 828–848 of the envelope glycoprotein of human immunodeficiency virus Type I with lipid bilayers. *Biochemistry* **32**, 3112–3118.

Gazit, E., and Shai, Y. (1993). Structural characterization, membrane interactions, and specific assembly within phospholipid membranes of hydrophobic segments from *Bacillus thuringiensis* var. *isrealensis* cytolytic toxin. *Biochemistry* **32**, 12363–12371.

Genest, D., and Wahl, P. (1983). Energy migration and fluorescence depolarization: Structural studies of ethidium bromide-nucleic acid complexes. In *Time-Resolved Fluorescence Spectroscopy in Biochemistry and Biology* (R. B., Cundall and R. E. Dale Eds.), Vol. 69, pp. 523–539. Plenum, New York.

Ghiggino, K. P., and Smith, T. A. (1993). Dynamics of energy migration and trapping in photoirradiated polymers. *Prog. React. Kinet.* **18**, 375–436.

Gohlke, C., Murchie, A. I. H., Lilley, D. M. J., and Clegg, R. M. (1994). Kinking of DNA and RNA helices by bulged nucleotides observed by fluorescence resonance energy transfer. *Proc. Natl. Acad. Sci. U.S.A.* **91**, 11660–11664.

Gösele, U., Hauser, M., Klein, U. K. A., and Frey, R. (1975). Diffusion and long-range energy transfer. *Chem. Phys. Lett.* **34**, 519–522.

Gösele, U., Hauser, M., and Klein, U. K. A. (1976). Eine neue, ausbaufähige Theorie der Statistik der weitreichenden Dipol-Dipol-Energieübertragung nach Förster. *Z. Phys. Chem. (Wiesbaden)* **99**, 81–96.

Greiner, W. (1986). *Theoretische Physik: Klassische Elektrodynamik*. Verlag Harri Deutsch, Frankfurt.

Gryczynski, I., Wiczk, W., Johnson, M. L., Cheung, H. C., Wang, C.-K., and Lakowicz, J. R. (1988). Resolution of end-to-end distance distributions of flexible molecules using quenching-induced variations of the Forster distance for fluorescence energy transfer. *Biophys. J.* **54**, 577–586.

Gutierrez-Merino, C. (1981). Quantitation of the Förster energy transfer for two-dimensional systems. I. Lateral phase separation in unilamellar vesicles formed by binary phospholipid mixtures. *Biophys. Chem.* **14**, 247–257.

Gutierrez-Merino, C., Munkonge, F., Mata, A. M., East, J. M., Levinson, B. L., Napier, R. M., and Lee, A. G. (1987). The position of the ATP binding site on the $(Ca^{2+} + Mg^{2+})$-ATPase. *Biochim. Biophys. Acta* **897**, 207–216.

Haas, E., and Katchalski-Katzir, E. (1978). Effect of the orientation of donor and acceptor on the probability of energy transfer involving electronic transitions of mixed polarization. *Biochemistry* **17**, 5064–5070.

Haas, E., and Steinberg, I. Z. (1984). Intramolecular dynamics of chain molecules monitored by fluctuations in efficiency of excitation energy transfer. *Biophys. J.* **46**, 429–437.

Haas, E., Katchalski-Katzir, E., and Steinberg, I. Z. (1978). Brownian motion of the ends of oligopeptide chains in solution as estimated by energy transfer between the chain ends. *Biopolymers* **17**, 11–31.

Haas, E., McWherter, C. A., and Scheraga, H. A. (1988). Conformational unfolding in the N-terminal region of ribonuclease A detected by nonradiative energy transfer:

Distribution of interresidue distances in the native, denatured, and reduced-denatured states. *Biopolymers* **27**, 1–21.

Haigh, E. A., Thulborn, K. R., and Sawyer, W. H. (1979). Comparison of fluorescence energy transfer and quenching methods to establish the position and orientation of components within the transverse plane of the lipid bilayer: Application to the Gramicidin A–bilayer interaction. *Biochemistry* **18**, 3525–3532.

Hanicak, A., Maretzki, D., Reimann, B., Pap, E., Visser, A. J. W. G., Wirtz, K. W. A., and Schubert, D. (1994). Erythrocyte band 3 protein strongly interacts with phosphoinositides. *FEBS Lett.* **348**, 169–172.

Hannan, L. A., Lisanti, M. P., Rodriguez-Boulan, E., and Edidin, M. (1993). Correctly sorted molecules of a GPI-anchored protein are clustered and immobile when they arrive at the apical surface of MDCK cells. *J. Cell Biol.* **120**, 353–358.

Hasselbacher, C. A., Street, T. L., and Dewey, T. G. (1984). Resonance energy transfer as a monitor of membrane protein domain segregation: Application to the aggregation of bacteriorhodopsin reconstituted into phospholipid vesicles. *Biochemistry* **23**, 6445–6452.

Hassoon, S., Lustig, H., Rubin, M. B., and Speiser, S. (1984). The mechanism of short-range intramolecular electronic energy transfer in bichromophoric molecules. *J. Phys. Chem.* **88**, 6367–6374.

Haugland, R. P., Yguerabide, J., and Stryer, L. (1969). Dependence of the kinetics of singlet–singlet energy transfer on spectral overlap. *Proc. Natl. Acad. Sci. U.S.A.* **63**, 23–30.

Hauser, M., Klein, U. K. A., and Gösele, U. (1976). Extension of Förster's theory of long-range energy transfer to donor-acceptor pairs in systems of molecular dimensions. *Z. Phys. Chem. (Wiesbaden)* **101**, 255–266.

Herman, B. A. (1989) Resonance energy transfer microscopy. *Methods Cell Biol.* **30**, 216–243.

Herman, B. A., and Fernandez, S. M. (1982). Dynamics and topographical distribution of surface glycoproteins during myoblast fusion: A resonance energy transfer study. *Biochemistry* **21**, 3275–3283.

Hirschfeld, T. (1976). Quantum efficiency independence of the time integrated emission from a fluorescent molecule. *Appl. Opt.* **15**, 3135–3139.

Hochstrasser, R. A., Chen, S.-M., and Millar, D. P. (1992). Distance distribution in a dye-linked oligonuleotide determined by time-resolved fluorescence energy transfer. *Biophys. Chem.* **45**, 133–141.

Holmes, A. S., Suhling, K., and Birch, D. J. S. (1993). Fluorescence quenching by metal ions in lipid bilayers. *Biophys. Chem.* **48**, 193–204.

Holowka, D., and Baird, B. (1983). Structural studies on the membrane-bound immunoglobulin E-receptor complex. 2. Mapping of distance between sites of IgE and the membrane surface. *Biochemistry* **22**, 3475–3484.

Horowitz, A. D., Baatz, J. E., and Whitsett, J. A. (1993). Lipid effects on aggregation of pulmonary surfactant protein SP-C studied by fluorescence energy transfer. *Biochemistry* **32**, 9513–9523.

Houston, P., and Kodadek, T. (1994). Spectrophotometric assay for enzyme-mediated unwinding of double-stranded DNA. *Proc. Natl. Acad. Sci. U.S.A.* **91**, 5471–5474.

Inacker, O., and Kuhn, H. (1974). Energy transfer from dye to specific singlet or triplet energy acceptors in monolayer assemblies. *Chem. Phys. Lett.* **27**, 317–321.

Inacker, O., Kuhn, H., Möbius, D., and Debuch, G. (1976). Manipulation in molecular dimensions. *Z. Phys. Chem.* (*Wiesbaden*) **101**, 337–360.

Jackson, J. D. (1962). *Classical Electrodynamics.* Wiley, New York.

Jovin, T. M. (1979). Fluorescence polarization and energy transfer: Theory and application in flow systems. In *Flow Cytometry and Sorting* (M. R. Melamed, P. F. Mullaney, and M. L. Mendelsohn Eds.), pp. 137–165. Wiley, New York.

Jovin, T. M., and Arndt-Jovin, D. (1989a). FRET microscopy: Digital imaging of fluorescence resonance energy transfer. Application in cell biology. In *Cell Structure and Function by Microspectrofluorometry* (E. Kohen, and J. G. Hirschberg, Eds.), pp. 99–117. Academic Press, San Diego, CA.

Jovin, T. M., and Arndt-Jovin, D. J. (1989b). Luminescence digital imaging microscopy. *Annu. Rev. Biophys. Chem.* **18**, 271–308.

Jovin, T. M., Marriott, G., Clegg, R. M., and Arndt-Jovin, D. J. (1989). Photophysical processes exploited in digital imaging microscopy: Fluorescence resonance energy transfer and delayed luminescence. *Ber. Bunsenges. Phys. Chem.* **93**, 387–391.

Kauzmann, W. (1957). *Quantum Chemistry.* Academic Press, New York.

Kellogg, R. E. (1970). Some aspects of dipole–dipole energy transfer. *J. Lumin.* **1**(2), 435–447.

Ketskeméty. I. (1962). Zwischenmolekulare Energieübertragung in fluoreszierenden Lösungen. *Z. Naturforsch.* **17A**, 666–670.

Kinnunen, P. K. J., Rytömaa, M., Kõiv, A., Lehtonen, J., Mustonen, P., and Aro, A. (1993). Sphingosine-mediated membrane association of DNA and its reversal by phosphatidic acid. *Chem. Phys. Lipids* **66**, 75–85.

Kirk, W. R., Wessels, W. S., and Prendergast, F. G. (1993). Lanthanide-dependent perturbations of luminescence in indolylethylenediaminetetraacetic acid–lanthanide chelate. *J. Phys. Chem.* **97**, 10326–10340.

Klafter, J., and Blumen, A. (1984). Fractal behavior in trapping and reaction. *J. Chem. Phys.* **80**, 875–877.

Klafter, J., and Drake, J. M. (1989). *Molecular Dynamics in Restricted Geometries.* Wiley, New York.

Klein, U. K. A., Frey, R., Hauser, M., and Gösele, U. (1976). Theoretical and experimental investigations of combined diffusion and long-range energy transfer. *Chem. Phys. Lett.* **41**, 139–142.

Kleinfeld, A. (1988). Tertiary structure of membrane proteins determined by fluorescence resonance energy transfer. In *Spectroscopic Membrane Probes* (L. M. Loew, Ed.), Vo. 1, pp. 63–92. CRC Press, Boca Raton, FL.

Koppel, D. E., Fleming, P. J., and Strittmatter, P. (1979). Intramembrane positions of membrane-bound chromophores determined by excitation energy transfer. *Biochemistry* **18**, 5450–5457.

Kost, S. H., and Breuer, H. D. (1991). Energy transfer in solid solutions and on fractal polymer surfaces. *Ber. Bunsenges. Phys. Chem.* **95**, 480–484.

Kubitscheck, U., Kircheis, M. Schweitzer-Stenner, R., Dreybrodt, W., Jovin, T. M., and Pecht, I. (1991). Fluorescence resonance energy transfer on single living cells: Application to binding of monovalent haptens to cell-bound immunoglobulin E. *Biophys. J.* **60**, 307–318.

Kubitscheck, U., Schweitzer-Stenner, R., Arndt-Jovin, D. J., Jovin, T. M., and Pecht, I. (1993). Distribution of type I Fc_ε-receptors on the surface of mast cells probed by fluorescence resonance energy transfer. *Biophys. J.* **64**, 110–120.

Kuhn, H. (1965). Versuche zur Herstellung einfacher organisierter Systeme von Molekülen. *Pure Appl. Chem.* **11**, 345–357.

Kuhn, H. (1968). On possible ways of assembling simple organized systems of molecules. In *Structural Chemistry and Molecular Biology* (A. Rich and N. Davidson, Eds.), pp. 566–572. Freeman, San Francisco.

Kuhn, H. (1970). Classical aspects of energy transfer in molecular systems. *J. Chem. Phys.* **53**, 101–108.

Kuhn, H. (1977). Energieübertragungsmechanismen. In *Biophysik* (W. Hoppe, W. Lohmann, H. Markl, and H. Ziegler, Eds.), pp. 187–197. Springer-Verlag, Berlin.

Kuhn, H., and Möbius, D. (1971). Systems of monomolecular layers-assembling and physiochemical behavior. *Angew. Chem. Int. Ed. Engl.* **10**, 620–638.

Kuhn, H., and Möbius, D. (1993). Monolayer assemblies. In *Investigations of Surfaces and Interfaces* (B. W. Rossiter, and R. C. Baetzold, Eds.), Vol. 9, pp. 375–542. Wiley, New York.

Kuhn, H., Möbius, D., and Bücher, H. (1972). Spectroscopy of monolayer assemblies. *Phys. Methods Chem.* **1**, 577–702.

Kusumi, A., Tsuji, A., Murata, M., Sako, Y., Yoshizawa, A. C., Kagiwada, S., Hayakawa, T., and Ohnishi, S. (1991). Development of a streak-camera-based time-resolved microscope fluorimeter and its application to studies of membrane fusion in single cells. *Biochemistry* **30**, 6517–6527.

Kwiatkowski, M., Samiotaki, M., Lamminmäki, U., Mukkala, V.-M., and Landegren, U. (1994). Solid-phase synthesis of chelate-labelled oligonucleotides: Application in triple-color ligase-mediated gene analysis. *Nucleic Acids Res.* **22**, 2604–2611.

Lakowicz, J. R. (1983). *Principles of Fluorescence Spectroscopy.* Plenum, New York.

Lakowicz, J. R. (1991). *Topics in Fluorescence Spectroscopy.* Plenum, New York.

Lakowicz, J. R., Kúsba, J., Wiczk, W., Gryczynski, I., and Johnson, M. L. (1990a). End-to-end diffusion of a flexible bichromophoric molecule observed by intramolecular energy transfer and frequency-domian fluorometry. *Chem. Phys. Lett.* **173**, 319–326.

Lakowicz, J. R., Szmacinski, H., Gryczynski, I., Wiczk, W., and Johnson, M. L. (1990b).

Influence of diffusion on excitation energy transfer in solutions by gigahertz harmonic content frequency-domain fluorometry. *J. Phys. Chem.* **94**, 8413–8416.

Lakowicz, J. R., Gryczynski, I., Kúsba, J., Wiczk, W., Szmacinski, H., and Johnson, M. L. (1994). Site-to-site diffusion in proteins as observed by energy transfer and frequency-domain fluorometry. *Photochem. Photobiol.* **59**, 16–29.

Lakowicz, J. R., Gryczynski, I., Wiczk, W., Kúsba, J., and Johnson, M. L. (1991a). Correction for incomplete labeling in the measurement of distance distributions by frequency-domain fluorometry. *Anal. Chem.* **195**, 243–254.

Lakowicz, J. R., Kúsba, J., Szmacinski, H., Gryczynski, I., Eis, P. S., Wiczk, W., and Johnson, M. L. (1991b). Resolution of end-to-end diffusion coefficients and distance distributions of flexible molecules using fluorescent donor–acceptor and donor–quencher pairs. *Biopolymers* **31**, 1363–1378.

Lamola, A. A. (1969). Energy transfer in solution: Theory and applications. In *Energy Transfer and Organic Chemistry* (A. A. Lamola, and N. J. Turro. Eds.), Vol. 14, pp. 17–132. Wiley (Interscience), New York.

Langlois, R. G., and Jensen, R. H. (1979). Interactions between pairs of DNA-specific fluorescent stains bound to mammalian cells. *J. Histochem. Cytochem.* **27**, 72–79.

Latt, S. A., Cheung, H. T., and Blout, E. R. (1965). Energy transfer: A system with relatively fixed donor–acceptor separation. *J. Am. Chem. Soc.* **87**, 995–1003.

Latt, S. A., Sahar, E., and Eisenhard, M. E. (1979). Pairs of fluorescent dyes as probes of DNA and chromosomes. *J. Histochem. Cytochem.* **27**, 65–71.

Laurence, D. J. R. (1957). Fluorescence techniques for the enzymologist. *Methods Enzymol.* **4**, 174–212.

LePecq, J. B., and Paoletti, C. (1967). A fluorescent complex between ethidium bromide and nucleic acids. Physical-chemical characterization. *J. Mol. Biol.* **27**, 87.

Liang, X. H., Volkmann, M., Klein, R., Herman, B., and Lockett, S. J. (1993). Co-localization of the tumor-suppressor protein p53 and human papillomavirus E6 protein in human cervical carcinoma cell lines. *Oncogene* **8**, 2645–2652.

Lilley, D. M. J., and Clegg, R. M. (1993). The structure of branched DNA species. *Q. Rev. Biophys.* **26**, 131–175.

Lin, S. H. (1973). On the theory of non-radiative transfer of electronic excitation. *Proc. R. Soc. London, Ser. A* **335**, 51–66.

Lin, S. H., Xiao, W. Z., and Dietz, W. (1993). Generalized Förster–Dexter theory of photoinduced intramolecular energy transfer. *Phys. Rev. E* **47**, 3698–3706.

Liu, Y. S., Li, L., Ni, S., and Winnik, M. A. (1993). Recovery of acceptor concentration distribution in a direct energy transfer experiment. *Chem. Phys.* **177**, 579–589.

Luan, P., and Glaser, M. (1994). Formation of membrane domains by the envelope proteins of vesicular stomatitis virus. *Biochemistry* **33**, 4483–4489.

Ludwig, M., Hensel, N. F., and Hartzman, R. J. (1992). Calibration of a resonance energy transfer imaging system. *Biophys. J.* **61**, 845–857.

Maher, R. J., Cao, D., Boxer, L. A., and Petty, H. R. (1993). Simultaneous calcium-dependent delivery of neutrophil lactoferrin and reactive oxygen metabolites to erythrocyte targets: Evidence supporting granule-dependent triggering of superoxide deposition. *J. Cell. Physiol.* **156**, 226–234.

Maliwal, B. P., Kúsba, J., Wiczk, W., Johnson, M. L., and Lakowicz, J. R. (1993). End-to-end diffusion coefficients and distance distributions from fluorescence energy transfer measurements: Enhanced resolution by using multiple acceptors with different Förster distances. *Biophys. Chem.* **46**, 273–281.

Mancheño, J. M., Gasset, M., Lacadena, J., Ramón, F., Martinez del Pozo, A., Oñaderra, M., and Gavilanes, J. G. (1994). Kinetic study of the aggregation and lipid mixing produced by α-sarcin on phosphatidylglycerol and phosphatidylserine vesicles: Stopped-flow light scattering and fluorescence energy transfer measurements. *Biophys. J.* **67**, 1117–1125.

Martin, I., Dubois, M.-C., Defrise-Quertain, F., Saermark, T., Burny, A., Brasseur, R., and Ruysschaert, J.-M. (1994). Correlation between fusogenicity of synthetic modified peptides corresponding to the NH_2-terminal extremity of simian immunodeficiency virus gp32 and their mode of insertion into the lipid bilayer: An infrared spectroscopy study. *J. Virol.* **68**, 1139–1148.

Mata, A. M., Stefanova, H. I., Gore, M. G., Khan, Y. M., East, J. M., and Lee, A. G. (1993). Localization of Cys-344 on the $(Ca^{2+} - Mg^{2+})$-ATPase of sarcoplasmic reticulum using resonance energy transfer. *Biochim. Biophys. Acta* **1147**, 6–12.

Matayoshi, E. D., Wang, G. T., Krafft, G. A., and Erickson, J. (1990). Novel fluorogenic substrates for assaying retroviral proteases by resonance energy transfer. *Science* **247**, 954–957.

Matkó, J., Szöllösi, J., Trón, L., and Damjanovich, S. (1988). Luminescence spectroscopic approaches in studying cell surface dynamics. *Q. Rev. Biophys.* **21**, 479–544.

Matkó, J., Jenei, A., Matyus, L., Ameloot, M., and Damjanovich, S. (1993). Mapping of cell surface protein-patterns by combined fluorescence anisotropy and energy transfer measurements. *J. Photochem. Photobiol. B: Biol.* **19**, 69–73.

Mátyus, L. (1992). Fluorescence resonance energy transfer measurements on cell surfaces: A spectroscopic tool for determining protein interactions. *J. Photochem. Photobiol.* **12**, 323–337.

McLean, L. R., and Hagaman, K. A. (1993). Kinetics of the interaction of amphipathic α-helical peptides with phosphatidylcholines. *Biochim. Biophys. Acta* **1167**, 289–295.

Mekler, V. M. (1994). A photochemical technique to enhance sensitivity of detection of fluorescence resonance energy transfer. *Photochem. Photobiol.* **59**, 615–620.

Mergny, J.-L., Boutorine, A. S., Garestier, T., Belloc, G., Rougée, M., Bulychev, N. V., Koshkin, A. A., Bourson, J., Lebedev, A. V., Valeur, B., Thuong, N. T., and Hélène, C. (1994). Fluorescence energy transfer as a probe for nucleic acid structures and sequences. *Nucleic Acids Res.* **22**, 920–928.

Mersol, J. V., Wang, H., Gafni, A., and Steel, D. G. (1992). Consideration of dipole orientation angles yields accurate rate equations for energy transfer in the rapid diffusion limit. *Biophys. J.* **61**, 1647–1655.

Miki, M., O'Donoghue, S., and dos Remedios, C. G. (1992). Structure of actin observed by fluorescence resonance energy transfer spectroscopy. *J. Muscle Res. Cell. Motil.* **13**, 132–145.

Morrison, L. E. (1988). Time-resolved detection of energy transfer: Theory and application to immunoassays. *Anal. Biochem.* **174**, 101–120.

Morrison, L. E., and Stols, L. M. (1993). Sensitive fluorescence-based thermodynamic and kinetic measurements of DNA hybridization in solution. *Biochemistry* **32**, 3095–3104.

Morrison, L. E., Halder, T. C., and Stols, L. M. (1989). Solution-phase detection of polynucleotides using interacting fluorescent labels and competitive hybridization. *Anal. Biochem.* **183**, 231–244.

Mugnier, J., Pouget, J., Bourson, J., and Valeur, B. (1985). Efficiency of intramolecular electronic energy transfer in coumarin bichromophoric molecules. *J. Lumin.* **33**, 273–300.

Murchie, A. I. H., Clegg, R. M., von Kitzing, E., Duckett, D. R., Dieckmann, S., and Lilley, D. M. J. (1989). Fluorescence energy transfer shows that the four-way DNA junction is a right-handed cross of antiparallel molecules. *Nature (London)* **341**, 763–766.

Mustonen, P., Lehtonen, J., Kõiv, A., and Kinnunen, P. K. J. (1993). Effects of sphingosine on peripheral membrane interactions: Comparison of adriamycin, cytochrome c, and phospholipase A_2. *Biochemistry* **32**, 5373–5380.

Narayanaswami, V., and McNamee, M. G. (1993). Protein–lipid interactions and *Torpedo californica* nicotinic acetylcholine receptor function. 2. Membrane fluidity and ligand-mediated alteration in accessibility of γ subunit cysteine residues to cholesterol. *Biochemistry* **32**, 12420–12427.

Narayanaswami, V., Kim, J., and McNamee, M. G. (1993). Protein–lipid interactions and *Torpedo californica* nicotinic acetylcholine receptor function. 1. Spatial disposition of cysteine residues in the γ subunit analyzed by fluorescence-quenching and energy transfer measurements. *Biochemistry* **32**, 12413–12419.

O'Donoghue, S. I., Hambly, B. D., and dos Remedios, C. G. (1992). Models of the actin monomer and filament from fluorescence resonance-energy transfer. *Eur. J. Biochem.* **205**, 591–601.

Oida, T., Sako, Y., and Kusumi, A. (1993). Fluorescence lifetime imaging microscopy (flimscopy). *Biophys. J.* **64**, 676–685.

Ozinskas, A. J., Malak, H., Joshi, J., Szmacinski, H., Britz, J., Thompson, R. B., Koen, P. A., and Lakowicz, J. R. (1993). Homogeneous model immunoassay of thyroxine by phase-modulation fluorescence spectroscopy. *Anal. Biochem.* **213**, 264–270.

Pap, E. H. W., Bastiaens, P. I. H., Borst, J. W., van den Berg, P. A. W., van Hoek, A., Snoek, G. T., Wirtz, K. W. A., and Visser, A. J. W. G. (1993). Quantitation of the interaction of protein kinase C with diacylglycerol and phosphoinositides by time-resolved detection of resonance energy transfer. *Biochemistry* **32**, 13310–13317.

Peng, L., Alcaraz, M.-L., Klotz, P., Kotzyba-Hibert, F., and Goeldner, M. (1994).

Photochemical labeling of membrane-associated and channel-forming domains of proteins directed by energy transfer. *FEBS Lett.* **346**, 127–131.

Perrin, J. (1927). Fluorescence et induction moléculaire par résonance. *C. R. Hebd. Seances Acad. Sci.*, **184**, 1097–1100.

Peters, R. (1971). Study of membrane thickness by energy transfer. *Biochim. Biophys. Acta* **233**, 465–468.

Poo, H., Fox, B. A., and Petty, H. R. (1994). Ligation of CD3 triggers transmembrane proximity between LFA-1 and cortical microfilaments in a cytotoxic T cell clone derived from tumor infiltrating lymphocytes: A quantitative resonance energy transfer microscopy study. *J. Cell. Physiol.* **159**, 176–180.

Remmers, A. E., and Neubig, R. R. (1993). Resonance energy transfer between guanine nucleotide binding protein subunits and membrane lipids. *Biochemistry* **32**, 2409–2414.

Rice, K. G., Wu, P., Brand, L., and Lee, Y. C. (1993). Modification of oligosaccharide antenna flexibility induced by exoglycosidase trimming. *Biochemistry* **32**, 7264–7270.

Ringsdorf, H., Sackmann, E., Simon, J., and Winnik, F. M. (1993). Interactions of liposomes and hydrophobically-modified poly-(N-isopropylacrylamides): An attempt to model the cytoskeleton. *Biochim. Biophys. Acta* **1153**, 335–344.

Robinson, G. W., and Frosch, R. P. (1962). Theory of electronic energy relaxation in the solid phase. *J. Chem. Phys.* **37**, 1962–1973.

Robinson, G. W., and Frosch, R. P. (1963). Electronic excitation transfer and relaxation. *J. Chem. Phys.* **38**, 1187–1203.

Rye, J. S., Drees, B. L., Nelson, H. C. M., and Glazer, A. N. (1993). Stable fluorescent dye–DNA complexes in high sensitivity detection of protein–DNA interactions. *J. Biol. Chem.* **268**, 25229–25238.

Saha, A. K., Kross, K., Kloszewski, E. D., Upson, D. A., Toner, J. L., Snow, R. A., Black, C. D. V., and Desai, V. C. (1993). Time-resolved fluorescnece of a new europium chelate complex: Demonstration of highly sensitive detection of protein and DNA samples. *J. Am. Chem. Soc.* **115**, 11032–11033.

Sanderson, A., Holmes, A. S., McLoskey, D., Birch, D. J. S., and Imhof, R. E. (1993). A single-photon timing fluorometer with liquid light guide sensing. *Proc. SPIE* **1885**, 466.

Schiff, L. I. (1968). *Quantum Mechanics.* McGraw-Hill, New York.

Schiller, P. W. (1975). The measurement of intramolecular distances by energy transfer. In *Biochemical Fluorescence: Concepts* (R. F. Chen and H. Edelhoch, eds.), pp. 285–303. Dekker, New York.

Schleicher, J., Hof, M., and Schneider, F. W. (1993). Determination of fractal dimensions of Xerogels via Förster energy transfer. *Ber. Bunsenges. Phys. Chem.* **97**, 172–176.

Scholes, G. D., and Ghiggino, K. P. (1994). Rate expressions for excitation transfer I. Radiationless transition theory perspective. *J. Chem. Phys.* **101**, 1251–1261.

Selvin, P. R. (1995). Fluorescence resonance energy transfer. *Methods Enzymol.* **246**, 300–334.

Selvin, P. R., and Hearst, J. E. (1994). Luminescence energy transfer using a terbium chelate: Improvements on fluorescence energy transfer. *Proc. Natl. Acad. Sci. U.S.A.* **91**, 10024–10028.

Selvin, P. R., Rana, T. M., and Hearst, J. E. (1994). Luminescence resonance energy transfer. *J. Am. Chem. Soc.* **116**, 6029–6030.

Shaklai, N., Yguerabide, J., and Ranney, H. M. (1977). Interaction of hemoglobin with red blood cell membranes as shown by a fluorescent chromophore. *Biochemistry* **16**, 5585–5592.

Silvius, J. R., and Zuckermann, M. J. (1993). Interbilayer transfer of phospholipid-anchored macromolecules via monomer diffusion. *Biochemistry* **32**, 3153–3161.

Sixou, S., Szoka, F. C., Jr., Green, G. A., Giusti, B., Zon, G., and Chin, D. J. (1994). Intracellular oligonucleotide hybridization detected by fluorescence resonance energy transfer (FRET). *Nucleic Acids Res.* **22**, 662–668.

Sklar, L. A., Doody, M. C., Gotto, A. M., Jr., and Pownall, H. J. (1980). Serum lipoprotein structure: Resonance energy transfer localization of fluorescent lipid probes. *Biochemistry* **19**, 1294–1301.

Speiser, S. (1983). Novel aspects of intermolecular and intramolecular electronic energy transfer in solution. *J. Photochem.* **22**, 195–211.

Stefanova, H. I., Mata, A. M., Gore, M. G., East, M. G., and Lee, A. G. (1993). Labeling the $(Ca^{2+}-Mg^{2+})$-ATPase of sarcoplasmic reticulum at Glu-439 with 5-(bromomethyl)fluorescein. *Biochemistry* **32**, 6095–6103.

Steinberg, I. A. (1971). Long-range nonradiative transfer of electronic excitation energy in proteins and polypeptides. *Annu. Rev. Biochem.* **40**, 83–114.

Steinberg, I. Z. (1968). Nonradiative energy transfer in systems in which rotatory Brownian motion is frozen. *J. Chem. Phys.* **48**, 2411–2413.

Steinberg, I. Z., and Katchalski, E. (1968). Theoretical analysis of the role of diffusion in chemical reactions, fluorescence quenching, and nonradiative energy transfer. *J. Chem. Phys.* **48**, 2404–2410.

Steinberg, I. Z., Haas, E., and Katchalski-Katzir, E. (1983). Long-range nonradiative transfer of electronic excitation energy. In *Time-Resolved Fluorescence Spectroscopy in Biochemistry and Biology* (R. B. Cundall and R. B. Dale, eds.), Vol. 69, pp. 411–450. Plenum, New York.

Steiner, R. F., Albaugh, S., and Kilhoffer, M.-C. (1991). Distribution of separations between groups in an engineered calmodulin. *J. Fluoresc.* **1**, 15–22.

Stratton, J. A. (1941). *Electromagnetic Theory.* McGraw-Hill, New York.

Stryer, L. (1960). Energy transfer in proteins and polypeptides. *Radiat. Res., Suppl.* **2**, 432–451.

Stryer, L. (1968). Fluorescence spectroscopy of proteins. *Science* **162**, 526–533.

Stryer, L. (1978). Fluorescence energy transfer as a spectroscopic ruler. *Annu. Rev. Biochem.* **47**, 819–846.

Stryer, L., and Haugland, R. P. (1967). Energy transfer: A spectroscopic ruler. *Proc. Natl. Acad. Sci. U.S.A.* **58**, 719–726.

Stryer, L., Thomas, D. D., and Meares, C. F. (1982). Diffusion-enhanced fluorescence energy transfer. *Annu. Rev. Biophys. Bioeng.* **11**, 203–222.

Sunderland, J. E., and Storch, J. (1993). Effect of phospholipid headgroup composition on the transfer of fluorescent long-chain free fatty acids between membranes. *Biochim. Biophys. Acta* **1168**, 307–314.

Szabò, G., Pine, P. S., Weaver, J. L., Kasari, M., and Aszalos, A. (1992). Epitope mapping by photobleaching fluorescence resonance energy transfer measurements using a laser scanning microscope system. *Biophys. J.* **61**, 661–670.

Szabò, G., Weaver, J. L., Pine, P. S., and Aszalos, A. (1993). Specific disengagement of cell-bound anti-LAM-1 (anti-L-selectin) antibodies by aurintricarboxylic acid. *Mol. Immunol.* **30**, 1689–1694.

Szöllösi, J., Trón, L., Damjanovich, S., Helliwell, S. H., Arndt-Jovin, D., and Jovin, T. M. (1984). Fluorescence energy transfer measurements on cell surfaces: A critical comparison of steady-state fluorimetric and flow cytometric methods. *Cytometry* **5**, 210–216.

Szöllösi, J., Damjanovich, S., Mulhern, S. A., and Trón, L. (1987a). Fluorescence energy transfer and membrane potential measurements monitor dynamic properties of cell membranes: A critical review. *Prog. Biophys. Mol. Biol.* **49**, 65–87.

Szöllösi, J., Mátyus, L., Trón, L., Balázs, M., Ember, I., Fulwyler, M. J., and Damjanovich, S. (1987b). Flow cytometric measurements of fluorescence energy transfer using single laser excitation. *Cytometry* **8**, 120–128.

ter Beest, M. A., and Hoekstra, D. (1993). Interaction of myelin basic protein with artificial membranes. Parameters governing binding, aggregation and dissociation. *Eur. J. Biochem.* **211**, 689–696.

Thomas, D. D., Carlsen, W. F., and Stryer, L. (1978). Fluorescence energy transfer in rapid-diffusion limit. *Proc. Natl. Acad. Sci. U.S.A.* **75**, 5746–5750.

Trón, L., Szöllösi, J., Damjanovich, S., Helliwell, S. H., Arndt-Jovin, D. J., and Jovin, T. M. (1984). Flow cytometric measurement of fluorescence resonance energy transfer on cell surfaces. Quantitative evaluation of the transfer efficiency on a cell-by-cell basis. *Biophys. J.* **45**, 939–946.

Ullman, E. F., and Khanna, P. L. (1981). Flourescence excitation transfer immunoassay (FETI). In *Immunochemical Techniques. Part C* (J. J. Langone and H. van Vunakis, Eds.) Vol. 74, pp. 28–60. Academic Press, New York.

Ullman, E. F., Bellet, N. F., Brinkley, J. M., and Zuk, R. F. (1980). Homogeneous fluorescence immunoassays. In *Immunoassays: Clinical Laboratory Techniques for the 1980s* (R. M. Nakamura, W. R. Dito and E. S. Tucker, Eds.), pp. 13–43. Alan R. Liss, New York.

Uster, P. S. (1993). *In situ* resonance energy transfer microscopy: Monitoring membrane fusion in living cells. *Methods Enzymol.* **221**, 239–246.

Uster, P. S., and Pagano, R.E. (1986). Resonance energy transfer microscopy: Observations of membrane-bound fluorescent probes in model membranes and in living cells. *J. Cell Biol.* **103**, 1221–1234.

Uster, P. S., and Pagano, R. E. (1989). Resonance energy transfer microscopy: Visual colocalization of fluorescent lipid probes in liposomes. *Methods Enzymol.* **171**, 850–857.

Valenzuela, C. F., Weign, P., Yguerabide, J., and Johnson, D. A. (1994). Transverse distance between the membrane and the agonist binding sites on the *Torpedo* acetylcholine receptor: a fluorescence study. *Biophys. J.* **66**, 674–682.

Valeur, B. (1989). Intramolecular excitation energy transfer In bichromophoric molecules—fundamental aspects and applications. In *Fluorescence Biomolecules: Methodologies and Applications* (D. M. Jameson and G. D. Reinhart, Eds.), pp. 269–303. Plenum, New York.

Vanderkooi, J. M., Ierkomas, A., Nakamura, H., and Martonosi, A. (1977). Fluorescence energy transfer between Ca^{2+} transport ATPase molecules in artificial membranes. *Biochemistry* **16**, 1262–1267.

van der Meer, B. W., Raymer, M. A., Wagoner, S. L., Hackney, R. L., Beechem, J. M., and Gratton, E. (1993). Designing matrix models for fluorescence energy transfer between moving donors and acceptors. *Biophys. J.* **64**, 1243–1263.

Vinogradov, A. E. (1994). Measurement by flow cytometry of genomic AT/GC ratio and genome size. *Cytometry* **16**, 34–40.

Walter, A., and Siegel, D. P. (1993). Divalent cation–induced lipid mixing between phosphatidylserine liposomes studied by stopped–flow fluorescence measurements: Effect of temperature, comparison of barium and calcium, and perturbation by DPX. *Biochemistry* **32**, 3271–3281.

Ward, R. J., Palmer, M., Leonard, K., and Bhakdi, S. (1994). Identification of a putative membrane-inserted segment in the α-toxin of *Staphylococcus aureus*. *Biochemistry* **33**, 7477–7484.

Watts, T. H., Gaub, H. E., and McConnell, H. M. (1986). T-cell-mediated association of peptide antigen and major histocompatibility complex protein detected by energy transfer in an evanescent wave-field. *Nature (London)* **320**, 179–181.

Wiczk, W. M., Gryczynski, I., Szmacinski, H., Johnson, M. L., Kruszynski, M., and Zboinska, J. (1988). Distribution of distances in thiopeptides by fluorescence energy transfer and frequency-domain fluorometry. *Biophys. Chem.* **32**, 43–49.

Wilkinson, F. (1965). Electronic energy transfer between organic molecules in solution. *Adv. Photochem.* **3**, 241–268.

Wolber, P. K., and Hudson, B. S. (1979). An analytic solution to the Förster energy transfer problem in two dimensions. *Biophys. J.* **28**, 197–210.

Wolf, H. C. (1967). Energy transfer in organic molecular crystals: A survey of experiment. *Adv. At. Mol. Phys.* **3**, 119–142.

Wu, P., and Brand, L. (1992). Orientation factor in steady-state and time-resolved resonance energy transfer measurements. *Biochemistry* **31**, 7939–7947.

Wu, P., and Brand, L. (1994). Review–Resonance energy transfer: Methods and applications. *Anal. Biochem.* **218**, 1–13.

Yguerabide, J. (1994). Theory for establishing proximity relations in biological membranes by excitation energy transfer measurements. *Biophys. J.* **66**, 683–693.

Young, R. M., Arnette, J. K., Roess, D. A., and Barisas, B. G. (1994). Quantitation of fluorescence energy transfer between cell surface proteins via fluorescence donor photobleaching kinetics. *Biophys. J.* **67**, 881–888.

Zheng, Y., Shopes, B., Holowka, D., and Baird, B. (1991). Conformations of IgE bound to its receptor $Fc_\varepsilon RI$ and in solution. *Biochemistry* **30**, 9125–9132.

Zheng, Y., Shopes, B., Holowka, D., and Baird, B. (1992). Dynamic conformations compared for IgE and IgG1 in solution and bound to receptors. *Biochemistry* **31**, 7446–7456.

Zumofen, G., Blumen, A., and Klafter, J. (1990). The role of fractals in chemistry. *New J. Chem.* **14**, 189–196.

CHAPTER

8

TWO-PHOTON EXCITATION MICROSCOPY

DAVID W. PISTON

Department of Molecular Physiology and Biophysics
Vanderbilt University
Nashville, Tennessee 37232-0615

8.1. INTRODUCTION

Two-photon excitation microscopy (TPEM) is a new form of laser scanning microscopy with advantages over both widefield and confocal laser scanning microscopy (CLSM) for the study of biological systems. Fluorescence micros-copy has become a powerful technique for studies of living cells due to its selec-tivity (directed by attaching probes to specific molecular sites) and sensitivity (detectability of individual photons). Confocal microscopy has increased the applicability of fluorescence methods by its ability to acquire data from three-dimensionally resolved volumes within cells and tissues (White et al., 1987; Chapter 6, this volume). However, confocal microscopy has some serious limitations (Sandison et al., 1994). For instance, the amount of photobleaching per detected fluorescence photon is increased over widefield fluorescence microscopy, because only photons originating from the focal plane are collected despite the fact that fluorescence is generated throughout the sample. In addition, the introduction of a detection pinhole leads to chromatic aberration (Sandison et al., 1994), which can cause a dramatic loss of signal, especially if UV excitation wavelengths are used (Ulfhake et al., 1991; Bliton et al., 1993; Niggli et al., 1994). TPEM offers an alternative technique that can help overcome these limitations.

Two-photon excitation as a single quantum event, where two photons combine their energies to cause the transition to an excited state of the chromophore, was predicted by Maria Göppert-Mayer in 1931. Because of the large intensities required, however, this phenomenon was not observed until the invention of the laser 30 years later (Kaiser and Garrett, 1991). Over the

Fluorescence Imaging Spectroscopy and Microscopy, edited by Xue Feng Wang and Brian Herman. Chemical Analysis Series, Vol. 137.
ISBN 0-471-01527-X © 1996 John Wiley & Sons, Inc.

last 30 years, many spectroscopic studies using two-photon excitation have been reported and reviewed (Friedrich and Mcclain, 1980; Friedrich, 1982; Birge, 1986). Nonlinear optical effects were first used in laser scanning microscopy to produce images of second harmonic generation in crystals (Hellwarth and Christiansen, 1974; Sheppard and Kompfner, 1978). Recently, two-photon excitation of fluorescence and photolysis in laser scanning microscopy was introduced, with the expectation of a powerful new tool for biophysical research (Denk et al., 1990).

This chapter will outline the principles of TPEM, including the basic physical parameters and instrumentation. The strengths and limitations of this technique will be discussed, and many of the applications of TPEM to biological samples will be described briefly. Finally, the outlook for future improvements and applications will be considered. Since a detailed discussion of the physical properties and limitations of TPEM has recently been presented elsewhere (Denk et al., 1994), the emphasis here will be on qualitative descriptions of TPEM and on the wide range of its use for biological studies.

8.2. TWO-PHOTON EXCITATION

Two-photon excitation is the simultaneous (as opposed to sequential) absorption of two photons that combine their energies (mediated by a virtual state) to produce an electronic excitation that is conventionally caused by a single photon with a shorter wavelength (Figure 8.1). For example, two photons of red light can excite an ultraviolet (UV)-absorbing fluorophore. Because two photons are required for each excitation, the excitation rate depends on the square of the instantaneous intensity. While in practice the two photons need not be of identical wavelength, for the experiments described here a single laser is used, so the two photons are similar. Selection rules dictate that excitation transitions allowed for one-photon absorption are not allowed for two-photon absorption (Figure 8.1) (Birge, 1986). However, the density of vibrational states for complex dye molecules in solution, often allows significant overlap between the allowed one-photon and two-photon transitions (McClain, 1971). Even with this overlap, though, differences between one- and two-photon excitation spectra are often observed. Fortunately, in practice, some two-photon excitation usually occurs at a particular wavelength, λ, when the one-photon excitation peak occurs at $\lambda/2$. As diagrammed in Figure 8.1, the emitted fluorescence properties (spectra, lifetime, environmental dependence, etc.) for all fluorophores studied so far with TPEM have been found to be the same whether excitation occurred via one- or two-photon absorption. This allows one to utilize well-established ratiometric and spectroscopic methods in TPEM.

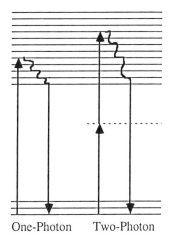

Figure 8.1. Jablonski diagram showing excitation by one or two photons. For two-photon excitation, there is a short-lived virtual state that mediates the absorption (shown by the dashed line). Despite the fact that one- and two-photon absorption populate different energy levels in the excited state, internal conversion (e.g., vibrational relaxation) to the lowest energy level available causes the fluorescence emission characteristics to be independent of the excitation pathway.

In the absence of saturation (ground-state depletion), the probability that a given fluorophore will absorb two photons simultaneously depends on both the spatial and temporal overlap of the incident photons. This is unlike one-photon absorption, in which the absorption cross section is independent of time. For one-photon absorption the number of excitations, N_{ex1}, is equal to the cross section, σ, times the excitation intensity, I; N_{ex1} depends on the number of photons that pass through the sample, but not on the rate at which the photons arrive (N_{ex1}, over time τ with I_0 is the same as N_{ex1} over $\tau/2$ with $2I_0$). A typical σ is $10^{-16}\,\text{cm}^2$. For two-photon absorption $N_{ex2} = \delta I^2$. δ arises from both the spatial extent of the chromophore and the required temporal overlap of incident photons, typically $\delta = 10^{-50}\,\text{cm}^4\,\text{s/photon}$. The two-photon excitation cross section can be understood by a simplified qualitative model. Assuming that each fluorophore sees the same cross section of the laser ($10^{-16}\,\text{cm}^2$), the photons must arrive within 10^{-18} s. This temporal overlap is consistent with the lifetime of the intermediate virtual state shown in Figure 8.1, which is generally 10^{-17} s.

The properties that make TPEM so useful for fluorescence microscopy are consequences of the quadratic dependence of absorption on excitation light intensity. While photon density (laser intensity) must be very high, significant two-photon excitation can be generated without collateral laser damage to the sample by concentrating the illumination in both time and space.

Temporal concentration by subpicosecond mode-locked laser pulses can be used since the probability of two photons being absorbed simultaneously depends on the square of the instantaneous light intensity. With these brief but intense pulses of laser light, a significant amount of two-photon absorption is generated, but the average power is kept relatively low. For biological applications of laser scanning mircoscopy, average excitation power must be minimal because one photon absorption occurs along the entire excitation path and is believed to be responsible for most of the heating and for some of the photodamage. In practice, TPEM is achieved by using 100 fs (10^{-13} s) pulses at a repetition rate of 100 MHz. The use of such short pulses and small duty cycles is essential to permit image acquisition in a reasonable time while preserving sample viability (described below). Note that although these pulses are called "ultrashort," 100 fs is still 4–5 orders of magnitude longer than the reaction time of two-photon absorption.

Spatial concentration of the excitation light is achieved by diffraction-limited focusing by a high numerical aperture (NA) microscope objective lens. In a focused excitation beam, as shown in Figure 8.2 (where the intensity varies as the square of the distance from the focal plane), two-photon absorption probability outside the focal region falls off with the fourth power of the distance along the optical axis (generally denoted the z-direction). For two-photon excitation of a uniform distribution of chromophores, 80% of the absorption occurs in a "focal volume" defined by an ellipsoid about 0.3 μm in diameter and 1 μm long (for $\lambda = 700$ nm) and an objective lens with a 1.4 NA (Sandison and Webb, 1944). This localization of excitation provides depth discrimination equivalent to that of an ideal one-photon confocal microscope

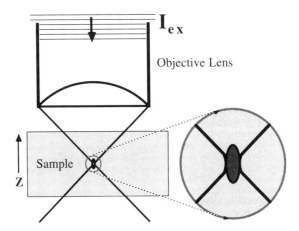

Figure 8.2. Spatial localization of two-photon excitation by high numerical aperture focusing. If the excitation. I_{ex}, is a plane wave into the back aperature of the objective lens, then the focal spot (at $Z = 0$) will be a diffraction-limited ellipsoid. As the area of the illumination cone above and below the focal grows as Z^2, the incident intensity decreases as Z^{-2}.

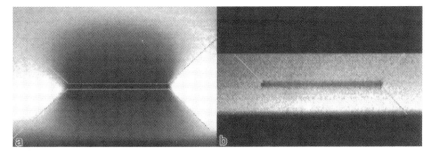

Figure 8.3. (a) XZ profile of the excitation pattern in a confocal microscope formed by repeatedly scanning a single XY optical section in a thick film of rhodamine-Formvar until fluorescence was completely bleached. The box represents the region from which data were collected by the confocal microscope for this optical section. Nearly uniform bleaching occurred both above and below the focal plane. (b) Same excitation pattern, but for two-photon excitation. The lines denote the path taken by the excitation light to reach the focal plane, but no excitation occurs outside the focal plane.

(point detector). In contrast to CLSM, which uses a confocal pinhole, three-dimensional (3-D) resolution is due entirely to the confinement of excitation to the focal volume. Therefore, out-of-focus photobleaching and photodamage do not occur. A further advantage can be gained since there is no attenuation of the excitation beam by out-of-focus absorption. This is demonstrated in Figure 8.3, which shows XZ-images through bleaching patterns generated by repeated scanning of a rectangle in a single XY plane in a thick rhodamine-stained Formvar layer. Figure 8.3a shows the localized photobleaching that results from two-photon excitation scanning; by comparison, Figure 8.3b shows an equivalent image generated by conventional single-photon scanning. Photobleaching in the latter case occurs throughout the sample, demonstrating that with one-photon excitation, background is excited all along the laser beam.

8.3. RESOLUTION AND SIGNAL-TO-BACKGROUND RATIO IN TPEM

Calculations of resolution and the signal-to-background ratio (S/B) can be performed using the same methods used in CLSM theory (Wilson and Sheppard, 1984) simply by plugging in the new excitation wavelength. Resolution in TPEM has been calculated using these methods (Sheppard and Gu, 1990; Nakamura, 1993). For most cell biological applications, the small differences in resolution between TPEM and CLSM are negligible. Significant resolution enhancement with TPEM, albeit at the expense of collection efficiency, can be achieved by using a detection spatial filter (Stelzer et al., 1994).

In order to gather information with 3-D spatial resolution, it is essential to differentiate between fluorescence signals originating inside and outside the focal volume. This differentiation can be done instrumentally, such as by a confocal microscope or TPEM, or by a computer deconvolution of a 3-D data set (Agard et at., 1989; Carrington et al., 1990). The ability of any technique to distinguish in-focus from out-of-focus photons is defined by the S/B, where S is the intensity of the photons arriving at the detector from emission within the focal volume and B is all of the intensity arriving from elsewhere (Sandison and Webb, 1994). For incident red light (~ 700 nm) and a 1.3 NA objective lens, the focal volume is an ellipsoid of diameter 0.3 μm in the plane of focus (X, Y) and 1.0 μm along the optic axis. In CLSM, high S/B is generated by rejection of B by the confocal pinhole, but in TPEM, S/B is inherently high, since very little B is excited. As mentioned for resolution above, these calculations for TPEM are identical to those for CLSM with an infinitely small confocal pinhole. For both TPEM and CLSM, S/B is typically 2 orders of magnitude larger than for widefield fluorescence microscopy of cells.

8.4. POTENTIAL PROBLEMS AND LIMITATIONS IN TPEM

8.4.1. Photodamage

Photodamage to cells can result from one- or two-photon absorption (Denk et al., 1994), depending on the excitation wavelength and on the intrinsic and extrinsic chromophores present. Photodamage may arise from both heating and chemical means. Since photochemical effects are, in general, poorly understood and vary widely among cell types, the discussion here will concentrate on heating effects. Heating in TPEM can arise by two mechanisms: one-photon absorption by water, and two-photon absorption by intrinsic and extrinsic fluorophores. In mammalian cells, there is practically no absorption by intrinsic chromophores at the infrared and near-infrared wavelengths used in TPEM. This may, of course, not be the case in nonmammalian cells (particularly plant cells).

For heating due to one-photon absorption by water, an approximation of the temperature rise can be made by assuming that absorption occurs all along the beam path. The calculation is the same as that used to analyze thermal lens effects (Kliger, 1983). For absorption by pure water, with a 10 μs pixel dwell time and 50 mW average power, the temperature rise is predicted to be 0.065 and 1.1 °C at 700 and 1000 nm, respectively (Denk et al., 1994). These calculations are in agreement with measurements of heating due to 1064 nm optical tweezer laser beams (Liu et al., 1994). For stationary beam applications, greater heating can occur, rising with the logarithm of time. Therefore, heating due to water absorption may sometimes limit the use of TPEM, especially with infrared (IR) excitation of visible fluorophores.

Heating due to two-photon absorption, on the other hand, is restricted to the focal region. The 3-D model, in which the release of heat occurs uniformly within a spherically symmetrical region centered at the focus, indicates that the effect of two-photon generated heating, even with large fluorophore concentrations, is negligible (Denk et al., 1994).

8.4.2. Cell Viability During Imaging

Viability of the biological sample during imaging is the most important constraint on the usefulness of any vital microscopic technique. Damage to cells due to interactions with light is poorly understood. In general, though, living cells interact less with the red-wavelength light (~ 700 nm) used in TPEM than with bluer wavelengths. Due to its lower quantum energy, red light is less able to break bonds, so it is expected to cause less photodamage. Any one-photon effects of the high laser powers used are likely to be due to heating (as discussed above). By contrast two-photon photodamage can arise from interactions with both the imaged fluorophore and other naturally occurring chromophores; in the focal plane, these effects are expected to be just as large for two-photon as for one-photon excitation. Still, any chemical reactions by the excited state of the chromophore persist only in the focal plane, not throughout the entire sample.

Viability tests of unstained samples with 700 nm laser excitation have yielded a rule of thumb for two-photon photodamage. If the content of reduced pyridine nucleotides (NADH, a naturally occurring fluorophore) in the cells is excited at a level sufficient for autofluorescence imaging (see the discussion of applications below), viability can be significantly compromised by TPEM. Under identical excitation condition, loss of viability has not been observed in any case when the focal plane was above or below the sample. In other words, water heating does not seem to be a problem for this wavelength, as is predicted by calculations (see above). Heating may be a problem in the case where 1μm laser light (which is absorbed by water much more strongly than is 700 nm light) is used to excite visible fluorophores, such as rhodamine or fluorescein.

Viability of stained cells can be compromised by exposure to two-photon excitation laser irradiation. However, in much the same fashion as confocal microscopy, viability can be maintained for extended periods by lowering excitation powers to the minimum required for imaging. Studies of tissue culture cell viability have indicated that damage is not strictly proportional to the amount of photobleaching (Ridsdale and Webb, 1993). Whatever the mechanism of damage, however, the localization of fluorophore excitations in TPEM should produce fewer damaging excitations than conventional illumination.

8.5. EXPERIMENTAL CONSIDERATIONS

Instrumentation for TPEM is almost identical to that of a laser scanning confocal microscope. The two differences are the type of laser light source and an increased number of options for detection. Scanning, data collection, and computer control are identical for one- and two-photon instruments. One extra feature of TPEM is its ability to perform time-resolved fluorescence measurements with the mode-locked laser required for TPEM (Piston et al., 1992).

8.6. LASERS FOR TPEM

Several types of laser are currently available. In making a choice, the main considerations are stability, tunability and range of wavelengths, and ease of use. Since two-photon excitation depends on the square of the excitation intensity, the fluorescence signal can show a larger effect of laser power fluctuations. However, the mode-locked lasers used in TPEM show only small fluctuations in amplitude. The small fluctuations in the two-photon-excited fluorescence ($\sim 2\%$) due to typical laser fluctuations ($< 1\%$) are almost always less than the photon shot noise in TPEM experiments ($< 5\%$). In this section, specifications of the usable laser sources are discussed.

8.6.1. Ti:Sapphire Laser

Currently, the self-mode-locked Ti:sapphire laser (Spence et al., 1991) is the preferred laser for most TPEM applications (Curley et al., 1992). This laser provides a large tuning range, from slightly below 700 to above 1000 nm, with pulse lengths of ~ 100 fs and sufficient power (> 100 mW throughout the tuning range) for saturation of two-photon excitation in most fluorophores. Several manufacturers offer this laser system, which requires little maintenance.

8.6.2. CPM Laser

The first two-photon images (Denk et al., 1990) were acquired using a colliding-pulse mode-locked (CPM) laser (Valdemanis et al., 1985). However,

CPM lasers have very limited tunability, give only 15 mW average power, and may require considerable maintenance of the absorber dye solution. However, a CPM allows access to short wavelengths that are not available from the Ti:sapphire laser (e.g., 630 nm).

8.6.3. Hybrid Mode-Locked Dye Laser

A third laser option for two-photon excitation is a hybrid mode-locked dye laser. These systems use actively mode-locked argon-ion or frequency-doubled Nd:YAG lasers to pump a laser dye that also contains an intracavity-saturable absorber jet. Such systems are rather expensive and difficult to operate. They yield > 100 mW and much better tunability than a CPM laser. These lasers, like CPMs, can access the spectral region below 700 nm, which is outside the tuning range of currently available ultrafast solid-state lasers.

8.6.4. Nd:YLF Laser

This entirely solid-state laser, generating light of 1054 nm, is exemplary of a new generation of lasers that are attractive light sources for TPEM in terms of stability, utility requirements, and cost. These systems are compact, and require less power and cooling resources than those described above (Malcolm et al., 1990). Currently, available pulse widths and powers are sufficient for two-photon excitation of visible of fluorophores, such as fluorescein and rhodamine, but are not able to reach fluorophore saturation. As these lasers mature over the next 10 years, however, it is expercted that a wide variety of wavelengths, with increased power and shorter pulse widths, will become available.

8.7. LASER INTENSITY CONTROL

Acousto-optic modulators (AOMs) can be used to control laser power and to shutter the laser beam in laser scanning microscopy experiments (Sandison et al., 1994). However, AOMs are difficult to use in TPEM. First, the deflection angle and efficiency depend on λ, which can spatially spread the ultrashort pulses used in TPEM. Second, efficient AOMs often are made of highly dispersive materials, which spread the laser pulse temporally. While the angular and temporal spreads of an AOM can be partially compensated for, an electro-optic modulator such as a Pockel's cell may provide an adequate, although difficult, solution for intensity control in two-photon excitation. These devices yield intensity extinction ratios of > 100:1 with no noticeable increase in the 80 fs pulse length. Such a linear extinction ratio translates into a 100,000-fold reduction of two-photon excitation, more than enough for most applications.

8.8.   OPTICAL ABERRATIONS

In TPEM, in spite of the strong NA dependence of the peak excitation rate, the total amount of two-photon excitation arising from a focused laser beam in a spatially homogeneous distribution of fluorophores is independent of NA as long as the focal spot maintains a diffraction-limited shape (Birge, 1986). This is because the reduction of the two-photon absorption probability on reducing the NA is exactly compensated for by the increased focal volume that is illuminated. For diffraction-limited focusing, the total amount of two-photon excitation arising from a focused laser beam is independent of NA [though the excitations per volume decrease at lower NA, the size increase of the focal spot directly compensates to keep the total number of excitations the same (Birge, 1986)]. However, in two-photon excitation, any change in the size and shape of the focal volume can affect not only the resolution of the microscopy but also the amount of fluorescence generated. In contrast to CLSM, the total two-photon-excited fluorescence can be severely reduced by optical aberrations and dispersion. Aberrations from the objective lens, and those introduced by focusing through the oil–coverslip and coverslip–water interfaces, also smear the central focal spot and shift the apparent focal point (Ling and Lee, 1984; Sheppard, 1988; Visser and Wiersma, 1991; Visser et al., 1992; Hell et al. 1993). Since these aberrations cause the focal spot to be spread out in a manner that does not maintain its diffraction-limited shape, the total amount of two-photon excitation is decreased. This effect probably reduces the amount of two-photon excited fluorescence $\sim 100\,\mu m$ into the sample to $<50\%$ of its maximal value just inside the coverslip (Bighouse, 1991).

8.9.   PULSE SPREADING DUE TO DISPERSION

The ultrashort pulses used in TPEM can be significantly changed by passing through any dielectric medium. In particular, pulses, are spread in time due to group velocity dispersion (GVD), and the shorter the pulse, the greater this effect becomes. This effect can be partially compensated for by "pre-chirping" the pulse by a prism or grating arrangement (Fork et al., 1984). However, spreading is minimal for pulse widths of 150 fs, which seems to be sufficiently short for TPEM. A complete discussion of GVD effects in TPEM is given elsewhere (Denk et al., 1994).

8.10. DETECTION

Because three-dimensional localization is accomplished in TPEM by excitation alone, increased flexibility in optical design and considerable improvement in fluorescence collection efficiency are possible compared to CLSM (Figure 8.4). The first design option is whether the emitted light passes back through the scanning mirrors (*descanned* detection) or whether the detector is sensitive to emitted light from the whole image area at all times (*nondescanned* detection).

Descanned detection is similar to a confocal arrangement in that the detected light is reflected off the scanning mirrors. This is useful to exclude room light with the use of a large confocal pinhole; to increase resolution at the cost of fluorescence signal collection efficiency (Sheppard and Gu, 1990; Sandison and Webb, 1994); or to allow detectors with small entrance apertures to be used, such as some avalanche photodiodes or spectrometers. However, use of a confocal pinhole to increase resolution produces a large drop in detection efficiency. Such losses in detection efficiency should be avoided since

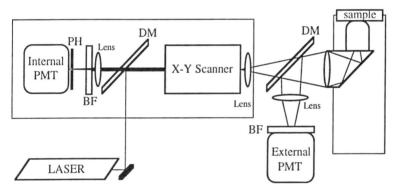

Figure 8.4. A schematic diagram of the two-photon laser scanning microscope, with both descanned and non-descanned detection shown. The incoming laser light is raster scanned ($X - Y$ scanner) and is focused on the sample by the objective lens. For descanned detection, fluorescence returns down the same path back through the scanner and dichroic mirror (DM). The signal is separated from the scattered laser light by a barrier filter (BF), and can be focused to pass through a confocal pinhole (PH), if desired (see text). This detector (internal) can be a photomultiplier tube (PMT), photodiode, or spectrometer. For nondescanned detection, a dichroic mirror is inserted between the objective lens and the scanner. The emitted signal is then refocused so that the back aperture of the objective is conjugate to the front face of a PMT (external PMT, as shown), or it can be refocused so that an image is projected onto an array detector, such as a charge-coupled device.

fluorescence imaging of living specimens is often limited by photobleaching and photodamage.

Nondescanned detection has been used successfully for several applications, including calcium imaging with Indo-1 (Piston et al., 1994). This method uses a dichroic mirror between the scanner and the microscope to isolate the fluorescence after a minimum number of optical surfaces in order to maximize fluorescence detection efficiency. The back aperture, a stationary pivot point of the scanning beam of the objective, is focused on the PMT. Nondescanned detection allows the highest collection efficiency but is vulnerable to contamination from ambient room light.

Another nondescanned detection strategy is to refocus the fluorescence to an image plane and use an array detector such as a CCD. Lateral resolution in this scheme is improved since it is determined by the emission wavelength, which in TPEM is shorter than the excitation wavelength.

One nonoptical detection scheme could become very useful due to the spatial localization of two-photon excitation. Two-photon scaning photochemical microscopy (Denk, 1994) generates images of receptor distributions by locally releasing agonists such as neurotransmitters from "caged" precursors and detecting the agonist induced ionic current in voltage-clamped cells.

8.11. APPLICATIONS OF TWO-PHOTON EXCITATION

8.11.1. Chromophores (Fluorophores and Caged Compounds)

Good chromophores for TPEM are defined as they would be for any other fluorescence microscopy technique: large absorption cross section, high quantum yield, low rate of photobleaching, and minimal toxicity to living cells. So far, the probes used for TPEM are those that have proven useful in conventional fluorescence microscopies.

It would be convenient if the two-photon absorption spectrum was identical to the one-photon spectrum (the same spectral shape and at exactly twice the wavelength). However, there can be significant differences between the one- and two-photon excitation spectra. An excellent example of this is the blue shift of the excitation peaks of tyrosine and phenylanine under two-photon excitation (Rehms and Callis, 1993). In contrast, the spectrum of tryptophan is largely unchanged. As a rule of thumb in dealing with symmetrical molecules (like tyrosine and phenylalanine), one expects the two-photon excitation peak to be less than twice the one-photon excitation peak.

Unfortunately, experimental measurements of two-photon cross sections are difficult to perfom. One cannot simply measure the absorbance, since the fraction of the incident power that is absorbed is rather small compared to the intensity fluctuations in the light sources. Therefore, it is easiest to measure the excitation spectrum. Due to variations in pulse widths and average power over the tuning range of two-photon excitation sources, the most promising technique for measurement of excitation spectra is to measure them in comparison to the spectrum from a reference compound with a known excitation cross section and spectrum (Kennedy and Lytle, 1986).

Two-photon absorption spectra of caged compounds are even more difficult to measure than those of fluorophores, since the amount of uncaged material generated is usually too small to be easily measured by analytical techniques. Similar to fluorophore spectra, uncaged excitation spectra can be measured as described above if fluorescence assays for the released agonist exist, as for caged ATP (Denk et al., 1990), or if the release product is fluorescent, as for caged fluorescein (Silberzan et al., 1993). A further limitation to the use of photorelease is the release rates can be very slow and sensitive to the chemical environment (Corrie and Trentham, 1993). The rate of release is particularly important in TPEM because diffusion of the released agonist tends to blur the "focal spot," which can prevent high-resolution localization of the photorelease. For example, a delay between excitation and photorelease of 1 ms allows the released agonist to diffuse a distance of about 1 μm.

8.11.2. Applications of TPEM to Biological Systems

8.11.2.1.   Intracellular Calcium Ion Activity

Advances in optical techniques for measuring $[Ca^{2+}]_i$ have allowed the study of subcellular calcium activity in many biological systems. To accurately determine Ca^{2+}-related cellular dynamics, it is necessary to measure three-dimensionally resolved $[Ca^{2+}]_i$ with sufficient temporal resolution to follow fast cellular responses that generate signal pulses and wave propagation. Dynamic quantitative imaging of intracellular calcium was undertaken, with particular emphasis on spatial heterogeneities in the calcium-induced calcium release wave associated with cardiac myocyte contraction. Because mycocytes are relatively thick ($> 10\,\mu m$), strongly scattering cells, conventional full field microscopy does not provide a sufficient S/B ratio to observe cellular dynamics. Visible excitation calcium indicators, such as fluo-3 (Minta et al., 1989), are not ratiometric, so they do not eliminate staining heterogeneities. Both of these problems can be solved by using the UV-excited calcium indicator dye indo-1 in TPEM (Grynkiewicz et al., 1985; Piston et al., 1994), as demonstrated in Figure 8.5. These results clearly show the transient calcium rises in both the cytoplasm and the nucleus on triggering of contraction.

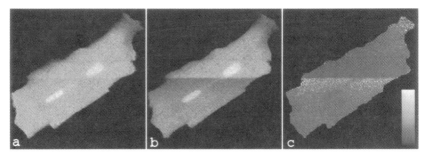

Figure 8.5. Fluorescence images of a rat cardiac myocyte acquired with two-photon excitation of indo-1. Approximately halfway through the scan, the cell is stimulated, causing a transient calcium rise. The two data channels, ~ 405 nm (panel a) and ~ 495 nm (panel b), correspond to the expected calcium-bound and free peaks of indo-1 with conventional excitation at ~ 355 nm. With two-photon excitation at 705 nm, image (a) is very close to an isospectic point of the indo-1 emission as a function of calcium. The ratio image of panels (a) and (b) is shown in (c). Each image is 100×100 µm.

Figure 8.6. A single optical section through a sea urchin embryo nucleus (metaphase) stained with Hoechst 33342. This image was taken ~ 30 µm into the embryo, with no excitation damage to cells above or below the focal plane. The entire chromosomal structure is about 10×5 µm in size, with individual chromosomes between 0.2 and 0.5 µm wide.

8.11.2.2.   Vital DNA Stains for the Study of Chromosome Patterns

Nuclear structures and division patterns of the sea urchin embryo have been visualized by two-photon excitation of the UV dye Hoechst 33342, as shown in Figure 8.6 (Piston et al., 1993) These results on living, developing embryos have confirmed the hypothesis, based on observations by CLSM on fixed embryos, that the fourth cells cleavage generates nonradial cleavage at the vegetal pole, contrary to the cononical textbook statements that the geometry of this division is radial (Summers et al., 1993).

8.11.2.3.   Cellular Structure and Dynamics by NADH Autofluorescence

The fluorescence from the natural reduced pyridine nucleotides (NADH) in cells is a potential indicator of cellular respiration (Chance and Thorell, 1959). However, NADH has a low quantum yield, and previous examinations of cellular dynamics by NADH autofluorescence have been limited by severe photobleaching and photodamage problems (Eng et al., 1989; Masters et al., 1993). By two-photon excitation of NADH, 3-D metabolic maps of cellular oxidative function have been obtained from many biological samples. As an example, the basal epithelium of the cornea is the source of renewal for the ocular surface through cell division, differentiation, and migration. Oxidative metabolism has been measured in this critical cell layer, as shown in Figure 8.7 (Piston et al., 1995). The images obtained were far superior to those previously recorded using a state-of-the-art Zeiss UV confocal microscope (Masters et al., 1993). Cellular NADH autofluroescence has also been imaged by TPEM from rat cardiac mycocytes (Niggli et al., 1994) and from individual β cells in intact pancreatic islets (Bennett et al., 1995) These autoflourescence imaging experiments could not be performed by UV confocal due to photobleaching and photodamage limitations.

8.11.2.4.   Other Applications

One powerful application of TPEM is its use in photoactivation of caged compounds (Silberzan et al., 1993; Denk, 1994). The inherent 3-D localization of excitation allows release in or near a single cell within a thick preparation. A related application is the use of photopolymerization to create a 3-D optical memory (Strickler and Webb, 1991). Since polymerized spots have a different index of refraction than unpolymerized media, 1 or 0 spots can be distinguished by Nomarski DIC differential interference contrast. Two-photon excitation has also been used in conjunction with 4Pi microscopy to achieve significant resolution improvements (Hell and Stelzer, 1992). Another application that utilizes the superb background-free fluorescence collection proper-

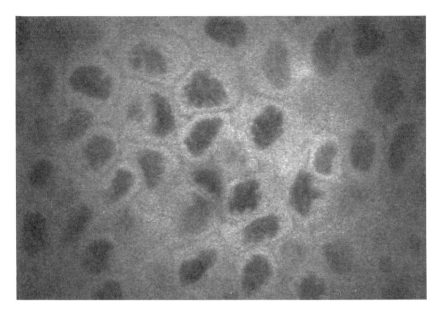

Figure 8.7. Detailed autoflourscence pattens seen in the basal epithelial cell layer of the rabbit cornea by two-photon excitation. The arrows point to invaginations of cytoplasm into the nuclei of two cells. Extremely pronounced cell borders are also visible throughout the image. The intensity of this NADH fluorescence can give a measure of metabolic energy stored within the cell. Image is 150 × 100 μm.

ties of TPEM is fluorescence decay time imaging (Piston et al., 1992; Guild et al., 1993). In this instrument, fluorescence lifetime, instead of intensity, is imaged to yield a new contrast mechanism. A final application of TPEM that is beginning to be explored is the imaging of visible absorbing fluorophores (e.g., fluorescein or rhodamine) with ~ 1 μm excitation. This may be a powerful method for performing 3-D reconstructions of easily photobleached samples, but in living biological systems, problems due to heat absorption by water may limit this application.

8.12. CONCLUSION

Two-photon excitation microscopy has been shown to work for dynamic imaging of living cells, including systems where UV imaging is not possible due to photobleaching or photodamage limitations. There are still many questions that must be answered to utilized the full quantitative potential of this

technique. The main impediment to the understanding of this form of microscopy is the difficulty in predicting or measuring two-photon absorption
spectra of the chromophores used. While it is possible to perform quantitative
TPEM imaging with dyes that were developed for one-photon, their behavior
may be different enough that changes must be made in the optics and calibrations in order to maximize the acquired information.
TPEM requires many of the same elements as CLSM but results in optical
simplifications because of the localization of the excitation. There are no
physical limitations to the implimentation of TPEM, and its use will most
likely increase as cheaper and more reliable ultrafast mode-locked lasers are
developed.

ACKNOWLEDGMENTS

The author wishes to thank Watt W. Webb and Winfried Denk for many helpful
discussions about two-photon excitation microscopy and Dave Sandison, Jeff Guild,
Brian Bennett, Guangtao Ying, Rebecca Williams, Mark Kirby, Heping Cheng, W. J.
Lederer, Andy Ridsdale, and Barry Masters for their help with experiments and
understanding of TPEM. Some of the data shown here were acquired at the Developmental Resource for Biological Imaging and Opto-Electronics at Cornell University
(NIH, 5P41-RR-04224) and the National Instrumentation Facility for Optical Microscopy (NSF, DIR-8800278). The author is a 1993 Beckman Young Investigator of the
Arnold and Mabel Beckman Foundation (Irvine, California).

REFERENCES

Agard, D. A., Hiraoka, Y., Shaw, P., and Sedat, J. W. (1989). Fluorescence microscopy
in three dimensions. In *Methods in Cell Biology*, Vol. 30 (D. L. Taylor and Y.-L.
Wang, Eds.), pp. 353–378. Academic Press, San Diego, CA.

Bennett, B. D., Ying, G., Jetton, T., Magnuson, M. A., and Piston, D. W. (1995).
Dynamics and distribution of glucose-induced NAD(P)H via two-photon excitation microscopy *Biophys. J.* **68**, A290. (Abstract.)

Bighouse, R. (1991). Measurement of the two-photon absorption spectra of several
fluorescent dyes. Master's Thesis in Applied and Engineering Physics, Cornell
University, Ithaca, NY.

Birge, R. R. (1986). Two-photon spectroscopy of protein-bound chromophores. *Acc.
Chem. Res.* **19**, 138–146.

Bliton, C., Lechleiter, J., and Clapham, D. E. (1993). Optical modifications enabling
simultaneous confocal imaging with dyes excited by ultraviolet- and visible-
wavelength light. *J. Microsc. (Oxford).* **169**, 15–26.

Carrington, W., Fogarty, K. E., and Fay, F. S. (1990). Three-dimensional imaging on confocal and wide-field microscopes. In *Handbook of Biological Confocal Microscopy* (J. Pawley Ed.), Chapter 14. Plenum, New York.

Chance, B., and Thorell, B. (1959). Localization and kinetics of reduced pyridine nucleotide in living cells by microfluorometry. *J. Biol. Chem.* **234**, 3044–3050.

Corrie, J. E. T. and Trentham, D. R. (1993). Caged Nucleotides and Neurotransmitters in Bioorganic Photochemistry (H. Morrison, Ed.), Vol. 2. Wiley, New York.

Curley, P. F., Ferguson, A. I., White, J. G., and Amos, W. B. (1992). Application of a femtosecond self-sustaining model-locked Ti:Sapphire laser to the field of laser scanning confocal microscopy. *Opt. Quantum Electron* **24**, 851–855.

Denk, W. (1994) Two-photon scanning photochemical microscopy: Mapping ligand-gated ion channel disctributions. *Proc. Nat Acad. Sci. U.S.A.* **91**, 6629–6633.

Denk, W., Piston, D. W., and Webb, W. W. (1994). Two-photon molecular excitation in laser scanning microscopy. In *Handbook of Biological Confocal Microscopy*, (J. Pawley, Ed.), 2nd Ed., Chapter 28. Plenum, New York.

Denk, W., Strickler, J. H., and Webb, W. W. (1990). Two photon laser scanning fluorescence microscopy. *Science* **248**, 73–76.

Eng, J., Lynch, R. M., and Balaban, R. S. (1989). Nicotinamide adenine dinucleotide fluorescence spectroscopy and imaging of isolated cardiac myocytes. *Biophys. J.* **55**, 621–630.

Fork, R. L., Martinez, O. E., and Gordon, J. P. (1984). Negative dispersion using pairs of prisms. *Opt. Lett.* **9**, 150.

Friedrich, D. M. (1982). Two-photon molecular spectroscopy. *J. Chem. Educ.* **59**, 472.

Friedrich, D. M., and McClain, W. M. (1980). Two-photon molecular electronic spectroscopy. *Annu. Rev. Phys. Chem.* **31**, 559–577.

Göppert-Mayer, M. (1931). Über Elementarake mit zwei Quantensprungen. *Ann. Phys. (Leipzig)* [5] **9**, 273–294.

Grynkiewicz, G., Poenie, M., and Tsien, R. Y. (1985). A new generation of indicators with greatly improved fluorescence properties. *J. Biol. Chem.* **260**, 3440–3450.

Guild, J. B., Piston, D. W., Sandison, D. R., and Webb, W. W. (1993). Fluorescence decay time imaging of rat basophillic leukemia cells by two-photon excitation microscopy. *Biophys. J.* **64**, A110.

Hell, S., and Stelzer, E. H. K. (1992). Fundamental improvement of resolution with a 4Pi-confocal fluorescence microscope using two-photon excitation. *Opt. Commun.* **93**, 277–282.

Hell, S., Reiner, G., Gremer, C., and Stelzer, H. K. (1993). Aberrations in confocal fluorescence microscopy induced by mismatches in refractive index. *J. Microsc. (Oxford)* **169**, 391.

Hellwarth, R., and Christiansen, P. (1974). Nonlinear optical microscopy examination of structure in polycrostalline ZnSe. *Opt. Commun.* **12**, 318–321.

Kaiser, W., and Garrett, G. G. B. (1961). Two-photon excitation in $CaF_2:Eu^{2+}$]. *Phys. Rev. Lett,* **7**, 229–231.

Kennedy, S. M., and Lytle. F. E. (1986) *p*-bis(*o*-Methylstyryl)benzene as a powersequared sensor for two-photon absorption measurements. *Anal. Chem.* **58**, 2643–2653.

Kliger, S. (1983). *Ultrasensitive Laser Spectroscopy.* Academic Press, New York.

Ling, H., and Lee, S. (1984). Focusing of electromagnetic waves through a dielectric interface. *J. Opt. Soc. Am. A* **1**, 965–973.

Liu, Y., Cheng, D. K., Sonek, G. J., Berns, M. W., and Tromberg, B. J. (1994). Microfluorometric technique for the determination of localized heating in organic particles *Appl. Phys. Lett.* **64**, 919–921.

Malcolm, G. P. A., Curley, P. F., and Ferguson, S. I. (1990). Additive-pulse mode-locking of a diode-plumped Nd-YLF Laser. *Opt. Lett.* **15**, 1305-1307.

Masters, B. R., Kriete, A., and Kukulies, J. (1993) Ultraviolet confocal fluorescence Microscopy of the *in vitro* cornea: Redox metabolic imaging. *Appl. Opt.* **32**, 592–596.

McClain, W. M. (1971). Excited state symmetry assignment through polarized two-photon absorption studies of fluids. J. Chem. Phys. **55**, 2789.

Minta, A., Kao, J. P. Y., and Tsien, R. Y. (1989). Fluorescent indicators for cytosolic calcium based on rhodamine and fluorescein chromophores. *J. Biol. Chem.* **264**, 8171–8178.

Nakamura, K. (1993). Three-dimensional imaging characteristics of laser scan fluorescence microscopy: Two-photon excitaton vs. single-photon excitation. *Optik* **93**, 39–45.

Niggli, E., Piston, D. W., Kirby, M. S., Cheng, H., Sandison, D. R., Webb, W. W., and Lederer, W. J. (1994). A confocal laser scanning microscope designed for indicators with ultraviolet excitation wavelengths. *Am. J. Physiol.* **266**, C303–C310.

Piston, D. W., Sandison, D. R., and Webb, W. W. (1992). Time-resolved fluorescence imaging and background rejection by two-photon excitation in laser scanning microscopy. *Proc. SPIE — Int. Soc. Opt. Eng.* **1640**, 379–390.

Piston, D. W., Summers, R. G., and Webb, W. W. (1993). Observation of nuclear division in living sea urchin embryos by two-photon fluorescence microcsopy. *Biophys. J.* **64**, A110.

Piston, D. W., Kirby, M. S., Cheng, H., Lederer, W. J., and Webb, W.W. (1994). Two-photon-excitation fluorescence imaging of three dimensional calcium-ion activity. *Appl. Opt.***33**, 662–669.

Piston. D. W., Masters, B. R., and Webb, W. W. (1995). Three-dimensionally resolved NAD(P)H cellular metabolic redox imaging of the *in situ* cornea with two-photon excitation laser scanning microscopy. *J. Microscopy. (Oxford)* **178**, 20–27.

Rehms, A. A., and Callis, P. R. (1993). Two-photon fluorescence excitation spectra of aromatic amino acids. *Chem. Phys. Lett.* **208**, 276–282.

Ridsdale, J. A., and Webb, W. W. (1993). The viability of cultured cells under two-photon laser scanning microscopy. *Biophys. J.* **63**, A109.

Sandison, D. R., and Webb, W. W. (1994). Background rejection and signal-to-noise optimization in the Confocal and alternative fluorescence microscopes. *Appl. Opt.* **33**, 603–615.

Sandison, D. R., Wells, K. S., Strickler, J., and Webb, W. W. (1994). Quantitative fluorescence imaging in laser scanning confocal microscopy. In *Handbook of Biological Confocal Microscopy* (J. Pawley, Ed.), 2nd ed., Chapter 3. Plenum, New York.

Sheppard, C. J. R. (1988). Aberrations in high aperture conventional and confocal imaging systems. *Appl. Opt.* **27**, 4782–4786.

Sheppard, C. J. R., and Gu, M. (1990). Image formation in two photon fluorescence microscopy. *Optik* **86**, 104–106.

Sheppard, C. J. R., and Kompfner, R. (1978). Resonant scanning optical microscope. *Appl. Opt.* **17**, 2879–2884.

Silberzan, I., Williams, R. M., and Webb, W. W. (1993). Fluorescence photoactivation by two-photon excitation: Kinetics of uncaging and three dimensional point diffusion measurements. *Biophys. J.* **64**, A109.

Spence, E. E., Kean, P. N., and Sibbett, W. (1991). 60-fsec pulse generation from a self-mode-locked Ti:sapphire laser. *Opt. Lett* **16**, 42.

Stelzer, E. H. K., Hell, S., and Lindek, S. (1994). Nonlinear absorption extends confocal fluorescence microscopy into the ultra-violet regime and confines the illumination volume. *Opt. Commun.* **104**, 223.

Strickler, J. H., and Webb, W. W. (1991). Three dimensional optical data storage in refractive media by two-photon point excitation. *Opt. Lett.* **16**, 1780–1782.

Summers, R. G., Morrill, J. B., Lieth, A., Marko, M., Piston, D. W., and Stonbreaker, A. T. (1993). A stereometric analysis of karyokinesis, cytokinesis and cell arrangements during and following fourth cleavage period in the sea urchin, *Lytechinus variegatus*. *Deve., Growh Differ.* **35**, 41–57.

Ulfhake, B., Carlsson, K., Mossberg, K., Arvidsson, U., and Helm, P. J. (1991). Imaging of fluorescent neurons labeled with fluoro-gold and fluorescent axon terminals labeled with AMCA (7-amino-4-methylcoumarine-3-acetic acid) conjugated antiserum using a UV-laser confocal scanning microscope. *J. Neurosci. Methods* **40**, 39–48.

Valdemanis, J. A., Fork, R. L., and Gordon, J. P. (1985). Generation of optical pulses as short as 27 femtoseconds directly from a laser balancing self-phase modulation, group-velocity dispersion, saturable absorption, and saturable gain. *Opt. Lett.* **10**, 112.

Visser, T. D., and Wiersma, S. H. (1991). Spherical aberration and the electromagnetic field in high-aperture systems. *J. Opt. Soc. Am. A* **8**, 1404–1410.

Visser, T. D, Oud, J. L., and Brakenhoff, G. J. (1992). Refractive index and axial distance measurements in 3-D microscopy. *Optik* **90**, 17–19.

White, J. B., Amos, W. B., and Fordham, M. (1987). An evaluation of confocal versus conventional imaging of biological structures by fluorescence light microscopy. *J. Cell Biol.* **105**, 41–48.

Wilson, T., and Sheppard, C. (1984). *Theory and Practice of Scanning Optical Microscopy*. Academic Press. New York.

CHAPTER

9

IMAGING APPLICATIONS OF TIME-RESOLVED FLUORESCENCE SPECTROSCOPY

JOSEPH R. LAKOWICZ and HENRYK SZMACINSKI

Center for Fluorescence Spectroscopy
Department of Biochemistry
University of Maryland at Baltimore
School of Medicine
Baltimore, Maryland 21201

9.1. INTRODUCTION

Time-resolved fluorescence spectroscopy is widely used as a research tool in biochemistry and biophysics, and has contributed immensely to our present understanding of the structure and function of biological macromolecules (1–3). However, time-resolved measurements require relatively complex instrumentation, and there have been relatively few reports of time-resolved measurements in fluorescence microscopy (4–6). In addition to the research applications of time-resolved fluorescence (1–3), there is a continuing effort to replace analytical methods based on radioactivity with fluorescence methods, as can be judged from the recent books and conferences on fluorescence sensing methods (7–11). These increasing clinical applications of fluorescence can be seen by the growth and introduction of improved methods for immunoassays, enzyme-linked immunoassays (ELISA), protein and DNA staining, and protein and DNA sequencing. In recent years, we have also witnessed an increasing use of fluorescence lifetimes in chemical sensing. Lifetime-based sensing is desirable because, in contrast to intensity, the fluorescence lifetime is mostly independent of the local probe concentration (12). It is now known that many fluorescence sensors display changes in lifetime in response to analytes, such as oxygen (13, 14), pH (15, 16), calcium (17–20), magnesium (21), and chloride (22). Importantly, lifetime-based

Fluorescence Imaging Spectroscopy and Microscopy, edited by Xue Feng Wang and Brian Herman. Chemical Analysis Series, Vol. 137.
ISBN 0-471-01527-X © 1996 John Wiley & Sons, Inc.

sensing circumvents the need for wavelength-ratiometric probes (23, 24), which, in turn, allows the use of longer wavelengths for excitation and emission. For instance, calcium imaging with wavelength-ratiometric probes requires ultraviolet excitation (23). In contrast, calcium imaging using lifetimes can be accomplished with visible-wavelength excitation (19).

An important factor in the evolution of fluorescence sensing is the introduction of long-wavelength probes that allow excitation with simple and reliable light sources such as laser diodes. The use of red-near infrared excitation also improves sensitivity because of the lower levels of autofluorescence observed with these long excitation wavelengths. Consequently, several laboratories are now developing long-wavelength probes that can be excited with laser diodes from 635 to 820 nm. Additionally, the increasing availability of picosecond (ps) and femtosecond (fs) lasers has resulted in the growing use of two-photon excitation, which allows excitation of UV or blue fluorescence with excitation wavelengths above 600 nm. In our opinion, lifetime-based sensing, long-wavelength sensing probes, ps and fs lasers, and time-resolved methods will result in a new generation of fluorescence microscopes that will provide a wide variety of chemical images based on lifetime imaging technology.

9.2. SCHEMES FOR FLUORESCENCE SENSING

Prior to describing the instrumentation for fluorescence lifetime imaging, it is informative to summarize the various possible schemes for fluorescence sensing (Figure 9.1). Most fluorescence measurements are based on the standard intensity-based methods, in which the intensity of the probe molecule changes in response to the analyte of interest. If the intensity of a probe varies in response to an analyte, or if the amount of signal is proportional to the analyte, then it appears simple and straightforward to relate this intensity to the analyte concentration (left). Intensity-based methods are initially the easiest to implement because many probe fluorophores change intensity and/or quantum yield in response to analytes. Additionally, collisional quenching processes, such as quenching by oxygen, iodide, or chloride, result in changes in intensity without significant shifts to the emission spectrum. While intensity measurements are simple and accurate in the cuvette, they are often inadequate for fluorescence microscopy. This is because one typically cannot control the amount of probe at each location in the microscopic sample. In fluorescence microscopy it is often impossible to know the probe concentration at each point in the image because the intensity changes continually due to photobleaching, phototransformation, and/or diffusive processes.

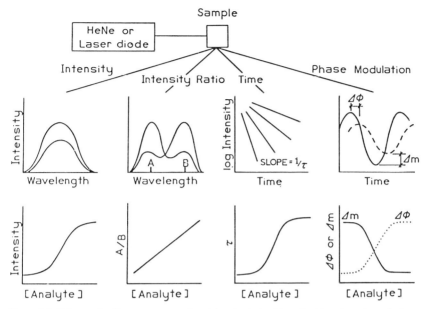

Figure 9.1. Schemes for fluorescence sensing: intensity, intensity ratio, time domain and phase modulation (from left to right).

The problems of intensity-based imaging can be avoided by using wavelength-ratiometric probes, that is, fluorophores that display spectral changes in the absorption or emission spectrum on binding or interaction with the analytes (Figure 9.1). In this case, the analyte concentration can be determined independently of the probe concentration by the ratio of intensities at two excitation or two emission wavelengths. Wavelength-ratiometric probes provide a straightforward means to avoid the difficulties of intensity-based imaging. However, few such probes are available, and it is clear that they are difficult to create (16, 23). For instance, in spite of the enormous interest in measuring intracellular Ca^{2+} concentrations, there appear to be no practical wavelength-ratiometric indicators for Ca^{2+} that allows visible-wavelength excitation. A recently synthesized visible- wavelength-ratiometric Ca^{2+} probe (24) remains to be tested in a fluorescence microscope. The two most widely used probes, fura-2 (Figure 9.2) and indo-1, both require UV excitation, with the associated problems of complex UV laser sources, the need for UV microscope optics, and the large amounts of cellular autofluorescence excited at these wavelengths. Attmepts to make long-wavelength Ca^{2+} probes have resulted in probes that change intensity but do not display spectral shifts in either the excitation or emission spectra, such as rhod-2, fluo-3 (25), and Calcium Color

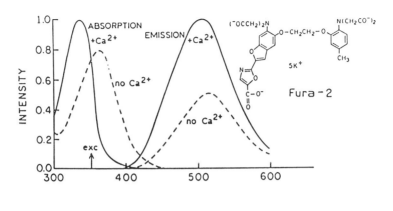

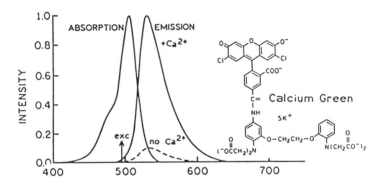

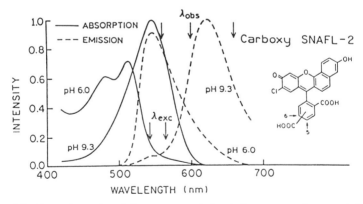

Figure 9.2. Absorption and emission spectra and chemical structures of representative fluorescence sensors for metal ion recognition. The excitation wavelength was 345 nm for fura-2, 488 nm for Calcium Green, and 543 nm for carboxy SNAFL-2.

Series (19). Wavelength-ratiometric probes for pH have only recently become available (16).

The difficulties of intensity-based imaging, and of the scarcity of wavelength-ratiometric probes, may be circumvented by the use of time-resolved or lifetime-based imaging. We reasoned that it would be easier to identify and/or synthesize probes that display changes in lifetime in response to analytes than to design and synthesize probes that display spectral shifts. Our opinion was based on the knowledge that a wide variety of quenchers and/or molecular interactions result in changes in the lifetimes of fluorophores, while changes in spectral shape were the exception rather than the rule. This prediction proved to be correct, as we now know that a number of probes for Ca^{2+} imaging

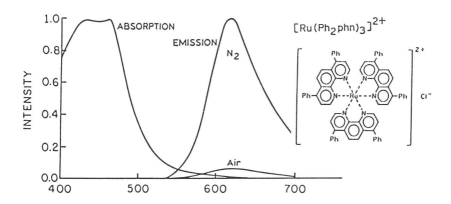

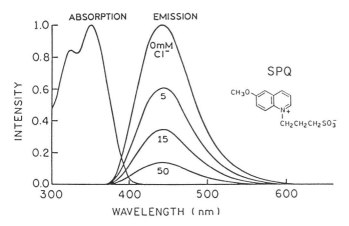

Figure 9.3. Absorption and emission spectra and chemical structures of fluorescent sensors based on collisional quenching by oxygen (top) and by chloride (bottom).

(17–20, 24) and the analogous Mg^{2+} probes (21) display lifetime changes in response to binding of their specific cations. Additionally, the pH probes of the SNAFL (seminaphthofluorescein) and SNARF (seminaphthorhodafluor) series display changes in lifetime on pH-induced ionization (15). The use of wavelength-ratiometric probes with lifetime imaging allows expansion of the analyte sensitivity range by judicious selection of the excitation and/or emission wavelength (15). Of course, collisional quenchers like Cl^-, O_2, and so on also cause changes in lifetimes (similar to the changes in intensities) for sensors such as $[Ru(Ph_2phn)_3]^{2+}$ and SPQ (Figure 9.3). However, in the last cases, there are no available specific chelating groups, so a wavelength-ratiometric probe for O_2, Cl^-, or other collisional quenchers does not presently seem possible. Also, it is more difficult to develop wavelength-ratiometric probes for Na^+ and K^+, as seen by the small spectral shift displayed by SBFI (sodium-binding benzofuran isophthalate) and PBFI (potassium-binding benzofuran isophthalate) fluorophores.

9.2.1. Molecular Mechanisms for Fluorescence Lifetime Imaging

We now consider the various factors that can alter the lifetimes of fluorophores. Such effects can be used for lifetime-based sensing or imaging. The fluorescence lifetime of a sample is the mean duration of time the fluorophore remains in the excited state. Following pulsed excitation, the intensity decays of many fluorophores are single exponentials

$$I(t) = I_0 \exp(-t/\tau) \tag{1}$$

where I_0 is the intensity at $t = 0$, and τ is the lifetime. A variety of molecular interactions can influence the decay time, as can be seen from the Jablonski diagram (Figure 9.4). The excited fluorophore can return to the ground state by the radiative (emission) pathway with a rate k_r. The inverse of this rate constant ($\tau_r = k_r^{-1}$) is usually called the intrinsic or radiative lifetime. The radiative decay rate k_r is generally of intramolecular origin, with only a modest dependence on the local environment.

There are many molecular interactions within the local environment that influence the fluorescence decay times. The measured fluorescence lifetime τ is usually shorter than the radiative lifetime τ_r because of other decay rates that depend on intermolecular interactions (Figure 9.4). These nonradiative decay rates depopulate the excited state and decrease the mean lifetime. The measured fluorescence lifetime (τ) is given by the inverse of the total rate of deactivation from the excited (mostly singlet S_1) state:

$$\tau = \frac{1}{k_r + k_{nr} + k_q[Q] + k_T} \tag{2}$$

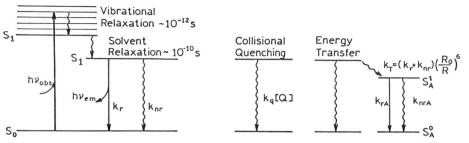

Figure 9.4. Modified Jablonski diagram for the processes of absorption and fluorescence emission (left), dynamic quenching (middle), and fluorescence resonance energy transfer (FRET) (right).

Nonradiative processes (k_{nr}) can occur with a wide range of rate constants. Molecules with high k_{nr} values display low quantum yields due to rapid depopulation of the excited state by this route. The lifetime in the absence of collisional $(k_q[Q])$ or energy transfer (k_T) quenching is usually referred to as τ_0 and is given by $\tau_0 = (k_r + k_{nr})^{-1}$.

One easily understood mechanism for lifetime-based sensing is collisional quenching (Figure 9.4). A variety of substances act as quenchers, including oxygen, heavy atoms, halides, acrylamide, and amines, to name only a few. By considering the lifetime in the absence (τ_0) and presence (τ) of collisional quenchers (no resonance energy transfer), one can easily see that such quenching decreases the lifetime of the excited state

$$\frac{\tau_0}{\tau} = 1 + \tau_0 k_q[Q] \tag{3}$$

where k_q is the biomolecular quenching constant. In fluid solution with efficient quenchers, k_q is typically in the range of 10^9 to $10^{10} \, M^{-1} \, s^{-1}$ Collisional quenching by oxygen has been used for intensity-based and lifetime-based sensing of oxygen using the long-lifetime transition metal complexes, resulting in large amounts of quenching by oxygen (13, 14). Collisional quenching can also be the result of electron transfer processes induced by excitation (26). In some cases, binding of cations can prevent electron transfer from the electron-donating group, preventing intramolecular quenching, which results in enhancement of quantum yield on binding of metal cations without observable spectral shifts (27).

Another mechanism for fluorescence intensity and lifetime sensing is fluorescence resonance energy transfer (FRET) (28). An energy transfer–based sensor consists of two kinds molecules: donor (fluorescent) and acceptor (fluorescent or nonfluorescent). For such a sensor, the donor need not be sensitive to the analyte, but the acceptor must display a change in its absorption spectrum as an extinction coefficient due to the analyte. Chemical sensing is accomplished by the changes in energy transfer efficiency from donor to acceptor due to the analyte-induced changes in acceptor absorption. Energy transfer efficiency can also be changed by changing the proximity of donor and acceptor. In this case, neither the donor nor the acceptor needs to be sensitive to the analyte. FRET requires partial overlap of the emission spectrum of the donor with the absorption spectrum of the acceptor. Since energy transfer depends on the local concentration of acceptor, lifetime imaging in the presence of donor- and acceptor-labeled macromolecules should allow proximity imaging in cells.

Lifetime-based imaging can also be based on the existence of two forms of the probe, the fraction of which depends on the analyte concentration This situation is presented in Figure 9.5, where it is assumed that both forms can be excited at the same wavelength. The sample displays two lifetimes (τ_F and τ_B) that are characteristic of the analyte-free (F) and analyte-bound (B) forms of the probe. The change in lifetime (from τ_F to τ_B) is due to binding of the analyte and may result from changes in the radiative and/or nonradiative decay rates. The fluorescence intensity decay of such probes is generally not a single exponential [Eq. (1)] but is given as a sum of the exponentials with decay times τ_F and τ_B. Nonetheless, measurement of the mean lifetime can be used to determine the analyte concentration, and it is not necessary to resolve the multiexponential decay for lifetime-based sensing or imaging.

Binding of the analyte to the probe may also result in spectral shifts. In Figure 9.5 we have assumed that the complexed form has blue-shifted spectra (higher energy) relative to that of the free form. This behavior is known to occur in several Ca^{2+}, Mg^{2+}, K^+, and pH probes, such as fura-2 (Figure 9.2), Mag-quin-2, PFBI, and the SNAFL probes (Figure 9.2). Some probes display larger spectral shifts in the absorption spectra than in the emission, like fura-2

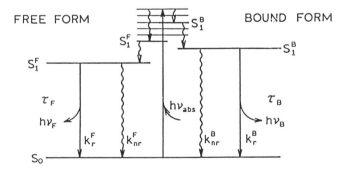

Figure 9.5. Jablonski diagram for two forms of a sensing probe.

(Figure 9.2). However, there are probes that do not display spectral shifts on complexation, like the Calcium Color Series, but that display significant changes in quantum yield and lifetimes due to different decay rates for free and bound forms (Calcium Green in Figure 9.2). In recent years, many lifetime sensors for ion recognition have been identified. This information is summarized in Table 9.1 (29–39). Given the relative newness of lifetime-based sensing, one can imagine an increasing number of lifetime probes for fluorescence microscoy.

9.2.2. Measurement of Fluorescence Lifetimes

There are two widely used methods for measuring fluorescence lifetimes, the time-domain and the frequency-domain or phase-modulation methods. The basic principles of time-domain (40) fluorometry and frequency-domain (41) fluorometry are described elsewhere in considerable detail (40–43). It is common to represent intensity decays of varying complexity in terms of the multiexponential model

$$I(t) = \sum_i \alpha_i \exp(-t/\tau_i) \tag{4}$$

where the α_i are the preexponential factors and the τ_i are the decay times. The fractional intensity of each form to the steady-state intensity is proportional to the preexponential factor and decay time:

$$f_i = \frac{\alpha_i \tau_i}{\sum_j \alpha_j \tau_j} \tag{5}$$

This fact can be understood by recognizing that the $\alpha_i \tau_i$ product represents the area under the decay curve which is due to the component with the lifetime τ_i. The fractional intensity may not be the same as the fractional number of molecules that result in each component of the emission because the relative absorption and quantum yields of the populations can be different.

A wide variety of molecular interactions can be responsible for a complex multiexponential or nonexponential decay. Depending on the molecular origin of the complex decay kinetics, the values of α_i and τ_i may have direct or indirect molecular significance. For instance, for a mixture of two fluorophores (complexed and free forms or protonated and unprotonated forms), each of which displays a single decay time, the τ_i are the two individual decay times and f_i are the fractional contributions of each fluorophore to the total steady-state emission. If a single fluorophore is in two different environments, such as protein

Table 9.1. Spectral Properties, Quantum Yields (Φ), and Mean Lifetimes (τ) of Free (F) and Bound (B) Forms of Probes for Ion Recognition and Their Dissociation Constants $(K_D)^a$

Probe	Absorption		Emission			Dissociation Constant (K_D)	Refs.
	$\lambda_F(\lambda_B)^b$ (nm)	$\varepsilon_F(\varepsilon_B)$ ($\times 10^{-3}\,M^{-1}\,cm^{-1}$)	$\lambda_F(\lambda_B)^b$ (nm)	$\Phi_F(\Phi_B)$	$\bar{\tau}_F(\bar{\tau}_B)$ (ns)		
Ca²⁺ Probes						(nM)	
Quin-2	356(336)	5.0(5.0)	500(503)	0.03(0.14)	1.35(11.6)	60,115[c]	17,18,29
Fura-2	362(335)	27(32)	518(510)	0.23(0.49)	1.09(1.68)	135,224[c]	23,30
Indo-1	349(331)	33	482(398)	0.38(0.56)	1.40(1.66)	250	23,31
CaG	506	74	534	0.06(0.75)	0.92(3.66)	189	19,32
CaO	555	80	576	0.11(0.33)	1.20(2.31)	328	19,32
CaC	588	108	610(612)	0.18(0.53)	2.55(4.11)	205	19,32
BOZ-crown	483(457)	—	642(574)	0.33(0.64)	2.1 (3.3)	79	33,34
Mg²⁺ Probes						(mM)	
Mag-quin-2	353(337)	4.2(4.0)	487(493)	0.003(0.07)	0.84(8.16)	0.8	35,36
MgG	506	76	532	0.04(0.42)	1.21(3.63)	0.9	21,35
BOZ-crown	483(469)	—	642(573)	0.33(0.48)	2.1 (3.3)	2.0	33,34
K⁺ Probe						(mM)	
PBFI	350(344)	42(42)	546(504)	0.024(0.072)	0.26(0.67)	8.0	35,37
Na⁺ Probes						(mM)	
NaG	506	—	536(538)	—	1.14(2.41)	6.0	35,38
SBFI	346(334)	42(47)	551(525)	0.045(0.083)	0.26(0.47)	7.4	35,37

282

pH Probes

						pK_a	15, 16
SNAFL-1	539(510)	51(28)	616(542)	0.093(0.33)	1.19(3.74)	7.7	
C. SNAFL-1	540(508)	52(29)	623(543)	0.075(0.32)	1.11(3.67)	7.8	
C. SNAFL-2	547(514)	48(25)	623(545)	0.054(0.43)	0.94(4.60)	7.7	
C. SNARF-1	576(549)	53(26)	638(585)	0.091(0.047)	1.51(0.52)	7.5	
C. SNARF-2	579(552)	60	633(583)	0.110(0.022)	1.55(0.33)	7.7	
C. SNARF-6	557(524)	50	635(559)	0.053(0.42)	1.03(4.51)	7.6	
C. SNARF-X	575(570)	42(18)	630(600)	0.163(0.069)	2.59(1.79)	7.9	
BCECF	503(484)	94(30)	528(514)	—	4.49(3.17)	7.0	
Resorufin acetate	571(484)	67(16)	587(578)	—	2.92(0.45)	~ 5.7	35
[Ru(bpy)$_2$(dcbpy)]$^{2+}$	458(480)	13(12)	650(675)		460(210)	3.9	39

[a] Abbreviations: CaG, Calcium Green; CaO, Calcium Orange; CaC, Calcium Crimson; BOZ, benzoxazine; MgG, Magnesium Green; PBFI, potassium-binding benzofuran isophthalate; NaG, sodium Green; SBFI, sodium-binding benzofuran isophthalate; C., carboxy; SNAFL, seminaphthofluorescein; SNARF, seminaphthorhodafluor; BCECF, 2′,7′-bis(2-carboxyethyl)-5(and 6)-carboxyfluorescein; [Ru(bpy)$_2$(dcbpy)]$^{2+}$, bis(2,2′-bipyridine)(2,2′-bipyridine-4,4′-dicarboxylic acid)ruthenium(II).

[b] λ refers to the wavelength at the maximum of the respective spectrum.

[c] In buffer with 1 mM Mg^{2+}.

283

bound and free, then the α_i values often represent the fraction of the molecules in each environment. This relationship occurs because the radiative decay rate (k_r) of the fluorophores is often insensitive to the local environment, whereas the quantum yield and/or lifetime is often strongly dependent on the environment. However, in many circumstances there is no obvious linkage between the α_i and τ_i values and the molecular heterogeneity of the samples. Additionally, there are many cases in which the intensity decays are nonexponential, such as resonance energy transfer, collisional quenching, and solvent relaxation. While it is of interest to interpret such complex decays in terms of molecular interactions within the sample, it is not necessary to have a complete understanding of the decay processes before a probe can be used for lifetime-based imaging. For imaging purposes, measurement of a mean lifetime is usually adequate to quantify the analyte. The mean lifetime ($\bar{\tau}$) is related to the multiexponential parameters by

$$\bar{\tau} = \frac{\sum_i \alpha_i \tau_i^2}{\sum_i \alpha_i \tau_i} \tag{6}$$

$$\bar{\tau} = \sum_i f_i \tau_i. \tag{7}$$

The mean lifetime can be used to determine the analyte concentration with a suitable calibration curve.

9.2.3. Instrumentation for Phase-Modulation Lifetime Measurement

The rapid timescale of fluorescence emission results in a significant technological challenge when measuring time-dependent decays. Instrumentation for lifetime measurements, using time-domain or frequency-domain methods, is complex and expensive. Present state-of-the-art instruments use ps laser sources and very fast photodetectors to provide resolution of multiexponential and nonexponential intensity decays. Consequently, time-resolved fluorescence spectroscopy has been primarily a research tool in biochemistry, biophysics, and chemical physics (1, 2). The instrumentation for lifetime imaging may be significantly simplified if the phase-modulation method is used. The phase angle and modulation can be measured with high accuracy, and a single modulation frequency is sufficient for imaging.

In phase-modulation fluorometry, the pulsed light source typical of time-domain measurements is replaced with an intensity-modulated source (Figure 9.6). Because of the time lag between absorption and emission, the emission is delayed in time relative to the modulated excitation. At each modulation

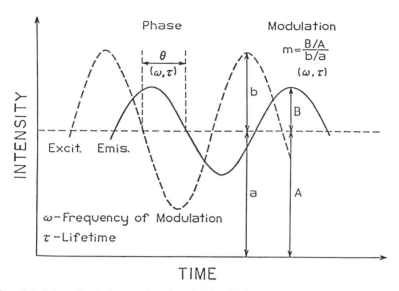

Figure 9.6. Schematic of phase angle and modulation lifetime measurements. ω = frequency of modulation, θ = phase angle, m = modulation, and τ = lifetime.

frequency ($\omega = 2\pi f$) this delay is described as the phase shift (θ_ω), which increases from $0°$ to $90°$ with increasing modulation frequency. The finite time response of the sample also results in demodulation of the emission by a factor m_ω, which decreases from 1.0 to 0.0 with increasing modulation frequency. The phase angle (θ_ω) and the modulation (m_ω) are separate measurements, each of which is related to the intensity decay parameters, α_i and τ_i and modulation frequency ω by (44)

$$\theta_\omega = \arctan(N_\omega/D_\omega), \qquad m_\omega = (N_\omega^2 + D_\omega^2)^{1/2} \tag{8}$$

where

$$N_\omega = \frac{1}{J}\sum_{i=1}^{n}\frac{\omega\alpha_i\tau_i^2}{1+\omega^2\tau_i^2}, \qquad D_\omega = \frac{1}{J}\sum_{i=1}^{n}\frac{\alpha_i\tau_i}{1+\omega^2\tau_i^2}, \qquad J = \sum_{i=1}^{n}\alpha_i\tau_i \tag{9}$$

Research instrumentation for phase-modulation fluorometry operates over a wide range of light modulation frequencies and excitation wavelengths. Consequently, instruments are somewhat expensive and complex. If the probe characteristics are known, the phase-modulation instrument can be designed to have a single-wavelength light source and to operate at just one light modulation frequency appropriate for the chosen probe. This specialization can result in decreased complexity.

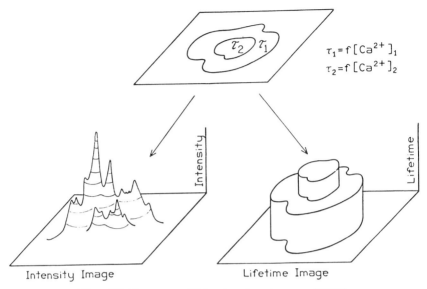

$$\tau_1 = f[Ca^{2+}]_1$$
$$\tau_2 = f[Ca^{2+}]_2$$

Intensity Image   Lifetime Image

Figure 9.7. Fluorescence lifetime imaging microscopy (FLIM).

9.3. FLUORESCENCE LIFETIME IMAGING MICROSCOPY (FLIM)

Fluorescence microscopy can be used to study the location and movement of intracellular species. In general, the fluorescence image reflects the location and concentration of the probe, or that amount of probe remaining in a photo-bleached sample (Figure 9.7, lower left). Consequently, quantitative fluo-rescence microscopy based on intensity is very difficult, except for those cases where wavelength-ratiometric probes are available. The ratio of intensities of two excitation or emission wavelengths is independent of local probe concentra-tion and reflects the microenvironment of the probe. Consider now that the lifetime of the probe is different in the two regions of the cell (Figure 9.7, top). If one could create a contrast based on the lifetime at each point in the image, one would resolve two regions of the cell, each with an analyte (e.g., Ca^{2+}) concentra-tion revealed by the lifetime image. What type of chemical imaging will be possible using FLIM technology? Based on our current understanding of FLIM, and of factors that affect fluorescence lifetimes, we can predict that lifetime imaging will provide information on a variety of cellular properties. These include imaging of ions, cofactor binding, probe binding, chromosomes, microviscosity, and proxim-ity imaging of associating macromolecules (Table 9.2).

The creation of such fluorescence lifetime images, in which the contrast is based on lifetimes, appeared to be a daunting challenge. Imagine performing 2.62×10^5

Table 9.2. Biomedical and Biological Applications of FLIM

Type of Imaging	Analyte or Property
Chemical imaging	$Ca^{2+}, Mg^{2+}, Cl^{-}, pH, CO_2, O_2, Na^{+}, K^{+}$
Ligand binding to proteins	NADH, TNS
Chromosome imaging	Acridine lifetimes depend on DNA base composition
Microviscosity imaging	Identify viscosity-lifetime probes
Proximity imaging by energy transfer	Protein–protein binding Protein–membrane association

lifetime measurements for a typical 512×512 image. Given the difficulties of measuring even a single lifetime in a cuvette, such a task seems nearly impossible. However, image intensifiers and charge-coupled device (CCD) camera technology now make this possible (17, 45). Figure 9.8 (right) shows the Ca^{2+} concentration image of COS (from Green monkey kidney) cells based on the lifetime of quin-2 (46), along with the intensity image (left). The intensity images show the expected spatial variations due to probe localization, and the Ca^{2+} (phase angle) image shows the expected uniform concentration of intracellular calcium. As

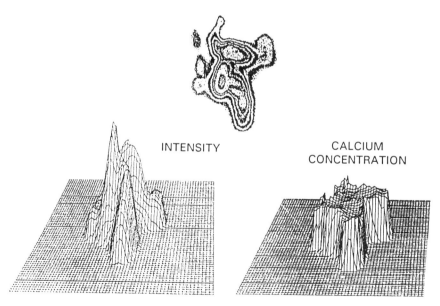

INTENSITY   CALCIUM
CONCENTRATION

Figure 9.8. Intensity (left) and calcium images (right) of quin-2 fluorescence in COS cells.

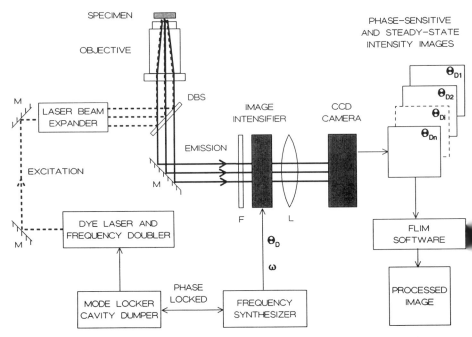

Figure 9.9. Instrumentation for FLIM. Excitation is provided by the frequency-doubled output of a cavity-dumped dye laser, which is synchronously pumped by a mode-locked Nd:YAG laser. The excitation light is expanded by a laser beam expander. The microscope is a Nikon Diaphot-TMD inverted fluorescence microscope with Nikon Fluor 40X, NA 1.3, and DM400 Nikon dichroic beam splitter (DBS). The gated image intensifier is positioned between the target and the CCD camera. The gain of image intensifier is modulated using output of a frequency synthesizer with a digital phase shift option. The detector is a CCD camera.

predicted, the lifetime imaging provides chemical imaging, which within limits is insensitive to the local probe concentration.

9.3.1. Apparatus for Fluorescence Lifetime Imaging

The cellular FLIM images in Figure 9.8 were obtained using moderately complex instrumentation, which consists of a ps dye laser, a gain-modulated image intensifier, and a slow-scan scientific grade CCD camera (Figure 9.9). However, the FLIM instruments in the future can be compact, even mostly solid-state devices. This possibility is shown in Figure 9.10, where we show that the light source can be a laser diode, assuming that long-wavelength probes become available. The image intensifier is a moderately simple device, but is delicate and requires high voltages. Reports have appeared on gatable CCD detectors (47). Presently available gatable CCDs are too slow (50 ns gating time). This time response is likely to be improved, and probes can be developed with

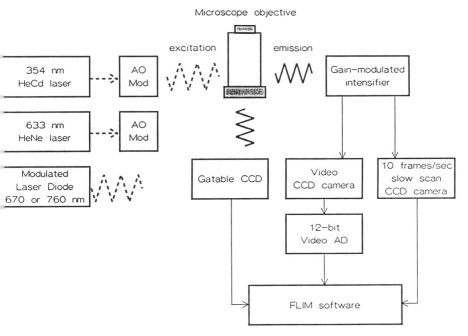

Figure 9.10. Future apparatus for FLIM.

longer decay times. Then the FLIM apparatus will consist of only modest additions to a standard fluorescence microscope.

We now describe in detail our approach to FLIM. The first lifetime measurements under a microscope were carried out almost 20 years ago (48–50) at selected sites in cells using standard time-domain methods. The first measurements were performed using a flash lamp (48) or nitrogen laser (49, 50) and the pulse sampling method for time-resolved detection. Single-photon counting was used later (51–53). Recent advances in high-repetition lasers for pulse excitation and time-correlated single-photon counting (TCSPC), together with the development of the frequency-domain method, have been used to obtain more reliable fluorescence lifetimes from microscopic samples. Several groups have developed time-resolved fluorescence microscopy (TRFM) using argon-ion pulsed lasers (54, 55), mode-locked ps dye lasers (56–59), and the TCSPC technique. The frequency-domain technique has also been utilized in fluorescence microscopy (60, 61). To increase temporal resolution over that of the single-photon counting method, a ps streak camera was adapted to a fluorescence microscope (62, 63). Compared to the photon counting technique and frequency domain, the streak camera exhibits the highest temporal performance; however, only limited time range can be examined.

An important point about all of these instruments is that fluorescence decays were measured more or less by conventional means on single points within a microscopic sample. It is, of course, possible to create fluorescence lifetime images by combining lifetime measurements and scanning techniques. However, for measurements of dynamic events, problems could arise due to the long measurement times involved in measuring lifetime at many pixels. Low photon-counting rates result in long observation times for mapping 2-D samples. To increase the counting rate and reduce the measurement time, multiphoton (64) and multichannel photon-counting methods have been developed (65–67). However, neither of these techniques provides the temporal resolution required to study cellular dynamics.

Until 1991 no attmepts were made to provide high-resolution images in which the contrast is determined by fluorescence lifetimes. As in time-resolved fluorescence spectroscopy, two methods were developed for lifetime imaging: time-domain and frequency-domain fluorometry. In time-domain imaging, gated detection was used to provide imaging of intensity in time windows (68–74). Since most of the fluorescent probes display multiexponential and short lifetimes (Table 9.1), time-domain FLIM has certain disadvantages (75, 76). However, the gated method is particularly useful for providing imaging of the intensity of long-lived emissions (69, 77). It has been shown that by using a confocal laser scanning microscope and gated methods, lifetime images can be obtained for relatively short lifetimes (78). Frequency-domain imaging has now been implemented in a number of laboratories (19, 45, 46, 75, 76, 79–83). A comprehensive review of current FLIM techniques with respect to lifetime resolution, formation of 2-D images, and measurement time has been published (75, 76).

9.3.2.   Lifetime Imaging Using Phase-Sensitive Image Detection

We now describe our apparatus that allows lifetime imaging with simultaneous measurement at all positions in the image. This method uses a gain-modulated image intensifier that converts the time-dependent image data into a steady-state "phase-sensitive" image, which is quantified using a slow-scan CCD camera. By using several phase-sensitive images collected with various electronics delays or phase shifts, it is possible to calculate the lifetime image of the object. This method of lifetime imaging has been evaluated using macroscopic samples in which the data were compared with standard lifetime measurements (17, 19, 45, 79). This method has been proved practical by obtaining Ca^{2+} lifetime images of cells loaded with a fluorescent calcium indicator such as Quin-2 (46).

In our apparatus, the fluorescence of the 2-D sample is excited by an intensity-modulated laser light at circular frequency ω and modulation degree, m_E:

$$E(t) = E_0(1 + m_E \sin \omega t) \tag{10}$$

The sinusoidally modulated emitted fluorescence is phase shifted and partially demodulated relative to the excitation:

$$F(r, t) = F_0(r)[1 + m_F(r) \sin(\omega t - \theta_F(r))] \tag{11}$$

where $F_0(r)$, $m_F(r)$, and $\theta_F(r)$ are the time-averaged, spatially dependent fluorescence intensity, modulation, and phase angle, respectively. The phase angle, $\theta_F(r)$, and the fluorescence modulation degree $m_F(r)$, depend on the lifetime at each position, r, and on the light modulation frequency:

$$\tan \theta_F(r) = \omega \tau_p(r) \tag{12}$$

$$m_F(r) = m_E[1 + \omega^2 \tau_m^2(r)]^{1/2} \tag{13}$$

where $\tau_p(r)$ and $\tau_m(r)$ are apparent phase and modulation lifetimes at each position, r. For single exponential decays, the phase and modulation lifetimes are equal and independent of modulation frequency. Generally, the emission kinetics are more complicated and the intensity decays are multiexponential or nonexponential, especially in polymers and biological systems. For multi-exponential decays, generally $\tau_p < \tau_m$, and both values decrease with increased modulation frequency. The heterogeneity may indicate the existence of many microscopically different environments in which fluorophore is located. Resolution of the intensity decay law from the phase shift (θ) and modulation (m) data requires multifrequency measurement (84, 85). In a FLIM experiment, the essential information is contained in the position-dependent phase and/or modulation of the emission. Irrespective of intensity decay heterogeneity, the phase angle and modulation can be measured with high accuracy, and a single modulation frequency is sufficient for chemical imaging (15, 19, 46).

The gain-modulated, high-speed gated image intensifier is used as an optical 2-D phase-sensitive detector (Figure 9.9). The fluorescence phase angle and modulation at each position, r, can be obtained from a series of phase-sensitive images. A schematic diagram of the instrumentation for FLIM using homodyne phase-sensitive detection is shown in Figure 9.9. The photocathode of the image intensifier converts the light image into an electron image, which is intensified by the microchannel plate (MCP), reconverted into an optical image on the phosphor screen, and recorded using a slow-scan CCD camera. The voltage between the photocathode and MCP input surface is varied at the desired frequency (limited by the time response of the image intensifier). The electronic gain is varied at a modulation frequency equal to the light modulation frequency or a harmonic of the pulse rate. This gain modulation signal is applied to the photocathode of the image intensifier, resulting in a time-varying gain $G(t)$, with

$$G(t) = G_0[1 + m_D \sin(\omega t - \theta_D)] \tag{14}$$

where m_D is the gain modulation degree and θ_D is the detector phase angle relative to the excitation phase angle. The time-dependent photocurrent [Eq. (11)] is detected with the time-dependent gain [Eq. (14)], resulting in a DC signal and high-frequency signals. However, due to the slow time response of the image intensifier screen ($\sim$ 1 ms), the high-frequency signals are averaged at the output screen. The time-averaged, phase-sensitive intensity from corresponding position r, is given by

$$I(r, \theta_D) = I_0(r)[1 + \tfrac{1}{2} m_D m(r) \cos(\theta(r) - \theta_D)] \tag{15}$$

The phase-sensitive intensity at each position, r, depends on the gain modulation of the detector, m_D; the modulated amplitude of the emission, $m(r)$; and the cosine of the phase angle difference between the gain modulation signal, θ_D, and the phase of the emission, $\theta(r)$. The phase-sensitive intensity images provide a constant intensity at each position, r, where values depend on the concentration of the local fluorophore $c(r)$ and the local lifetime $\tau(r)$. A value of $\theta_D = 0$ results in maximum intensity for a zero lifetime [$\theta(r) = 0$], i.e., scattered light. This procedure is analogous to the method of phase-sensitive or phase-resolved fluorescence (86, 87). However, these earlier measurements of phase-sensitive fluorescence were performed electronically on the heterodyne low-frequency cross-correlation signal, whereas, here, homodyne detection is performed electro-optically on the high-frequency modulated emission.

It is not possible to calculate the fluorescence phase angle $\theta(r)$ or modulation $m(r)$ images from the single phase-sensitive image. However, the phase angle and modulation of fluorescence can be determined from a series of phase-sensitive images by varying the phase angle of the detector, θ_D [Eq. (15)]. This can be accomplished by a series of electronic delays or by using the digital phase shift option of the synthesizer. The phase-sensitive intensity images can be collected using a slow-scan CCD camera. There are many advantages to using a slow-scan cooled CCD camera, such as high resolution, high sensitivity, wide dynamic range, photometric accuracy, geometric stability, and the ability to integrate the image directly.

In our measurements, we collected series of phase-sensitive images in which θ_D was varied over $360°$. A series of phase-sensitivity images of a standard fluorophore, POPOP(p-bis[2-(5-phenyloxazazolyl)]benzene) in propylene

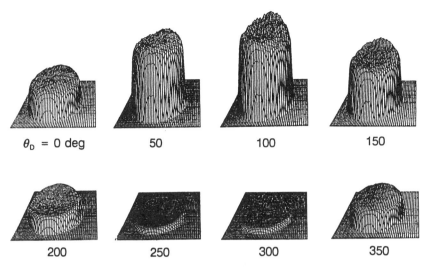

θ_D = 0 deg   50   100   150

200   250   300   350

Figure 9.11. Phase-sensitive intensity images of the fluorescence standard (POPOP) obtained at a modulation frequency of 49.53 MHz using instrumentation described in Figure 9.9.

glycol, between two glass coverslips are shown in Figure 9.11. The phase-sensitive images in Figure 9.11 contain information about the decay time of the fluorescence at each position. In this case, fluorescence decay times are similar throughout the image because a uniform sample was used. Maximum (in-phase) intensity is observed at about $\theta_D = 100°$ and minimum (out-of-phase) intensity between $250°$ and $300°$. The fluorescence phase angle is related to the detector phase, θ_D, which includes a constant instrumental phase shift. A reference sample is required to determine the instrumental phase angle and modulation, which are needed to calculate the apparent lifetimes (Eqs. (12) and (13)].

The data sets for FLIM are rather large (in our case, 512×512 pixels, resulting in about 520 kbyte storage per image), which can result in time-consuming data storage, retrieval, and processing. To allow rapid calculation of images, we developed an algorithm that uses each phase-sensitive image only once. The goal is to use a set of images taken at different detector phase angles (Figure 9.11) and generate the lifetime image. Our algorithm yields three images. The first image is of the phase of the fluorescence, $\theta(r)$ in Eq. (15). The second image is the modulated amplitude of the fluorescence at the particular detector modulation frequency (i.e., the AC component or $m(r)$ in Eq. (15)]. The third is an image of the steady-state or DC component of the fluorescence (Figure 9.12). This procedure, including corrections for the position-dependent response of the image intensifier, has been described in detail (46).

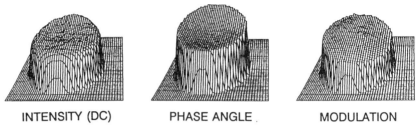

INTENSITY (DC)   PHASE ANGLE   MODULATION

Figure 9.12. Calculated images: intensity (left), phase angle (middle), and modulation (right).

9.3.3. Intracellular Calcium Concentration Imaging

We used COS cells to work out the technical requirements for FLIM in a cellular system. The COS cells were selected for the initial FLIM studies because these cells are relatively flat in terms of their biological architecture, easy to grow, and adhere moderately well to the glass surface. The cells were loaded with calcium probe quin-2 by exposure to quin-2 AM. The fluorescence lifetime of quin-2 is strongly dependent on Ca^{2+} [17, 18, 20]. The calcium-induced change in the lifetime of quin-2 results in a dramatic change in phase angle and modulation vs. free calcium concentration, ranging from 0 to 600 nM. Phase-sensitive images were collected using the instrumentation described in Figure 9.9 at a modulation frequency of 49.53 MHz.

The fluorescence intensity image (Figure 9.8 top) shows the area, shape, and orientation of the COS cell (or cell cluster). Figure 9.8 (left) shows that the local intensity of quin-2 varies dramatically throughout the cell. The phase angle and modulation images (Figure 9.13) display relatively constant values

PHASE ANGLE   MODULATION

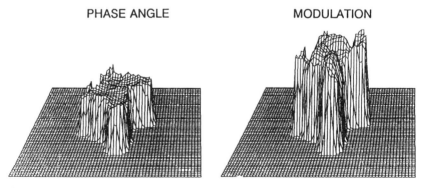

Figure 9.13. Phase angle and modulation images of quin-2 fluorescence in a COS cell at 49.53 MHz.

throughout the cell, suggesting that the Ca^{2+} concentration is uniform (Figure 9.8 right). This constancy is in agreement with the result of ratiometric imaging in diverse cells (88). The calcium concentrations determined from the phase angle or modulation image are not dependent on the intensity signal. The variations in the intensity image (Figure 9.8) are due to the different quin-2 concentrations and/or different thicknesses of the cell. In contrast, the phase and modulation images are relatively constant throughout the cell. The spatial variation in phase and modulation appears to be inversely correlated. This result is expected because an increase in the local lifetime results in an increase in the local phase angle and a decrease in the local modulation. The use of both the phase and modulation information is likely to increase the accuracy and range of the FLIM. More detailed information about these cellular imaging experiments has been reported elsewhere (46).

9.3.4.   Phase Suppression Imaging

Creation of a lifetime image requires considerable data collection and image processing. This is because of the nature of the technological challenge and because in our initial experiments we chose to use "complete" data sets that in fact overdetermine the answer. However, one can imagine many circumstances in which it is desirable to obtain less fully resolved images more quickly or with fewer computational demands. For example, one can collect only two images, with a θ_D difference of 90° to calculate the phase angle or lifetime image, from the ratio of these two images. Alternatively, one may be interested in quickly identifying regions of the cell where the lifetime exceeds some threshold value. The phase-modulation imaging method is unique in this regard. By selecting two detector phase angles, one can eliminate the contribution (suppress) of any desired lifetime. Then only regions of longer (or shorter) lifetime appear as positive values in the difference image. This procedure is described elsewhere in more detail (45, 79).

9.4.   ADVANCED APPLICATIONS OF FLUORESCENCE SPECTROSCOPY

Time-resolved imaging is now an existing technology, and lifetime-based sensing is moving toward analytical and clinical applications (9). Hence it is natural to wonder which topics of current interest in basic research are likely to have applications in imaging and clinical chemistry. One could describe many possibilities, such as the present efforts to perform medical imaging based on the time-dependent migration of light in tissues (89–94), the uses of FRET to recover the structure and dynamics of biological molecules (95, 96), and the use of oriented systems to obtain the information content of the fluorescence spectra data (97). In this chapter, we emphasize two areas of special promise, both of which take advantage of the intense pulses available from modern laser sources: two-photon-induced fluorescence and light quenching.

9.4.1. Two-Photon-Induced Fluorescence (TPIF)

Most of us are familiar only with one-photon-induced fluorescence (OPIF), which is the common form in our experiments. With intense laser sources, it is possible to observe the emission resulting from the simultaneous absorption of two long-wavelength photons (Figure 9.14). For instance, tryptophan in proteins, which normally absorbs light at 290 nm, can be excited by the simultaneous absorption of two 580 nm photons. This remarkable phenomenon occurs only with high light intensity because the two photons must be in the same place at the same time to allow simultaneous absorption.

This dependence on the square of the intensity is shown in Figure 9.15, which shows the spectra of human serum albumin (HSA) excited at 295 or 590 nm. Also shown is the emission intensity of HSA, which is linearly dependent on the incident light intensity at 295 nm and quadratically dependent on the intensity at 590 nm (98). TPIF has also been observed for fluorophores bound to membranes (99) and nucleic acids (100).

With OPIF, the violet light is absorbed according to Beer's law, with light absorption starting immediately at the surface of the sample. With TPIF, absorption occurs primarily at the point of highest intensity—in this case, where the red incident light is focused at the center of the sample. The fact that

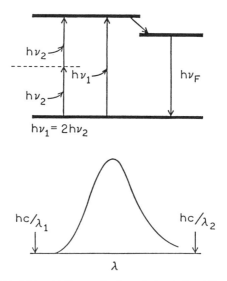

Figure 9.14. Jablonski diagram for two-photon-induced fluorescence.

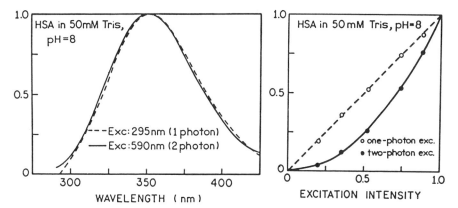

Figure 9.15. Emission spectra of HSA for one- and two-photon excitation (left) and dependence of emission intensity on incident light intensity (right).

TPIF depends on the intensity squared provides an important opportunity for fluorescence microscopy. In fluorescence microscopy, confocal optics are often used to eliminate fluorescence from outside the focal plane of the lenses (101, 102). Removal of this out-of-focus light provides remarkable improvement in image quality because in OPIF the fluorescence occurs from the entire thickness of the sample. Much of this emission is devoid of spatial information and only serves to degrade image contrast and resolution.

Professor Webb and colleagues at Cornell University recognize that the intensity-squared dependent of TPIF provided the opportunity for intrinsic "confocal" excitation (103, 104). The sample can be excited only at the desired depth (Figure 9.16), and the signal comes only from this region. Perhaps more important, the fluorophores that are not in the focal plane are not excited, consequently are not photobleached, and are thus available for imaging when the focal plane is moved (Figure 9.16). There are additional advantages of two-photon microscopy, such as the greater availability of optical components and increased transmission of the optics for the longer wavelengths. It is possible that sample autofluorescence will be lower with two-photon excitation, but at present, we do not know if the endogenous fluorophores in cells will display high or low cross sections for two-photon excitation. It is already known that some Ca^{2+} probes display good two-photon absorption (31), and lifetime images with two-photon excitation have already been reported (105). Hence, we can now imagine the creation of 3-D chemical images of cells (Figure 9.16), which could display the local Ca^{2+} concentration as seen for the 2-D FLIM imaging in Figure 9.8.

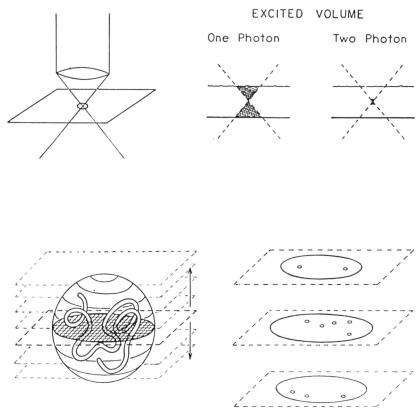

Figure 9.16. Intrinsic confocal excitation using TPIF in microscopy (top) and 3-D optical imaging (bottom).

9.4.2. TPIF and Time-Resolved Fluorescence

We were surprised by the ability to observe TPIF with ps laser sources. The experiments with two-photon microscopy (105) used an fs laser, and it was generally assumed that these exotic lasers were necessary to obtain adequate intensity for TPIF. (At the same average power, a 10 fs pulse is expected to result in 10^6-fold more TPIF than a 10 ps pulse.) These observations will allow the use of TPIF in many other laboratories. In the next several years, we expect to see increasing use of the time-resolved emission resulting from TPIF. One reason will be the increased photo-selected orientation that results from two-photon excitation (Figure 9.17). For OPIF, the probability of absorption depends on $\cos^2 \theta$, where θ is the angle between the electric vector of the

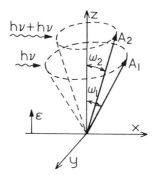

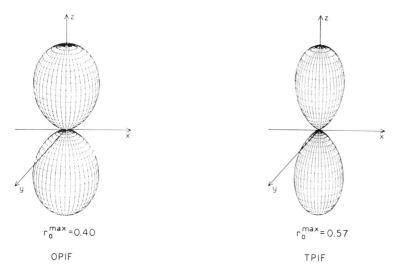

$r_0^{max} = 0.40$ $r_0^{max} = 0.57$

OPIF   TPIF

Figure 9.17. Increased photo-selected orientation for TPIF. The lower surfaces represent the orientation of the excited state, assuming that $r_0 = 0.4$ and no rotational diffusion exists.

polarized excitation and the transition moment of the fluorophores. The fact that the maximal observable anisotropy for a random distribution of fluorophores is 0.4 is a direct consequence of the $\cos^2 \theta$ dependence of light absorption.

Simultaneous absorption of two photons can result in an apparent $\cos^4 \theta$, which results in a more highly oriented excited-state population. This effect is illustrated in the lower panel of Figure 9.17, which shows the excited-state population immediately following light absorption. Consequently, the maxi-

mum anisotropy for TPIF is near 0.57, which may provide improved resolution of complex rotational motions of macromolecules in solution. Another
important aspect of TPIF is that the cross section (absorption spectra) for
two-photon absorption can be different from the cross section for one-photon
absorption (after correcting for λ and 2λ). Such differences have already been
observed for indole and tryptophan (106), which appear to be the result of
different relative cross sections for the ^1L_a and ^1L_b transitions for OPIF and
TPIF. Hence, different information may be available from the OPIF and
TPIF spectral data.

9.4.3. Light Quenching of Fluorescence

Measurements of time-resolved fluorescence, particularly TPIF, require the
use of intense laser sources. The use of these ps laser sources allows observation of the phenomena of stimulated emission. If a fluorophore is illuminated
at a wavelength that overlaps its emission spectra, the fluorophore can be
stimulated to return to the ground state (Figure 9.18). Since the stimulated
photon travels parallel to the "quenching beam," and since the emission is
generally observed at right angles to the illumination, the emission appears to
be quenched.

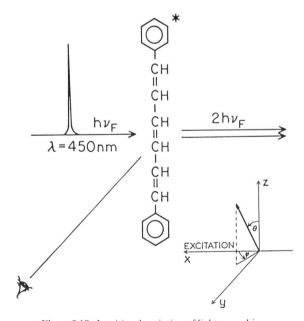

Figure 9.18. Intuitive description of light quenching.

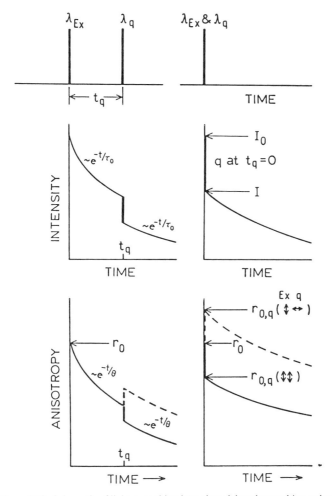

Figure 9.19. Schematic of light quenching by a time-delayed quenching pulse.

Of course, light quenching or, stimulated emission was predicted by Einstein in 1917 for atoms in the gas phase (107). Historically, light quenching has only been observed using the very intense pulses from Q-switched ruby lasers (108–110). Since we now know that light quenching can be observed wth modern ps lasers, there are numerous opportunities for novel fluorescence experiments (111–113).

Consider that the sample is excited with one pulse, followed by a second longer wavelength quenching pulse (Figure 9.19). The quenching pulse can result in an instantaneous change in the excited-state population. It is

important to recognize that this change in population should be nondestructive, and we are not depleting the ground state or bleaching the sample. Hence,
the experiment may be repeated numerous times for improved signal-to-noise
ratios, if needed to measure small effects.

One remarkable property of light quenching (and there are other properties
that are not described in this chapter) is that it displays the same $\cos^2 \theta$
dependence as does light absorption (109–113). This means that not only is the
total excited-state population altered by the quenching pulse but also the
selectively oriented parts of the excited-state population are quenched. Consequently, depending on the polarization of the quenching light, the polarization
of the emission can be altered from 1.0 to -1.0, (Figure 9.20), resulting in

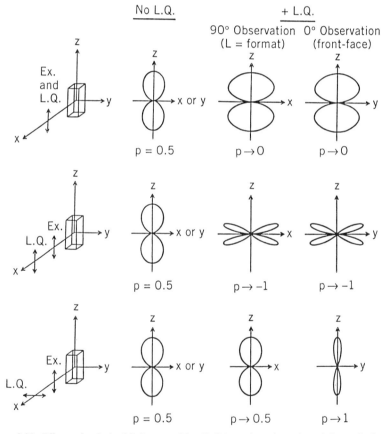

Figure 9.20. Effects of polarized light quenching (L.Q.) on the orientation of the excited-state
population.

a high degree of orientation of the excited-state population. In contrast, one-photon excitation of randomly oriented fluorophores can only result in polarization values of 0.5 to -0.33. It may even be possible to break the z-axis symmetry, which heretofore has been pervasive in the optical spectroscopy of randomly oriented solutions. In our opinion, the phenomena of light quenching can result in a new class of fluorescence experiments in which the sample is excited with one pulse and the excited-state population is modified by the quenching pulse(s) prior to measurement.

9.5. CONCLUSION

In closing, we wish to reiterate our opening statement. Time-resolved fluorescence is now positioned to move out of the biophysics research laboratory and into the world of fluorescence microscopy, cell biology, and clinical chemistry (114, 115). Advances in laser sources, CCD detection, and other technologies are resulting in the possibility of simple instrumentation for previously complex measurements. The increasing availability of intense ps and fs lasers, as well as laser systems that provide multiple time-delayed pulses, will result in the increased use of two-photon excitation and stimulated emission to control and/or modify the excited-state population.

In our opinion, the applications of time-resolved fluorescence to microscopy are limited not by the instrument technology but by the available probes. There are only a limited number of conjugatable long-wavelength probes and none that display specific analyte sensitivity. What is needed is an arsenal of probes, all of which can be excited with laser diodes or light-emitting diodes (LEDs), and that are specifically sensitive to cations, anions, and other analytes. While several laboratories are working in this area, the total effort is minor in comparison to the number of scientists engaged in instrument development, technology development, theory, or applications. The development of this arsenal of probes is crucial for expanded applications of fluorescence to cell biology.

ACKNOWLEDGMENTS

The concepts and results described in this chapter were developed over a number of years. Dr. Joseph R. Lakowicz wishes to thank the National Institutes of Health (RR-08119, RR-07510, GM 39617, GM 35154) and the National Science Foundation (MCB 8804931, BIR 9319032, DIR 8710401) for their continued support. The authors thank Dr. Ignacy Gryczynski and Dr. Józef Kuśba for use of the results on two-photon excitation and light quenching. Dr. Lakowicz expresses appreciation to the Medical Biotechnology Center for its support.

REFERENCES

1. Dewey, T. G. (Ed.). (1991). *Biophysical and Biochemical Aspects of Fluorescence Spectroscopy.* Plenum, New York.

2. Lakowicz, J. R. (Ed.), (1994). *Time-resolved laser spectroscopy in biochemistry IV. SPIE—Int. Soc. Opt. Eng.* **2137**, 806 pp.

3. Taylor, D. L., Waggoner, A. S., Lanni, F., Murphy, R. F., and Birge, R. R. (Eds.). (1986). *Applications of Fluorescence in the Biomedical Sciences.* Alan. R. Liss, New York.

4. Tian, R., and Rodgers, M. A. J. (1991). Time-resolved fluorescence microscopy in new techniques in optical microscopy and spectrophotometry. *Top. Mol. Struct. Biol.* **15**, 177–198.

5. Schneckenburger, H., Siedlitz, H. K., and Eberz, J. (1988). Time-resolved fluorescence in photobiology. *J. Photochem. Photobiol. B: Biol.* **2**, 1–19.

6. Morgan, C. G., Mitchell, A. C., and Murray, J. G. (1990). Nanosecond time-resolved fluorescence microscopy: Principles and practice. *Trans. R. Microsc. Soc.* **1**, 463–466.

7. Wolfbeis, O. S. (Ed.). (1993). *Fluorescence Spectroscopy: New Methods and Applications.* Springer-Verlag, New York.

8. Schulman, S. G. (1985). *Molecular Luminescence Spectroscopy: Methods and Applications,* Vol. 1, Wiley, New York.

9. Lakowicz, J. R., and Thompson, R. B. (1993). *Advances in fluorescence sensing technology. Proc. SPIE—Int. Soc. Opt. Eng.* **1885**, 480 pp.

10. Van Dyke, K., and Van Dyke, R. (Eds.). (1990). *Luminescence Immunoassay and Molecular Applications.* CRC Press, Boca Raton, FL.

11. Wolfbeis, O. S. (Ed.). (1993). Proceedings of the First European Conference on Optical Chemical Sensors and Biosensors. *Sensors Actuators B* **11**, 1–3.

12. Lakowicz, J. R., and Szmacinski, H. (1993). Fluorescence lifetime-based sensing of pH, Ca^{2+}, K^+ and glucose. *Sensors Actuators B* **11**, 133–143.

13. Bacon, J. R., and Demas, J. N. (1987). Determination of oxygen by luminescence quenching of a polymer immobilized transition metal complex. *Anal. Chem.* **59**, 2780–2785.

14. Lippitsch, M. E., Pasterhofer, J., Leiner, M. J. P., and Wolfbeis, O. S. (1988). Fibre-optic oxygen sensor with the fluorescence decay time as the information carrier. *Anal. Chim. Acta* **205**, 1–6.

15. Szmacinksi, H., Lakowicz, J. R. (1993). Optical measurements of pH using fluorescence lifetimes and phase-modulation fluorometry. *Anal. Chem.* **65**, 1668–1674.

16. Whitaker, J. E., Haugland, R. P., and Prendergast, F. G. (1991). Spectral and photophysical studies of benzo[*c*]xanthene dyes: Dual emission pH sensors. *Anal. Biochem.* **194**, 330–344.

17. Lakowicz, J. R., Szmacinski, H., Nowaczyk, K., and Johnson, M. L. (1992). Fluorescence lifetime imaging of calcium using Quin-2. *Cell Calcium* **13**, 131–147.

18. Hirshfeld, K. M., Toptygin, D., Packard, B. S., and Brand, L. (1993). Dynamic fluorescence measurements of two-state systems: Applications to calcium- chelating probes. *Anal. Biochem.* **209**, 209–218.

19. Lakowicz, J. R., Szmacinski, H., and Johnson, M. L. (1992). Calcium imaging using fluorescence lifetimes and long-wavelength probes. *J. Fluoresc.* **2**(1), 47–62.

20. Miyoshi, N., Hara, K., Kimura, S., Nakanishi, K., and Fukuda, M. (1991). A new method of determining intracellular free Ca^{2+} concentration using Quin-2 fluorescence. *Photochem. Photobiol.* **53**, 415–418.

21. Szmacinski, H., and Lakowicz, J. R. (1993). Optical measurements of Mg^{2+} using fluorescence lifetimes and phase-modulation fluorometry. *Biophys. J.* **64**, A108.

22. Verkman, A. S. (1990). Development and biological applications of chloride sensitive fluorescent indicators. *Am. J. Physiol.* **259**, C375–C388.

23. Grynkiewicz, G., Poenie, M., and Tsien, R. Y. (1985). A new generation of Ca^{2+} indicators with greatly improved fluorescence properties. *J. Biol. Chem.* **260**, 3440–3450.

24. Akkaya, E. U., and Lakowicz, J. R. (1993). Styryl-based wavelength ratiometric probes: A new class of fluorescent calcium probes with long wavelength emission and a large Stokes' shift. *Anal. Biochem.* **213**, 285–289.

25. Minta, A., Kao, J. P. Y., and Tsien, R. Y. (1989). Fluorescent indicators for cytosolic calcium based on rhodamine and fluorocein chromophores. *J. Biol. Chem.* **264**, 8171–8178.

26. Bissel, R. A., Prasanna de Silva, A., Gunaratne, H. Q. N., Lynch, P. L. M., McCoy, C. P., Maguire, G. E. M., and Sandanayake, K. R. A. s. (1993). Fluorescent photoinduced electron-transfer sensors. *ACS Symp. Ser.* **538**, 45–58.

27. de Silva, A. P., and de Silva, S. A. (1986). Fluorescent signalling crown ethers; 'switching on' of fluorescence by alkali metal ion recognition and binding in situ. *J. Chem. Soc., Chem. Commun.* **20**, 1709–1710.

28. Förster, Th. (1948). Intermolecular energy migration and fluorescence. *Ann. Phys. (Leipzig)* [6] **2**, 55–75 (Translation 1974 by R. S. Knox).

29. Tsien, R. Y. (1980). New calcium indicators and buffers with high selectivity against magnesium and protons: Design, synthesis, and properties of prototype structures. *Biochemistry* **19**, 2396–2404.

30. Szmacinski, H., and Lakowicz, J. R. (1995). Possibility of simultaneous measuring low and high calcium concentration using Fura-2 and lifetime-based method. *Cell Calcium* **18**, 64–75.

31. Szmacinski, H., Gryczynski, I., and Lakowicz, J. R. (1993). Calcium-dependent fluorescence lifetimes of indo-1 for one- and two-photon excitation of fluorescence. *Photochem. Photobiol.* **58**, 341–345.

32. Kuhn, M. A. (1993). 1,2-bis(2-Aminophenoxy)ethan-N,N,N',N',-tetraacetic acid conjugate used to measure intracellular Ca^{2+} concentration. *ACS Symp. Ser.* **538**, 147–161.

33. Frey-Forgues, S., Le Bris, M.-T., Guette, J.-P., and Valeur, J. B. (1988). Ion-responsive fluorescent compounds. I. Effect of cation binding on photophysical

properties of a benzoxazinone derivative linked to monoaza-15- crown-5. *J. Phys. Chem.* **92**, 6233–6237.

34. Valeur, B., Bourson, J., and Pouget, J. (1993). Ion recognition detected by changes in photoinduced charge or energy transfer. *ACS Symp. Ser.* **538**, 25–44.

35. Szmacinski, H., and Lakowicz, J. R., unpublished data of lifetimes.

36. Haugland, R. P. (1992–1994). *Handbook of Fluorescent Probes and Research Chemicals* (K. D. Larison, Ed.). Molecular Probes, Eugene, OR.

37. Minta, A., and Tsien, R. Y. (1989). Fluorescent indicators for cytosolic sodium. *J. Biol. Chem.* **264**, 19449–19457.

38. Molecular Probes (1993). *BioProbes 18*, p. 20. Molecular Probes, Eugene. OR.

39. Schimidzu, T., Iyoda, T., and Izaki, K. (1985). Photochemical properties of bis(2,2'-Bipyridine)(4,4''-dicarboxy-2,2'-bipyridine)ruthenium(II) chloride, *J. Phys. Chem.* **89**, 642–645.

40. Birch, D. J. S., Imhof, R. E. (1991). Time-domain fluorescence spectroscopy using time-correlated single-photon counting. In *Topics in Fluorescence Spectroscopy*. (J. R. Lakowicz, Ed.), Vol. 1, pp. 1–95. Plenum, New York.

41. Lakowicz, J. R., and Gryczynski, I. (1991). Frequency-domain fluorescence spectroscopy. In *Topics in Fluorescence Spectroscopy* (J. R. Lakowicz, Ed.), Vol. 1, pp. 293–355. Plenum, New York.

42. Demas, J. N. (1983). *Excited State Lifetime Measurements*. Academic Press, New York.

43. O'Connor, D. V., and Phillips, D. (1984). *Time-Correlated Single Photon Counting*. Academic Press, London.

44. Lakowicz, J. R., Laczko, G., Cherek, H., Gratton, E., and Limkeman, M. (1984). Analysis of fluorescence decay kinetics from variable-frequency phase shift and modulation data. *Biophys. J.* **46**, 463–477.

45. Lakowicz, J. R., Szmacinski, H., Nowaczyk, K., Berndt, K., and Johnson, M. L., (1992). Fluorescence lifetime imaging. *Anal. Biochem.* **202**, 316–330.

46. Lakowicz, J. R., Szmacinski, H., Nowaczyk, K., Lederer, W. J., Kirby, M. S., and Johnson, M. L., (1994). Fluorescence lifetime imaging of intracellular calcium in COS cells using quin-2. *Cell Calcium* **15**, 1–21.

47. Riech, R. K., Mountain, R. W., McGonagle, W. H., Huang, C. M., Twichell, J. C., Kosicki, B. B., and Savoye, E. D., (1991). An integrated electronic shutter for back-illuminated charge-coupled devices. *IEEE* **91**, 171–174.

48. Loeser, C. N., Clark, E., Maher, M., and Tarkmeel, H. (1972). Measurement of fluorescence decay time in living cells. *Exp. Cell Res.* **72**, 480–484.

49. Sacchi, C. A., Svelto, O., and Prenna, G. (1974). Pulsed tunable lasers in cytofluorometry. *Histochem. J.* **6**, 251–258.

50. Bottirolli, G., Prenna, G., Andreoni, A., Sachhi, C. A., and Svelto, O. (1979). Fluorescence of complexes of quinacrine mustard with DNA. I. Influences of the DNA base composition on the decay time in bacteria. *Photochem. Photobiol.* **29**, 23–28.

51. Kinoshita, K., Jr., Mitaku, S., Ikegami, A., Ohbo, N., and Kunii, T. L., (1976). Construction of nanosecond fluorometry system for applications to biological samples at cell and tissue levels. *Jpn. J. Appl. Phys.* **15**, 2433–2440.

52. Herman, B. A., and Fernandez, S. F. (1978). Changes in membrane dynamics associated with myogenic cell fusion. *J. Cell. Physiol.* **94**, 253–264.

53. Arndt-Jovin, D. J., Latt, S. A., Striker, G., and Jovin, T. M. (1979). Fluorescence decay analysis in solution and in a microscope of DNA and chromosomes stained with quinacrine. *J. Histochem. Cytochem.* **27**, 87–95.

54. Schneckenburger, H., Reuter, B. W., and Schoberth, S. M. (1984). Time-resolved fluorescence microscopy for measuring specific coenzymes in methanogenic bacteria. *Anal. Chim. Acta* **163**, 249–255.

55. Minami, T., Kawahigashi, M., Sakai, Y., Shimamoto, K., and Hirayama, S. (1986). Fluorescence lifetime measurement under a microscope by the time-correlated simple-photon counting technique. *J. Lumines.* **35**, 247–253.

56. Rodgers, M. A. J., and Firey, P. A. (1985). Instrumentation for fluorescence microscopy with picosecond time resolution. *Photochem. Photobiol.* **42**, 613–616.

57. Schneckenburger, H., Pauker, F., Unsöld, E., and Jocham, D. (1985). Intracellular distribution of retention of the fluorescent compounds of Photofrin II. *Photobiochem. Photobiophys.* **10**, 61–67.

58. Ramponi, R., and Rodgers, M. A. J. (1987). An instrument for simultaneous acquisition of fluorescence spectra and fluorescence lifetimes from single cells. *Photochem. Photobiol.* **45**, 161–165.

59. Keating, S. M., and Wensel, T. G. (1991). Nanosecond fluorescence microscopy. Emission kinetics of Fura-2 in single cells. *Biophys. J.* **59**, 186–202.

60. Murray, J. G., Cundall, R. B., Morgan, C. G., Evans, G. B., and Lewis, C. (1986). A single-photon-counting Fourier transform microflurometer. *J. Phys. E* **19**, 349–355.

61. Verkman, A. S., Armijo, M., and Fushimi, K. (1991). Construction and evaluation of a frequency-domain epifluorescence microscope for lifetime and anisotropy decay measurements in subcellular domains. *Biophys. Chem.* **40**, 117–125.

62. Schneckenburger, H., Frenz, M., Tsuchiya, Y., Denzer, U., and Schleinkofer, L. (1987). Picosecond fluorescence microscopy for measuring chlorophyll and porphyrin compounds in conifers and cultured cells. *Laser Life Sci.* **1**(4), 299–307.

63. Kusumi, A., Tsuji, A., Murata, M., Sako, Y., Yoshizawa, A. C., Kagiwada, S., Hayakawa, T., and Ohnishi, S. (1991). Development of a streak-camera-based time-resolved microscope fluorimeter and its application to studies of membrane fusion in single cells. *Biochemistry* **30**, 6517–6527.

64. Pauker, F., Schneckenburger, H., and Unsöld, E. (1986). Time-resolved multiphoton counting. *J. Phys. E* **19**, 240–244.

65. Iwata, T., Uchida, T., and Minami, S. (1985). A nanosecond photon counting fluorometric system using a modified multichannel vernier chronotron. *Appl. Spectrosc.* **39**, 101–109.

66. Wang, X. F., Kitajima, S., Uchida, T., Coleman, D. M., and Minami, S. (1990).Time-resolved fluorescence microscopy using multichannel photon counting. *Appl. Spectrosc.* **44**, 25–30.

67. Wang, X. F., Tsuji, T., Uchida, T., and Minami, S. (1992). Multichannel photon-counting fluorometric system using optical fiber dynamic memory. *Proc. SPIE— Int. Soc. Opt. Eng.* **1640**, 271–277.

68. Minami, T., and Hirayama, S. (1990). High quality fluorescence decay curves and lifetime imaging using an elliptical scan streak camera. *Photochem. Photobiol. A: Chem.* **53**, 11–21.

69. Marriott, G., Clegg, R. M., Arndt-Jovin, D. J., and Jovin, T. M. (1991). Time-resolved imaging microscopy. Phosphorescence and delayed fluorescence imaging. *Biophys. J.* **60**, 1347–1387.

70. Wang, X. F., Uchida, T., Coleman, D. M., and Minami, S. (1991). A two dimensional fluorescence lifetime imaging system using a gated image intensifier. *Appl. Spectrosc.* **45**, 360–366.

71. Cubeddu, R., Taroni, P., Valentini, G., and Canti, G. (1992). Use of time-gated fluorescence imaging for diagnosis in biomedicine. *Photochem. Photobiol.* **12**; 109–113.

72. Cubeddu, R., Canti, G., Taroni, P., and Valentini, G. (1993). Time-gated fluorescence imaging for diagnosis of tumors in a murine model. *Photochem. Photobiol.* **57**, 480–485.

73. Oida, T., Sako, Y., and Kusumi, A. (1993). Fluorescence lifetime imaging microscopy. *Biophys. J.* **64**, 676–685.

74. Schneckenburger, H., König, K., Kunzi-Rapp, K., Westphal-Frösch, C., and Rück, A. (1993). Time-resolved in vivo fluorescence of photosensing porphyrins. *Photochem. Photobiol. B* **21**, 143–147.

75. Wang, X. F., Periasamy, A., and Herman, B. (1992). Fluorescence lifetime imaging microscopy (FLIM): Instrumentation and applications. *CRC Crit. Rev. Anal. Chem.* **23**, 369–395.

76. Gadella, T. W. J., Jr., Jovin, T. M., and Clegg, R. M. (1993). Fluorescence lifetime imaging (FLIM): Spatial resolution of microstructures on the nanosecond time scale. *Biophys. Chem.* **48**, 221–239.

77. Seveous, L., Väisälä, M., Syrjänen, S., Sandberg, M., Kuusisto, A., Harju, R., Salo, J., Hemmilä, I., Kojola, H., and Soini, E. (1992). Time-resolved fluorescence imaging of europium chelate label in immunohistochemistry and in situ hybridization. *Cytometry* **13**, 329–338.

78. Buurman, E. P., Sanders, R., Draauer, A., Gerritsen, H. C., Van Veen, J. J. F., Houpt, P. M., and Levine, Y. K. (1992). Fluorescence lifetime imaging using a confocal laser scanning microscope. *Scanning* **14**; 155–159.

79. Szmacinski, H., Lakowicz, J. R., and Johnson, M. L. (1994). Fluorescence lifetime imaging microscopy (FLIM). Homodyne technique using high-speed image intensifier. *Methods Enzymol.* **240**, 723–748.

80. Piston, D. W., Sandison, D. R., and Webb, W. W. (1992). Time-resolved fluo-

rescence imaging and background rejection by two-photon excitation in laser scanning microscopy *Proc. SPIE—Int. Soc. Opt. Eng.* **1640**, 379–389.

81. Morgan, C. G., Mitchell, A. C., and Murray, J. G. (1992). Prospects for confocal imaging based on nanosecond fluorescence decay time. *J. Microsc. (Oxford)* **165**, 49–60.

82. French, T., Gratton, E., and Maier, J. (1992). Frequency-domain imaging of thick tissues using a CCD. *Proc. SPIE—Int. Soc. Opt. Eng.* **1640**, 254–261.

83. Clegg, R. M., Fedderson, B., Gratton, E., and Jovin, T. M. (1992). Time-resolved imaging fluorescence microscopy. *Proc. SPIE—Int. Soc. Opt. Eng.* **1640**, 448–460.

84. Gratton, E., Limkeman, M., Lakowicz, J. R., Maliwal, B. P., Cherek, H., and Laczko, G. (1984). Resolution of mixtures of fluorophores using variable-frequency phase and modulation data. *Biophys. J.* **46**, 479–486.

85. Alcala, J. R., Gratton, E., and Prendergast, F. G. (1987). Resolvability of fluorescence lifetime distributions using phase fluorometry. *Biophys. J.* **51**, 587–596.

86. Lakowicz, J. R., and Cherek, H. (1981). Phase sensitive fluorescence spectroscopy. A new method to resolve fluorescence lifetimes or emission spectra of components in a mixture fluorophores. *J. Biochem. Biophyss. Methods* **5**, 19–35.

87. Lakowicz, J. R., and Cherek, H. (1981). Resolution of heterogeneous fluorescence from proteins and aromatic amino acids by phase-sensitive detection of fluorescence. *J. Biol. Chem.* **256**, 6348–6353.

88. Berlin, J. R., Wozniak, M. A., Cannel, M. B., Bloch, R. J., and Lederer, W. J. (1990). Measurement of intracellular Ca^{2+} in BC3H-1 muscle cells with Fura-2. *Cell Calcium* **11**, 371–384.

89. Grunbaum, F. A., Kohn, P., Latham, G. A., Singer, J. R., and Zubelli, J. P. (1991). Diffuse tomography. *Proc. SPIE—Int. Soc. Opt. Eng.* **1431**, 232–238.

90. Spears, K., Serafin, J., Abramson, N. H., Zhu, X., and Bjelkhagen, H. (1989). Chromo-coherent imaging for medicine. *IEEE Trans. Biomed. Eng.* **BME-36**, 1210–1221.

91. Andersson-Engels, S., Berg, R., Svanberg, S., and Jarlman, O. (1990). Time-resolved illumination for medical diagnosis. *Opt. Lett.* **15**, 1179–1181.

92. Wang, L., Ho, P. P., Liu, C., Zhang, G., and Alfano, R. R. (1991). Ballistic 2-D imaging through scattering wells using an ultrafast optical Kerr gate. *Science* **253**, 769.

93. Wilson, B. C., Sevick, E. M., Patterson, M. S., and Chance, B. (1993). Time-dependent optical spectroscopy and imaging for biomedical applications. *Proc. IEEE* **80**, 918–930.

94. Sevick, E. M., Frisoli, J. K., Burch, C. L., and Lakowicz, J. R. (1993). Localization of absorbers in scattering media using frequency-domain measurements of time-dependent photon migration. *Apl. Opt.* **33**(16), 3562–3571.

95. Cheung, H. C., (1991). Resonance energy transfer. In *Topics in Fluorescence Spectroscopy* (J. R. Lakowicz, Ed.), Vol. 2, pp. 127–176. Plenum, New York.

96. Uster, P. S., and Pagano, R. E. (1986). Resonance energy transfer microscopy: Observations of membrane-bound fluorescent probes in model membranes and in living cells. *J. Cell Biol.* **103**, 1221–1234.

97. Burghardt, T. P., and Ajtai, K. (1991). Fluorescence polarization from oriented systems. In *Topics in Fluorescence Spectroscopy* (J. R. Lakowicz, Ed.), Vol. 2, pp. 307–365. Plenum, New York.

98. Lakowicz, J. R., and Gryczynski, I. (1993). Tryptophan fluorescence intensity and anisotropy decays of human serum albumin resulting from one-photon and two-photon excitation. *Biophys. Chem.* **45**, 1–6.

99. Lakowicz, J. R., Gryczynski, I., Kusba, J., and Danielsen, E. (1992). Two-photon-induced fluorescence intensity and anisotropy decays of diphenylhexatriene in solvents and lipid bilayers. *J. Fluoresc.* **2**, 247–258.

100. Lakowicz, J. R., and Gryczynski, I. (1992). Fluorescence intensity and anisotropy decay of the 4′,6-diamidino-2-phenylindole-DNA complex resulting from one-photon and two-photon excitation. *J. Fluoresc.* **2**, 117–122.

101. Wilson, T. (Ed.). (1990). *Confocal Microscopy*. Academic Press, London.

102. Pawley, J. B. (Ed.). (1990). *Handbook of Biological Confocal Microscopy*. Plenum, New York.

103. Denk, W., Strickler, J. H., and Webb, W. W. (1990). Two-photon laser scanning fluorescence microscopy. *Science* **248**, 73–76.

104. Webb, W. W. (1990). *Micro'90*, pp. 445–450

105. Piston, D. W., Sandison, D. R., and Webb, W. W. (1992). Time-resolved fluorescence imaging and background rejection by two-photon excitation in laser scanning microscopy. *Proc. SPIE—Int. Soc. Opt. Eng.* **1640**, 379–389.

106. Lakowicz, J. R., Gryczynski, I., Danielsen, E., and Frisoli, J. K. (1992). Unexpected anisotropy spectra of indole and N-acetyl-L-tryptophanamide observed for two-photon excitation fluorescence. *Chem Phys. Lett.* **194**, 282–287.

107. Einstein, A. (1917). On the quantum theory of radiation. *Phys. Z.* **18**, 121. (This paper has been reprinted in *Laser Theory* by F. S. Barnes, IEEE Press, New York, 1972.)

108. Galanin, M. D., Kirsanov, B. P., and Chizhikova, Z. A. (1969). Luminescence quenching of complex molecule in a strong laser field. *Sov. Phys.—JETP Lett.* (*Engl. Transl.*) **9**(9), 502–507.

109. Mazurenko, Y. T., Danilov, V. V., and Vorontsova, S. I. (1973). Depolarization of photoluminescence by powerful excitation. *Opt. Spectrosc.* (*Engl. Transl.*) **35**(1), 107–108.

110. Butko, A. I., Voropai, E. S., Zholnerevic, I., Saechnikov, V. A., and Sarzhevskii, A. M. (1978). Spectral distribution of polarization under light quenching. *Izve. Akad. Nauk SSSR* **42**(3), 626–630.

111. Lakowicz, J. R., Gryczynski, I., Kuśba, J., and Bogdanov, V. (1994). Light quenching of fluorescence: A new method to control the excited state lifetime and orientation of fluorophores. *Photochem. Photobiol.* **60**(6), 546–562.

112. Gryczynski, I., Bogdanov, V., and Lakowicz, J. R. (1993). Light quenching of tetraphenylbutadiene fluorescence observed during two-photon excitation. *J. Fluoresc.* **3**, 85–92.

113. Lakowicz, J. R., Gryczynski, I., Bogdanov, V., and Kuśba, J. (1994). Light quenching and fluorescence depolarization of rhodamine B and applications of this phenomenon to biophysics. *J. Phys. Chem.* **98**, 334–342.

114. Lakowicz, J. R., Koen, P. A., Szmacinski, H., Gryczynski, I., and Kusba, J. (1994). Emerging biomedical and advanced applications of time-resolved fluorescence spectroscopy. *J. Fluoresc.* **4**(1), 117–136.

115. Szmacinski, H., and Lakowicz, J. R. (1994). Fluorescence lifetime-based sensing. In *Topics in Fluorescence Spectroscopy* (J. R. Lakowicz, Ed.), Vol. 4, pp. 295–334. Plenum, New York.

TIME-RESOLVED FLUORESCENCE LIFETIME IMAGING MICROSCOPY: INSTRUMENTATION AND BIOMEDICAL APPLICATIONS

XUE FENG WANG, AMMASI PERIASAMY, PAWEL WODNICKI, GERALD W. GORDON, AND BRIAN HERMAN

Department of Cell Biology and Anatomy
University of North Carolina at Chapel Hill
Chapel Hill, North Carolina 275997090

10.1. INTRODUCTION

Fluorescence microscopic imaging is a powerful tool for a wide range of biological applications (1–3). Recently, quantitative image measurements using fluorescence microscopy have been achieved by using sensitive image detectors [e.g., charge coupled device (CCD) camera] and fluorescent probes (e.g., fura-2 as a Ca^{2+} indicator), as well as computer image processing (e.g., ratio imaging) (4–6). However, currently most fluorescence microscopic imaging is performed as measurements of emission or excitation spectra. These conventional fluorescence microscopic imaging measurements have several limitations: (1) generally, fluorescence imaging can be difficult to quantify, since there is no easy way to determine quantum yields across the sample (especially for turbid biological samples); (2) fluorescence intensity measurements cannot normally be used in a quantitative fashion because nonuniform fluorescent probe labeling or staining alters fluorescence intensity in a microscopically heterogeneous sample; (3) because conventional low-speed image detectors are used, fast dynamic events in cell physiology and biology cannot be studied; (4) fluorescence emitted by the sample under investigation may be complex, and a number of different molecular species may contribute to the overall observed signal, which cannot be individually analyzed using only spectral information; and (5) autofluorescence of intrinsic cellular components and scattered light affect detection sensitivity. To address these limitations and

Fluorescence Imaging Spectroscopy and Microscopy, edited by Xue Feng Wang and Brian Herman. Chemical Analysis Series, Vol. 137.
ISBN 0-471-01527-X © 1996 John Wiley & Sons, Inc.

to provide the ability to obtain additional novel information, a new approach based on imaging of spatial variations in fluorescence lifetime has been developed.

Like excitation and emission spectra, the fluorescence lifetime is an important reflection of a molecule's structure, dynamics, and environment, and generally falls in the range of 1–100 ns (7–9). Different molecules have different lifetimes, and even the same molecule in different environments can display distinct lifetimes. Moreover, molecules with similar spectra may have significant differences in their lifetimes, which makes lifetime analysis a powerful analytical tool. Importantly, the fluorescence lifetime is related to the quantum yield of the fluorescence. Measurements of lifetimes are independent of probe concentration (within a certain range), resulting in less photobleaching and providing more quantitative and accurate fluorescence measurements. By combining the power of fluorescence lifetime measurements with fluorescence microscopy, it is now possible to quantitatively determine dynamic and structural information from microscopic samples with high sensitivity and specificity (10). Because fluorescent lifetimes are not affected by scattering or decay characteristics of the background, measurements of fluorescent lifetimes provide more sensitive and quantitative information from complex structures. Unfortunately, the expense, complexity, and limitations of instrumentation have prevented lifetime measurements from being performed in two or three dimensions.

Recently, key advances in (1) laser light sources, (2) high-speed and high-sensitivity image detection devices, (3) fluorescence microscopy, (4) sensitive and specific fluorescent probes, and (5) image processing techniques have allowed the development of fluorescence lifetime imaging microscopy (FLIM) (11–17). FLIM permits the measurement of the two-dimensional (2-D) fluorescence lifetime distribution, thus providing 2-D structural and dynamic information about microscopic samples. FLIM has enormous potential applications in biomedical research. For example, for the cell biologist, FLIM is an extremely important advance, as it allows, for the first time, the sensitivity of the fluorescence lifetime to environmental parameters to be monitored in spatial and temporal manner in single living cells (18). In this chapter, we discuss the concept, principles, and instrumentation of FLIM. We describe in detail a time-resolved FLIM system that uses pulsed excitation in combination with a gated image intensifier and a confocal FLIM based on frequency-domain lifetime determination techniques. We also discuss numerous biological applications of FLIM.

10.2. PRINCIPLES AND BACKGROUND

Fluorescence lifetime measurements have been used for a wide variety of research applications (7–10) because of the following features:

1. *Specificity:* Fluorescence lifetimes are usually characteristic of a molecule's structure and composition. Different molecules usually display distinct lifetimes that can be selectively used for the analysis of complex mixtures of molecular species.

2. *Sensitivity:* Detection of fluorescence lifetime can be carried out with only small numbers of fluorescent molecules. By discrimination of autofluorescence from specific fluorescence (based on differences in lifetimes), highly sensitive fluorescence detection is possible.

3. *Quantitation:* Fluorescence lifetimes provide truly quantitative measurements because the fluorescence lifetime is directly related to the fluorescence quantum yield of the fluorophore. Quantification is feasible at relatively low concentrations because of the inherent sensitivity associated with emission as opposed to absorption processes.

4. *Environmental sensitivity:* Fluorescence lifetimes are extremely sensitive to the immediate physical and chemical environments; fluorophores have been designed for sensing environmental factors (pH, viscosity, etc.) and other related parameters (e.g., Ca^{2+}).

5. *High temporal resolution:* Fluorescence lifetime measurements can be used to detect very fast chemical and molecular changes in specimens. Chemical and biological processes occurring on the 10^{-12} to approximately 10^{-7} s timescale can be detected and measured.

6. *High spatial resolution:* Fluorescence lifetimes can be measured from single molecules if the molecules contain a sufficient number of fluorophores. Cellular components with dimensions below the diffraction-limited resolution of the light microscope and their interactions can be visualized by using fluorescence resonance energy transfer techniques.

7. *Three-dimensional (3-D) resolution:* Since 3-D fluorescence images can be constructed by conventional or confocal microscopy, in principle 3-D fluorescence lifetime imaging information can be obtained.

So far, most fluorescence lifetime measurements have been carried out in bulk samples or using single-point detection in cells. Why do we need fluorescence lifetime imaging? The utility of fluorescence lifetime imaging is based on the fact that most biological systems are not homogeneous, but rather exist as 3-D, spatially heterogeneous structural and functional complexes. Fluorescence lifetime imaging provides useful and unique information in terms of deciphering the 3-D, spatially heterogeneous structural and functional complex of the cell.

The basic configuration of a FLIM instrument is shown in Figure 10.1. To obtain fluorescence lifetime images in a 2-D sample, there are two options for imaging: (1) the scanning mode, in which the lifetime image is built up using

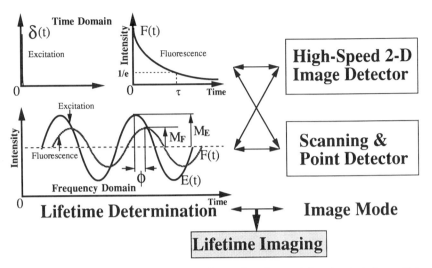

Figure 10.1. FLIM combines fluorescence lifetime detection, and imaging techniques. It is possible to measure fluorescence lifetime images by the combination of lifetime determination (time- and frequency-domains) techniques with high-speed 2-D detectors and scanning techniques such as mechanical stage scanning, laser beam scanning, or electronic (e.g., image dissector, streak-camera) scanning methods.

a scanning device such as a laser confocal scanner with a single detector or (2) the parallel image detection mode, in which a high-speed 2-D imaging detector is used for lifetime imaging. There are also two options for excitation and detection of fluorescence lifetimes: (a) time-domain pulse methods (19, 20) and (b) frequency-domain or phase-resolved methods (7). In the time-domain method, employing pulsed excitation, lifetimes are measured directly from the fluorescence signal or by photon-counting detection. In the frequency-domain or phase-resolved method, sinusoidally modulated light is used as an excitation source, and the lifetimes are determined from the phase shift or modulation depth of the fluorescence emission signal. It is possible to obtain fluorescence lifetime images by a combination of the lifetime determination techniques (time and frequency domains) with high-speed 2-D detectors and scanning techniques such as mechanical stage scanning, laser beam scanning, or electronic (e.g., image dissector, streak-camera) scanning methods. Recently, two-photon excitation techniques have been introduced to the fluorescence spectroscopic community that have improved spatial and lifetime detection resolution (21).

FLIM is a new imaging technology and in its infancy. The first paper about fluorescence lifetime imaging was published in 1989 and was based on the use of a high-speed, sensitive image dissector tube (IDT) coupled with frequency-domain lifetime detection (22). By heterodyning the modulated fluorescence and reference signals at the blanking electrode of the IDT, high-frequency responses of up to 1 GHz were obtained, and significant improvement of time resolution was observed. Two-dimensional lifetime and intensity images were obtained by electronically scanning images within the IDT. A FLIM that combined fluorescence microscopy with multichannel photon-counting detection was discribed in 1990 (11). Multichannel photon detection was used to enhance the rate of data collection and improve temporal resolution for 2-D lifetime images. Subsequently, several groups have reported prototype FLIM instruments capable of spatially mapping the excited-state lifetimes of dye molecules (13–17, 23–25). Recently, fluorescence lifetime imaging with single-molecule sensitivity, picosecond temporal resolution, and a spatial resolution beyond the optics diffraction limit using near-field optics has been developed (26).

In this chapter, we discuss the design and application of two FLIM systems: (1) a time-domain FLIM system using a high-speed, highly sensitive gated image intensifier and (2) a frequency-domain confocal FLIM. The other chapters in this volume contain detailed discussions of frequency-domain FLIM using 2-D image devices (see Chapter 9) and two-photon FLIM (see Chapters 8 and 11). Several types of FLIM systems have been developed by different groups. Each type of FLIM has its advantages and disadvantages, which depend mainly on the biological events to be monitored, as well as the light sources and microscopic techniques used in the experiments.

10.3. TIME-RESOLVED FLIM

In time-resolved fluorescence spectroscopy, the time-correlated single photon-counting (SPC) technique is currently the most widely used method (20). By combining SPC with scanning techniques, it would be possible to perform lifetime imaging. However, measurements of dynamic events using this approach would be problematic due to the long measurement time involved in collecting photons at each point. To improve photon-counting efficiency and reduce measurement time for photon counting, multichannel photon-counting techniques have been developed (11). A corresponding reduction in measurement time makes 2-D lifetime mapping practical. However, the scanning time for 2-D lifetime imaging still limits temporal resolution in studies of cellular dynamics. This problem can be solved by using highly sensitive, high-speed 2-D image detectors (23).

10.3.1. Image Detectors for FLIM

As an alternative to scanning, an imaging detector can be used for time-resolved FLIM with a pulsed laser system. Depending on the timescales of the applications, different imaging detectors can be used. High-speed frame cameras have very high time responses, usually ranging from picoseconds to milliseconds (27). With a single-frame camera tube and single-pulse shot, two to four time-resolved fluorescence images can be obtained at different locations inside the tube. Based on the time-resolved fluorescence images detected by the frame camera, it is possible to perform lifetime imaging. However, the extremely high price and low detection sensitivity of high-speed frame cameras have prevented their usage. In addition, because of distortion of corresponding pixels in different images, it is difficult to obtain accurate, spatially registered lifetime images. Recently, more sensitive and gatable CCD cameras have become available (28). However, because of limits in gating speed (which is usually microseconds to milliseconds), it is not possible to use these CCDs for lifetime imaging.

Newer advances in gated image intensifier techniques allow time gating to a few nanoseconds or even subnanoseconds (29). One kind of gated image intensifier is produced by Hamamatsu Photonic Systems (Bridgewater, New Jersey) (30). At present, the practical choice for time-resolved FLIM is to use such a gated image intensifier as an image detector. When these types of detectors are used the decay time resolution attainable is limited by the gate time and gating jitter of the detector. A functional description of a gated image intensifier–CCD combination is shown in Figure 10.2. In the gated operation mode, the photocathode of the intensifier is biased to a cutoff positive potential ($+150$ V) relative to the MCP-In potential, while keeping V_{MCP} and V_S at their normal operating potentials. When a high negative (-300 V) gate pulse is applied to the photocathode, an intensified image can be observed on the phosphor screen. The phosphor screen luminance is typically around 4×10^{-2} ft-L at about 100 ns gate time. The output image of the intensifier on the phosphor screen is focused at unity magnification onto a CCD camera. Time-resolved images can be accumulated on the CCD chip by repeatedly pulsed light excitation. Following the production of an image with a sufficient signal-to-noise ratio (SNR), the image is read out to a computer. For weak fluorescence emission, the target integration in the CCD is effective in enhancing the SNR of detected images. In a given situation, the available fluorescence light level determines the integration time required to obtain acceptable SNRs. Cooling the CCD to reduce the dark current to negligible levels allows integration times for very long times. In Table 10.1, the electrical and optical characteristics of the gated image intensifier and CCD camera we have used are listed.

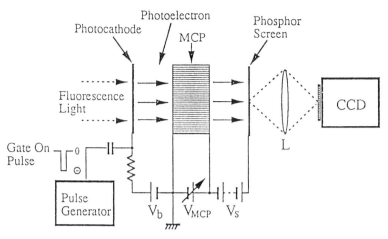

Figure 10.2. The principle of operation of the gated image intensifier used for time-resolved FLIM.

10.3.2. Time-Resolved FLIM System

Using a gated image intensifier, we developed a time-resolved FLIM. The configuration of the time-resolved FLIM is shown in Figure 10.3. Basically, the system has five main components: (1) the pulsed light source; (2) the image detection system (a gated image intensifier); (3) the timing control unit; (4) a fluorescence microscope; and (5) an SPC detection system. In our system, a picosecond pulsed light source (Coherent, Inc., Santa Clara, California) is used, consisting of a mode-locked YAG laser (Antares 76-S), a dye laser (700 Dye Laser) with a third harmonic generator (model 7950), and a cavity-dumper (model 7220). Picosecond pulses with tunable wavelengths from the UV to the IR at rates of single shot to 76 MHz can be generated. The pulsed excitation light is guided to a fluorescence microscope (Diaphot 300, Nikon, Inc., Garden City, New York) using a multimode optical fiber system (Newport Co, Irvine, California). The optical fiber output is focused to illuminate the entire field of view. The objective lens used is a Nikon 40 × Fluor [oil, numerical aperture (NA) = 1.3]. The fluorescence microscope has two output ports, one for time-resolved FLIM and another for SPC detection.

Time-resolved fluorescence microscopic images are obtained using a high-speed, gated MCP image intensifier (minimum gate width: 3.0 ns, C2925-01, Hamamatsu Photonic Systems, Bridgewater, New Jersey) that is installed at one of the microscope's detection ports. The principal features of the gated image intensifier are very small image distortion, wide spectral response, compact size, and high sensitivity, gain, and spatial resolution. The input gate

Table 10.1. Specifications of a Gated Image Intensifier Used for Time-Resolved FLIM

High-Speed Gated Image Intensifier (C2925-01)

MCP	Single stage
Spectral response	150–850 nm
Gate width	2.5 ns–100 ms
Gate jitter	200 ps max
Gate pulse repetition rate	10 kHz
Photocathode	
Window material	Synthetic silica
Effective diameter	17.7 mm
Cathode sensitivity	
Luminous sensitivity	150 µA/lm typ.
Radiant sensitivity at 430 nm	57 mA/W typ.
Quantum efficiency at 400 nm	14% typ.
Phosphor screen material	P-20
Image intensifier gain	(MCP 900 V) 10000 ft-l/ft-C min
Spatial resolution (MTF)	23 lp/mm min

Cooled CCD Camera (C4880)

Spectral response	340–1000 nm
No. of effective pixels	1000(H) × 1018(V)
Pixel size	12 × 12 µm square pixels
Frame rate	
High-precision mode	4 s/frame max.
High-speed mode	7 Hz max.
Mean readout noise	
High-precision mode	12 electron rms
High-speed mode	50 electron rms
Mean dark current	0.1 electron/pixel/s (at −40 °C)
A/D converter resolution	10–14 bits
Cooling method	Thermoelectronic + water cooling

signal for the gated image intensifier is delayed relative to the excitation laser pulse using a digital delay/pulse generator (DG535, Stanford Research, Sunnyvale, California). The digital delay/pulse generator has a total of nine output channels and can be operated by both external and internal triggers, with jitter of 100 ps. By using the pulse signal from the cavity-dumper driver of the laser system as an external trigger starting pulse, the measurement timing can be synchronized and the gating timing and gate width for the gated image intensifier can be determined. In our system, all the timing and gating functions for the experiments are automatically controlled by computer via

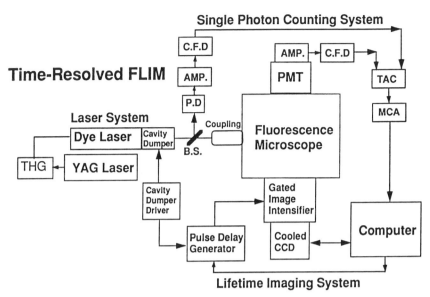

Figure 10.3. Time-resolved FLIM system developed in our laboratory. There are five main sections in the system: (1) pulsed light source, (2) image detection system using gated image intensifier, (3) timing control unit, (4) fluorescence microscope, and (5) SPC system.

a GPIB bus. The gated image intensifier operates maximally at a frequency of 10 kHz; thus, the laser pulse frequency is fixed at 10 kHz at the cavity dumper. To enhance the SNR of detected images, target integration on the CCD is effective. In our system, a slow-scan cooled CCD camera (1000 × 1018 pixels, C4880, Hamamatsu Photonic Systems, Bridgewater, New Jersey) is used to reduce dark current to negligible levels, allowing prolonged integration times.

A personal computer (Pentium 90, Gateway 2000, North Sioux City, South Dakota) interfaced to an image processing board (Image 1280, Matrox Electronic Systems, Ltd., Dorval, Quebec, Canada) and a GPIB board are used to control the experiments, to analyze the time-resolved fluorescence images, and to reconstruct the fluorescence lifetime images. The fluorescence lifetime images are calculated pixel by pixel by assuming a single exponential decay using a rapid lifetime determination method. In principle, the fluorescent decay may be multiexponential, but, in practice, one component usually represents most of the fluorescence signal and is often sufficient to observe relative changes. By using the rapid lifetime determination method, lifetime images can be obtained a short time, which is critical for biological applications.

In our system, an SPC detection system [Tennelec, Oak Ridge, Tennessee, time-to-amplitude converter (TAC): TC 864; constant fraction discriminator (CFD): TC 454; delay line: TC 412A] is used for sensitive single-point detection with high temporal resolution. For detection, a highly sensitive and ultrafast MCP-PMT tube (R3809U, Hamamatsu Photonic Systems) is used. A wideband, high-frequency amplifier (8447F, Hewlett-Packard) is used to amplify the signal from the photomultiplier (PMT), which is then fed into the TAC. In most cases, we have used reverse SPC mode for detection (i.e., the stop signal for the TAC comes from the cavity-dumper driver, while the start pulse for the TAC comes from the photon-counting signal). By adding SPC capabilities to the system, high spatial and temporal resolution measurements at a single point can be obtained that generate more detailed information (e.g., multiexponential data analysis). Thus the design of the FLIM system provides flexibility for different applications.

10.3.3.   Rapid Lifetime Determination for FLIM

To obtain lifetime images at rapid rates, we have developed a rapid lifetime determination method for multigate detection. A functional description of lifetime imaging using this lifetime determination method and the gated MCP image intensifier is shown in Figure 10.4. The time-resolved fluorescence image of the sample at t_d is obtained by applying a sampling gate pulse (duration ΔT) to the photocathode of the intensifier at time t_d following sample excitation. Several sampling time positions can be selected during fluorescence decay (i.e., multgate detection). By means of rapid lifetime computational procedures for multigate detection, fluorescence lifetime images can be derived (8). For a single exponential decay of fluorescence after pulsed light excitation, the fluorescence intensity with time change is described as

$$I(t) = A \exp(-t/t) \qquad (1)$$

Fluorescence decay can be detected at two different delay times t_1 and t_2, with the gate width ΔT. The gated fluorescence signals (D_1 and D_2) can be described as

$$D_1 = \int_{t_1}^{t_1 + \Delta T} A \exp(-t/\tau)\,dt \qquad (2)$$

$$D_2 = \int_{t_2}^{t_2 + \Delta T} A \exp(-t/\tau)\,dt \qquad (3)$$

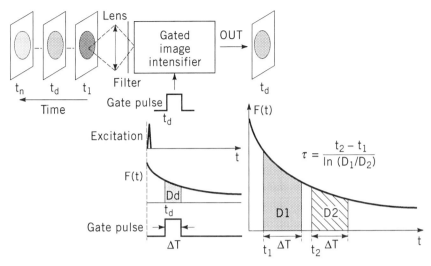

Figure 10.4. A functional description of rapid lifetime determination for time-resolved fluorescence lifetime imaging using a gated MCP image intensifier.

The lifetime τ and pre-exponential factor A can be extracted as follows:

$$\tau = \frac{t_2 - t_1}{\ln(D_1/D_2)} \tag{4}$$

$$A = \frac{D_1}{\tau \exp(-t_1/\tau)[1 - \exp(-\Delta T/\tau)]} \tag{5}$$

The lifetime depends on the D_1/D_2 ratio and the selection of t_1 and t_2 time points; it is not dependent on the gated image intensifier time resolution (3.0 ns). The D_1/D_2 ratio, is an important feature for correct lifetime calculation since this ratio decreases the effect of jittering of the detector and other timing electronics on the lifetime calculation. This method also results in extremely short calculation times. Fluorescence lifetimes and pre-exponential factors can be calculated directly from only four parameters (D_1, D_2, t_1, and t_2) without fitting a large number of data points, as required by conventional least-squares methods. With a gated image intensifier that is capable of multigate detection, both the observation time and the calculation time can be reduced.

In theory, optimum fitting regions exist for the rapid ratio lifetime determination method that depend on the lifetime value and SNR of the fluorescent signal (31). For the gate width ΔT, the optimum value occurs at $\Delta T = 2.5\tau$, and the region around the optimum selection is quite stable and is not sensitive to the choice of $\Delta T/\tau$. When $\Delta T/\tau$ become either small (<1) or large (>4), calculation precision decreases rapidly. In practice, fluorescence lifetimes can usually be predicted (but not precisely), so that the correct gate width and gating timing needed to carry out lifetime imaging experiments can be determined.

10.3.4. Procedures for FLIM Experiments

The general procedure for time-resolved FLIM experiments can be outlined as follows:

1. *Experimental setup:* For different samples, the setup for experiments will need to be adjusted. The adjustments include integration time for the CCD camera and gain for the gated image intensifier, gate width, and gate timing. An important parameter for time-resolved FLIM is the determination of the time when the sample is excited by the pulsed laser (zero start point). By using short lifetime probes (e.g., Rhodamine) and short gate widths, several gated time-resolved images can be obtained. The time point at which image intensity is maximum should be close to the zero point. Based on the start point image, the integration time and gain of the gated image intensifier can be determined to obtain high SNR images. The gate width and timing can also be determined using the optimal lifetime determination method (31). Before acquiring time-resolved fluorescence images from samples, a series of corresponding background images need to be taken. The background of the sample can be a buffer solution or a scattering medium (e.g., casein suspension). These background images at different t_d values are taken automatically.

2. *Lifetime imaging for samples:* After loading samples on the microscope stage, several time-resolved images are taken automatically with the same settings used for the background images. After subtracting the appropriate background image from each time-resolved image, lifetime images are calculated from any pair of gated images. When the range of lifetime values is known. The optimal lifetime image can be determined.

To ensure proper system performance, it is necessary to calibrate the FLIM system. FLIM calibration experiments can be carried out with solutions of well-known standard probes (e.g., Rhodamine B: 2.8 ns; FITC: 4.2 ns) and fluorescent beads whose diameter varies from 1.0 to 10 μm are used. The calculated lifetime values from lifetime images of fluorophore solutions are used to calibrate the system, while the measurements with fluorescent beads are used to characterize spatial resolution (32).

10.4. CONFOCAL FLIM

Recent advances in scanning and computer technology have allowed the development of confocal fluorescence microscopy, providing remarkably detailed 2-D and 3-D images with new standards of resolution and contrast (33, 34). Confocal microscopes reject light from out-of-focus planes, and by moving the specimen up and down, a 3-D image can be recorded. Just as confocal microscopy has provided information unobtainable with conventional fluorescence microscopy, the extension of conventional FLIM to confocal FLIM (CFLIM) should provide a more detailed description of cell activity.

10.4.1. Principle of CFLIM

In principle, CFLIM can be based on either time- or frequency-domain lifetime methods. For time-domain methods, the fluorescence lifetime would be generated on a point-to-point basis, using a pulsed laser with SPC detection or gated integration. However, at a typical spatial resolution of 256×256 or 512×512 pixels per image, a high laser repetition rate is needed to acquire an image in a reasonable time since each pixel will require many measurements to generate the decay curve. It is not reasonable to increase the laser power density to improve sensitivity, since this can alter the photochemistry of the samples. Low-cost pulsed nitrogen and nitrogen-pumped dye lasers are limited to low repetition rates (e.g., 100 Hz), while mode-locked laser sources have high repetition rates but are expensive and require skilled technical support. The only reasonable approach for time-resolved CFLIM is to use a conventional continuous wave (CW) laser and an external light modulator to produce short pulses of light with high repetition rates for excitation. By coupling to time-resolved gated detection, lifetimes can be determined within a short time; the addition of scanning would allow 2-D lifetime imaging to be undertaken. Based on this principle, time-resolved CFLIM using an electro-optic light modulator and gated sampling detection has been developed (15). This system provides reasonable time resolution and can be used for some applications.

In the frequency-domain method, high-frequency modulated laser excitation provides the possibility of high temporal resolution CFLIM. One possible approach is to use an image dissector tube (IDT) for CFLIM. A confocal fluorescence microscope using IDT has been developed (35). In principle, using IDT, it is possible to combine confocal and lifetime imaging detection. However, for many reasons, the IDT has recently become less popular. CFLIM systems using an IDT involve complicated system design and adjustment, which has limited their development and applications. Recently, two-photon frequency-domain CFLIMs have been developed (36). However, the high cost of the required laser system has limited its availability. The practical choice for CFLIM is to use an inexpensive, compact CW laser and employ external modulation methods to supply light modulation. Compared with time-domain detection for confocal microscopy, frequency-domain detection provides more flexible and accurate measurements. Since frequency-domain detection has been well developed and commercialized, it is relatively easy to adapt frequency-domain detection to the confocal microscope.

10.4.2. CFLIM System

A schematic of our CFLIM is shown in Figure 10.5. It is based on a laser scanning confocal microscope (LSCM: Olympus LSM-GB200) and frequency-domain lifetime detection equipment. In the LSM-GB200, the excitation optics are based on an optical fiber system, making input of light sources relatively easy. Moreover, because of three-channel detection on the emission side, it is possible to utilize one channel (or two) for fluorescence lifetime imaging measurements and, at the same time, use the other two channels (or one) for conventional confocal fluorescence imaging measurements. With this design, the CFLIM is multifunctional and flexible for many different applications.

In the frequency-domain method, high-frequency (MHz–GHz), sinusoidally modulated radiation is used to excite the sample. Following excitation, fluorescence lifetimes are determined from phase shifts or amplitude demodulation factors. In our CFLIM, an inexpensive CW laser (e.g., Ar^{2+} ion

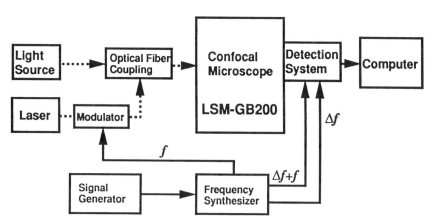

Figure 10.5. Schematic of CFLIM based on the use of phase-resolved fluorescence lifetime detection and a laser scanning confocal microscope.

laser) is used as the light source. The intensity of laser light is modulated sinusoidally with an extracavity electro-optic modulator (Pockels cells, Fast-pulse Technology, Inc., Saddle Brook, New Jersey). Two polarizers are used to increase the polarization of the laser light and modulation efficiency. The modulator is driven by a high-frequency signal from a frequency synthesizer (model K310, ISS, Inc.) driven by a signal generator (model K309, ISS, Inc., Champaign, Illinois). The frequency synthesizer can also be used to generator other phase-locked, high-frequency reference signals. Both the frequency generator and the synthesizer have a frequency range from 10 kHz to 1 GHz and resolution of 1 Hz. The modulation frequency (f) is determined by the measured lifetimes of the samples (usually around 30 to 300 MHz). The output of the laser light modulator is directed into the LSM-GB200 through a fiber optic coupling system.

10.4.3. Lifetime Determination of CFLIM

When the sample in a single point under study is excited with time-dependent, sinusoidally modulated light $E(t)$,

$$E(t) = A[1 + M_e \cos(\omega t)] \qquad (6)$$

where A is the dc intensity component of the exciting light, $\omega = 2\pi f$ where f is the modulation frequency, and M_e is the ratio of the amplitude of the ac intensity to the dc intensity component. The resulting wavelength-dependent fluorescence $F(t)$, for a single-component sample with exponential decay, will have frequency f but will be phase shifted by an angle Φ which is unique to the particular fluorescent species:

$$F(t) = F_0[1 + M_f \cos(\omega t - \Phi)] \qquad (7)$$

where F_0 is the average fluorescence intensity. The phase shift Φ and the modulation factor M are related to the fluorescence lifetime τ by

$$\tan \Phi = \omega \tau \qquad (8)$$

$$M = M_f/M_e = 1/[1 + (\omega \tau)^2]^{1/2} \qquad (9)$$

To extract fluorescence lifetimes from frequency-domain data, the phase shift and demodulation of the fluorescence with respect to the excitation beam can be determined by measuring the phase, ac component, and dc component. Lifetime measurements using phase shifts alone have been extensively employed for time-resolved fluorescence spectroscopy and require only one phase shift parameter to be measured. The use of the phase shift Φ for lifetime determination is also advantageous because it can be detected by an electronic circuit as it occurs, which allows real-time confocal fluorescence lifetime imaging. To detect lifetimes efficiently, the heterodyne detection method is used, in which high-frequency signals are transformed into low-frequency signals containing the same phase and modulation information as the original high-frequency signal. The frequency synthesizer generates a sinusoidal electronic signal $C_r(t)$ at $f + \Delta f$, slightly different from the modulation frequency (f). This signal is amplified and applied to the second dynode of the PMT (Hamamatsu R928) as the detection modulation signal, which results in a multiplication of the fluorescence signal for heterodyne detection.

$$C_r(t) = C_0[(1 + M_C \cos(\omega_c t - \Phi_c)]$$ (10)

The resulting output photocurrent $I_o(t)$ of the PMT is given as

$$I_o(t) = F_0 C_0[1 + M_f \cos(\omega t - \Phi) + M_C \cos(\omega_c t - \Phi_c)$$
$$+ M_f M_C \cos(\omega t - \Phi)\cos(\omega_c t - \Phi_c)]$$ (11)

The last term of Eq. (11) can be rewritten as

$$\tfrac{1}{2} M_f M_C [\cos(\Delta \omega t - \Delta \Phi) + \cos(\omega t + \omega_c t - \Phi - \Phi_c)]$$ (12)

where $\Delta \omega = \omega_c - \omega$ and $\Delta \Phi = \Phi_c - \Phi$. When ω_c is very close to ω, Eq. (11) simplifies to four terms: a constant, $\Delta \omega (= 2\pi \Delta f)$, ω, and a high-frequency term (2ω). The $\Delta \omega$ term contains all the phase and modulation information of the original fluorescence signal that is necessary to determine fluorescence lifetimes. The constant term contains the initial fluorescence intensity information. The high-frequency terms are then electronically filtered out, leaving the constant and $\Delta \omega$ terms. When these terms and a high-speed phase detection circuit are used, fluorescence lifetimes are measured in one pixel, and then the LSCM scans the laser beam pixel by pixel over the specimen mounted on an inverted microscope. Fluorescence lifetimes are calculated at each pixel and then combined to yield 2-D lifetime images. The focal plane is changed and the process repeated to generate 3-D confocal lifetime images.

A functional description of heterodyne detection in the PMT is shown in Figure 10.6. The interrelationship between the aforementioned signals is shown for the measurement of one fluorescent data point. A reference signal $C_r(t)$ of frequency Δf is used for phase shift Φ detection. Usually, in heterodyne detection, Δf ranges from tens to hundreds of hertz. At this slow frequency, it is not possible to detect phase signals with high enough speed for real-time

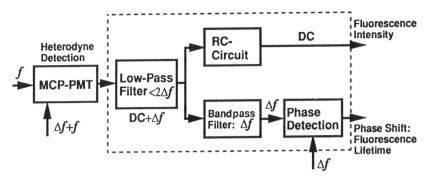

Figure 10.6. Functional description of high-speed heterodyne phase detection circuitry.

CFLIM. To detect lifetimes in each pixel on a microsecond time scale, a high-speed phase detection system is being developed based on a high-speed electronic phase detection circuit at high frequency Δf ($\Delta f \Rightarrow 1$ MHz). This frequency is useful for high-speed phase detection because it is below the limiting frequency (which is determined by the PMT transit spread time). To obtain fluorescence intensity and lifetime information, it is essential that the dc and phase shift components from the PMT heterodyne output signal be measured. We are developing high-speed phase shift detection circuitry to measure lifetimes simultaneously at each pixel, allowing us to make CFLIM measurements in real time. We believe that in most cases, one-component lifetime measurements will provide suitable data for fluorescence resonance energy transfer (FRET) and Ca^{2+} imaging experiments. If resolution of multi-component lifetimes is required, data must be obtained at several different modulation frequencies, and more complicated image processing software using global data analysis techniques will be needed (37).

10.5. BIOMEDICAL APPLICATIONS OF FLIM

FLIM is a new technology, and its future relies heavily on the development of applications. The rationale for employing FLIM in biomedical studies is based on the following advantages:

1. Because fluorescent molecules with similar spectra often display distinct lifetimes, multiparameter imaging of cellular structures is possible (38). With steady-state fluorescence microscopy, the number of spectral components that can be resolved is limited by spectral overlap of the absorption and emission of

the different dyes used. Thus, the number of analytical parameters that can be examined is increased by FLIM even at the same wavelength (39).

2. FLIM can discriminate autofluorescence from living cells from true fluorescence on the basis of distinct lifetime differences, leading to increased contrast and sensitivity in images.

3. Fluorescence lifetime probes are sensitive to numerous chemical and environmental factors such as pH, oxygen, and Ca^{2+}. FLIM can directly image the environmental and structural environments immediately surrounding the probe. For example, FLIM provides quantitative information regarding microviscosity in living cells through measurements of time-resolved emission anisotropy. FLIM can also provide data on the rotational and translational dynamics of cellular constituents (40).

4. FLIM is an effective technique for quantifying the interaction, binding, or association of two types of molecules using FRET spectroscopy (41–43). Because FRET decreases in proportion to the inverse sixth power of the distance between the donor and acceptor probes, this phenomenon is effective when the donor and acceptor molecules are within 10–100 Å of each other. This method has the advantages of superior spatial resolution, high sensitivity, and applicability in complex systems. FRET imaging is particularly useful in examining temporal and spatial changes in the distribution of fluorescent molecules in living cells by monitoring donor lifetime imaging.

In the following sections, we provide some examples of how FLIM is being used for biomedical research in our laboratory.

10.5.1. Quantitative Calcium Imaging

Ionized calcium (Ca^{2+}) is an important signal transduction element in cells ranging from bacteria to specialized neurons (44). With the introduction of fluorescent probes (e.g., fura-2, fluo-3), Ca^{2+} imaging has been intensively undertaken in a number of different cell types (45). Quantitative Ca^{2+} imaging has been pursued by using fluorescent ratiometric probes (e.g., fura-2, indo-1) (46). However, in practice, there are still problems with quantitative analysis using these probes. Existing ratiometric probes are usually excited at ultraviolet (UV) wavelength, which could generate photobleaching and other photodamage in living cells. Moreover, there is no proper ratio probe for visible confocal laser scanning microscopy studies; therefore, nonuniform loading of the Ca^{2+} fluorescent probe concentration can affect quantitative confocal Ca^{2+} imaging. Although some confocal microscopes have modified optics and are adapted for use with UV excitation, the available UV laser wavelengths are limited (e.g., it is not possible to perform Ca^{2+} measurements using fura-2, which requires two UV excitation wavelengths), and the interactions between UV excitation and the natural chromophores in living cells can generate photobleaching and photodamage as well as autofluorescence. To address these problems, new types of fluorescent probes have been developed (47). Some of these can be used as fluorescence lifetime probes for Ca^{2+} imaging. Fluorescence lifetime probes are equilibrium probes that respond to the calcium concentration independently of probe concentration and photobleaching (48). Fluorescence lifetime probes eliminate the need for ratiometric measurements using UV-wavelength Ca^{2+}-sensitive probes. This is advantageous, as recently nonratiometric probes for Ca^{2+} such as Calcium Green, Calcium Crimson, and Calcium Orange have been commercialized and can be used to measure Ca^{2+} (49). These nonratiometric probes are extremely important for imaging the calcium concentration in cells using confocal laser scanning microscopy for 3-D Ca^{2+} imaging. This is due to the limited visible-excitation wavelengths available from most lasers employed in confocal microscopes. Another important feature of fluorescence lifetime sensing is that lifetime measurements provide information on microenvironmental factors that can be used to elucidate the role of environmental factors in quantitative Ca^{2+} measurements.

Our group has investigated the role and regulation of ligand-stimulated spatial and temporal alterations in cellular Ca^{2+} (50). We have applied time-resolved FLIM using a gated image intensifier (as described above) and fluorescence lifetime probes to determine (1) the role of Ca^{2+} in platelet-derived growth factor (PDGF)-stimulated mitogenesis (51) and (2) whether recently reported differences in nuclear and cytosolic free Ca^{2+} are real or artifactual (52). We have employed Calcium Green and Calcium Crimson (Molecular Probes, Eugene, Oregon) as lifetime probes for Ca^{2+} imaging since these two probes have relatively long lifetimes and good quantum yield. We have performed lifetime measurements for these two probes in standard Ca^{2+} buffer solutions with different Ca^{2+} concentrations (Molecular Probes) using both lifetime imaging and SPC detection. The SPC measurement decay data were analyzed using a program provided by Dr. J. Beechem, which includes a global data analysis package and nonlinear least-squares curve fitting. The accuracy of time-resolved FLIM and SPC detection for determination of fluorescence lifetimes was assessed using the well-known lifetime probe rhodamine B in solution. The lifetime of rhodamine B determined by FLIM was 2.71 ns (compared to 2.88 ns by SPC), which indicates proper functioning of the FLIM system. The fluorescence lifetime measurement data for Ca^{2+} standard solutions using FLIM and SPC are shown in Figure 10.7. With increases in Ca^{2+} concentration, fluorescence lifetimes increase. The lifetime values measured by SPC and FLIM are close. Calcium Green and Calcium Crimson lifetime values determined by FLIM were 1.05 and 2.56 ns,

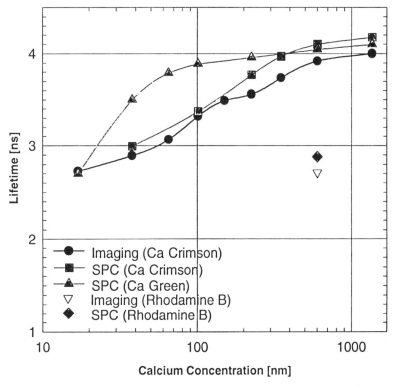

Figure 10.7. The fluorescence lifetimes measured by FLIM and SPC for different Ca^{2+} concentration standard solutions.

respectively, at zero Ca^{2+} and 3.60 and 4.10 ns, respectively, at saturating (1039 nM) Ca^{2+}. From the data, we found that Calcium Crimson has better linearity and less saturation at high Ca^{2+} levels. We have used Calcium Crimson as a lifetime probe for quantitative Ca^{2+} determination. After the calibration of fluorescence lifetime measurements in both the lifetime imaging mode and single-photon detection, we applied both lifetime detection modes to measure and quantitative Ca^{2+} in living cells.

As one example, time-resolved fluorescence images and lifetime images of Ca^{2+} in BALB fibroblasts are shown in Figure 10.8 (see Color Plates). A BALB/C-3T3 cell was labeled with 10 µm Calcium Crimson-AM and observed by FLIM. The intensity of the time-resolved images decreased as the time increased between the excitation pulse and turning on of the gated intensifier. A fluorescence lifetime image calculated from images A and D is shown on the right. We wish to emphasize the following:

1. To study the role and response of Ca^{2+} in biological events, fast temporal resolution for lifetime imaging is required. In our FLIM, it took 0.2 s to acquire each time-resolved image, and within 1 s, a lifetime image can be obtained.

2. In the time-resolved intensity images (A–D), the images show a difference in intensity between the nuclear and cytoplasmic regions (lower in the nucleus). However, fluorescence lifetime imaging indicates that there is no difference in Ca^{2+} between the nucleus and cytoplasm. The finding of a similar Ca^{2+} concentration measured by lifetime imaging demonstrates the uniqueness and usefulness of FLIM. Lifetime detection provides correct, reliable information, while intensity-based measurements may not.

We have also employed Calcium Crimson–salt (by microinjection) for Ca^{2+} imaging, as shown in Figure 10.9 (see Color Plates). Two BALB/c-3T3 fibroblasts were microinjected with 20 mM Calcium Crimson–salt and imaged using our FLIM. Time-resolved fluorescence intensity images at various times (0, 2, and 4 ns) after pulsed laser excitation are shown in Figure 10.9A. The lifetime image calculated from the 0 and 4 ns time-resolved images is shown in Figure 10.9B. Unlike the AM form of the Calcium Crimson probe, microinjection of the salt form caused the fluorescence intensity in the nucleus to be higher than in the cytoplasm. However, the fluorescence lifetime image displays a similar lifetime in both the nucleus and the cytoplasm, which indicates the same level of Ca^{2+} in both regions. Based on standard Ca^{2+} solution lifetime measurements, we have calculated that the Ca^{2+} concentration in the nucleus of 64 nM $\pm$ 20% vs. 68 nM $\pm$ 25% in the cytoplasm, indicating similar Ca^{2+} levels in the cytoplasm and nucleus. These results demonstrate the usefulness of FLIM for quantitative Ca^{2+} imaging. To further study the effects of environmental factors in the nucleus and cytoplasm on the observed Ca^{2+} level, we have applied SPC detection to measure the lifetimes of Ca^{2+} Crimson in the nucleus and cytoplasm, respectively. With fluorescence lifetime analysis, it is possible to study environmental effects on Ca^{2+}-sensing probes and therefore to investigate and quantitate real Ca^{2+} levels.

FLIM can be used to measure other types of cellular ions. The fluorescence lifetime of a probe can depend on pH, oxygen, intracellular ion concentrations (e.g., Mg^{2+}, Na^{+}, K^{+}), and a variety of other substances. For example, the fluorescence lifetime of the electron carrier nicotinamide adenine dinucleotide (reduced form) (NADH) increases after binding to proteins, allowing one to image the free and protein-bound NADH in cells (53). Moreover, the time resolution of present calcium imaging systems is insufficient to study rapid changes (e.g., receptor-mediated changes) in living cellular calcium levels in real time (54, 55). Such changes are expected to be very rapid and not assessable using currently available instrumentation. Use of FLIM would enable examination of these transient calcium signal changes on a nanosecond time scale in living cells.

10.5.2. Fluorescence Resonance Energy Transfer Imaging

Fluorescence resonance energy transfer (FRET) imaging (56, 57) is a microscopic technique for quantifying the distance between two molecules conjugated to different fluorophores. In principle, if one has a donor molecule whose fluorescence emission spectrum overlaps the absorbance spectrum of a fluorescent acceptor molecule, then the donor and acceptor will exchange energy through a nonradiative dipole–dipole interaction. This energy transfer is manifested by quenching of donor fluorescence in the presence of the acceptor and in sensitized emission of acceptor fluorescence. In theory, the distance (R) between the donor and acceptor is obtained from the equation

$$R = R_0(1/E_{\mathrm{T}} - 1)^{1/6} \tag{13}$$

where E_{T} is the efficiency of energy transfer and R_0 is the critical distance at which $E_{\mathrm{T}} = 0.50$. R_0 can be determined from the following equation:

$$R_0 = 9786[K^2\eta^{-4}q_0 J(\lambda)]^{1/6} \quad (\text{Å}) \tag{14}$$

where K^2 is the orientation factor, η is the refractive index of the medium, q_0 is the quantum yield of the donor in the absence of the acceptor, and $J(\lambda)$ the overlap integral.

If E_{T} could be measured, one could calculate the distance between the donor and acceptor. By measurements of donor lifetime in the presence and absence of the acceptor, E_{T} can be measured accurately:

$$E_{\mathrm{T}} = \tau_{\mathrm{DA}}/\tau_{\mathrm{D}} \tag{15}$$

where τ_{D} is the lifetime of the donor in the absence of the acceptor and τ_{DA} is the lifetime of the donor in the presence of the acceptor.

Because FRET decreases in proportion to the inverse sixth power of the distance between the two probes, this is effective in measuring separation of the donor- and acceptor-labeled molecules when they are within 10–100 Å of each other (41, 42). Recently, two review papers on FRET theory, relevant literature, practical suggestions, applications, and new developments in FRET techniques have been published (57, 58). In these papers extensive bibliographies are provided, as well as a table containing R_0 values for a variety of different donor–acceptor pairs. Because energy transfer occurs over distances of 10–100 Å, a FRET signal corresponding to a particular field (or pixels) within a microscope image provides additional information beyond the microscopic limit of resolution down to the molecular scale. The ability to measure interactions and distances of molecules provides the principal and unique benefits of FRET for microscopic imaging (56–59). FRET imaging is particularly useful in examining temporal and spatial changes in the distribution of fluorescently conjugated biological molecules in living cells.

In practice, there are two ways to perform FRET microscopy imaging: steady–state or time-resolved methods. In steady-state FRET imaging, an increase in the proximity of the donor and acceptor fluorophores results in nonradiative transfer of the donor excitation energy to the acceptor molecules, which results in a decrease in donor emission (I_D) and an increase in the intensity of emission of the acceptor $(I_A$; sensitized fluorescence). By ratio imaging of I_A/I_D (sensitized ratio FRET imaging), it is possible to obtain 2-D FRET information. FRET imaging can be performed using a standard fluorescence microscope equipped with excitation/emission filters and sensitive video cameras that are currently used for the popular technique of ratio imaging. Unlike conventional ratio imaging, in which one measures the emission at a single wavelength when the fluorphore is excited at two different wavelengths (i.e., fura-2), in FRET imaging one typically measures the emission at two wavelengths: that of the donor and the acceptor. Since FRET results in an increase in acceptor emission (i.e., sensitized emission) and a decrease in donor emission, the ratio of the intensity of the acceptor-to-donor fluorescence emission when excited at donor excitation can be taken as a convenient experimental measure of FRET (56). The value of this ratio depends on the average distance between donor–acceptor pairs. Since this method is simple and easy to perform, it has been used in a variety of FRET imaging applications (60, 61). The drawback of using steady-state FRET imaging is that it is difficult to perform rigorous quantitative measurements since there are several sources of distortion that need to be corrected for (e.g., polarization effects, overlap of the donor and acceptor emission spectra). Time-resolved FRET can be used to solve these problems. Time-resolved FRET imaging allows the quantitative measurement of donor–acceptor separation distances and is based on measuring the lifetime of the donor in the presence and absence of the acceptor. Time-resolved FRET imaging can be performed using FLIM. The advantage of time-resolved FRET imaging is that the donor–acceptor distance can be mapped in a more accurate and quantitative manner.

Successful FRET imaging requires that several factors be considered:

1. The donor and acceptor fluorophore concentrations need to be tightly controlled. To achieve high SNRs, high fluorescence signals are preferred.

However, too high a concentration of dye can cause self-quenching and disordered biological function (62).

2. Photobleaching needs to be prevented. Almost all fluorescent molecules are sensitive to this problem. Photobleaching can alter the donor–acceptor ratio and therefore the value of FRET. To prevent photobleaching, highly sensitive image devices and lower levels of excitation intensity are preferred.

3. Ideally, the donor emission spectrum should substantially overlap the absorption spectrum of the acceptor.

4. There should be relatively little direct excitation of the acceptor at the excitation maxima of the donor.

5. The emissions of both the donor and the acceptor occur in a wavelength range in which the detector has maximum sensitivity.

6. There should be little if any overlap of the donor absorption and emission spectra, thus minimizing donor–donor self-transfer.

7. The emission of the donor should ideally result from several overlapping transitions and thus should exhibit low polarization. This will minimize uncertainties associated with the K^2 factor.

8. When antibodies are used for FRET imaging, it is necessary to determine whether the antibody reagent itself may be affecting the FRET measurements. For example, because antibody labeling may affect the original molecule's structure, and therefore distance information, the correct selection of the labeling procedure and antibody needs to be seriously evaluated. Quantitative analysis of FRET data depends on the availability of correct controls and on the ability to choose the best method for FRET imaging measurements.

There are increasing applications for FRET imaging (see Chapter 7 in this volume). However, there are only few reports of studies using time-resolved FRET imaging. Recently, studies of epidermal growth factor (EGF) receptors have been carried out using time-resolved FRET imaging (63). Using FLIM and FRET, EGF receptor clustering during signal transduction was monitored, and a stereochemical model for the tyrosine kinase of the EGF receptor was investigated. Time-resolved FRET imaging has also been applied to study the extent of membrane fusion of individual endosomes in single cells (17). With the use of time-domain FLIM and FRET, the extent of fusion and the number of fused and unfused endosomes were clearly visualized and quantitated.

Our group is working on the application of time-resolved FRET imaging to the study of PDGF signal transduction mechanisms and the role of PDGF receptor interactions in signal transduction. PDGF is recognized as a major mitogen in serum for mesenchymally derived cells. Recent findings indicate that PDGF exists as three isoforms (AA, BB, and AB) that bind to at least two distinct cell surface receptors (α and β). The mechanism by which PDGF binding to its receptors is transduced into biological activity is not clear (64). Based on many different approaches, a general model has been proposed for the formation of high-affinity PDGF-binding sites. In these models, functional PDGF receptors require dimerization of the α and β subunits, resulting in the formation of three receptors that differ in their ability to bind the different PDGF isoforms. However, conflicting data have been presented on whether α–β receptor dimerization is required for PDGF signal transduction. A particular weakness in the current studies has been the inability to monitor sequential PDGF-inducible events at the molecular level in the same cell. With the development of FRET imaging microscopy, which provides the ability to monitor several cellular functions at the molecular level of a single intact living cell, it is now possible to strive to overcome this limitation.

In our studies, using donor- and acceptor-labeled monoclonal Fab fragments specific for the α or β receptor, FLIM is applied to determine (1) whether all PDGF isoforms cause receptor dimerization; (2) whether receptors preexist in a dimerized state or are induced to dimerize after PDGF addition; and (3) whether PDGF-AB causes dimerization of one α and one β receptor subunit, two β subunits, or binds to just the β receptor in monomer form. These studies involve different labeling combinations for the α and β receptors and are performed in MG-63 cells containing or lacking the α-type PDGF receptor. Fab fragments of monoclonal antibodies produced against the extracellular domain of the α and β PDGF receptors (but which do not inhibit PDGF binding to the receptor) are labeled with fluorescein (donor) and Lissamine Rhodamine B or Texas Red (acceptor). By measuring the lifetime of the donor in the absence of the acceptor and the lifetime of the donor in the presence of the acceptor, we are able to quantitate FRET and obtain direct information about receptor interactions during PDGF signal transduction. In combination with the Ca^{2+} imaging techniques described above, we can also determine whether interaction of distinct PDGF receptors is required to cause alterations in Ca^{2+}. Recently, green fluorescent protein (GFP) has been introduced as a marker for gene expression (65). GFP is a unique protein that can also be used for FRET measurements. By inserting the GFP cDNA into the donor molecule cDNA using molecular biology techniques, direct labeling of the donor molecule can be achieved. It also has been reported that a rhodamine-like fluorescent protein exists. Thus, it may be possible to obtain directly labeled donor and acceptor molecules in which the label is part of the molecule itself.

10.5.3. Clinical Imaging

Advances in fluorescence microscopy and the availability of bioreagents such as monoclonal antibodies and nucleic acid probes have opened up new possibilities for the localization and analysis of proteins and nucleic acid sequences in cells, tissues, and chromosomes. Fluorescence assays are increasingly utilized for a variety of clinical applications because fluorescent labels are very sensitive, specific, and quantitative. Fluorescence can also be used for the simultaneous detection of multiple antigens. For instance, fluorescently labeled antibodies and nucleic acid probes allow the quantitative measurement of multiple disease markers in individual cells of patient specimens. DNA-binding fluorophores are also widely used for high-resolution chromosome analysis (66), and luminescent labels are increasingly used for clinical immunoassays (67). With the advent of genetic engineering and recombinant DNA technology, fluorescently labeled DNA and RNA probes are becoming important tools for ultra-low-level detection of specific nucleotide sequences in specimens (68). With the development of lifetime imaging techniques, it is possible to utilize the lifetime as a parameter for fluorescence-based clinical imaging applications (69,70).

There are two main advantages of lifetime imaging techniques for clinical applications. First, using time-resolved fluorescence imaging, detection sensitivity can be significantly improved. In practice, theoretically obtainable detection sensitivity is never achieved due to nonspecific autofluorescence, fixative-induced fluorescence of cells and tissues, and autofluorescence of the optical components in the microscopic system. Usually, background and autofluorescence signals are rapidly decaying processes with lifetimes in the range of 1 to 100 ns, whereas phosphorescence and delayed fluorescence have lifetimes in the range of 1 μs to 10 ms. Therefore, if a sample using a long lifetime probe could be excited with a pulse of light, and detection delayed to the micro- or millisecond range, the emitted intensity could be observed without interference from autofluorescence and scattering light, leading to a substantial incease in detection sensitivity. The ability to discriminate between specific types of fluorescence and autofluorescence has improved the sensitivity of diagnostic tests such that they perform comparably to or even more sensitively than radioisotopic assays. Second, by introducing fluorescence lifetime-based probes and lifetime imaging to clinical applications, simultaneous detection of multiple antigens in clinical samples becomes a reality. In clinical practice, correct diagnosis relies on the detection of several related parameters rather than only one parameter (e.g., specific protein or DNA). Usually, detection of multiple probes is based on the difference of each probe's fluorescence emission wavelength. However, many existing probes generate spectral overlap, which limits the number of probes that can be used for detection and diagnosis. By introducing lifetime probes, it is possible to distinguish each probe in the time domain (based on the difference in fluorescence lifetime) and thus increase the number of parameters that can be assessed in a single live sample. In principle, the combination of spectral and temporal separation using FLIM would allow detection of 10 different probes at the same time.

As one example, we have been interested in the role of human papillomavirus (HPV) in cervical cancer and its diagnosis; to that end, we have been developing microscopic imaging and fluorescent *in situ* hybridization (FISH) techniques to genotype and quantitative the amount of HPV present at the single-cell level in cervical Pap smears (71, 72). However, we have found that low levels of HPV DNA are difficult to detect accurately because of background signal coming from the sample, nonspecific staining, and optical detection system. In addition, the absorption stains used for Pap smears are intensely autofluorescent. We have developed a time-resolved fluorescence microscope technique to improve the sensitivity of detection of specific molecules of interest in slide-based specimens (73,74). This technique is based on FLIM in conjunction with the use of long-lifetime fluorescent probes. Several long-lifetime probes have been made and are commercially available (75–77). The phosphor yttriumoxisulfide activated with europium (Y_2O_2S: Eu) emits maximally at 620 nm and has a half-life of 700 µs. This phosphor has strong luminescence, displays minimal photobleaching, and is not significantly influenced by pH or temperature. The phosphors Zn_2SiO_4: MnAs and ZnS: Ag emit in the green and blue, respectively, permitting simultaneous multiparameter measurements. The chelate 4,7-bis(chlorosulfophenyl)-1,10-phenanthroline-2,9-dicarboxylic acid (BCPDA) complexes with lanthanide ions (Eu, Tb, Dy, and Sm) and can be used to label antibodies, biotin, and streptavidin. Amplification is also possible by multiple labeling. Recent advances in long-lifetime probe design and synthesis have improved several aspects of their practical use (e.g. water solubility, high quantum yield).

In our studies (78), the stable fluorescent europium chelate of 4-(4-isothiocynatophenylethynyl)-2,6-[N, N-bis(carboxymethyl)aminomethyl]-pridine was used as a long-lifetime probe. Lanthanide ions (Eu^{3+}) bound to streptavidin were used to detect bound DNA HPV 16 probes. This probe has a strong luminescence that is not significantly influenced by pH and temperature. The lifetime of the europium chelate is about 700 µs. Excitation for Eu^{3+} is at 355 nm and emission is at 610 nm. To observe europium emission independent of background autofluorescence, we detected that the Eu^{3+} signal emitted > 10 µs after excitation. A full-length HPV-16 DNA insert (8.0 kb) was labeled with biotin-11-dUTP by the random priming method and then used for FISH. At the last step, 10 µL of europium-conjugated avidin (Wallace Inc., Gaithersburg, Maryland) was placed on each slide, incubated in an *in situ* incubator, and then washed. After finishing the FISH procedure for HPV DNA 16, we carried out experiments using conventional fluorescence microscopy and time-resolved fluorescence microscopy.

The utility of the long-lifetime probe and TRFM for highly sensitive HPV DNA 16 detection is shown in Figure 10.10 (see Color Plates). In Figure 10.10A, a FISH image of HPV DNA 16 detected by conventional fluorescence microscopy is shown. For detection of HPV DNA, the CCD integration time was set at 1 s. In Figure 10.10B, a time-resolved fluorescence image for the same sample as in Figure 10.10A, detected with a gated image intensifier with a gate width 250 μs and turned on at 30 μs after the excitation pulse, is shown. The time-resolved image was detected with 10 s integration time in the CCD. In (A), the use of conventional microscopy detected nonspecific fluorescence surrounding the DNA probes. Because of nonspecific fluorescence (e.g., autofluorescence) and other scattering, correct DNA distribution and quantitation could not be obtained. By using a long-lifetime probe and TRFM, the correct DNA distribution was observed, as shown in (B). Because background and nonspecific signals are not present, an increase in detection sensitivity, specificity, quantitative accuracy, and correct DNA distribution was obtained. These results indicate that the use of TRFM and long-lifetime probes increases detection sensitivity by removing autofluorescence and scattering. Since highly sensitive detection of DNA in clinical samples using FISH imaging is useful for the diagnosis of many other diseases, the system we have develped should find numerous applications for diagnosis of disease states.

10.5.4. Fluorescence Lifetime Imaging for DNA Sequencing

The development of novel instrumentation for high-sensitivity, high-speed, high-throughput, low-cost DNA sequencing is important in achieving the goals of the Human Genome Project (79). Currently available technology for DNA sequencing by capillary electrophoresis (CE) is based on the spectral emission characteristics of the individual dideoxynucleotides labeled with different fluorescent probes in the gel during or after electrophoretic separation (80). Detection of fluorescence requires the sensitivity to detect the fluorescence intensity from each of the reporter molecules coupled to the dideoxynucleotides, as well as sufficient spectral resolution to be able to distinguish A, C, T, and G. Attempts to increase sequencing speed have led to the development of four-dye/one-lane capillary arrangements (81). The four-spectral-channel technique allows simultaneous sequencing in a single gel column, ease of automation, and high throughput (82,83). On the other hand, the different dye molecules used impart nonidentical electrophoretic mobilities to the DNA fragments. Further, in addition to the imperfections of the enzymatic DNA sequencing method itself, the emission spectra of the four different dyes employed substantially overlap. This overlapping spectral interference results in peaks composed of more than one dye and increases the uncertainty of DNA sequence base interpretation (84).

Recently, the fluorescence lifetimes of four different fluorescently conjugated primers associated with four different A-, T-, G-, and C-dideoxy-nucleotides used in the chain-terminating reaction for DNA sequencing— fluorescein isothiocyanate (FITC), 4-chloro-7-nitrobenzo-2-oxa-1,3-diazole (NBD), tetramethylrhodamine isothiocyanate (TMRITC), and sulfo-rhodamine 101 acid chloride (Texas Red)—have been measured (85). The mean fluorescence lifetimes of these four dyes were found to be 3.5, 1.1, 2.1, and 4.3 ns, respectively. The difference in their lifetimes is about 1 ns. These data clearly demonstrate that different DNA dyes have different (distinct) fluorescence lifetimes that can be measured with a reasonably inexpensive instrument.

In recent years, high-field, ultrathin slab gel electrophoresis and CE have been introduced for automated DNA sequencing based on the detection of fluorescently labeled nucleotides. CE coupled with fluorescence detection is a promising technique, not only because capillary gels are fast and easy to handle but also because a high electric field can be applied to a capillary without resulting in a large amount of Joule heating or a significant temperature gradient. Although CE provides rapid analysis, its throughput is limited because only one capillary is used at a time (86). The feasibility of performing capillary zone electrophoresis using time-resolved fluorescence detection has been demonstrated (87). However, in this study, only single-point detection was undertaken and protein, not DNA, was sequenced. This work, however, clearly demonstrates the possibility of fluorescence lifetime measurements for DNA sequencing using CE.

By using fluorescence lifetime imaging and multiple CE, it is possible to overcome the constraints of the current instruments used for DNA sequencing and achieve high-speed, high-sensitivity, high-throughput, low-cost, automated DNA sequencing. Of particular importance is the fact that the fluorescent lifetime (1–100 ns) of a fluorescence molecule is independent of its intensity. In addition, molecules that have similar spectral properties often display distinct lifetimes, making it possible to use a larger variety of fluorescent reporter molecules in which nucleotide discrimination would be based on differences in lifetime rather than differences in absorption and emission properties. Thus, the problems of low spectral resolution and sensitivity, as well as photobleaching and other factors affecting the detection of fluorescence signals, could be overcome by the measurement of fluorescence lifetimes.

The concept of FLIM for DNA sequencing is illustrated in Figure 10.11. When multiple lanes (four-dye/one-lane configuration) are used for DNA sequencing, 2-D fluorescence detection is needed. To simplify the case, in

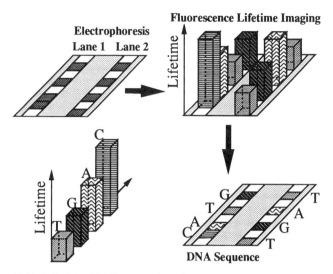

Figure 10.11. Principle of DNA sequencing using FLIM and multiple CE techniques.

Figure 10.11 two-lane electrophoresis DNA sequencing is shown. Instead of discrimination of fluorescence emission/excitation spectra, the fluorescence lifetime can be employed to distinguish different DNA bases. Suppose that the fluorescence lifetimes of four different A-, T-, G-, and C-dideoxynucleotides for DNA sequencing are 3.5, 1.1, 2.1, and 4.3 ns for FITC, NBD, TMRITC, and Texas Red, respectively (85). From 2-D FLIM, we will be able to identify each DNA base, based on its characteristic lifetime, and therefore obtain two-lane DNA sequencing information, as shown in Figure 10.11. Such 2-D parallel fluorescence lifetime image detection without moving parts would result in very high speed DNA sequencing because the 2-D image device (a gated image intensifier) would "look at" all the capillaries at all times, with data rates fast enough for high-speed DNA sequencing (at > 1 base/s lane). In addition, the use of long-lifetime (μs to ms) fluorescent probes would provide yet another advantage for DNA sequencing; when long-lifetime fluorescence probes are employed, the fluorescence signal from dideoxynucleotides will be obtained without interference from autofluorescence, scattering, and other nonspecific background, which will significantly enhance the SNR.

Another exciting application of fluorescence lifetime imaging techniques concerns the new DNA sequencing technology known as sequencing by hybridization (SBH) (88), in which sequence information is obtained by hybridization of small (e.g., eight nucleotides) probes to the target to be sequenced. By arraying probe oligomers on a support and hybridizing the target sequence to the array, the hybridization pattern becomes representative of the original target DNA sequence. Probe arrays can be custom designed for specific DNA diagnostic applications, such as mutation detection, HLA typing, and genetic identification. These arrays could be very powerful for the sequence of short, nonrepetitive DNA fragments and, when used in conjunction with primary sequence data derived by other methods, could provide a rapid means of confirming and correcting sequence data. Support-bound oligonucleotide arrays of limited complexity can be made manually by simply arranging suitably derivatized oligonucleotides on a surface; preparation of large combinatorial arrrays requires more sophisticated approaches involving photolithography and microfabrication (89). For this new genetic imaging technology, FLIM (both conventional and confocal) may provide a unique solution to discriminate perfect DNA matches from mismatches by using lifetime information. In addition, the use of long-lifetime probe can increase the SNR and substantially enhance detection sensitivity for SBH imaging.

10.5.5. Other Applications

There are other applications for FLIM. One area is in membrane biology. Current membrane studies employ single-point detection (steady-state and time-resolved) or steady-state imaging. Because of growing evidence that membrane structure, composition, and function are heterogeneous, FLIM should provide a sensitive technique capable of obtaining data relative to both the dynamics and the heterogeneous nature of membrane components related to plasma membrane fluidity, transportation, and fusion (90,91). FLIM can also be used for quantitative oxygen imaging of tissue and cells. By means of phosphorescence quenching and lifetime imaging techniques, quantitative oxygen distribution has been measured (92). By quantitatively detecting oxygen levels, localization of tumors and evaluation of their state of oxygenation are possible (93). Similarly, pH concentration distribution can be determined using FLIM (94). Another application for FLIM is in near-field fluorescence microscopy, allowing visualization of molecular information beyond the diffraction limit. Since fluorescent molecules are limited at one detection point of the near field, conventional CW excitation in one small spot generates substantial photobleaching and photodamage. The use of pulsed light excitation can prevent such problems and, at the same time, obtain lifetime information. By scanning at each point, near-field lifetime imaging can be generated for 2-D dynamic imaging and enhanced image contrast.

10.6. CONCLUSIONS

Research on FLIM is in its early stage. The principles, instrumentation, and some applications of FLIM have been demonstrated. However, the real challenge is in the next step. Current FLIM systems are too expensive because of the high price of laser systems. The temporal resolution of lifetime imaging and lifetime resolution need to be improved. Another important aspect of FLIM is fluorescent lifetime probes. So far, only limited lifetime probes have been used. The lifetime probes we have used were not designed for lifetime measurement purposes. The future of FLIM depends on meeting these challenges. Future developments of laser technology should result in the development of inexpensive FLIM systems based on compact, inexpensive solid-state laser systems. CFLIM could be used for 2-D and 3-D lifetime imaging. When ultrafast pulse laser technology becomes mature and inexpensive, two-photon FLIM may be the best choice for FLIM. Progress in biotechnology is another important aspect of the development of new chemical and immunological fluorescent probes for FLIM. We expect FLIM technology to play an important role in future biomedical research. As these instruments become available commercially, numerous applications in biomedicine will be developed.

ACKNOWLEDGMENTS

The work on time-resolved FLIM projects was supported by grants from the National Science Foundation, American Cancer Society, and North Carolina Biotechnology Center. The CFLIM project is supported by the Whitaker Foundation and a UNC Research Council award. We thank Dr. Pamela A. Diliberto for useful discussions and Mr. Seongwook Kwon for microinjection work. We also gratefully acknowledge support from Hamamatsu Photonic Systems, the Olympus Co., and ISS, Inc., for our FLIM research. The research on fluorescence lifetime imaging systems was originally guided by Dr. S. Minami and Dr. T. Uchida in the Department of Applied Physics, Osaka University, Japan. Valuable discussions and technical assistance from them are gratefully acknowledged.

REFERENCES

1. Arndt-Jovin, D. J., Robert-Nicoud, M., Kaufman, S. J., and Jovin, T. M. (1985). Fluorescence digital imaging microscopy in cell biology. *Science* **230**, 247–256.

2. Inoue, S. (1987). *Video Microscopy.* Plenum, New York.

3. Herman, B., and Jacobson, K. (1990). *Optical Microscopy for Biology.* Wiley-Liss, New York.

4. Taylor, D. L., and Wang, Y.-L. (1989). *Fluorescence Microscopy of Living Cells in Culture: Parts, A and B.* Academic Press, San Diego, CA.

5. Jovin, T. M., and Arndt-Jovin, D. J. (1989). Luminescence digital imaging microscopy. *Annu. Rev. Biophys. Biophys. Chem.* **18**, 271.

6. Herman, B., and Lemasters, J. J. (1992). *Optical Microscopy: Emerging Methods and Applications.* Academic Press, San Diego, CA.

7. Lakowicz, J. R. (1983). *Principles of Fluorescence Spectroscopy.* Plenum, New York.

8. Cundall, R. B., and Dale, R. E. (1983). *Time-Resolved Fluorescence Spectroscopy in Biochemistry and Biology.* Plenum, New York.

9. Bayley, P. M., and Dale, R. E. (1985). *Spectroscopy and the Dynamics of Molecular Biological Systems.* Academic Press, San Diego, CA.

10. Tian, R., and Rodgers, M. A. J. (1991). Time-resolved fluorescence microscopy. In *New Techniques in Optical Microscopy and Spectrophotometry* (R. J. Cherry, Ed.), pp. 312–351. CRC Press, Boca Raton, FL.

11. Wang, X. F., Kitajima, S., Uchida, T., Coleman, D. M., and Minami, S. (1990). Time-resolved fluorescence microscopy using multichannel photon counting. *Appl. Spectrosc.* **44**, 25.

12. Morgan, C. G., Mitchell, A. C., and Murray, J. G. (1990). Nanosecond time-resolved fluorescence microscopy: Principle and practice. *Trans. R. Microsc. Soc.* (*Micro '90*), pp. 463–466.

13. Clegg, R. M., Feddersen, B., Gratton, E., and Jovin, T. M. (1991). Time resolved imaging microscopy. *Proc. SPIE—Int. Soc. Opt. Eng.* **1640**, 448–460.

14. Lakowicz, J. R., and Berndt, K. W. (1991). Lifetime-selective fluorescence imaging using an RF phase-sensitive camera. *Rev. Sci. Instrum.*, **62**, 3653.

15. Buurman, E. P., Sanders, R., Draaijer, A. Van Veen; J. J. F., Houpt, P. M., and Levine, Y. K. (1992). Fluorescence lifetime imaging using a confocal laser scanning microscope. *Scanning* **14**, 155–159.

16. vandeVen, M., and Gratton, E. (1992). Time-resolved fluorescence lifetime imaging. In *Optical Microscopy: Emerging Methods and Applications* (B. Herman and J. J. Lemasters, Eds.), pp. 373–402. Academic Press, San Diego, CA.

17. Oida, T., Sato, Y., and Kusumi, A. (1993). Fluorescence lifetime imaging microscopy (flimscopy): Methodology development and application to studies of endosome fusion in single cells. *Biophys. J.* **64**, 676–685.

18. Wang, X. F., Periasamy, A., Coleman, D. M., and Herman, B. (1992). Fluorescence lifetime imaging microscopy: Instrumentation and applications. *CRC Crit. Rev. Anal. Chem.* **23**, 369–395.

19. Demas, J. N. (1983). *Excited State Lifetime Measurements.* Academic Press, New York.

20. O'Connor, D. V., and Phillips, D. (1984). *Time Correlated Single Photon Counting.* Academic Press, New York.

21. Denk, W., Strickler, J. H., and Webb, W. W. (1990). Two-photon laser scanning fluorescence microscopy. *Science* **248**, 73–76.

22. Wang, X. F., Uchida, T., and Minami, S. (1989). A fluorescence lifetime distribution measurement system based on phase-resolved detection using an image dissector tube. *Appl. Spectrosc.* **43**, 840.

23. Wang, X. F., Uchida, T., Coleman, D. M., and Minami, S. (1991). A two-dimensional fluorescence lifetime imaging system using a gated image intensifier. *Appl. Spectrosc.* **45**, 360.

24. Morgan, C. G., Mitchell, A. C., and Murray, J. G. (1992). In-situ fluorescence analysis using nanosecond decay time imaging. *Trends Anal. Chem.* **11**(1), 32–41.

25. Lakowicz, J. A., Szmacinski, H., Nowaczyk, K. and Johnson, M. L. (1992). Fluorescence lifetime imaging. *Anal. Biochem.* **202**, 316–330.

26. Xie, X. S., and Dunn, R. C. (1994). Probing single molecule dynamics. *Science* **265**, 361; Kopelman, R., and Tan, W. (1993). Near-field optics: Imaging single molecules. *Ibid.* **262**, 1382–1384.

27. Young, P. E., Kilkenny, J. D., Phillion, D. W., and Campbell, E. M. (1988). Four-frame gated optical imager with 120-ps resolution. *Rev. Sci. Instrum.* **59**, 1457–1460.

28. CCD Camera Application Note 1/2 (1994). Princeton Instrument Inc.

29. Thomas, R. S., Shimkunas, A. R., and Manger, P. E. (1992). Sub-nanosecond intensifier gating using heavy and mesh cathode underlays. *Proc. Int. Congr. High Speed Photo. Photon 19th, 1992* (1984).

30. C2925-01, C4078-01 High-Speed Gated Image Intensifier Units, Instruction Manual, Hamamatsu Photonics Systems, Bridgewater, New Jersey.

31. Woods, R. J., Scypinski, S., Love, L. J. C., and Ashworth, H. A. (1984). Transient digitizer for the determination of microsecond luminescence lifetime. *Anal. Chem.* **56**, 1395–1400.

32. Wang, X. F., Periasamy, A., Gordon, J., and Herman, B. (1994). Fluorescence lifetime imaging microscopy (FLIM) and its applications. *Proc. SPIE* **2137**.

33. Pawley, J. B. (Ed.). (1990). *Handbook of Biological Confocal Microscopy*. Plenum, New York.

34. Matsumoto, B. (1993). *Cell Biological Applications of Confocal Microscopy*. Academic Press, San Diego, CA.

35. Goldstein, S. R., Hubin, T., and Washburn, C. (1990). A video rate image dissector confocal microscope. In *Optical Microscopy for Biology* (B. Herman, and K. Jacobson, Eds.), pp. 59–72. Wiley-Liss, New York.

36. Williams, R., Piston, D. W., and Webb, W. W. (1994). Two-photon molecular excitation provides intrinsic 3-D resolution for laser-based microscopy and microphotochemistry. *FASEB J.* **8**, 804–13.

37. Gadella, T. W. J., Clegg, R. M., and Jovin, T. M. (1995). Fluorescence lifetime imaging microscopy: Pixel-by-pixel analysis of phase-modulation data. *Bioimaging* **2**, 139–58.

38. McGown, L. B. (1989). Fluorescence lifetime filtering. *Anal. Chem.* **61**, 839A.

39. Wang, X. F., Uchida, T., Maeshima, M., and Minami, S. (1991). Fluorescence pattern analysis based on time-resolved ratio method. *Appl. Spectrosc.* **45**, 560.

40. Verkman, A. S., Armijo, M., and Fushimi, K. (1991). Construction and evaluation of a frequency-domain epifluorescence microscope for lifetime and anisotropy decay measurements in subcellular domains. *Biophys. Chem.*. **40**(1), 117–125.

41. Forster, T. (1948). *Ann. Phys. (Leipzig)* [6] **2**, 55–75; in *Modern Quantum Chemistry* (O. Sinanoglu, Ed.), pp. 93–137. Academic Press, New York, 1966.

42. Stryer, L. (1978). Fluorescence energy transfer as a spectroscopic rule. *Annu. Rev. Biochem.* **47**, 819.

43. Dewey, T. G. (Ed.). (1991). *Biophysical and Biochemical Aspects of Fluorescence Spectroscopy.* Plenum, New York.

44. Clapham D. E. (1995). Calcium signaling-review. *Cell (Cambridge, Mass.)* **80**, 259–268.

45. Tsien, R. Y. (1994). Fluorescence imaging creates a window on the cell. *Chem. Eng. News* **72**(29), 34–44.

46. Tsien, R. Y. (1989). Fluorescence probes of cell signaling. *Annu. Rev. Neurosci.* **12**, 227–253.

47. Haughland, R. P. (1992). *Handbook of Fluorescent Probes and Research Chemicals.* Molecular Probes, Eugene, OR.

48. Lakowicz, J. A., Szmacinski, H., Nowaczyk, K., and Johnson, M. L. (1992). Fluorescence lifetime imaging of calcium using Quin-2. *Cell Calcium* **13**, 131–147.

49. Lakowicz, J. A., Szmacinski, H., and Johnson, M. L. (1992). Calcium imaging using fluorescence lifetimes and long-wavelength probes. *J. Fluoresc.* **2**, 47–62.

50. Diliberto, P. A., Wang, X. F., and Herman, B. (1994). Confocal imaging of [Ca^{2+}] in living cells. *Methods Cell Biol.* **40**, 243–262.

51. Periasamy, A., Wang, X. F., Wodnicki, P., Gordon, G., and Herman, B. (1995). High-speed fluorescence microscopy: Lifetime imaging in the biomedical science. *Microsc. Soc. Am. Bull.* **1**(1), 13–23.

52. Al-Mohanna, F. A., Caddy, K. W. T., and Bolsover, S. R. (1994). The nucleus is insulted from large cytosolic calcium ion changes. *Nature (London)* **367**, 745–750.

53. Lakowicz, J. A., Szmacinski, H., Nowaczyk, K. and Johnson, M. L. (1992). Fluorescence lifetime imaging of free and protein-bound NADH. *Proc. Natl. Acad. Sci. U.S.A.* **89**, 1271–1275.

54. Hibino, M., Shigemori, M., Itoh, H., Nagayama, K., and Kinosita, K. (1991). Membrane conductance of an electroporated cell analyzed by submicrosecond imaging of transmembrane potential. *Biophys. J.* **59**, 209.

55. Monck, J. R., Robinson, I. M., and Fernandez, J. M. (1994). Pulsed laser imaging of rapid Ca^{2+} gradients in excitable cells. *Biophys. J.* **67**, 505.

56. Herman, B. (1989). Resonance energy transfer microscopy. *Methods Cell Biol.* **30**, 219–243.

57. Jovin, T. M., and Arndt-Jovin, D. (1989). FRET microscopy: Digital imaging of

fluorescence resonance energy transfer. In *Cell Structure and Function by Micro-spectrofluorometry* (E. Kohen and J. Hirschberg, Eds.), Academic Press, San Diego, CA.

58. Wu, P., and Brand, L. (1994). Review: Resonance energy transfer: Methods and applications. *Anal. Biochem.* **218**, 1–13.

59. Uster, P. S. (1993). In situ resonance energy transfer microscopy: Monitoring membrane fusion in living cells. In *Membrane Fusion Techniques. Part B* (N. Duzgunes, Ed.), Vol. 221, pp. 239–246. Academic Press, San Diego, CA.

60. Adams, S. R., Harootunian, A. T., and Tsien, R. Y. (1991). Fluorescence ratio imaging of cyclic AMP in single cells. *Nature (London)* **394** (21), 694.

61. Wang, X. F., Lemasters, J. J., and Herman, B. (1993). Plasma membrane architecture during hypoxic injury in rat hepatocytes measured by fluorescence quenching and resonance energy transfer imaging. *Bioimaging* **1**, 30–39.

62. Selvin, P. R. (1995). Fluorescence resonance energy transfer. *Methods Enzymol.* **246**, 300–334.

63. Gadella, T. W. J. J., Jovin, T. M., and Clegg, R. M. (1994). Fluorescence lifetime imaging microscopy (FLIM): Spatial resolution of microstructures on the nanosecond time scale. *Biophys. Chem.* **48**, 221–239.

64. Heldin, C. H. (1995). Dimerization of cell surface receptors in signal transduction. *Cell (Cambridge, Mass.)* **80**, 213–223.

65. Chalfie, M., Tu, Y., and Prasher, D. C. (1994). Green fluorescent proteins as a marker for gene expression. *Science* **263**, 802–805.

66. Dahlen, P., Hurskainen, P., Lovgren, T., and Hyypia, T. (1988). Time resolved fluorometry for the identification of viral DNA in clinical specimens. *J. Clin. Microbiol.* **26**; 2434–2436.

67. Diamandis, E. P., and Christopulos, T. K. (1990). Europium chelate labels in time resolved fluorescence immunoassays and DNA hybridization assays. *Anal. Chem.* **62**, 1149A; Stahlberg, T., Markela, E., Mikolw, H., Motram, P., and Hemmila, I. (1993). Europium and samarium in time-resolved fluoroimmunoassays. *Am. Lab.* pp. 15–20.

68. Soini, E., and Lovgren, T. (1987). Time-resolved fluorescence of lanthanide probes and applications in biotechnology. *CRC Crit. Rev. Anal. Chem.* **18**, 105–154.

69. Batishko, C. R., Stahl, K. A., Erwin, D. N., and Kiel, J. (1990). A quantitative luminescence imaging system for biochemical diagnostics. *Rev. Sci. Instrum.* **61**(9), 2289–2295.

70. Seveus, L., Vaisala, M., Sandberg, M., Kuusisto, A., Hemmila, I., Kojola, H., Soini, E. (1992). Time-resolved fluorescence imaging of europium chelate label in immunohistochemistry and in situ hybridization. *Cytometry* **13**, 329–338.

71. Saidat-Pajouh, M., Periasamy, A., Ayscue, A. H., Moscicki, A. B., Palefsky, J. M., Walton, L., DeMars, L. R., Power, J. D., Herman, B., and Lockett, S. J. (1993). Detection of human papillomavirus type 16/18 DNA in cervicovaginal cells by fluorescence based in situ hybridization and automated image cytometry. *Cytometry* **15**, 245–258.

72. Siadat-Pajouh, M., Ayscue, A. H., Periasamy, A., Lockett, S. J., and Herman, B. (1994). Introduction of a fast and sensitive fluorescent *in situ* method for single copy detection of human papillomavirus (HPV) genome. *J. Histochem. Cytochem.* **42**, 1503–1512.

73. Wang, X. F., Periasamy, A., Wondnicki, P., Siadat-Pajouh, M., and Herman, B. (1995). Highly sensitive detection of human papillomavirus type 16 DNA using time-resolved fluorescence microscopy and long lifetime probes. *Proc. SPIE* **2137**, 162–168.

74. Periasamy, A., Wang, X. F., Wodnick, P., Gordon, G., Kwon, S., Diliberto, P., and Herman, B. (1995). High-speed fluorescence microscopy: Lifetime imaging in the biomedical sciences. *J. Microsc. Soc.* **1**, 13–23.

75. Suonpaa, M., Markela, E., Stahlberg, T., and Hemmila, I. (1992). Europium-labeled strepavidin as a highly sensitive universal label: Indirect time resolved fluorescence immunofluorometry of FSH and TSH. *J. Immunol. Methods.* **149**, 247–253.

76. Mottram, P. S. (1990). Applications of lanthanide chelates for time-resolved fluoroimmunoassay. *Am. Clin. Lab.* **9**(5), 34–38.

77. Beverloo, H. B., van Schadewijk, A., van Gelderen-Boele, S., and Tanke, H.J. (1990). Inorganic phosphors as new luminescent labels for immunocytochemistry and time-resolved microscopy. *Cytometry* **11**, 784–792.

78. Periasamy, A., Siadat-Pajouh, M., Wodnick, Wang, X. F., and Herman, B. (1995). Time-gated fluorescence microscopy for clinical imaging. *Microscopy Anal.*, March, pp. 33–35.

79. Collins, F., and Galas, D. (1993). A new five-year plan for the U.S. human genome project. *Science* **262**, 43–46.

80. Smith, L. M., Sanders, J. Z., Kaiser, R. J., Hughes, P., Dodd, C., Connell, C. R., and Hood, L. E. (1986). Fluorescence detection in automated DNA sequence analysis. *Nature, (London)* **321**, 674–679.

81. Van Ranst, M., Fiten, P., and Opdenakker, G. (1992). Comparison of three non-isotopic automated DNA sequence analysis systems. *Clin. Chem. (Winston-Salem, N.C.)* **38**, 465.

82. Mathies, R. A., and Huang, X. C. (1992). *Nature (London)* **359**, 167.

83. Swerdlow, H., Zhang, J. Z., Chen, D. Y., Dovichi, N. J., and Fuller, C. (1991). Three DNA sequencing methods using capillary gel electrophoresis and laser-induced fluorescence. *Anal. Chem.* **63**, 2835–2841.

84. Stevenson, C. L., Johnson, R. W., and Vo-Dinh, T. (1994). Synchronous luminescence: A new detection technique for multiple fluorescent probes used for DNA sequencing. *BioTechniques* **16**, 1104–1111.

85. Chang, K., and Force, R. K. (1993). Time-resolved laser-induced fluorescence study on dyes used in DNA sequencing. *Appl. Spectrosc.* **47**, 24–29.

86. Monnig, C. A., and Keenedy R. T. (1994). Capillary electrophoresis. *Anal. Chem.* **66**, 280R.

87. Miller, K. J., and Lytle, F. E. (1993). Capillary zone electrophoresis with time-

resolved fluorescence detection using a diode-pumped solid-state laser. *J. Chromotogr.* **648**, 245.

88. Drmanac, R., Drmanac, S., Strezoska, Z., Hood, L., and Crkvenjakov, R. (1993). DNA sequence determination by hybridization: A strategy for efficient large-scale sequencing. *Science* **260**, 1949 1652.

89. Pease, A. C., Solas, D., Sullivan, E. J., Cronin, M. T., Holmes, C. P., and Fodor, S. P. A. (1994). Light-generated oligonucleotide arrays for rapid DNA sequence analysis. *Proc. Natl. Acad. Sci. U.S.A.* **91**, 5022 5026.

90. Thevenin, B. J., Periasamy, N., Shohet, S. B., and Verkman, A. S. (1994). Segmental dynamics of the cytoplasmic domain of erythrocyte band 3 determined by time-resolved fluorescence anisotropy: Sensitivity to pH and ligand binding. *Proc. Natl. Acad. Sci. U.S.A.* **91**(5), 1741 1745.

91. Bicknese, S., Periasamy, N., Shohet, S. B., and Verkman, A. S. (1994). Cytoplasmic viscosity near the cell plasma membrane: Measurement by evanscent field frequency-domain microfluorimetry. *Biophys. J.* **65**(3), 1272 1282.

92. Shonat, R. D., Wilson, D. F., Riva, C. E., and Pawlowski, M. (1992). Oxygen distribution in the retinal and choroidal vessels of the cat as measured by a new phosphorescence imaging method. *Appl. Opt.* **31**, 3711 3718.

93. Wilson, D. F., and Cerniglia, G. J. (1992). Localization of tumors and evaluation of their state of oxygenation by phosphorescence imaging. *Cancer Res.* **52**, 3988 3993.

94. Draxler, S., Lippitsch, M. E., and Leiner, M. J. P. (1993). Optical pH sensors using fluorescence decay time. *Sensors Actuators B* **11**, 421 424.

CHAPTER

11

TWO-PHOTON FLUORESCENCE MICROSCOPY: TIME-RESOLVED AND INTENSITY IMAGING

P. T. C. SO, T. FRENCH, W. M. YU, K. M. BERLAND, C. Y. DONG, and E. GRATTON

Laboratory for Fluorescence Dynamics
Department of Physics
University of Illinois at Urbana–Champaign
Urbana, Illinois 61801

11.1. INTRODUCTION

A new, powerful fluorescence microscope has been developed by combining two-photon excitation and fluorescence time-resolved imaging. Scanning microscopy with two-photon excitation is an alternative to conventional confocal scanning for studying the internal structure of cells (1–3). This method produces a three-dimensional (3-D) sectioning effect similar to that of conventional confocal microscopy, but with additional advantages: (1) photobleaching and photodamage are localized; (2) samples labeled with UV probes can be visualized without UV lasers or expensive quartz optics; and (3) superior background rejection can be obtained. Fluorescence time-resolved measurements provide an additional contrast-enhancing mechanism for microscopic imaging (4–10). Multiple labeling of cellular organelles using fluorescent probes with similar excitation/emission spectra but different lifetimes can be distinguished by time-resolved methods. Lifetime measurements can also monitor the local environment of fluorescent probes, such as cellular calcium concentration and oxygen distribution.

In this chapter, we present the fundamentals of two-photon scanning microscopy for both steady-state (intensity) and time-resolved imaging techniques. The special considerations for successful integration of these two methods are discussed. We also characterize the performance of this new system in terms of spatial and time resolution. Finally, applications will be

Fluorescence Imaging Spectroscopy and Microscopy, edited by Xue Feng Wang and Brian Herman. Chemical Analysis Series, Vol. 137.
ISBN 0-471-01527-X © 1996 John Wiley & Sons, Inc.

demonstrated in three studies of cellular systems: (1) Laurdan-labeled cell membranes using two-photon intensity imaging; (2) time-resolved imaging of cells labeled with multiple dyes; and (3) quantitative measurement of intracellular calcium concentration with a ratiometric indicator, Calcium Green.

11.2. TWO-PHOTON EXCITATION

Two-photon molecular excitation is a nonlinear process in which a chromophore simultaneously absorbs two incident photons. The energy of the photon pair should be equal to the energy required for excitation (11–14). In the simplest case, each photon will provide half of the energy. Two-photon cross sections are small, typically on the order of 10^{-48} [p-bis(o-methylstyryl)-benzene] to 10^{-50} (rhodamine B) $cm^4 s$ photon^{-1} molecule^{-1} (14–16). The low-absorption cross section implies that the excitation of probe molecules is negligible unless they are located in a region of extremely high photon flux. The development of mode-locked, high-peak-power laser sources with femtosecond to picosecond pulses has made two-photon excitation a practical method in microscopy. CPM dye lasers were used in the pioneering work of two-photon microscopy (1). Femtosecond titanium (Ti):sapphire lasers are now the preferred choice due to their stability, high peak output power, and ease of operation and maintenance. By focusing these light sources to a diffraction-limited spot through a high numerical aperture (NA) objective, sufficiently high photon flux at the focal point can be generated. For a mode-locked laser source with average power P_0, repetition rate f_p, pulse width τ_p, and wavelength λ, focused with an objective with an NA of A, the number of photon pairs absorbed for each pulse n_a is given by (1, 17)

$$n_a \approx \frac{p_0^2 \delta}{\tau_p f_p^2} \left(\frac{\pi A^2}{hc\lambda} \right)^2 \tag{1}$$

where δ is the chromophore two-photon excitation cross section, h is Planck's constant, and c is the speed of light in the medium. When a Ti:sapphire laser with 150 fs pulse width and a repetition rate of 80 MHz is used, saturation of most chromophores occurs when the average power delivered by a 1.25 NA objective reaches about 10–50 mW.

Two photon excitation provides depth discrimination similar to that of confocal microscopy without the use of emission pinholes. The z-sectioning is due to the quadratic dependence of the fluorescence intensity on the incident photon flux. The contribution of out-of-focus planes to total fluorescence intensity decreases rapidly with increasing distance from the focal plane. When a 1.25 NA objective with 780 nm excitation is used, over 80% of total fluorescence intensity is confined to within $\pm 1 \mu m$ of the focal plane, which demonstrates the strong depth discrimination property of two-photon excitation (Figure 11.1a). In contrast, for a uniform sample under one-photon excitation, assuming negligible attenuation of the excitation light, the fluorescence intensity contribution from each z-section is independent of focal plane position. This behavior is a consequence of the conservation of energy (18). Thus, one-photon excitation without a confocal pinhole does not provide depth discrimination.

The resolution of two-photon microscopy has been compared to that of one-photon confocal microscopy (19, 20). For the same excitation wavelength, it can be shown that one-photon confocal and two-photon excitation systems have very similar spatial resolution (19, 20). However, if one uses the same fluorescence probe for confocal and two-photon imaging, the excitation light required will be twice the wavelength in the two-photon case. In this case, the

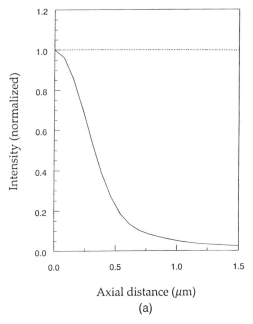

Axial distance (μm)

(a)

Figure 11.1. [pp. 353 and 354] Depth discrimination and two-photon excitation profiles. This simulation was calculated assuming 760 nm excitation and 1.25 NA objective. (a) The intensity contribution from different z-sections as a function of their distance from the focal plane in a uniform sample with one- and two-photon excitation; (b) the axial intensity profile of one- and two-photon excitation; (c) the radial intensity profile of one- and two-photon excitation. The dotted line is one-photon excitation and the solid line is two-photon excitation.

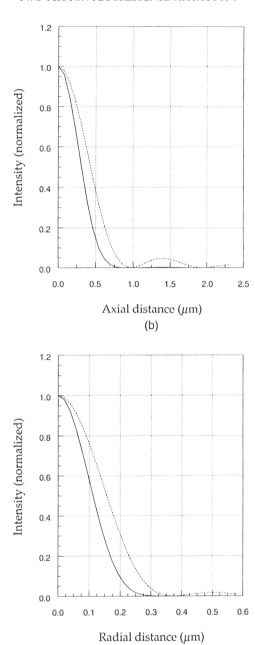

Axial distance (μm)

(b)

Radial distance (μm)

(c)

Figure 11.1 (*Contd.*)

confocal method will have better radial and axial resolution by a factor of 2. When a 1.25 NA objective with 960 nm excitation is used, two-photon axial resolution of 0.9 μm and radial resolution of 0.3 μm can be achieved. (Figure 11.1b, c). Greater suppression of the higher order diffraction rings of the Airy disk is achieved with two-photon excitation.

11.3. FLUORESCENCE TIME-RESOLVED IMAGING

Intensity imaging with two-photon excitation is a powerful technique; combining it with time-resolved imaging opens up many exciting possibilities. Time-resolved fluorescence measurement techniques for bulk solutions are well developed in both frequency and time domains (21). However, the application of these techniques to the study of cellular systems has its own technical challenges. The field of time-resolved microscopy has seen rapid development since unique information about cellular systems was obtained using single-point, time-resolved microscopy (4, 5). Extending these single-point measurements to real-time acquisition of lifetime images is an important technological advance. Time-resolved imaging in the frequency domain using charge-coupled device (CCD) cameras coupled to gain-modulated microchannel plates has been the most popular approach. Both heterodyning (22–24) and homodyning (6, 9, 25–28) techniques have been developed. Time-domain photon-counting systems (8, 10) have also been constructed. The confocal method has added three-dimensional (3-D) capability to time-resolved microscopes (8). Frequency-domain, time-resolved microscopy using two-photon excitation was first developed by Piston et al. (2, 3) and was shown to be another promising method for obtaining 3-D time-resolved images in cellular systems (29). The application of time-resolved measurements to near-field microscopy has further extended the time-resolved technique to nanometer scales (30–32). In this chapter, we will only consider two-photon, time-resolved techniques.

The two-photon, time-resolved microscope constructed in our laboratory employs the well-established frequency-domain heterodyning technique. The detection of fluorescence decay in the frequency domain requires the use of intensity-modulated light sources. When the intensity of the excitation light is sinusoidally amplitude modulated, the fluorescence emission will also be sinusoidal, with a characteristic phase delay and demodulation resulting from the finite lifteime of each chromophore (Figure 11.2a). Pulse sources can be considered a sum of the sinusoids according to the Fourier theory. The phase shift and demodulation are functions of modulation frequency and can be used to determine the fluorescence decay modes. For a single exponential decay, the phase shift, ϕ, and demodulation, M, are related to the fluorescence lifetime

τ and the angular modulation frequency ω as follows:

$$\tan(\phi) = \omega\tau$$

$$M = \frac{1}{\sqrt{1 + \omega^2\tau^2}} \qquad (2)$$

Since typical modulation frequencies are on the order of tens to hundreds of megahertz, direct measurement of the phase shift and demodulation is

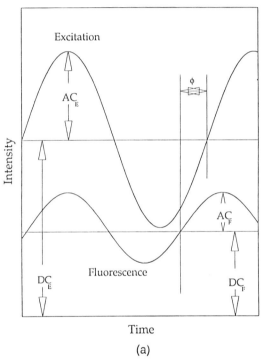

(a)

Figure 11.2. Principle of frequency heterodyning. (a) The definitions of DC (intensity), AC, phase (ϕ), and modulation $M = AC/DC$; (b) the digitization of the cross-correlation waveform, and determination of phase and modulation by fast Fourier transform. The figure indicates that four points per waveform are digitized. The intensity, phase, and modulation values can be calculated from the four digitized values:

$$DC = \frac{V_1 + V_2 + V_3 + V_4}{2} \qquad AC = \frac{[(V_0 - V_2)^2 + (V_3 - V_1)^2]^{1/2}}{2}$$

$$\phi = \tan^{-1}\left(\frac{V_3 - V_1}{V_0 - V_2}\right) \qquad M = \frac{AC}{DC}$$

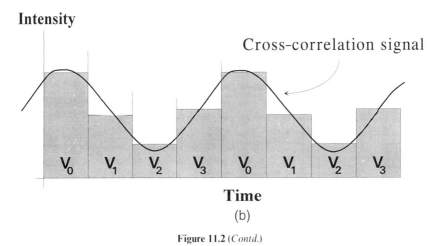

Intensity

Cross-correlation signal

V_0 V_1 V_2 V_3 V_0 V_1 V_2 V_3

Time

(b)

Figure 11.2 (*Contd.*)

difficult. Heterodyning techniques modulate the gain of the photodetector at a frequency $\omega + d\omega$ slightly above the modulation frequency ω, typically on the order of tens of hertz. The fluorescence and detector modulation frequencies are mixed in the detector, and the phase and demodulation information is translated into the signal at the beat frequency $d\omega$, also called the cross-correlation frequency. Since the cross-correlation signal is at a much lower frequency than the excitation source, it can easily be isolated by an electronic low-pass filter. This low-frequency signal can then be digitized, and the lifetime information can be recovered via Fourier transform techniques (Figure 11.2b).

11.4. DESIGN OF THE TWO-PHOTON, TIME-RESOLVED MICROSCOPE

The design schematic of the two-photon, time-resolved microscope developed in our laboratory is presented in Figure 11.3.

11.4.1. Laser Light Sources

The laser used in this microscope is a Ti: sapphire Mira 900 laser pumped by an Argon-Ion Innova 310 laser (Coherent Inc., Palo Alto, California). The Ti: sapphire laser generates 150 fs pulses with a high average output power of 250–1500 mW. The instantaneous peak power of this laser is about 100 times that of a 100 mW dye laser with 2 ps pulse width. The intensity stability of the Mira is excellent, with typical fluctuations of less than 0.5% of the total power.

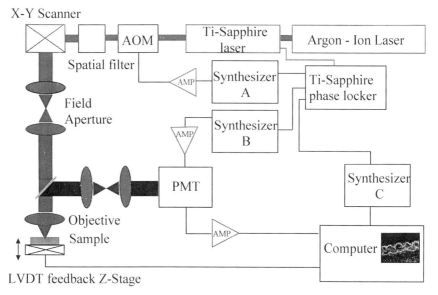

Figure 11.3. Schematic of the two-photon, time-resolved microscope.

This is important for uniform illumination of the sample in a scanning beam microscope. The output of this laser is tunable from 720 to 1000 nm, which provides two-photon excitation for probes with absorption from 360 to 500 nm. This wavelength range covers most near-UV, blue, and green probes commonly used in fluorescence microscopy.

Currently, most two-photon microscopy studies are performed in the range of 720–820 nm. The longer wavelength region is rarely explored. In recent experiments, we have obtained efficient excitation of probes in the absorption range of 450–490 nm (e.g., fluorescein), as well as a satisfactory result with some probes in the absorption range of 490–520 nm (e.g., rhodamine) (see Section 11.5). Furthermore, preliminary results indicate that using longer wavelength excitation slows down the process of photodamage to cells and reduces cellular autofluorescence. This result is expected since absorption of cellular proteins in this wavelength range is less than that in the near-UV region. Two drawbacks of operating the two-photon microscope in the longer wavelength region are lower power and reduced stability of the laser. This situation is due to the reduced fluorescence efficiency of the Ti: sapphire crystal and the rapidly rising infrared absorption by water vapor in the laser cavity. Additional studies are now underway to characterize two-photon excitation in this wavelength regime.

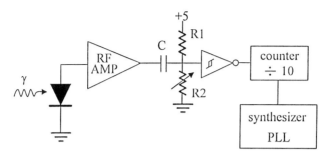

Figure 11.4. Phase-lock circuitry for the synchronization of the Ti:sapphire laser. The RF amplifier used is ZHL-32A; the Schmitt trigger inverter is 74F14; the 100 MHz counter is 74F579; the capacitor C is 1 nF; and the resistors R1 and R2 are 1 kΩ.

The Coherent Ti:sapphire laser is a passive mode-locked system. It generates a self-sustained oscillation at 80 MHz, stable to within 20 Hz. This repetition rate can be manually adjusted by varying the cavity length of the laser. Daily drift is typically less than 1 kHz. To use this laser as the excitation light source in a frequency-domain instrument, the other frequency synthesizers used in the instrument (PTS 500, Programmed Test Sources, Inc., Littleton, Massachusetts; HP 3325B, Hewlett Packard, Inc., Santa Clara, California) must be phase locked (synchronized) with the laser. Phase locking of this laser can be accomplished with a simple and inexpensive circuit (Figure 11.4). The Coherent Ti:sapphire laser is equipped with an internal fast photodiode that monitors the laser pulse train. The output of the diode is accessible from the exterior of the laser. The photodiode signal that tracks the laser pulses is first amplified by a high-frequency amplifier (ZHL-32A, Minicircuit, Inc., Brooklyn, New York). The output of the amplifier is shaped into a standard transistor-to-transistor logic (TTL) pulse using a Schmitt trigger inverter. By means of an 8 bit binary counter, the 80 MHz signal is divided down to 10 MHz, which serves as a master oscillator; the signal is accepted by the phase-lock circuitry of most commercial synthesizers. All electronic components used are F-series TTL (National Semiconductor, Santa Clara, California). This phase-locking scheme has been tested and found to be accurate to one part per million.

11.4.2. Additional Modulation of Laser Pulses

The Ti: sapphire laser is operated at a fixed repetition rate of 80 MHz. Due to its narrow (150 fs) pulse width, the laser pulse train has harmonic content up to 220 GHz (3 dB bandwidth). However, the ability to obtain multiple modulation frequencies below 80 MHz is critical for resolving slower fluorescence decay modes. To do so, a custom-made, fused quartz standing wave

acousto-optic modulator (AOM) (Intra-action, Belwood, Illinois) is placed in
the beam path of the Ti: sapphire laser. This type of AOM was chosen for its
low intensity loss (typical transmittance is 30–50%) compared with other
modulators. Furthermore, this modulator causes minimal broadening of the
150 fs pulses and maintains the high peak power. The AOM can be modulated
from about 30 to 120 MHz at discrete resonances 330 kHz apart. The modulation frequency of the AOM is mixed with the 80 MHz repetition frequency of
the laser and its higher harmonics (160, 240, 320 MHz, etc.). New frequencies
are generated at the sums and differences of the AOM modulation frequencies
and the laser harmonics. We are able to obtain good light modulation in the
range of kilohertz to gigahertz using this mixing scheme.

11.4.3. Scanner and Optics

After spatial filtering and beam expanding, the laser excitation is incident on
a galvanomotor-driven x-y scanner (Cambridge Technology, Watertown,
Massachusetts). The scanner motor driver circuitry provided by Cambridge
Technology accepts both analog and digital control. The digital driver is the
more natural choice for performing raster scans and is used here. Both the
x and y scanners have an angular range of $\pm 60°$, and their positions can be
specified by a 16 bit binary number. Additional synchronization and control
circuitry have been constructed to interface the scanner driver with the master
data acquisition computer. This interface circuit utilizes a low-cost single-
board computer (New Micros, Inc., Dallas, Texas). The scanning region is
stored as digital coordinates in the single-board computer memory. The
master computer synchronizes the scan by delivering a TTL trigger to this
interface circuit. On triggering, the interface circuit outputs the proper 16 bit
number to step the scanner to the next pixel position. This circuit allows us to
be very flexible in specifying the scan region. The array of positions stored in
the digital memory can be easily modified to allow scan regions of different
sizes, shapes, and resolutions to be specified. The scanner has a maximum
scan rate of 500 Hz over its full range of $\pm 60°$. The minimum time required to
scan a typical 256×256 pixel region is 0.524 s with a pixel residence time of
8 μs.
    The x–y scanner steers the laser excitation within the epiluminescence light
path of a Zeiss Axiovert 35 microscope (Zeiss, Inc., Thronwood, New York).
The epiluminescence light path has been modified such that a 10X eyepiece,
used as the scan lens, can be mounted. The scan lens is positioned such that the
x–y scanning mirror is at its eye point and the field aperture plane is at its focal
point. Through the eyepiece, the angular deviation of the input beam, generated by the scanner, is linearly translated into lateral translation of the focal
point position on the field aperture plane. Since the field aperture plane is

telecentric to the object plane of the microscope objective, the movement of the focal point on the object plane is also proportional to the angular deviation of the scanned beam (33).

The z-scanning is accomplished by varying the separation between the objective and the sample. The objective is driven by a geared stepper motor coupled to its standard manual height adjustment mechanism. The distance between the objective and the sample is monitored by a linear variable differential transformer (LVDT; Schaevitz Engineering, Camden, New Jersey). The LVDT and its signal conditioning circuitry are designed to have a position resolution of 0.2 μm over a total range of 200μm. The signal from the LVDT is monitored by a 8052 microprocessor-based single-board computer (Iota System, Inc., Incline Village, Nevada) that dynamically controls the stepper motor and maintains the distance between the objective and the sample. This dynamic feedback control system has a bandwidth of 10 Hz.

Because the Zeiss Axiovert microscope uses infinity-corrected optics, a tube lens is needed to recollimate the light. The excitation is then reflected by the dichroic mirror to the objective. The dichroic mirrors used are custom-made short-pass filters (Chroma Technology, Inc., Brattleboro, Vermont) instead of the long-pass filters commonly used in one-photon confocal systems.

We most often use a well-corrected, aberration-free Zeiss 60X Plan-Neofluar objective because of its high NA of 1.25. However, with the excitation flux (sometimes in excess of 500 mW) used, costly damage may occur to the objective. For less critical applications, a more economical Zeiss 100X CP-Achromat objective (1.25 NA) provides satisfactory performance with slight degradation of resolution and fluorescence excitation efficiency.

The fluorescence signal from the sample is collected by the same objective, transmitted through the dichroic mirror, and refocused on the detector. Since the power of excitation is much higher than the fluorescence signal, a short-pass filter, in addition to the dichroic mirror, further attenuates the transmitted light. Two-photon excitation has the advantage that the excitation and emission wavelengths are well separated by about 400 nm. Suitable short-pass filters can easily be found to eliminate most of the residual excitation, with minimal attenuation of the fluorescence signal. Dielectric filters are adequate, but a properly chosen liquid filter can provide improved performance (34). In the excitation wavelength region of 720–820 nm, a 1.5-in.-long liquid filter filled with 2 M $CuSO_4$ has an optical density (O.D.) of over 20 for the infrared while retaining over 90% transmittance of the fluorescence signal. In the excitation wavelength region of 900–1000 nm, a liquid filter filled with 1 M IRA980 absorber dye dissolved in methanol (Exciton Inc., Dayton, Ohio) also works well.

11.4.4. Signal Detection and Noise Reduction

The fluorescence signal at each pixel is detected by a photomultiplier tube (PMT), and its analog output is fed into the digitizer of the data acquisition computer. For steady-state measurements, a cooled, high-sensitivity R1104 PMT (Hamamatsu, Bridgewater, New Jersey) is used. For time-resolved measurements, we use a R928 PMT (Hamamatsu) modified for gain modulation, as previously described (34).

The cross-correlation frequency is chosen such that a complete period of the cross-correlated waveform can be collected at each pixel. As discussed previously, the minimum pixel residence time imposed by the mechanical response of the scanner is about 8 μs. The actual pixel residence time is determined by two competing requirements. First, since the quality of a microscope image is typically limited by the number of available photons, pixel dwell times should be long enough to collect a sufficient number of photons to obtain good signal-to-noise ratios. Second, the pixel residence time must be short enough that pictures (256 × 256 pixels) can be acquired in reasonable time. Typically, acquiring one frame in less than 1 min is acceptable to most microscope operators. Shorter pixel dwell times also minimize photodamage to the sample caused by localized heating. We found that a cross-correlation frequency of 25 kHz (40 μs period) is a good compromise.

The cross-correlation rate of 25 kHz is much higher than the typical cross-correlation frequency ($\sim$ 80 Hz) used in conventional fluorometers. However, with the availability of low-noise, fast digitizers, the speed of data acquisition poses no additional difficulties. A 100 kHz, 12 bit sampling digitizer (A2D-160, DRA Laboratories, Sterling, Virginia) is used in this instrument. The Shannon Sampling Theorem dictates that at least two points per waveform need to be acquired to determine a sinusoidal signal. We typically digitize four points per waveform to reduce the effect of total harmonic distortion (see Figure 11.26). As will be discussed shortly, the most efficient way to reduce noise is to average over a few waveforms at each pixel. In our apparatus, at least four waveforms are averaged, which brings the pixel dwell time to 160 μs with a reasonable frame rate of 10 s.

Three forms of digital filtering and averaging are applied during data acquisition. These techniques were developed for conventional frequency-domain fluorometry (35) and are equally applicable here. The goal of digital filtering is to isolate the signal at the cross-correlation frequency from both random and harmonic noise. The first two filters operate at the pixel level. The first is an in-phase averaging filter. A number of the cross-correlated waveforms are collected, and the data points at the corresponding phase of each waveform are averaged. This filter is very efficient in removing random noise; it strongly rejects noise that is not at the cross-correlation frequency and its harmonics. It has been shown that the increase in the signal-to-noise ratio is proportional to the number of waveforms averaged. The second filter consists of a fast Fourier transform (FFT). This process separates the desired signal at the cross-correlation frequency from the higher harmonic noise. These two processes are typically applied in succession to obtain filtered phase and modulation values at each pixel. To further improve the signal-to-noise ratio, multiple images can be collected and the intensity, phase, and modulation values at each pixel averaged. This is a less efficient filter, and the noise reduction obeys Poisson statistics.

A dedicated program has been developed for data acquisition in this time-resolved microscope. The program coordinates the data acquisition process, performs the necessary digital filtering, and displays the time-resolved images in real time. The 3-D images are manipulated and analyzed by importing the data into a 3-D image analysis package, ANALYZE (Mayo Clinic, Rochester, Minnesota), running on an IBM RS-6000 computer.

11.5. RESULTS AND DISCUSSIONS

It is important to characterize this new two-photon, time-resolved microscope in terms of its spatial and time resolutions. By understanding its strengths and limitations, we can better define the biological areas where this technology can be successfully applied. We will first present measurements of the spatial and temporal instrumental resolution. Three studies will then be described that utilize the unique capabilities of this microscope.

11.5.1. Microscope Performance and Characterization

11.5.1.1. Image Depth Discrimination and Resolution

We have imaged mouse fibroblast cells with their cytoplasm loaded with blue fluorescent 0.014 μm diameter latex spheres to demonstrate the 3-D imaging capability of this microscope. Thirty-five z-sections of the cell were imaged with a z-scan step of 0.7 μm. Each z-section is 256×256 pixels with scan steps of 0.14 μm. These images were 3-D reconstructed and are presented in Figure 11.5a (see Color Plates). The images indicate that the latex spheres are in the cytoplasm but are excluded from some cellular organelles, providing good contrast between the regions. Three typical $x-y$ (Figure 11.5b), $y-z$ (Figure 11.5c), and $z-x$ (Figure 11.5d) cross-sectional images are also presented. As shown, the $x-y$ resolution is significantly higher than that of the $y-z$ and $z-x$ sections. This is expected from the two-photon excitation profile with finer radial resolution compared to axial resolution.

To further characterize the spatial resolution of the system, yellow fluorescent latex spheres (Molecular Probe, Eugene, Oregon) 0.03 µm in diameter were immobilized between a coverslip and a flat microscope slide with viscous Fluoromount-G compound (Fisher Scientific, Pittsburgh, Pennsylvania). The size of these latex spheres is very uniform and has been characterized by the manufacturer using an electron microscope. At 960 nm, using a 64 × 1.25NA objective, we have imaged 12 z-sections across the sample with a z-scan step of 0.03 µm. The x–y scan steps are 0.035 µm. For all the spheres, the intensity at the radial direction drops off rapidly, with a full width at half-maximum of 0.3 µm. The intensity at the axial direction decreases slowly, with a full width at half-maximum of 0.9 µm (Figure 11.6a, b). The measured radial and axial dimensions are consistent with the Fraunhofer diffraction prediction. Equipped with the point spread function of this sytem, the resolution of more complex cellular images obtained with two-photon excitation may be further improved using 3-D deconvolution methods (36). Additional work is underway in our laboratory to use the point spread function and to sharpen complex images.

11.5.1.2. Quantitative Lifetime Measurement with Fluorescent Latex Spheres

This time-resolution measurement was obtained with a Ti:sapphire laser tuned to a wavelength of 960 nm at a modulation frequency of 80 MHz. The cross-correlation frequency was 25 kHz. Eight waveforms were in-phase averaged at each pixel. The images presented are an average of four frames. A solution of fluorescein was imaged as a lifetime reference.

Latex spheres containing two different chromophores were imaged to demonstrate the quantitative time resolution of this microscope. Orange fluorescent latex spheres (2.3 µm in diameter) were mixed with Nile red fluorescent latex spheres (1.0 µm diameter) and were immobilized with Fluoromount compound. The lifetimes of these spheres have been independently measured in a conventional cuvette fluorometer (ISS, Inc., Champaign, Illinois). The orange spheres exhibit a single exponential decay with a lifetime of 4.3 ns. The fluorescent decay of the red spheres can be fit to a single exponential decay of 2.9 ns. However, a better fit is obtained using a double exponential decay with a major lifetime component (58.8%) of 3.8 ns and a second lifetime component (41.2%) of 2.0 ns. Since the imaging experiment has been performed only at a single modulation frequency, we can not resolve a double exponential decay. The average values obtained from the single exponential decay data are sufficient for our analysis.

Figure 11.7a (see Color Plates) presents the intensity image of two orange spheres (the larger ones) and three red spheres (the smaller ones). Note that the

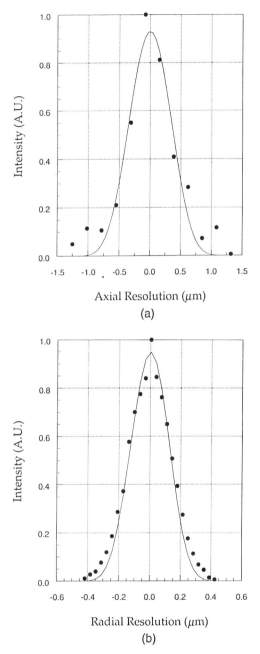

Figure 11.6. Determination of the two-photon excitation point spread function by imaging 0.03 μm fluorescent latex spheres. (a) The axial profile; (b) the radial profile. The point spread function full width at helf-maximum in the axial and radial directions are 0.9 and 0.3 μm, respectively. The experimental data (·) are fitted with Fraunhofer diffraction theory (solid line). The fitting parameters used are a scale factor and a translation along the position axis.

Table 11.1. Lifetimes of Fluorescent Latex Spheres

	τ_p	τ_m	$\langle \tau \rangle$	τ_f
Orange spheres	4.2 ± 0.4	3.8 ± 0.4	4.0 ± 0.4	4.3
Red spheres	2.4 ± 0.3	2.8 ± 0.3	2.6 ± 0.3	2.9

Note: τ_p and τ_m are the lifetimes derived from phase and modulation data, respectively; $\langle \tau \rangle$ is the average of the phase and modulation data; τ_f is the lifetime value obtained in a conventional fluorometer.

small sphere at the left edge of the picture has the same intensity as the large orange spheres but is considerably less intense than the red ones. This observation stresses the fact that intensity is a poor parameter for discerning the property of the specimen in a fluorescence image because intensity differences can be affected by experimental factors such as the uniformity of the illumination or the uniformity of the detector response. In contrast, the phase and modulation pictures (Figure 11.7b, c; see Color Plates), show that the two different colored spheres each have their own distinct phase and modulation since they contain different chromophores.

From the phase and modulation pictures, we can compute the lifetimes of these fluorescent spheres (Table 11.1). By averaging the lifetime data derived from the phase- and modulation-resolved images, a quantitative time-resolved image of the spheres can be generated (Figure 11.8a, see Color Plates). These values of 4.0 ns and 2.6 ns are in good agreement with the values obtained in the conventional fluorometer.

This result demonstrates the possibility of measuring quantitative dynamic fluorescence parameters in microscopy. While this study uses a single modulation frequency, future studies with multiple frequencies should allow us to resolve multiple fluorescence decay modes. Multiple modulation frequency capability has been built into the design of this microscope. No technical difficulty is expected, provided that the cellular samples can survive the longer laser exposure times that multiple frequency measurements require.

Figure 11.8b (see Color Plates) shows a histogram of the number of pixels in the image with a particular lifetime. Two peaks at 4.0 and 2.6 ns are observed corresponding to the two populations of spheres. These peaks have a full width at half-maximum of 1 ns. This indicates that our system can resolve fluorescence components with a lifetime difference of 500 ps for chromophores with lifetimes in the range of 1 to 10 ns commonly used in microscopy. This lifetime resolution can be further improved by increasing the data acquisition time for additional averaging.

11.5.2. Applications of Two-Photon Scanning and Time-Resolved Imaging to Cellular Studies

11.5.2.1. Laurdan-Labeled Imaging by Two-Photon Excitation

The preceding discussion demonstrates that while two-photon excitation provides excellent spatial resolution, confocal microscopy often performs better in this respect. However, if resolution is not the only concern, two-photon excitation has other significant advantages. One of the most important advantages of two-photon excitation is that photodamage and photobleaching are localized to the submicron volume of excitation. One common membrane fluorescent probe that presents difficulty in a traditional confocal microscope is Laurdan. Laurdan is used to measure the fluidity of lipid membrane in cells and purified membrane system (37–39). It is rarely used in microscopy due to its high susceptibility to photobleaching. In the fluorescence confocal microscope, as one collects one z-section image, the out-of-focus planes are photobleached simultaneously. This problem makes 3-D reconstruction difficult, and no satisfactory confocal images of Laurdan-labeled cells have been obtained so far. When two-photon excitation is used, this problem can be avoided.

Mouse T-cell membranes were labeled with Laurdan using a procedure previously described (39). A series of z-section images of the cell was taken with 780 nm two-photon excitation (Figure 11.9, see Color Plates). The x–y pixel size is 0.14 µm, and each x–y picture shown here is separated in the axial direction by 2.4 µm. We found that the rate of photobleaching at each plane is very slow, with 50% intensity reduction in 5 min providing ample time to obtain a reasonable image at each z-section. The actual frame rate for the data presented here was 10 s. As can be seen, the plasma and nuclear membrane are brightly labeled. This provides excellent contrast between the cell membrane and cytoplasmic structures that are slightly autofluorescent. Given the excellent quality of the data obtained in this experiment, measuring the membrane fluidity and cholesterol content of live cells using Laurdan G-polarization measurements appears to be a promising experiment.

11.5.2.2. Time-Resolved Multiple-Dye Imaging

In fluorescence microscopy, it is frequently necessary to simultaneously image and distinguish various organelles in a single cell. Traditionally, these distinct strutures are specifically labeled with fluorescent probes of sufficiently well separated absorption spectra such that each structure can be selectively excited by picking the proper excitation wavelength. By collecting a series of pictures with multiple excitation wavelengths, these images can be superimposed to generate an image in which the spatial relationship between these organelles can be studied. This process not only requires multiple excitation light sources, achromatic optics, and filter sets, but the sample must also be subjected to multiple exposures, which may cause significant photodamage. With the ability to quantitatively resolve fluorescence lifetimes as demonstrated in the previous section, we propose to simplify this process by labeling these different organelles with dyes that have similar absorption spectra but distinct fluorescence lifetimes. The goal is to excite and image all the labeled organelles at a single excitation wavelength and to distinguish these individual structures by their unique lifetime values.

We have double-labeled mouse fibroblast cells with ethidium bromide (EtBr) and 3,3'-dioctadecyloxacarbocyanines perchlorate (DiO) (Molecular Probes, Eugene, Oregon) to demonstrate the time-resolved multiple-labeling method. Both of these probes can be excited at 960 nm. The lifetimes of EtBr and DiO are known. EtBr has a lifetime of 1.7 ns when free in solution and a lifetime of 24 ns when bound to nucleic acid. The lifetime of DiO-labeled fibroblast cells was determined independently using this time-resolved microscope. A lifetime of 3 ± 1 ns was measured.

Eight pictures were averaged to generate the images presented (Figure 11.10, see Color Plates). The intensity image (Figure 11.10a) shows that the most strongly fluorescent regions are the two nucleoli and a few small regions on the plasma membrane. A less intense fluorescent background throughout the cell is also observed. The contrast in the intensity image is clearly dominated by differential dye concentration but does not reflect the functional difference of the cellular region.

On the other hand, the phase and modulation pictures (Figure 11.10b, c; see Color Plates) provide a more informative image. Functionally distinct regions are seen to have different phase and modulation values independent of dye concentration. Assuming single exponential decay, we have converted the modulation picture into a lifetime resolved picture. Five distinct lifetime regions are observed, and the image can be sectioned according to these distinct lifetime ranges (Figure 11.11, see Color Plates): Since DiO does not permeate the plasma membrane of fixed cells, the fluorescence seen inside the cell must have come from the presence of EtBr. The lifetimes of EtBr molecules are different in the cytoplasm, the nucleus, and the two nucleoli. The lifetimes observed in all three regions lie between the free and bound EtBr lifetimes. The differences in these three regions can be explained by the different ratios of coexisting bound and free EtBr molecules. EtBr molecules in the nucleus and the nucleoli regions have significantly longer lifetimes than those in the cytoplasm because of the much higher nucleic acid concentration in the nucleus. The slightly shorter lifetimes of the two nucleoli may indicate the presence of a larger excess of free EtBr compared to the rest of the nucleus.

The plasma membrane has the typical lifetime of DiO. In the localized, highly fluorescent region of the plasma membrane, a significantly less than normal lifetime is observed. One plausible explanation may be that there is a significant difference in the lipid compositions of these regions that affects the lifetime of DiO molecules bound there. Quantitative understanding of the fluorescence lifetime distribution inside a complex cellular system is difficult; the preceding explanation is conjectural and requires further experimental confirmation. However, as a contrast-enhancing mechanism, time-resolved imaging provides a unique lifetime label for the distinct structures in the cell and allows them to be distinguished. By properly choosing a series of fluorescent dyes with different lifetimes but similar absorption spectra, multiple regions in a cell can be selected and imaged simultaneously.

11.5.3. Quantitative Calcium Concentration Measurement Using Time-Resolved Methods

Calcium imaging is an important application of optical microscopy to cellular systems. Accurate measurement of the free calcium concentration is possible using ratiometric dyes such as the Fura and Indo series (40, 41). When ratiometric methods are used, the calcium concentration in the cells can be measured independently of the probe distribution in the cell. This is important since the probe distribution is not an experimentally controllable parameter. We propose an alternative method in which we measure the lifetime of the calcium probe molecules that are sensitive to their surrounding free calcium distribution but independent of the concentration of the probe molecules. Lifetime imaging shows great promise in allowing quantitative cellular calcium concentration measurement with a large variety of non-ratiometric calcium probes (42–44). Furthermore, the development of this methodology should be important for measuring the concentration of biologically relevant molecules such as Mg^{2+} or O_2, which have no associated ratiometric probes (45, 46).

We have imaged mouse fibroblast cells loaded with Calcium Green (Molecular Probes, Eugene, Oregon) (Figure 11.12, see Color Plates). Calcium Green is a very attractive probe for calcium imaging since it can be excited in the blue–green wavelength range, where there is significantly reduced cellular autofluorescence (41). Reduced autofluorescence not only allows a more accurate determination of cellular calcium content but also reduces photodamage to the cell. Since Calcium Green is a nonratiometric intensity probe, its use in the quantitative determination of the cellular calcium content has been limited compared to UV ratiometric probes such as fura-2. In this study, we demonstrate that with time-resolved imaging methods, quantitative calcium imaging can be achieved using Calcium Green.

The image presented in Figure 11.12 (see Color Plates) was acquired in 160 s. The pixel step size is 0.24 μm. The intensity image (Figure 11.12a) shows three fibroblast cells labeled with Calcium Green. In all three cells, the nuclei are the most brightly fluorescent region in the cells, typically two to three times more fluorescent than the cytoplasm. If one assumes that these intensity differences are entirely due to a variation in calcium concentration in the cell, this image indicates that the calcium concentration in the nucleus is an order of magnitude greater than that in the cytoplasm. Furthermore, the intensity difference between the cells indicates that the calcium concentration also varies greatly from cell to cell. However, this result is contrary to observations made with ratiometric dyes, indicating that the calcium concentrations in cells at rest are roughly constant (40).

This conflict can be resolved by obtaining phase- and modulation-resolved pictures (Figure 11.12b, c; see Color Plates). In sharp contrast to the intensity

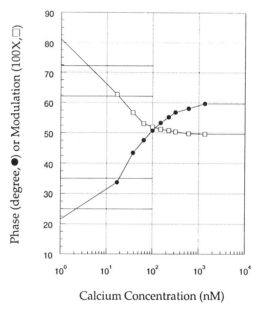

Calcium Concentration (nM)

Figure 11.13 Intracellular calcium concentration of mouse fibroblast cells measured with the calcium indicator Calcium Green. As a calibration, the Calcium Green fluorescence phase (solid circles) and modulation (open squares) as a function of calcium concentration have been measured independently *in vitro*. When the range of phase values (between the lower two horizontal lines) and the modulation values (between the upper two horizontal lines) extracted from the time-resolved microscope images are used, the range of calcium concentration in the fibroblast cells can be determined to be about 12 nM.

picture, the phase and modulation values are roughly uniform within each cell. Independent *in vitro* two-photon measurement of the Calcium Green lifetime as a function of the free calcium concentration has been performed, and the phase and modulation values can be directly translated into the calcium concentration inside the cell (Figure 11.13). We found that the calcium concentration in the cytoplasm is about 10 nM, and the calcium concentration in the nucleus is only slighly higher at 12 nM (47–49). These values are in reasonable agreement with published observations with ratiometric dyes. The calibration curves of phase and modulation values as a function of calcium concentration are obtained in buffer solution. Similar to the case of fura-2, there are indications that *in vivo* calibration procedures may be necessary to yield a more accurate determination of cellular calcium concentration. Nevertheless, lifetime imaging of nonratiometric dyes provides a viable alternative method for quantitative calcium concentration measurement in cells.

11.6. CONCLUSION

Two-photon fluorescence microscopy has seen rapid development in the past two years. As the technical aspects of both time-resolved and intensity-imaging modes approach maturity, the most important issue appears to be locating unique biological applications for this microscope. We have demonstrated three such applications: Laurdan intensity imaging, calcium time-resolved imaging, and time-resolved multiple-dye labeling.

ACKNOWLEDGMENTS

We thank Dr. David Piston for his advice on the techniques of two-photon excitation. We also thank Dr. Matt Wheeler, Dr. Laurie Rund, and Ms. Melissa Izard for providing mouse fibroblast cells and Dr. David Kranz for providing mouse T cells. This work was supported by the National Institute of Health (RR03155).

REFERENCES

1. Denk, W., Strickler, J. H., and Webb, W. W. (1990) Two-photon laser scanning fluorescence microscopy. *Science* **248**, 73–76.
2. Piston, D. W., Sandison, D. R., and Webb, W. W. (1992). Time-resolved fluorescence imaging and background rejection by two-photon excitation in laser scanning microscopy. *Proc. SPIE—Int. Soc. Opt. Eng.* **1640**, 379–389.
3. Piston, D. W., Kirby, M. S., Cheng, H., Lederer, W. J., and Webb, W. W. (1994).

Two photon-excitation fluorescence imaging of three-dimensional calcium-ion-activity. *Appl. Opt.* **33**, 662–669.

4. Dix, J. A., and Verkman, A. S. (1990). Pyrene eximer mapping in cultured fibro-blasts by ratio imaging and time-resolved microscopy. *Biochemistry* **29**, 1949–1953.

5. Keating, S. M., and Wensel, T. G. (1990). Nanosecond fluorescence microscopy: Emission kinetics of Fura-2 in single cells. *Biophys. J.* **59**, 186–202.

6. Lakowicz, J. R., Szmacinski, H., Nowaczyk, K., Berndt, K. W., and Johnson, M. L. (1992). Fluorescence lifetime imaging. *Anal. Biochem.* **202**, 316–330.

7. vande Ven, M., and Gratton, E. (1993). *Optical Microscopy: Emerging Methods and Applications*, pp. 373–402. Academic Press, New York.

8. Buurman, E. P., Sanders, R., Draaijer, A., Gerritsen, H. C., Van Veen, J. J. F., Houpt, P. M., and Levine, Y. K. (1992). Fluorescence lifetime imaging using a confocal laser scanning microscope *Scanning* **14**, 155–159.

9. Gadella, T. W. J., Jr., Jovin, T. M., and Clegg, R. M. (1993). Fluorescence lifetime imaging microscopy (FLIM): Spatial resolution of microstructures on the nanosecond time scale. *Biophys. Chem.* **48**, 221–239.

10. Oida, T., Sako, Y., and Kusumi, A. (1993). Fluorescence lifetime imaging micros-copy (filmscopy). *Biophys. J.* **64**, 676–685.

11. Friedrich, D. M. (1992). Two-photon molecular spectroscopy. *J. Chem. Educ.* **59**, 472–481.

12. Birge, R. R. (1983). One-photon and two-photon excitation spectroscopy. In *Ultrasensitive Laser Spectroscopy* (D. S. Kliger, Ed.), pp. 109–174. Academic Press, New York.

13. Birge, R. R. (1985). Two-photon spectroscopy of protein-bound chromophores. *Acc. Chem. Res.* **19**, 138–146.

14. Kennedy, S. M., and Lytle, F. E. (1986). p-bis(o-Methylstyryl)benzene as a power squared sensor for two-photon absorption measurements between 537 and 694 nm. *Anal. Chem.* **58**, 2643–2647.

15. Jiang, S. (1989). Two-photon spectroscopy of biomolecules. *Prog. React. Kinet.* **15**, 77–92.

16. Monson, P. R., and McClain, W. M. (1970). Polarization dependence of the two-photon absorption of tumbling molecules with application to liquid 1-chloronaphthalene and benzene. *J. Chem. Phys.* **53**, 29–37.

17. Jones, R. D., and Callis, P. R. (1988). A power-squared sensor for two-photon spectroscopy and dispersion of second-order coherence. *J. Appl. Phys.* **64**, 4302–4305.

18. Wilson, T., and Sheppard, C. (1984). *Theory and Practice of Scanning Optical Microscopy*. Academic Press, New York.

19. Sheppard, C. J. R., and Gu, M. (1990). Image formation in two-photon fluo-rescence microscopy. *Optik* **86**, 104–106.

20. Nakamura, O. (1992). Three-dimensional imaging characteristic of laser scan

fluorescence microscopy: Two-photon excitation vs. single-photon excitation. *Optik* **93**, 39–42.

21. Gratton, E., and Limkeman, M. (1983). A continuously variable frequency cross-correlation phase fluorometer with picosecond resolution. *Biophys. J.* **44**, 315–324.

22. French, T., Gratton E., and Maier, J. (1992). Frequency domain imaging of thick tissues using a CCD. *Proc. SPIE—Int. Soc. Opt. Eng.* **1640**, 220–229.

23. Mantulin, W. W., French, T., and Gratton, E. (1993). Optical imaging in the frequency domain. *Proc. SPIE—Int. Soc. Opt. Eng.* **1892**, 158–166.

24. So, P. T. C., French, T., and Gratton, E. (1994). A frequency domain time-resolved microscope using a fast-scan CCD camera. *Proc. SPIE—Int. Soc. Opt. Eng.* **2137**, 83–92.

25. Marriott, G., Clegg, R. M., Arndt-Jovin, D. J., and Jovin, T. M. (1991). Time resolved imaging microscopy. *Biophys. J.* **60**, 1347–1387.

26. Lakowicz, J. R., Szmacinski, H., and Johnson, M. L. (1992). Calcium imaging using fluorescence lifetimes and long-wavelength probes. *J. Fluoresc.* **2**, 47–62.

27. Lakowicz, J. R., Szmacinski, H., Nowaczyk, K., and Johnson, M. L. (1992). Fluorescence lifetime imaging of calcium using Quin-2. *Cell Calcium* **13**, 131–147.

28. Morgan, C. G., Mitchell, A. C., and Murray, J. G. (1991). Prospects for confocal imaging based on nanosecond fluorescence decay time. *J. of Microsc. (Oxford)* **165**, 49–60.

29. So, P. T. C., Dong, C. Y., Berland, K. M., French, T., and Gratton, E. (1994). A two-photon confocal lifetime microscope. *Biophys. J.* **66**, 276a.

30. Betzig, E., and Chichester, R. J. (1993). Single molecules observed by near-field scanning optical microscopy. *Science* **262**, 1422–1425.

31. Xie, X. S., and Dunn, R. C. (1994). Probing single molecule dynamics. *Science* **265**, 361–364.

32. Ambrose, W. P., Goodwin, P. M., Martin, J. C., and Keller, R. A. (1994). Alternation of single molecular fluorescence lifetimes in near-field optical microscopy. *Science* **265**, 364–367.

33. Stelzer, E. H. K. (1989). The intermediate optical system in confocal microscopes. In *Handbook of Biological Confocal Microscopy* (J. Pawley, Ed.), pp. 83–91. IMR Press, Madison, WI.

34. Alcala, J. R., Gratton, E., and Jameson, D. M. (1985) A multi frequency phase flurometer using the harmonic content of mode-locked laser. *Anal. Instrum.* **14**, 225–250.

35. Feddersen, B. A. (1993), Digital frequency domain fluorometry and the study of Hoechst 33258 dye-DNA interactions. Ph.D. Thesis, University of Illinois at Urbana-Champaign.

36. Carrington, W. A., Fogarty, K. E., Lifschitz, L., and Fay, F. S. (1989). Three-dimensional imaging on confocal and wide-field microscopes. In *Handbook of Biological Confocal Microscopy* (J. Pawley, Ed.), pp. 137–147. IMR Press, Madison, WI.

37. Parasassi, T., Stefano, M. D., Loiero, M., Ravagnan, G., and Gratton, E. (194). Influence of cholesterol on phospholipid bilayers phase domains as detected by Laurdan fluorescence. *Biophys. J.* **66**, 120–132.

38. Parasassi, T., Ravagnan, G., Rusch, R., and Gratton, E. (1993). Modulation and dynamics of phase properties in phospholipid mixtures detected by Laurdan fluorescence. *Photochem. Photobiol.* **57**, 403–410.

39. Parasassi, T., Stefano, M. D., Ravagnan, G., Sapora, O., and Gratton, E. (1992). Membrane aging during cell growth ascertained by Laurdan generalized polarization. *Exp. Cell Res.* **202**, 432–439.

40. Tsien, R. Y., and Poenie, M. (1986). Fluorescence ratio imaging: A new window into intracellular ionic signaling. *Trends Biochem. Sci.* **11**, 450–455.

41. Piston, D. W., and Webb, W. W. (1991). Three dimensional imaging of intracellular calcium activity, using two-photon excitation of the fluorescent indicator dye INDO-II in laser scanning microscopy. *Biophys. J.* **59**, 156a.

42. Eberhard, M., and Erne, P. (1991). Calcium binding to fluorescent calcium indicators: Calcium Green, Calcium Orange and Calcium Crimson. *Biochem. Biophys. Res. Commun.* **180**, 209–215.

43. Grynkiewicz, G., Poenie, M., and Tsien, R. Y. (1985). A new Generation of Ca^{2+} indicators with greatly improved fluorescence properties. *J. Biol. Chem.* **260**, 3440–3450.

44. Minta, A., Kao, J. P. Y., and Tsien, R. Y. (1989). Fluorescent indicators for cytosolic calcium based on rhodamine and fluorescein chromophores. *J. Biol. Chem.* **264**, 8171–8178.

45. Rink, T. J., Tsien, R. Y., and Pozzan, T. (1982). Cytoplasmic pH and free Mg^{2+} in lymphocytes. *J. Cell Biol.* **95**, 189–196.

46. Carrero, J., French, T., and Gratton, E. (1992). Oxygen imaging in tissues. *Biophys. J.* **59**, 167a.

47. Berlin, J. R., and Konishi, M. (1993). Ca^{2+} transients in cardiac myocytes measured with high and low affinity Ca^{2+} indicators. *Biophys. J.* **65**, 1632–1647.

48. Olivera, A., Zhang, H., Carlson, R. O., Mattie, M. E., Schmidt, R. R., and Spiegel, S. (1994). Stereospecificity of sphingosine-induced intracellular calcium mobilization and cellular proliferation. *J. Biol. Chem.* **269**, 17924–17930.

49. Gonzalez-Gronow, M., Gawdi, G., and Pizzo, S. V. (1993). Plasminogen activation stimulates an increase in intracellular calcium in human synovial fibroblasts. *J. Biol. Chem.* **268**, 20791–20795.

CHAPTER

12

LASER MANIPULATION AND FLUORESCENCE SPECTROSCOPY OF MICROPARTICLES

KEIJI SASAKI and HIROSHI MASUHARA

Department of Applied Physics
Osaka University
Osaka 565, Japan

12.1. INTRODUCTION

Micrometer-sized particles such as polymer latexes, liquid droplets, microcapsules, microcrystals, catalysts, and colloids have generated much interest in the fields of photophysics and photochemistry owing to their characteristic properties, which are different from those of bulk material (1). Since the surface area-to-volume ratio of a micrometer particle is 10,000 times larger than that of a centimeter of matter, the contribution of the surface and interface to the physical/chemical properties increases for the microparticle (2). Phenomena peculiar to surface/interface such as adsorption, adhesion, creation of electric double layers, surface tension, and mass transfer across interfaces sometimes dominate the nature of microparticles. In addition, the diffusional motion of reactive molecules in particles is confined within micrometer volumes; therefore, the reaction dynamics may deviate from those in a free space. Besides their physical/chemical characteristics, microspherical particle possess a unique optical property of acting as optical cavities in which intense resonant field are created along the spherical surfaces (3). The enhancement of the electromagnetic fields is also expected to influence the photophysical and photochemical processes occurring in microparticles.

Various physical/chemical analyses on microparticles in solution have been performed; however, most of them have been based on the measurement of an ensemble of the particles. The obtained results were therefore always the sum or the average of those for a number of microparticles, which concealed the characteristic properties of individual particles. Even if an optical microscope

Fluorescence Imaging Spectroscopy and Microscopy, edited by Xue Feng Wang and Brian Herman. Chemical Analysis Series, Vol. 137.
ISBN 0-471-01527-X © 1996 John Wiley & Sons, Inc.

is employed for the measurement of a single microparticle, its vigorous Brownian motion makes selective observation difficult. To obtain a clear picture of the physics and chemistry of microparticles, advanced technologies for controlling Brownian motion of particles in solution and for measuring individual particles are indispensable.

Optical trapping is one of the potential candidates for noncontact and nondestructive manipulation of a single microparticle. This technique is based on radiation pressure induced by refraction, reflection, and absorption of light in/on particles. This force can be theoretically derived from Maxwell's equations, and the result shows that the radiation pressure is extremely weak, i.e., on the order of pico-Newtons (pN), which is comparable to the force detected by an atomic force microscope. Although macroscopic objects cannot be moved by such weak forces, the motion of microparticles is appreciably influenced by the radiation pressure. The gravity force and the viscous drag exerted on a micrometer-sized particle are much smaller than the radiation pressure. Hence, one can observe a small object being levitated and transferred by irradiation of light under a microscope.

Optical trapping of a microparticle, based on radiation pressure, was first demonstrated by Ashkin in 1970 (4). In his experiment, a polystyrene latex particle was placed between two opposing laser beams. The Gaussian beam attracts a particle towards the high-intensity region at the center of the beam and pushes it in the direction of beam propagation, so that the particle is trapped in stable equilibrium at the point of symmetry of the opposing beams. Ashkin's group also succeeded in levitating a particle by a single vertically directed laser beam, just like lifting a ball by a fountain of water (5). Since the upward radiation pressure exerted on the levitated particle should be balanced by the gravity force on the particle, the radiation pressure can be precisely estimated based on this experiment. Indeed, Ashkin et al. observed the wavelength dependence of the levitation force and clarified the relationship between the radiation pressure and optical resonances in a particle (6). In 1986, they proposed a single-beam gradient force trapping method in which a laser beam was focused on a particle (7). The radiation pressure is directed to the focal point of a laser beam, so that the particle is three-dimensionally trapped in the vicinity of the focused beam spot. Conceptually and practically, this method is one of the simplest and most flexible optical trapping technique (8–10).

To apply the trapping technique to a wide variety of physical and chemical studies, all kinds of microparticles are required to be trapped; in addition, simultaneous manipulation of more than one particle is expected to open up a new field of science and technology. A particle to be trapped by Ashkin's method, however, has to be transparent at the wavelength of laser light, and its refractive index must be higher than that of the surrounding medium. Furthermore, the conventional trapping technique is essentially limited to single-particle manipulation. Although the number of manipulated particles can be increased by installing multiple laser beams, instrumental restriction determines the maximum number of beams.

In this chapter, we described a new micromanipulation technique that enables us to optically trap and manipulate a low refractive index droplet or a metal particle. Simultaneous trapping and spatial patterning of several particles are also made possible by this technique. In addition, by combining the trapping technique with picosecond time-resolved fluorescence spectroscopy and confocal fluorescence microscopy, photophysical and photochemical phenomena occurring in individual microparticles can be elucidated. Details of the laser micromanipulation and microspectroscopy system and its applications to particle analyses are introduced in this chapter.

12.2. LASER TRAPPING OF A SINGLE MICROPARTICLE

12.2.1. Radiation Pressure

A single photon with wavelength λ and frequency v possesses momentum given as h/λ (h: Planck's constant), as well as the energy of hv. When a photon flux is refracted or reflected by matter, the momentum of each photon is changed in direction and magnitude. Absorption of matter removes both the momentum and energy from a photon. Based on the conservation law of momentum, photon momentum is transferred from the incident light to the matter, which causes the force on the matter. This force is called radiation pressure.

The radiation pressure exerted on a microparticle under the irradiation of a focused laser beam is schematically illustrated in Figure 12.1. The laser light incident on a particle is refracted twice on the boundaries when it enters and exits the particles. Besides propagation direction, the wavelength is changed in the different refractive index medium resulting in photon momentum change. As shown in Figure 12.1, the difference between the momentums before and after refraction, i.e., $\Delta P = P_1 - P_0$, is transferred to the particle, so that the radiation pressure is exerted in the direction opposite to that of ΔP (i.e., $-\Delta P$). If the refractive index of the particle (n_p) has no imaginary part (no absorption) and a larger real part than the surrounding medium (n_m), the sum of the force at each point of the particle is directed to the high-intensity region at the focal point. Hence, the particle is attracted to the focused beam and then three-dimensionally trapped in the vicinity of the focal point against the thermal Brownian motion, gravity, and convection.

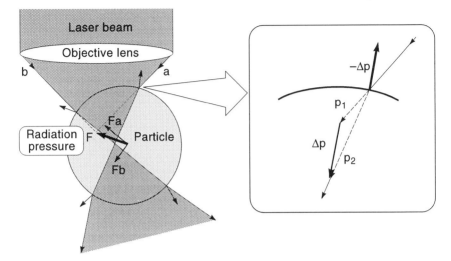

Figure 12.1. Radiation pressure exerted on a microparticle by a focused laser beam.

This explanation of optical trapping is based on geometrical optics, which can be easily understood with the photon momentum change. The theory, however, cannot be applied to a particle whose size is small compared to its wavelength. For such a particle, the path of beam propagation within the particle cannot be represented as a ray, and the focal spot is not a point but rather a wavelength-sized spot. Hence, the use of wave optics is indispensable for understanding the radiation pressure exerted on small particles.

According to the Rayleigh scattering theory, a particle whose diameter is much smaller than its wavelength works as a single electric dipole. The dipole experience Lorenz's force induced by the optical electromagnetic field. This force corresponds to the radiation pressure, which can be theoretically expressed as

$$F = \frac{1}{2}\alpha\Delta E^2 + \alpha\frac{\partial}{\partial t}(E \times B) \tag{1}$$

where E and B are the electric field and the magnetic flux density, respectively, and Δ represents the gradient operator with respect to the spatial coordinates; α is the polarizability of a particle, which is given by

$$\alpha = r^3\frac{(n_p/n_m)^2 + 1}{(n_p/n_m)^2 + 2} \tag{2}$$

where r is the radius of a particle. The first term of Eq. (1) is an electrostatic force acting on the dipole in the inhomogeneous electric field, which is called gradient force. When $n_p > n_m$, polarizability α is given as a positive value, so that the gradient force is directed to the higher electric field intensity region. The second term is derived from the change in the direction of the pointing vector, which is called scattering force. Since the gradient force is usually much stronger than the scattering force, the radiation pressure attracts the particle to the high intensity region, which is the same phenomenon operating in the large particle. Hence, any particles with an arbitrary size can be trapped at a focal spot of a laser beam under the condition $n_p > n_m$.

12.2.2. Single-Particle Manipulation

Figure 12.2. shows laser trapping of a polystyrene (PSt) latex particle (4 μm) in ethylene glycol (11). A CW Nd:YAG laser (54 mW) was focused on a particle (indicated by the arrows in Figure 12.2) perpendicular to the plane of the photograph. Another particle was transferred by moving a microscope stage in the z direction, resulting in the blur of the contours (Figure 12.2b), while the irradiated particle was always in focus. In addition, the particle was fixed at the same position when the stage was scanned in the lateral (x and y) directions even when another particle collided with the trapped particle.

Laser trapping of a single microparticle such as a poly(methyl methacryrate) (PMMA) particle, a toluene droplet, a liquid–paraffin droplet, a melamine–resin microcapsule containing a toluene solution of pyrene, and a swollen micelle in water was also successful. PSt and PMMA latex particles could be captured in ethylene glycol and diethylene glycol. Furthermore, nonspherical particles such as titanium dioxide (needle-like). *Salmonella typhymurium* (ellip-

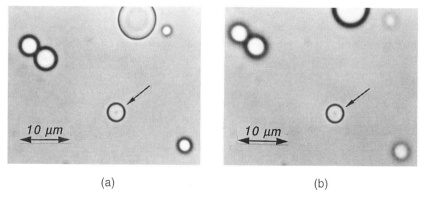

        (a)   (b)

Figure 12.2. Laser trapping of a PSt latex particle (4 μm) in water. A sample stage was moved in the z direction (a and b).

Table 12.1. Optical Trapping of Various Microparticles in Water and Alcohols

Particle	Medium
Poly(methyl methacryrate) latex	Water, ethylene glycol, diethylene glycol
Polystyrene latex	Water, ethylene glycol, diethylene glycol
Toluene droplet	Water
Benzyl alcohol droplet	Water
cis-Decalin droplet	Water
Silicon oil droplet	Water
Liquid paraffin droplet	Water
Silica gel	Water
Glass bead	Water
Titanium dioxide	Water, ethylene glycol
Salmonella typhymurium	Water
Calf thymus DNA	Water
Water droplet	Liquid paraffin
Ethylene glycol droplet	Liquid paraffin
Iron particle	Water
Aluminum particle	Water
Carbon black	Water

soid), and calf thymus DNA (rod-like) were optically trapped in water. The results are summarized in Table 12.1 (11). Laser trapping of transparent dielectric particles was always successful in relatively low refractive index media. Ashkin et al. showed that theoretically the minimum particle size that can be trapped with a laser power of 1.5 W at room temperature is 14 nm. In fact, trapping of a 26 nm PSt latex particle in water has been reported (7).

12.2.3.   Measurement of Radiation Pressure

An optically trapped particle experiences viscous flow in a medium (i.e., viscous drag) when a sample stage is driven at constant velocity. With increasing velocity, the particle is slightly shifted, so that the trapping force increases to keep balance with the viscous drag. When the viscous drag overcomes the maximum radiation pressure, the particle is released from the trapping. Thus, we could quantitatively determine the magnitude of radiation pressure by measuring the maximum velocity (v_0) at which a particle is detrapped. The viscous drag (F) exerted on a particle at the flow velocity v_0 is given by Stokes law:

$$F = 6\pi\eta r v_0 \tag{3}$$

where η is the viscosity of the medium and r is the radius of the particle.

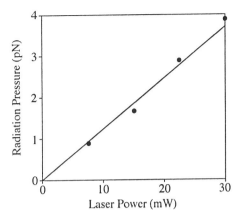

Figure 12.3. Trapping force in the z direction as a function of laser power.

Figure 12.3 shows the laser power dependence of the trapping force in the z direction. The laser beam irradiated on a PMMA particle (6.8 μm) in ethylene glycol ($\eta = 17.3$ cP at 20 °C) and a sample stage was scanned by a piezo actuator under a microscope. The maximum radiation pressure was on the order of pN and was proportional to the laser power, which can be explained by the fact that the number of photons interacting with a particle determines the trapping force. Figure 12.4 is a plot of trapping force as a function of particle size. The radiation pressure strengthened as the diameter increased, that is, the larger particle was trapped in the deeper potential well. Besides laser power and particle size, refractive indices of particles and media,

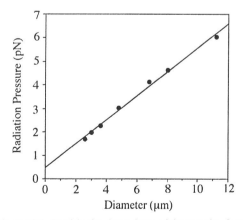

Figure 12.4. Particle size dependence of the trapping force.

optical conditions such as numerical aperture (NA) and beam profile are important factors in determining radiation pressure.

Theoretical comparison of the trapping potential and thermal energy of a particle indicates that a 10 µm particle can be trapped against Brownian motion by a 30 mW laser beam even when the temperature is 10^6 K. Hence, the pN force is sufficiently strong to trap micrometer-sized particles at a temperature below the damage threshold of particles. At room temperature, the walking distance of a thermally moving particle in the trapping potential well is ~ 3.5 nm, which can be reduced in inverse proportion to the square root of the laser power. Thus, laser trapping can provide nanometer positioning control of a single microparticle.

12.3. OPTICAL TRAPPING OF METAL
AND LOW REFRACTIVE INDEX PARTICLES

The relationship between the refractive indices of a particle (n_p) and the medium (n_m) is important in laser manipulation. As mentioned in the previous section, the gradient force trapping method is based on attractive radiation pressure, which is obtained only for a transparent particle with $n_p > n_m$. When $n_p < n_m$, the situation is reversed, so that the direction of the radiation pressure

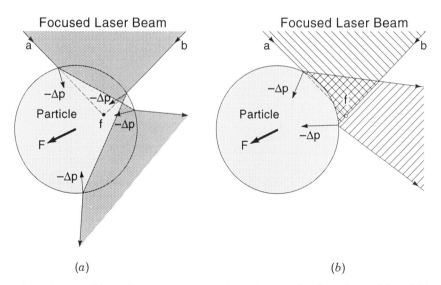

(a) (b)

Figure 12.5. Repulsive radiation pressure exerted on (a) a low refractive index particle and (b) a highly reflective particle.

is opposite to that of the laser beam, as shown in Figure 12.5a. Hence, the particle experiences repulsive force. Wave optics also indicates that the polarizability α given by Eq. (2) is negative when $n_p < n_m$, so that the particle is repelled by the laser beam towards the lower intensity region. For simplicity, we shall call such a force *repulsive radiation pressure*. For example, a water droplet ($n_p = 1.33$) in liquid paraffin ($n_m = 1.46$–1.47) cannot be optically trapped by a single focused laser beam. Another class of particles that cannot be trapped by the conventional technique is those with a high reflection coefficient at the wavelength of an incident laser beam. Simple geometrical optics in Figure 12.5b indicate that such particles experience repulsive radiation pressure analogous to the pressure of particles with $n_p < n_m$. Hence, metal particles are pushed to the outside of the laser beam. [The exception is particles much smaller than skin depth (nanometer order); such particles can be categorized as transparent scattering objects.] Although laser trapping of micrometer-sized, low refractive index droplets and highly reflective particles is required for chemical applications, a single beam trapping technique is not applicable to these particles.

To trap such particles, we have proposed a scanning laser manipulation technique (12). The principle of this technique is explained by the trapping potential shown in Figure 12.6. The potential of the radiation pressure exerted

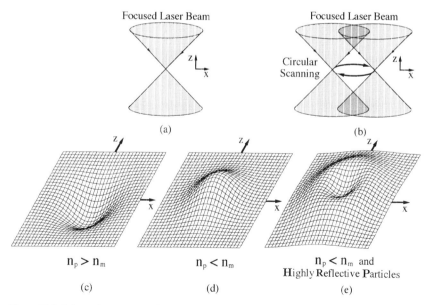

Focused Laser Beam   Focused Laser Beam

                            Circular
                            Scanning

(a)   (b)

$n_p > n_m$ $n_p < n_m$ $n_p < n_m$ and
                                          Highly Reflective Particles
(c)   (d)   (e)

Figure 12.6 Schematic representation of the spatial intensity distributions of a focused laser without and with beam scanning (a and b), and the relevant potentials of the radiation pressure (c–e).

on a microparticle with $n_p > n_m$ is shown in Figure 12.6c, which has the potential well at the focal spot. This indicates that the particle is attracted to and trapped in this potential well, as explained above. On the other hand, in the case of $n_p < n_m$ or a high reflective particle (Figure 12.6d), there is no potential well for trapping, so the particle is pushed out by the repulsive force. The key to our technique is the repetitive scanning of a focused laser beam in a sample space. If the repetition rate of the scanning is much higher than the cutoff frequency for the mechanical response of a particle in the medium, the particle will experience the radiation pressure given by the time-averaged intensity distribution. When a focused laser beam is scanned circularly around a particle with $n_p < n_m$ or with a high reflection coefficient, as shown in Figure 12.6b, the potential of the radiation pressure calculated by the averaged intensity distribution can be shown, as in Figure 12.6e. This demonstrates that the circular scanning of a focused laser beam produces a relatively high, circular potential barrier. A particle that experiences repulsive radiation pressure is expected to be caged inside this potential barrier if the repetition rate of the beam scanning is faster than the mechanical motion of the particle. In addition, the focused laser beam produces high-intensity regions over and under the particle, so that the potential barrier surrounds the particle three-dimensionally. Thus, the particle will be trapped at the position where the repulsive radiation pressure is balanced in all directions.

Figure 12.7 shows a schematic diagram of the scanning manipulation system. A trapping beam of a CW Nd:YAG laser was spatially modulated by

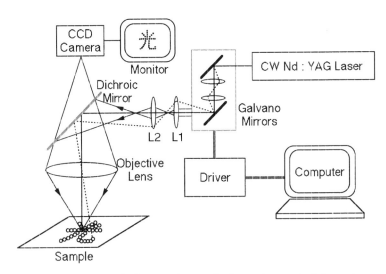

Figure 12.7. A schematic diagram of a scanning laser micromanipulation system.

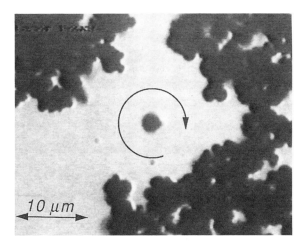

Figure 12.8. Optical trapping of an iron particle (3 μm) in water. The circle shown in the figure represents the locus of the scanning laser beam.

two galvano mirrors, which were operated by a driver and controlled by a microcomputer. The modulated laser beam was introduced into a microscope through lenses L1 and L2, which matched the beam diameter to the numerical aperture of the microscope and imaged the galvano mirror surfaces in the plane of an aperture diaphragm. In the microscope, the laser beam was focused on a ∼ 1 μm spot on a sample by an oil-immersion objective lens (NA = 1.30). The micromanipulation process was monitored and recorded by a change-coupled device (CCD) camera and a video recording system.

Figure 12.8 shows optical trapping of an iron particle (3 μm) in water (12). A focused beam of CW Nd: YAG laser (145 mW) was scanned along the circle shown in the figure. The particle inside the laser cage was optically trapped, and it could be manipulated by moving the position of the cage. Since repulsive radiation pressure is also exerted on particles outside the cage, the particles neither enter nor come close to the cage, that is, stable trapping of a single particle is achieved. For conventional trapping of a particle with $n_p > n_m$, prolonged irradiation of a focused beam frequently induces the aggregation of trapped particles.

A further sophisticated experiment involved laser manipulation of water ($n_p = 1.33$) containing red dye (eosin) and ethylene glycol (EG; $n_p = 1.43$) droplets in liquid paraffin ($n_m = 1.46$–1.47) (12). The laser beam was separated into two beams by a beam splitter (320 mW for each beam), and these beams were scanned independently by using two sets of galvano mirrors. As clearly

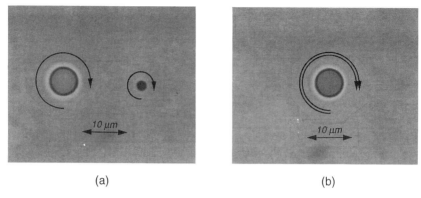

(a)   (b)

Figure 12.9. Optical trapping of low refractive index particles. (a) Water and ethylene glycol droplets were manipulated by two laser cages and (b) were fused into one droplet.

seen in Figure 12.9a, double-beam scanning manipulation succeeded in trapping the dye water and EG droplets independently. Three-dimensional manipulation of each droplet was also attained by controlling the galvano mirrors and/or the optics outside the microscope. The dye water and EG droplets were manipulated in the lateral direction and fused into one droplet by causing the positions of the two laser cages to coincide (Figure 12.9b). The dye–water droplet was diluted by the EG droplet, and the color of the droplet turned to pale red.

Besides the scanning laser manipulation, a TEM_{01}^* mode laser beam, which has the intensity minimum on the beam axis, also provides the potential well for a particle with $n_p < n_m$ or with a high reflection coefficient, similarly to Figure 12.6e. Indeed, Ashkin et al. reported the levitation of a hollow dielectric sphere by the TEM_{01}^* beam (13). However, the potential well given by the TEM_{01}^* beam is only two-dimensional, that is, there is no well in the z direction. Hence, three-dimensional trapping cannot be performed with the TEM_{01}^* beam. Furthermore, TEM_{01}^* beam trapping is restricted by the size and/or shape of particles. On the other hand, the laser caging technique can be applied to nonspherical particles by adapting the scanning pattern and its size, which can be easily controlled by a computer.

By means of this technique, laser caging of iron, aluminum, and carbon black particles in water has been demonstrated. Optical trapping techniques based on attractive and repulsive radiation pressures are complementary, and any microparticles can be optically manipulated by scanning laser trapping or conventional trapping, depending on the nature of the particle and the surrounding medium.

12.4. SPATIAL PATTERNING BY SCANNING MICROMANIPULATION

The optical trapping techniques have been widely used as the noncontact and nondestructive positioning methods for single-particle manipulation. On the other hand, we demonstrated that several particles are simultaneously trapped at the positions of intensity maxima of the standing wave field formed by a laser beam (14). The particles can be aligned on concentric circles or line patterns by adjusting an interference optical system. This technique has the ability to organize new functional materials and systems composed of various reactive particles. The spatial patterns produced by the interference method are, however, essentially limited to simple fringe patterns. Another possible approach is the use of photomask, as is widely applied in photolithographic technology. Unfortunately, the projected image is usually degraded by speckle and/or unexpected interference fringes due to the high coherence of the laser. Furthermore, since the laser beam is greatly attenuated by a photomask, a high-power laser is required to achieve spatial patterning, which is likely to damage the microscope optics and the mask.

In a new approach to manipulation of several particles, we applied the scanning laser manipulation technique, which makes it possible to align microparticles on arbitrary patterns without coherent noises and loss of laser power (15). When a focused laser beam is repetitively scanned in a sample space, the transparent particles with $n_p > n_m$ are attracted to the beam spot but they cannot follow the scanning beam, so that these particles can be simultaneously trapped and aligned along a path defined by the scanning of a laser beam.

Figure 12.10 demonstrates spatial patterning of PSt latex particles (1 μm) in EG along an italic-like "μm" (15). The sample solution containing the particles was placed between two quartz plates separated by a spacer of 100 μm and mounted on the microscope stage. A focal spot was scanned along the pattern at the bottom of the liquid layer with a repetition rate of 13 Hz. The spatial pattern of "μm" was formed by ~ 100 particles under beam irradiation of 145 mW power, that is, a ~ 1.5 mW/particle. When the laser was switched on, particles were rarely observed in the ocular field. After several tenths of a second, the radiation pressure successfully attracted the dispersed particles and created the spatial pattern on the quartz plate. The formed pattern could be transported in the lateral and longitudinal directions without deformation of the pattern. After the laser was switched off, the particles immediately disappeared from the pattern.

Figure 12.11 shows spatial alignment of titanium dioxide particles (~ 0.5 μm) in EG in the geometrical figure of a star. The laser power and repetition rate of the beam scan were 145 mW and 24 Hz, respectively. Since thermal motion depends on the size of the particles and the viscosity of the

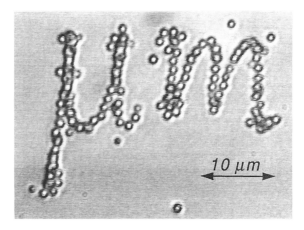

Figure 12.10. Spatial patterning of PSt latex particles (1μm) in ethylene glycol along the italic-like "μm."

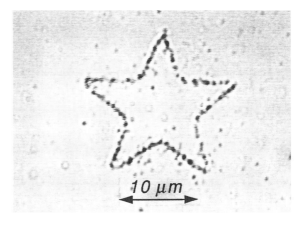

Figure 12.11. Scanning micromanipulation of titanium dioxide (~ 0.5 μm) in ethylene glycol along the star.

medium, the repetition rate was optimized for the given particles and medium, as well as for the complexity of the pattern to be produced.

Scanning laser manipulation is novel since arbitrary spatial pattern of particles can be produced by a single laser beam. In addition, the pattern can be easily constructed and distracted by switching the laser on or off, respectively, and is continuously varied by programming the sequence. Furthermore, the pattern formation is based on incoherent imaging, as in confocal scanning

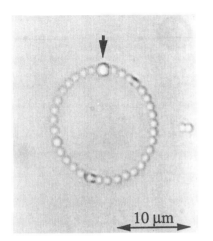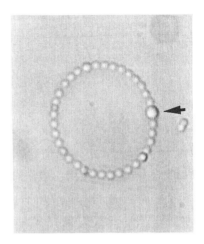

Figure 12.12. Optical transport of PSt latex particles (1 μm) along the circle in 1-pentanol.

microscopy, so that the present technique can be further extended to three-dimensional patterning of particles. Indeed, we succeeded in creating the spatial pattern of particles at a specific distance from the quartz plate, that is, levitation of the aligned particles.

In addition to pattern formation, the scanning micromanipulation technique has the ability to transport all the trapped particles along the produced pattern at constant velocity (16). This is based on the fact that careful adjustment of the experimental conditions results in a slight driving force, in addition to the trapping force, on particles, Since the driving force can be varied by the scan speed and laser power, the velocity of particle flow is controllable. The motion of particles can be expressed as Newton's equation with the coefficients of viscosity and friction.

Figure 12.12 illustrates circular patterning of PSt latex particles (1 μm) in 1-pentanol ($n_m = 1.41$) (16). In this experiment, the particles were not trapped at fixed positions but were moved on the quartz plate, forming the circular pattern. Figure 12.12 shows sequential images recorded at intervals of 0.6 s. The circle with a diameter of 13.4 μm was created by repetitive scanning of the focused beam at the repetition rate of 15 Hz in the right-handed rotation. The slightly larger particle, marked by an arrow, followed the same direction. All the particles moved together in an orderly fashion at a flow velocity of 12.2 μm/s, and the scan speed of the laser beam (120 mW) was 642 μm/s.

Figure 12.13 shows the velocity of particle flow as a function of scan speed (16). It is worth noting that particle flow becomes slower as scan speed is increased. Since the mechanical response function of the particles in viscous

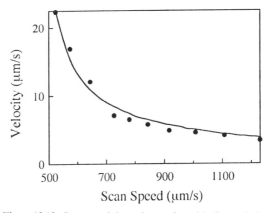

Figure 12.13. Scan speed dependence of particle flow velocity.

solution is lower in the high-frequency region, the driving force also decreases in high-speed scanning. The solid curve in Figure 12.13 was calculated from Newton's equation, which fitted well with the experimental data. From this fitting, we could estimate the friction coefficient between the particles and the quartz plate, which is useful for analyzing mechanical characteristics in a micrometer-sized space.

12.5. FLUORESCENCE SPECTROSCOPY OF A MANIPULATED MICROPARTICLE

Time-resolved fluorescence spectroscopy with the use of short-pulse lasers and high-speed detectors is a powerful method of elucidating photophysical and photochemical phenomena, such as excitation energy relaxation, electron as well as proton transfer, molecular vibrational relaxation, and isomerization. Its temporal resolution has improved considerably, from microsecond to femtosecond. By combining laser micromanipulation with time-resolved fluorescence spectroscopy, the dynamic processes occurring in/on individual microparticles can be observed.

We have developed a space- and time-resolved fluorescence spectroscopy and laser manipulation system (17) based on the confocal fluorescence microscope and single-photon timing, as well as the focused beam–trapping technique. The confocal fluorescence microscope is composed of point-excitation and point-detection systems combined with a scanning mechanism (18). The sample or laser beam is mechanically or optically scanned, so that fluorescence is measured point by point. The detected signal is sequentially processed on a digital or analog computer to form the images. In contrast to conventional fluorescence microscopes, the confocal one provides three-dimensional imaging, that is, longitudinal (depth) resolution can be obtained in addition to lateral resolution. This three-dimensional resolution is indispensable for the measurement of internal structures and/or surface properties of individual microparticles. Furthermore, the confocal microscope provides clear images free from the speckle and interference fringe caused by coherence of the laser. Flare and scattered light are also negligibly weak due to the use of the point detector. On the other hand, single-photon timing has the advantages of high sensitivity and wide dynamic range (19) compared with the frequency up-conversion technique (20) and the streak camera (21) that have been used widely for time-resolved fluorescence spectroscopy. These advantages of single-photon timing are appropriate for weak fluorescence measurements in the application of microscopic spectroscopy.

Figure 12.14 is a schematic diagram of the three-dimensional space- and time-resolved fluorescence spectroscopy and laser manipulation system (17). A cavity-dumped dye laser, synchronously pumped by the second harmonics of a CW mode-locked Nd:YLF laser, was used as a light source. Its wavelength was tunable from 560 to 620 nm by using a rhodamine 6G dye. The pulse width and repetition rate were 2 ps (FWHM) and 3.8 MHz, respectively. An ultraviolet pulse (280 ~ 310 nm) produced by a second harmonic generator was condensed by a lens and introduced into a pinhole P1 of a microscope. A zoom lens was used to match the beam diameter with the numerical aperture

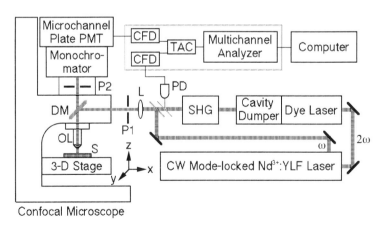

Figure 12.14. A schematic diagram of a three-dimensional space- and time-resolved fluorescence spectroscopy and manipulation system. L, lens; P1 and P2, pinhole; DM, dichroic mirror; OL, objective lens; S, sample; PD, photodiode; CFD, constant-fraction discriminator; TAC, time-to-amplitude converter.

of the microscope. In the microscope, the laser light was reflected by a dichroic mirror and focused onto a sample by an oil-immersion objective lens. This objective was made of quartz for ultraviolet excitation; its magnification and NA were 100 and 1.25, respectively. Fluorescence emitted from the sample was collected by the same objective lens and imaged on a pinhole P2. Its diameter of 40 µm corresponds to 0.25 µm on the sample, as the magnification was 160. Since fluorescence from the focal spot is condensed on the second pinhole, most of its energy goes through, while fluorescence from out-of-focus positions is defocused on the pinhole plate, as shown by the dotted lines in Figure 12.14, so that most of its energy is cut off. Therefore, the observed fluorescence is ascribed to the three-dimensionally minute volume. The sample stage was driven with steps of 0.25 µm in the lateral direction and 0.1 µm in the depth direction for measuring the three-dimensional structure. Fluorescence passing the pinhole P2 was spectrally resolved by a monochromator and detected by a microchannel-plate photomultiplier. The output signals of the photomultiplier and a PIN photodiode detecting the excitation pulse were processed by constant fraction discriminators. Their outputs were time correlated by a time-to-amplitude converter and processed by a multichannel pulse-height analyzer. A microcomputer controlled the sample stage, the monochromator, and the single-photon timing apparatus, and the data analyses were performed with a workstation.

The performance of the system, which was specified by measuring thin liquid layers of pyrene and rhodamine B, provides three-dimensional space resolutions of 0.3 µm (lateral) and 0.5 µm (depth) and a temporal resolution of 2 ps [full width at half-maximum (FWHM) of the instrumental response function, 33 ps], in addition to spectral resolution of 1 nm (the observable wavelength range, 300 ∼ 1000 nm) (17).

The present fluorescence spectroscopy system can be extended to space- and time-resolved fluorescence depolarization measurement in a selected small volume (22), which is useful for the analysis of rotational relaxation processes of molecules in inhomogeneous systems. For polarization spectroscopy, the excitation laser beam is passed through a Babinet–Soleil compensator to adjust the polarization direction. The polarized fluorescence is selected by an analyzer and passed through a depolarizer, both of which are set in front of the monochromator. By setting the excitation polarization parallel and perpendicular to that of the analyzer, two respective decay curves are measured in each small volume. Unfortunately, the conventional method for calculating the anisotropy decay from the two decay curves is not applicable to the present system because the excitation light is far from the plane wave due to strong condensation by the high numerical aperture objective lens. The fluorescence from the sample is also collected at a large solid angle, so that various directions of polarization are mixed together on the detector plane. In addition, chromatic and spherical aberrations of the objective lens and the optical imperfection of the polarizer and analyzer lead to deviations of the experimental data from ideal values. To solve these problems, we have derived a practical theory of fluorescence depolarization analysis for microscopic measurement (22). In this theory, the changes in polarization caused by the microscope optics are represented as linear equations with two parameters. The parameters can be evaluated experimentally and the estimates used to calculate precise anisotropy decays. Since the parameters are characteristic of the developed system, their estimates can be used in the anisotropy analysis for any sample.

Here some problems with the fluorescence microspectroscopy system are considered. The most serious problem with microspectroscopy is chromatic aberration. Since the focal length of an objective lens depends on the wavelength, the depth coordinates on fluorescence images observed at different wavelengths are shifted away from each other. Therefore, chromatic aberration causes distortion of spectra obtained at small volumes. In the developed system, chromatic aberration is automatically compensated for by varying the position of both the sample stage and the zoom lens as a function of wavelength (17). The other problem is sample damage caused by the high-intensity excitation. The laser beam is condensed onto the submicrometer spot, so that the excitation pulse energy often rises to over tens of mJ/cm^2. Such an intense light may damage most molecular materials. Therefore, the intensity of the excitation laser has to be attenuated and adjusted to be kept below the damage threshold of the sample. Under the latter condition, the observed fluorescence usually becomes weak. Fortunately, the system has the advantage of high detection efficiency due to the high numerical aperture of the objective lens, which overcomes the weakness of the fluorescence intensity.

One application of the system, the quantitative concentration estimation of dye molecules in small volumes, is described (17). Figure 12.15 shows depth-resolved fluorescence curves of PMMA films. Pyrene fluorescence at 385 nm was observed along the longitudinal axis perpendicular to the films. The fluorescence intensities inside the films vary; this can be explained in terms of attenuation of the excitation light in the films. The intensity of the excitation light decreases as the light goes deeply because it is absorbed by pyrene. The logarithmic plots of the fluorescence intensities are well fitted to linear lines, which indicates that pyrene was homogeneously distributed, and the Lambert–Beer law held for the attenuation in the films. The pyrene concentration determined on the basis of the homogeneous distribution has a good relation to the gradient of the logarithmic curves, so that the molar extinction coefficient at 293 nm (excitation wavelength) can be calculated.

Figure 12.16 shows the depth profile of a PMMA latex particle (7.2 μm). The PMMA latex was soaked in a methanol solution of pyrene, washed with

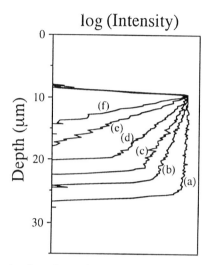

Figure 12.15. Concentration dependence of depth profiles of pyrene-doped PMMA films. The thickness of the films was adjusted to decrease the concentration increases. The estimated concentrations in the films are (a) 1.6×10^{-3} M, (b) 2.1×10^{-2} M, (c) 4.0×10^{-2} M, (d) 7.4×10^{-2} M, (e) 1.8×10^{-1} M, and (f) 3.4×10^{-1} M.

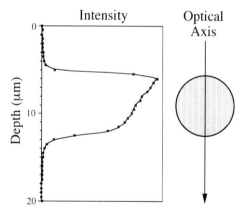

Figure 12.16. A fluorescence depth profile of pyrene-doped PMMA latex particles (7.2 μm).

cold water, and then dispersed in pure water. The use of the molar extinction coefficient given by the curves in Figure 12.15 makes it possible to estimate the pyrene concentration in the latex particle from the depth proile. The obtained concentration was 4.9×10^{-2} M. It is noteworthy that this concentration estimation is very powerful since it is not affected by surrounding materials, even if the surroundings absorb or scatter the light.

12.6. EXCIMER FORMATION DYNAMICS
IN A SINGLE MICROCAPSULE

We have applied the laser manipulation and spectroscopy system to the analysis of fluorescence dynamics in various kinds of microparticles. Here we describe one application: observation of excimer formation dynamics in a single microcapsule dispersed in aqueous solution (23). Since the excimer formation kinetics of aromatic molecules in solution depends on concentration, solvent viscosity, and so forth, these physical/chemical conditions in individual microparticles are expected to be clarified by precise spectroscopic measurement and analysis of a single capsule. Microcapsules that possess unique geometrical structures with solvents encapsulated by thin polymer resin walls are widely used for industrial applications. Their physical and chemical natures are determined by the chemical composition of the polymer, contained molecules, and their concentrations.

The samples we prepared were melamine resin wall microcapsules containing pyrene in toluene (8.1×10^{-3} M) as the inner solution. The diameters were several micrometers to tens of micrometers. The microcapsules were dispersed in pure water, and the solution was placed between two quartz plates. As a preliminary experiment, the pyrene fluorescence spectrum of the solution containing a number of microcapsules was observed by spatially unresolved (conventional) spectroscopy, which showed both monomer and excimer fluorescence with an excimer (at 475 nm) and monomer (at 384 nm) fluorescence intensity ratio (I_E/I_M) of 2.4. This value corresponds to the concentration of pyrene of 1.14×10^{-2} M, as estimated by the concentration dependence of I_E/I_M in an air-saturated bulk toluene solution (separate experiment). This value was slightly higher than that of the mother solution, which could be due to evaporation of toluene during the synthetic procedure of the capsules.

Figure 12.17 shows the excimer fluorescence dynamics of pyrene in the microcapsules observed under spatially unresolved conditions. Since the sample solution was not deaerated, the excited pyrene is quenched by oxygen, showing relatively fast decay. Actually, the excited pyrene monomer in diluted solution showed decay with a time constant of 17.5 ns. In homogeneous solution, pyrene excimer formation is known to proceed via the Birks kinetics model, where the rise and decay of both monomer and excimer are characterized by double exponential curves with the same time constants. However, the curve fitting to the data of Figure 12.17 resulted in failure (large χ^2 and a small Durbin–Watson parameter), which phenomenologically indicated that the excimer formation observed for microcapsules by spatially unresolved spectroscopy cannot be explained on the basis of the Birks kinetics model.

We studied the pyrene fluorescence dynamics for individual microcapsules by the three-dimensional space- and time-resolved fluorescence spectroscopy

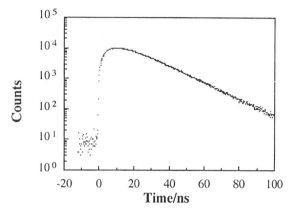

Figure 12.17. A fluorescence rise/decay curve of pyrene excimer in a microcapsule observed by spatially unresolved spectroscopy.

system. The fluorescence spectrum and the dynamics of a single microcapsule were measured at the center of the capsule. Fluorescence spectra of three different microcapsules are shown in Figure 12.18 (23). It was clearly demonstrated that the efficiency of excimer formation was quite different for the various capsules and is independent of particle size. Pyrene excimer rise/decay curves relevant to the spectra in Figure 12.18 are shown in Figure 12.19 (23). The microcapsule with relatively large I_E/I_M decayed faster than the microcapsule with smaller I_E/I_M. In contrast to the results of Figure 12.17, furthermore, the rise/decay curves for individual microcapsules were well fitted with

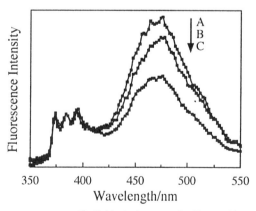

Figure 12.18. Fluorescence spectra of individual microcapsules dispersed in water. The diameters of capsules A, B, and C were 19.2, 6.5, and 6.8 μm, respectively.

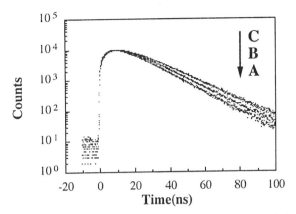

Figure 12.19. Excimer fluorescence dynamics of individual microcapsules A, B, and C corresponding to those in Figure 12.18.

double-exponential functions. Analogous results were obtained for a number of microcapsules. These results clearly indicate that pyrene excimer formation in a single microcapsule can be explained by the Birks kinetics model, as found in homogeneous solution. The origin of the multiexponential behavior observed by spatially unresolved spectroscopy (Figure 12.17) is due to the observation of the sum of various microcapsules with different fluorescence dynamics.

Here possible explanations for the variation in excimer formation efficiency (I_E/I_M) in the individual capsules are considered. The factors influencing excimer formation are the viscosity of the inner solution and the concentration of pyrene in the capsule. The viscosity of the inner toluene solution might increase with decreasing size of the capsule owing to the surface forces between the melamine-resin wall and toluene. To test this possibility, we selected microcapsules of various sizes and determined the I_E/I_M of each capsule. The I_E/I_M was not correlated with the size of the capsule, indicating that the viscosity of the inner toluene solution is not a factor governing the I_E/I_M of each capsule. The most probable explanation for the variation in I_E/I_M and the fluorescence dynamics with the capsules is scattering of the pyrene concentration.

Assuming that the variation in I_E/I_M of the capsules is primarily ascribable to the pyrene concentration, we analyzed the monomer decay and excimer rise/decay profiles based on the Birks kinetic model. Double-exponential fittings (24) were successful ($\chi^2 \approx 1$ and large Durbin–Watson parameter) for every capsule. Two time constants (λ_1 and λ_2) and their pre-exponential factors (A_1 and A_2, respectively) were thus obtained for three microcapsules. λ_1

and λ_2 determined at 475 nm (excimer) coincide fairly well with the corresponding values observed for monomer fluorescence at 384 nm for a given capsule. Furthermore, A_1 and A_2 are in good agreement with the expectations of the Birks kinetics model; $(A_1 + A_2) \sim 1.0$ and $(A_1 - A_2) \sim 0$ for the monomer and excimer curves, respectively.

Using the pyrene concentration for each capsule, which was estimated from I_E/I_M based on the concentration dependence of I_E/I_M in toluene (separate experiment), we determined the rate parameters for excimer formation dynamics. The estimated values for every capsule are comparable to those determined for a homogeneous toluene solution of pyrene. The variation in I_E/I_M with the capsules is, therefore, reasonably well explained by the variation in the pyrene concentration in the capsule. Excimer formation dynamics in each microcapsule are concluded to be similar to those in a homogeneous bulk solution.

One possible origin of the concentration distribution between the capsules may be inhomogeneous location of pyrene, which could be determined during the synthetic procedure of the microcapsules. Evaporation of toluene during vigorous stirring of a pyrene–toluene solution in water accounts for the higher than average I_E/I_M values (> 2.3). For the capsules with $I_E/I_M < 2.3$, on the other hand, we suspect partition of water into toluene droplets, leading to the change in pyrene concentration. The evaporation rate of a toluene droplet and partition of water into the toluene layer are also influenced by the polymerization rate and the wall thickness of the capsule that are scattered among the capsules since the emulsion polymerization of reactants proceeds at the toluene–water interface in the inhomogeneous solution. All these factors are related to each other and result in the present fluorescence characteristics of microcapsules.

12.7. LASER OSCILLATION IN A TRAPPED MICROPARTICLE

The focused laser beam used in the present microspectroscopy system can excite molecules at extremely high intensity, which easily induces various nonlinear phenomena in microparticles. One example is laser oscillation within a microspherical particle (3). When a dye-doped microparticle is irradiated by intense excitation light, an inverted population of dye molecules can be induced in the particle. The fluorescence emitted at the particle–medium boundary with an angle larger than the critical angle is reflected back inside the particle if the refractive index of the particle is higher than that of the surrounding medium. This emission is reflected repeatedly and propagates circumferentially along the boundary, which forms an optical cavity with a high quality factor ($Q > 10^8$). When the emission returns to the starting point with the same phase, except for some interger multiple of 2π, oscillation is induced, so that the emission is amplified by the population-inverted dye molecules, laser oscillation. Since the surface of the microsphere curved rather than flat, the emission is not totally reflected at the boundary, that is, the evanescent wave does not completely return to the inside of the sphere. Because of this leakage, the laser light moves from the particle boundary toward the tangents of the sphere. Thus, the laser emission is observed on the rim of the microsphere (25).

The space- and time-resolved microspectroscopy and manipulation system can be applied to precise analyses of spectral and temporal characteristics of the microspherical laser oscillation with no disturbances such as thermal Brownian motion, gravity, or convection (26). Three-dimensional trapping also avoids optical interactions of a lasing particle with its surroundings, such as glass plates and other particles, which reduce the quality factor of a microspherical cavity and affect the lasing process. We confirmed the lasing of a rhodamine B (RhB)-doped PMMA particle (26). The particle was dispersed in water and optically manipulated by a focused 1064 nm beam under a microscope. RhB in the particle was uniformly pumped by a second harmonic pulse from a Q-switched, mode-locked Nd:YAG laser (532 nm, $\sim$ 40 ps, 10 Hz). In addition to spontaneous orange emission from RhB molecules, more intense emission was observed near the particle–water boundary. This ring-like emission could be ascribed to laser light from the microspherical cavity.

Emission spectra from an RhB-doped PMMA particle (26 μm) are shown in Figure 12.20 (26). In this experiment, the emission from the particle was

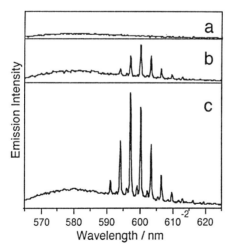

Figure 12.20. Emission spectra of an RhB–PMMA microparticle (26 μm) pumped with the laser power of (a) 1.1, (b) 3.0, and (c) 9.7 mJ cm^{-2} pulse^{-1}.

detected by a polychromator and a double-intensified, multichannel photo-diode array detector (spectral resolution $= 0.4$ nm). When the pumping laser power is low, the emission spectrum is broad and structureless (Figure 12.20a), which corresponds to a spontaneous emission spectrum of RhB. As pumping intensity increases, distinct resonance peaks appear and their intensities increase (Figure 12.20b, c). Several pairs of adjacent high- and low-intensity peaks, corresponding to TE and TM modes, respectively, are spaced at the constant-wavelength interval. This ripple structure is attributed to the whispering-gallery mode resonances. The interval of the TE mode peaks is 3.1 nm, which is the same as that of the TM mode and well agrees with the calculated value of 3.06 nm based on the Mie–Debye scattering theory.

A microparticle acts as a short cavity with a micrometer resonator length, so that a picosecond lasing pulse can be produced by a single pulsed pumping, one of the characteristic properties of microspherical lasing (27). Temporal profiles of laser emission from an RhB-PMMA particle (21 µm) were observed with a streak camera (temporal resolutions $= 10$ ps). When pumping intensity is low (Figure 12.21a), the emission slowly decays, corresponding to the fluorescence decay of RhB in PMMA (lifetime $= 3.3$ ns). At a pumping inten-sity of 8 mJ cm^{-2} pulse^{-1} (Figure 12.21b), short pulsed emission appears and its relative contribution increases compared with that of fluorescence. This fast decay clearly indicates that the stimulated emission process is induced in the microparticle. As the pumping power increases (Figure 12.21c), the intensity of the pulsed emission increases nonlinearly, so that the decay curve includes no

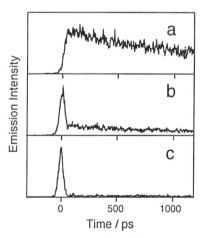

Figure 12.21. Temporal emission profiles of an RhB–PMMA microparticle (21 µm). The pump-ing intensities were (a) 4, (b) 8, and (c) 16 mJ cm^{-2} pulse^{-1}.

appreciable fluorescence component. The pulse width of the laser emission was determined to be ~ 40 ps (FWHM).

The pulse shape depends on the lasing wavelength and the dye concentration (27). This dependence can be explained by the cavity gain provided by stimulated emission of RhB and by the cavity loss due to reabsorption of the emission in the particle. The rise curve of the lasing pulse depends on the quality factor of the cavity and the inverted population of RhB, and the exponential decay is mainly determined by the absorption of the laser emission by RhB itself. Therefore, the shorter pulse can be produced by the higher gain and loss of the microspherical cavity, while the low cavity loss—that is, the high quality factor—is required for studies of the resonance effect on photochemical processes in a microparticle, as mentioned later.

Rise and decay curves of the pulsed laser oscillation provide valuable information on the characteristic molecular dynamics in a microparticle, as well as the mechanism of microspherical lasing. Some photophysical phenomena interact with the high-Q resonances within a microparticle so that the dynamics of the microspherical lasing and photophysical processes are influenced by each other. Thus, the efficiency of the processes in microparticles is sometimes quite different from that of bulk materials (28). For example, we determined the effect of transient absorption of one molecule on the lasing process of another molecule within a microspherical particle. The photon energy of laser dye molecules can be transferred to transient species produced with excitation light different from the pumping light for lasing. This transient absorption induces cavity loss in the particle, resulting in suppression of microspherical lasing. The transient absorbance of the particle is amplified by the high-Q resonance effect, so that the lasing dynamics sensitively varies, depending on the concentration of transient species. Based on this phenomenon, the resonance peak intensity and pulse shape of microspherical lasing can be varied by optically controlling the quality factor of a microspherical cavity with the irradiation of an excitation pulse.

Figure 12.22 shows emission spectra of a PMMA particle (46 μm) doped with RhB and 9,10-diphenylanthracene (DPA) (29). Lasing of RhB was induced by a second harmonic pulse (532 nm) from a Q-switched, mode-locked Nd: YAG laser (Figure 12.22a). DPA absorbs neither 532 nm laser light nor emission of RhB. When DPA was excited by the third harmonic pulse (355 nm), transient absorption was induced, whose spectrum (top curve in Figure 12.22a) overlapped with the emission band of RhB. Hence, the intensity of laser emission from the microparticle was appreciably reduced, as shown in Figure 12.22b. By cutting off the 355 nm excitation light, the laser emission intensity was immediately recovered. Furthermore, it was confirmed that the intensity of microspherical lasing increased as the excitation intensity decreased.

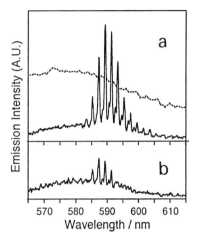

Figure 12.22. Emission spectra of an RhB–DPA–PMMA microparticle (46 μm) in water. (a) Lasing of RhB was induced by a 532 nm light and (b) quenched by transient absorption of DPA, which was induced by a 355 nm laser pulse.

By contrast, spontaneous emission intensity exhibits no appreciable difference between the two spectra in Figures 12.22a and 12.22b. This can be explained by the fact that the optical path length for the straight ray in the microparticle is no more than tens of a micrometer, so that the transient absorbance is negligibly small under the present experimental conditions. However, the laser emission goes around the particle, so that the effective path length reaches millimeter or centimeter orders, which is hundreds or thousands of times longer than that of the straight path. Hence, the transient absorbance in the microparticle is enhanced for laser emission.

Based on this enhancement, high-sensitive absorption measurements can be realized [intracavity laser absorption spectroscopy (30)], which is indispensable for studies on absorption dynamics of micrometer-sized particles (31). In addition, the picosecond pulse of the microspherical lasing makes it possible to extend the technique to time-resolved spectroscopy. The 532 nm pumping pulse for lasing is optically delayed to the 355 nm excitation pulse, so that the temporal variation of transient absorption can be obtained with the picosecond time-resolution.

Figure 12.23 shows lasing spectra of an RhB–DPA–PMMA particle (31 μm) observed by varying the delay time between the pumping and excitation pulses (31). When lasing of RhB was induced before the excitation of DPA (curve a), intense lasing peaks were observed with no influence of transient absorption. When the pumping pulse was temporarily overlapped with the excitation pulse, the laser oscillation was quenched (curve b), and then the

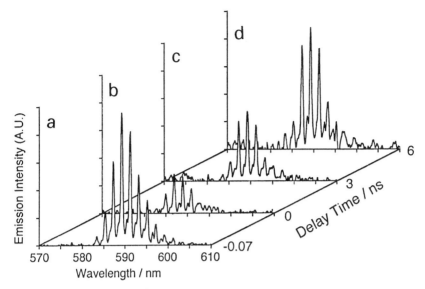

Figure 12.23. Lasing spectra of an RhB–DPA–PMMA microparticle (31μm) observed at the delay times of − 70 ps and 0.3 and 6 ns.

peak intensity recovered gradually with increasing delay time of the lasing to the excitation (curves c and d). By plotting the relative intensity change as a function of the delay time, a decay curve of transient absorption of DPA could be obtained. The curve exponentially decayed with a time constant of ∼ 7 ns, which almost fitted with the lifetime of excited DPA molecules in bulk solutions (14). By comparison, a DPA-doped PMMA film with a thickness of ∼ 10 μm was measured by the conventional time-resolved absorption spectroscopy system; however, transient absorbance was too small to be detected. These result demonstrated the high sensitivity and high time resolution of the present technique.

The spatial characteristics of microspherical lasing can also be studied by the micromanipulation and spectroscopy system (26). The Mie–Debye light scattering theory indicates that the electric field formed by the whispering-gallery mode resonance is localized in the vicinity of the particle-surrounding boundary, that is, the surface wave (3). Hence, the intracavity transient absorption measurement mentioned in the previous section can be applied in analyzing photochemical dynamics at surfaces and interfaces. In addition, the microspherical resonance forms an evanescent field around a particle, so that lasing dynamics is also sensitive to changes in absorption and in the refractive index of the surroundings just outside the particle (32). We have employed

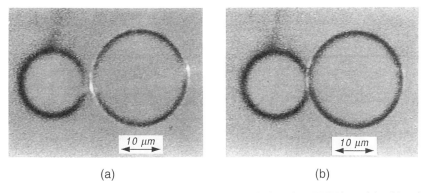

(a)   (b)

Figure 12.24. Optical interaction between lasing and nonlasing RhB–PMMA particles (25 and 17 μm, respectively). (a) The right particle was excited in a limited region, so that lasing was observed in two narrow parts. (b) Quenching of laser oscillation by photon tunneling with a nonlasing particle (left).

a multibeam laser manipulation technique to determine the optical interaction between lasing and nonlasing particles.

Two RhB–PMMA particles (25 and 17 μm) were independently trapped and manipulated by two laser beams, as shown in Figure 12.24 (26). Only a limited region of the right edge in the particle (Figure 12.24a right) was pumped, so that laser emission could be observed in two narrow parts at the left and right sides. The cavity for lasing might be formed along a circular path through these two spots. When the distance between the two particles is large enough (> 1 μm), the nonlasing particle (Figure 12.24a left) does not influence lasing of the other particle. If the nonlasing particle is positioned close to (< 100 nm, noncontact) the lasing particle by optical manipulation, laser oscillation in the RhB–PMMA particle is suppressed, as seen in Figure 12.24b, which may be ascribed to reduction of the quality factor of the relevant modes for lasing. The results will be understood as photon tunneling from the lasing to nonlasing particles through the evanescent field.

Although we have not determined the absolute distance between the two particles in Figure 12.24b, our estimation suggests that the distance between two particles can be controlled within 10 nm by optical trapping. Therefore, three-dimensional manipulation of a lasing particle will play an important role in studies of the mechanism of a laser oscillation process in a microspherical cavity and on the distance dependence of photon tunneling between particles. We also expect that lasing microparticles can be utilized as a probe for a photon-mode scanning tunneling microscope or a near-field scanning optical microscope. By scanning a lasing particle on a sample substrate, the variation of the emission intensity and the lasing spectrum provide information on microstructures of the sample and spectroscopic properties at each position. If a very small particle is attached to a microparticle to be lased by the reported technique, the spatial resolution of the method as a spectroscopic microscope will be further improved.

12.8. CONCLUSION

We have described scanning laser micromanipulation, which made it possible to simultaneously trap several particles and manipulate metal and low refractive index particles. Furthermore, photophysical and photochemical phenomena occurring in individual microparticles can be elucidated by this system combined with three-dimensional space- and time-resolved fluorescence spectroscopy. In addition to particle spectroscopy, chemical reactions such as photopolymerization, gel phase transition, and laser ablation can be induced in/on individual particles by this system, which are applicable to formation (33), dissolution, assembling (34), modification, and fabrication (11) of individual microparticles. These applications of the scanning laser micromanipulation and spectroscopy system are expected to play a major role in physical and chemical studies of fine particles, as well as in micro-optical and microelectronic devices, and micromachines.

ACKNOWLEDGMENTS

This review is based on the work performed in Microphotoconversion Project, ERATO Program, Research Development Corporation of Japan. The authors wish to express their sincere thanks to their colleagues, Professor N. Kitamura, Professor H. Misawa, Mr. M. Koshioka, and Mr. K. Kamada.

REFERENCES

1. Masuhara, H. (Ed.) (1994). *Microchemistry*. Elsevier, Amsterdam.
2. Israelachvili, J. N. (1985). *Intermolecular and Surface Forces*. Academic Press, London.
3. Barber, P. W., and Chang, R. K. (Eds). (1988). *Optical Effects Associated with Small Particles*. World Scientific, Singapore.
4. Ashkin, A. (1970). *Phys. Rev. Lett.* **24**, 156.
5. Ashkin, A., and Dziedzic, J. M. (1972). *Appl. Phys. Lett.* **19**, 283.
6. Ashkin, A., and Dziedzic, J. M. (1977). *Phys. Rev. Lett.* **38**, 1351.
7. Ashkin, A., Dziedzic, J. M., Bjorkholm, J. E., and Chu, S. (1986). *Opt. Lett.* **11**, 288.

8. Sleubing, R. W., Cheng, S., Wright, W. H., Numajiri, Y., and Berns, M. W (1990). *Proc. SPIE—Int. Soc. Opt. Eng.* **1202**, 272.

9. Ashkin, A., Dziedzic, J. M., and Yamane, T. (1987). *Nature (London)* **330**, 769.

10. Block, S. M., Blair, D. F., and Berg, H. C. (1989). *Nature (London)* **338**, 514.

11. Misawa, H., Koshioka, M., Sasaki, K., Kitamura, N., and Masuhara, H. (1991). *J. Appl. Phys.* **70**, 3829.

12. Sasaki, K., Misawa, H., Koshioka, M., Kitamura, N., and Masuhara, H. (1992). *Appl. Phys. Lett.* **60**, 807.

13. Ashkin, A., and Dziedzic, J. M. (1974). *Appl. Phys. Lett.* **24**, 586.

14. Misawa, H., Koshioka, M., Sasaki, K., Kitamura, N., and Masuhara, H. (1991). *Chem. Lett.* 469.

15. Sasaki, K., Koshioka, M., Misawa, H., Kitamura, N., and Masuhara, H. (1991). *Jpn. J. Appl. Phys.* **30**, L907.

16. Sasaki, K., Koshioka, M., Misawa, H., Kitamura, N., and Masuhara, H. (1991). *Opt. Lett.* **16**, 1463.

17. Sasaki, K., Koshioka, M., and Masuhara, H. (1991). *Appl. Spectrosc.* **45**, 1041.

18. Wilson, T. (Ed.) (1990) *Confocal Microscopy*, Academic Press, London.

19. O'Connor, D., and Phillips, D., (1985). *Time-Correlated Single-Photon Counting.* Academic Press, London.

20. Kahlow, M. A., Jarzeba, W., DuBruil, T. P., and Barbara, P. F. (1988) *Rev. Sci. Instrum.* **59**, 1098.

21. Masuhara, H., Eura, S., Fukumura, H., and Itaya, A. (1989). *Chem. Phys. Lett.* **156**, 446.

22. Koshioka, M., Sasaki, K., and Masuhara, H. (1995) *Appl. Spectrosc.* **49**, No. 2.

23. Koshioka, M., Misawa, H., Sasaki, K., Kitamura, N., and Masuhara, H. (1992) *J. Phys. Chem.* **96**, 2909.

24. Sasaki, K., and Masuhara, H. (1991). *Appl. Opt.* **30**, 977.

25. Tzeng, H. M., Wall, K. F., Long, M. B., and Chang, R. K. (1984). *Opt. Let.* **9**, 499.

26. Sasaki, K., Misawa, H., Kitamura, N., Fujisawa, R., and Masuhara, H. (1993). *Jpn. J. Appl. Phys.* **32**, L1144.

27. Kamada, K., Sasaki, K., Misawa, H., Kitamura, N., and Masuhara, H. (1993). *Chem. Phys. Lett.* **210**, 89.

28. Folan, L. M., Arnold, S., and Druger, S. D. (1985). *Chem. Phys. Lett.* **11**, 322.

29. Sasaki, K., Kamada, K., and Masuhara, H. (1994). *Jpn. J. Appl. Phys.* **33**, L1413.

30. Harris, S. J. (1984). *Appl. Opt.* **23**, 1311.

31. Kamada, K., Sasaki, K., and Masuhara, H. (1994). *Chem. Phys. Lett.* **229**, 559.

32. Fuller, K. A. (1991). *Appl. Opt.* **30**, 4716.

33. Ishikawa, M., Misawa, H., Kitamura, N., and Masuhara, H. (1993). *Chem. Lett.* 481.

34. Misawa, H., Sasaki, K., Koshioka, M., Kitamura, N., and Masuhara, H. (1992). *Appl. Phys. Lett.* **60**, 310.

CHAPTER

13

NANOSCALE IMAGING AND SENSING BY NEAR-FIELD OPTICS

WEIHONG TAN

Department of Chemistry and Brain Institute
University of Florida
Gainesville, Florida 32611-7200

RAOUL KOPELMAN

Department of Chemistry
University of Michigan
Ann Arbor, Michigan 48109

13.1. INTRODUCTION

Conventional optical techniques are based on focusing elements such as a lens. The sample is usually positioned at a relatively large distance from the light source. In such far-field optics, the standard rules of interference and diffraction lead to the Abbè diffraction limit (1873) (1) on the resolution of optical microscopes. This limit is approximately $\lambda/2$, where λ is the wavelength. The rapidly developing physical and biological sciences demand better spatial resolution. The realization of better resolution by subwavelength light sources has led to the concept of near-field optics (NFO) (2–10). NFO makes it possible to bypass the optical diffraction limit through the use of a small light source that effectively focuses photons through an aperture that may be as small as $\lambda/50$ (5). Near-field scanning optical microscopy (NSOM) is another form of scanning probe microscopy that has recently been developed and is generating considerable interest (2–10). NFO has enabled researchers to examine optically a variety of specimens without being limited in resolution to one half of the wavelength of light. It has been applied in microscopy, spectroscopy, and the nanofabrication and application of subwavelength optical biochemical sensors (2–8). NFO is realized by subwavelength optical

Fluorescence Imaging Spectroscopy and Microscopy, edited by Xue Feng Wang and Brian Herman. Chemical Analysis Series, Vol. 137.
ISBN 0-471-01527-X © 1996 John Wiley & Sons, Inc.

407

light sources and probes that are leading to new technologies and techniques ranging from electronics to biomedical research devices. The high demand for NFO probes has also led to the development of active light sources (4, 6, 8). Much more light emanates from the same-sized active light source than from a passive one (2, 3). We note that traditional NSOM can also be carried out with active light sources based on exciton transport and quenching, which is called molecular exciton microscopy (MEM) (11). In this case, exciton transport and quenching occur only inside the source, while the sample is exposed only to virtual photons. The different regimes of optical and exciton microscopy and spectroscopy are shown in Table 13.1; the advantages of NSOM and MEM are evident.

Optical microscopy and spectroscopy are key techniques in medicine, biology (12), chemistry, and materials science. Among their advantages are the following:

1. *Universality:* All meterials and samples attenuate light and have spectroscopic states.

2. *Energy and chemical state resoultion:* The obvious advantages of spectroscopy and photochemistry, at ambient temperature can be easily added to those of the optical methods. By contrast, this is not easily accomplished with other techniques such as electron, microscopy or X-ray crystallograpy.

3. *Noninvasiveness:* Most often, the sample is not altered in a microscopic and/or spectroscopic investigation. Moreover, biological samples can usually be studied in their native environment. Most chemical reactions are not perturbed by light of long enough wavelength.

4. *Real-time observation:* Biological phenomena, chemical reactions, crys-

Table 13.1. Comparison of Optical Microscopies

Optical Technique	Interaction	Resolution Limit (Å)	Implementation
Far-field optics	Conventional diffraction-limited	2500–5000	Lens
Near-field optics	Evanescent wave intensity-limited	100–200	Fiber-optic tip; micropipette tip
Molecular exciton microscopy	Excitation transport	3.5–10	Molecular donor supertip
	Spin-orbit coupling	2.5	Molecular sensor supertip

tallization and so on can be observed under the microscope as they happen *in situ* (even with the eyes); spectroscopic measurements can be performed on line in an industrial process and in other fields.

5. *Safety:* Optical and spectroscopic analyses are usually very safe, and precautions are mostly limited to wearing optically protective eyeglasses.

6. *Low price:* Optical microscopy is much cheaper than, say, electron microscopy; optical spectroscopy is usually a bargain compared with, say, nuclear magnetic resonance machines. Obviously, there are exceptions.

7. *Contrast:* For optical microscopy, several contrast mechanisms, such as fluorescence, absorbance, polarization, and diffraction, can be easily applied to increase chemical identification abilities.

8. *Speed, zoom, and human factors:* Optical techniques are usually fast and can be extended even into the femtosecond domain. They can be used for distances ranging from astronomical to microscopic. Preliminary or concomitant observations can be made using the most highly developed sense—sight (and in "living color")—even without the brokerage services of an analog or digital interface.

One can obviously add to the list of advantages the use of optical fiber probing techniques. On the negative side there are two major factors:

1. *Spatial resolution:* There is a well-known diffraction limit, as mentioned above, which is on the order of half a wavelength. This has opened the way to competing technologies such as electron microscopies and the recently developed scanning probe microscopies.

2. *Attenuation:* Penetration of samples is limted, opening the way to X-ray, magnetic resonance, acoustical, and other imaging techniques.

This chapter describes some newly emerging techniques of NFO that overcome the diffraction limit and retain the advantages of conventional optics. Eventually, with the aid of optical nanoprobes and molecular engineering, this may even lead to techniques that can penetrate deeply into the human body and perform measurements with molecular precision.

13.2. NEAR-FIELD OPTICS AND ITS REALIZATION

13.2.1. Near-Field Optics

Conventional (far-field) optical techniques are diffraction limited (1–5). However, light can be *apertured* down to much smaller sizes, with no obvious

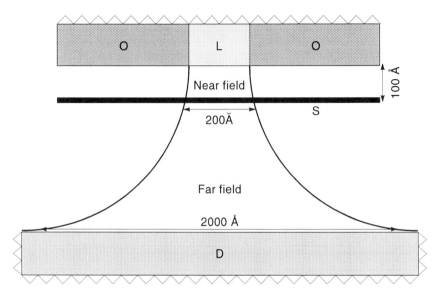

Figure 13.1. Illustration of NFO. Key: O, opaque material; L, active or passive light source; S, sample (support not shown); D, far-field detection system (e.g., lens).

theoretical limit. The simplest example is the passage of light through a small hole. In NFO the diffraction limit is overcome by using subwavelength light sources and samples positioned very close to them (i.e., in the near field). The realization of NFO is through optical nanoprobes. The principle underlying this concept is schematically shown in Figure 13. 1. The near-field apparatus consists of a near-field light source, a sample in the near field, and a far-field detector. To form a subwavelength optical probe, light is directed to an opaque screen containing a small aperture. The radiation emanating through the aperture and into the region beyond the screen is first highly collimated, with a dimension equal to the aperture size, which is independent of the wavelength of the light employed. The region of collimated light is known as the near-field region. The highly collimated emissive photons occur only in the near-field regime. To generate a high-resolution image, a sample has to be placed within the near-field region of the illuminated aperture. The aperture then acts as a subwavelength-sized light probe that can be used as a scanning tip to generate an image. That is why this optical microscopy is called near-field scanning optical microscopy (2–5). The NFO principle has been discovered and rediscovered several times [in 1928 (13), 1956 (14), 1972 (15), and the 1980s (2, 3)]. Actually, the photon "scanning probe technique" preceded all others, such as Scanning tunneling microscopy (STM). However, only in the 1980s was the principle followed by optical experiments [the 1972 work demonstrating it with microwave radiation (15)]. As is well known, scanning probes give an image due to raster scanning over a sample. Today's optical methods have borrowed some electromechanical and computer control scanning techniques from STM, AFM (atomic force microscopy) and other systems (16–18).

Unlike STM or AFM, imaging in NSOM occurs via the interaction of light with the surface by either a simple refraction/reflection contrast or absorption and fluorescence mechanisms. The advantages of NSOM are its noninvasive nature, its ability to look at nonconducting and soft surfaces, and the addition of a spectral dimension; the last does not exist in either STM (at room temperature) or AFM. This potential for extracting spectroscopic information from a nanometer-sized area makes it particularly attractive for biomedical research and materials science. The resolution of an NSOM image is limited by the size of the light probe. Thus, the light source is the heart of the NFO technique. It has to be small, intense, durable, and spatially controlled. As pointed out above, its size determines the resolution, *provided that it can be scanned close to the sample*, as close as its size (technically, 20–80% of it).

What are the techninal difficulties? One is that we need a subwavelength light source with enough intensity. Another is that the sample has to be scanned closely and quickly. The latter requirement is not too difficult today (magnetic memories are scanned extremely quickly at even closer distances). However, the first requirement of a tiny, intense light source has been a problem. Originally, tiny nanofabricated orifices (holes) were used as light sources (2,7). However, the photon throughput is very limited. Very recently, with the advent of *active* subwavelength light sources (4–6) it has been possible to get very high light throughput. There are other problems, such as the need for a "feedback" mechanism to avoid physical contact and damage to the source. There are also wonderful recent solutions, like combined NFO and force or combined NFO and STM operation (20–22) (see below). These further enrich the contrast mechanisms of NFO: refraction, reflection, polarization, luminescence, lateral force interactions, and so on.

13.2.2. Realization of NFO

The original idea NFO depended on light transmission through a subwavelength "hole," i.e., a nanofabricated hole in a thin metal sheet or film. Obviously, a flat sheet or film (with a nano-hole) can be used as a scanning probe only with completely flat samples. This led, among other things, to the idea of using a glass micropipette such as the one invented by the Nobel prize winners Neher and Sackmann (23) for intracellular electrical measurements and used for microelectrochemical sensors (24). The first step in the probe nanofabrication process is the pulling of micropipette and fiber optic tips of appropriate size and shape. The second step is the metal coating of such tips. This is followed by crystal (or polymer) growing if active optical or excitonic probes are desired. Pulling such micropipettes is mostly done today with a commercial puller (e.g., Sutter P-87 and P-2000, Novato, California). Recently, this process has been improved with computerized multistep control of the pulling procedure (25) and with infrared lasers replacing the electric heater strips. Thus, one can now reproducibly pull robust and efficient micropipettes with orifices as small as 10 nm (inner diameter), optically streamlined (e.g., short "shank"), and clean enough on the outside to facilitate the metal coating. The latter procedure (vacuum deposition of metals) is well known but far from trivial. Here we illustrate the nanofabrication of optical fiber tips. Very similar techniques have been applied in the fabrication of micropipettes (25–26).

Optical fiber tips have been used in many areas (5, 6, 27) and can be fabricated either by heating and stretching or by chemical etching. Our apparatus for fiber tip pulling, as shown schematically in Figure 13.2, consists of a micropipette puller and a 25 W CO_2 infrared laser. The CO_2 infrared laser beam replaces the electric filament in the puller to heat the optical fiber for the pulling process. The laser beam is reflected by a mirror and directed to heat the optical fiber, which is fixed on the puller. The details of the pulling setup and procedures can be found in several references (5, 19, 25, 27). Most of the tips used in our NFO application range from 0.05 μm to about 0.3 μm in diameter (5, 6, 8–10, 19). After pulling, the optical fiber tip is coated with aluminum by vapor deposition to form a small aperture. A specially built high-vacuum chamber is employed for coating these pulled fiber tips; only the fiber tip sides are coated with aluminum, leaving the end face as a transmissive aperture. To make it into a light source, a visible or UV laser beam is coupled to the opposite end of the pulled tip. This probe delivers light very efficiently, since most of the radiation is bound to the core up to a few microns away from the

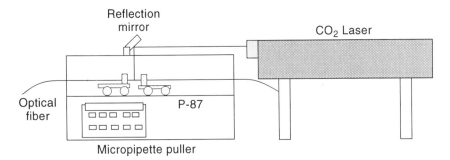

Figure 13.2. Schematic drawing of optical fiber tip pulling setup.

tip. A randomly chosen $0.2\,\mu m$ optical fiber probe gives 10^{12} photons per second (19).

When the same puller but with electric film heating is used, glass micropipettes with different diameters can be pulled. To pull short-shank micropipettes, we developed a multistep pulling program (25). In multistep pulling, the general idea is to use initial puller settings to produce a rapid taper over a desired distance, allowing the operation to occur over one or more pulling cycles and to form an ultrafine tip during the final pulling cycle. By varying the parameters in each step and the glass type and tube dimensions, a wide variety of micropipette shapes and sizes were achieved. According to scanning electron microscopy (SEM) micrographs, the smallest tips used in our crystal growth experiments for active light sources are about $500\,\text{Å}$ inner diameter with a usable shank. Tips pulled by multistep pulling programs show great promise for crystal growth inside them (25, 28–30).

13.2.3. Optical Nanoprobes

Two major probes are used in NFO: metal-coated glass micropipettes and nanofabricated optical fiber tips (2–10). Both probes are easily fabricated to sizes of approximately 50 nm; the smallest nanofabricated optical fiber tip reported to date is about 20 nm (5). The fabrication of miniaturized optical probes has keyed the development and application of NFO in a wide variety of fields. NFO nanoprobes can be classified into three kinds: (1) passive optical probes, such as coated micropipettes or small holes on a screen (2, 3); (2) semiactive light sources such as optical fiber tips (5); and (3) active light sources such as nanometer crystal light sources (4, 6, 28–32). Micrographs of the first two light sources are shown in Figure 13.3. Both optical fiber tips and micropipettes have been used in NFO applications. The micropipette approach has been extended in two ways: (1) The tip has been filled with a photoactive material (4, 28–32); (2) the "pre-tip" region is specially bent to double as a "force probe" (22, 32). These extensions have kept the micropipette tip in competition with the optical fiber tip. The problem in using micropipettes is the conflict between smallness and light intensity (see below). Thus, micropipettes gained limited usage in NSOM after their initial application in NFO.

Compared to a hollow micropipette tip, a nanofabricated optical fiber tip is a semiactive photon tip (5, 8). Generally, it is orders of magnitude brighter, easily coupled to an optical source, and at least as mechanically sturdy as a micropipette. But it is not transparent to short-wave radiation (deep UV or soft X-rays). Also, its preparation is a bit trickier. Not only does the higher melting point silica require the temperature provided by an infrared laser (5, 33), but the metallization process is more demanding, both on the pulling

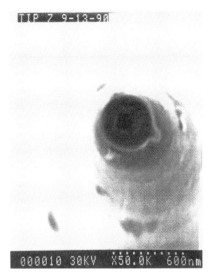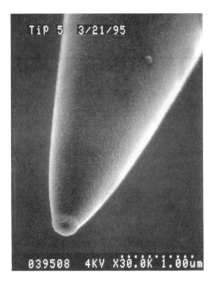

Figure 13.3. SEM micrographs of optical nanoprobes. Left: micropipette probe, right: optical fiber probe.

process and on the deposition process. For instance, to avoid metallizing (and thus blocking) the photon tip, the tip's surface has to be essentially orthogonal to the pulled fiber wall. Furthermore, the deposition angle has to be carefully adapted to the tip geometry. It is interesting to notice that the top of a fiber tip is strongly resistant to breakage. The mechanical stability of optical tips is excellent (lasting from days to months). The photochemical stability of optical fiber tips is also excellent; under very intense illumination, it is heat that damages the aluminum coating at the tip. The price of a tip is reasonable. The spectral range extends from about 300 to 2000 nm (depending on optical fiber quality). Both micropipette and fiber probes have been made, with a diameter of about 500 Å, without difficulty, in applications as light sources.

Using a classical optics description (34), we note that the higher index of refraction (n) of the fiber material reduces the photon wavelength inside it ($\lambda = \lambda_0/n$) compared to the wavelength in vacuum or air (λ_0). This reduces significantly the diffraction of the light at the orifice. In principle, as λ approaches the optical absorption of the dielectric, n increases and eventually becomes a complex quantity (35, 36). Alternatively, one can use a quantum approach and consider the exciton–polariton resonance or quasi-resonance

(35, 36). Thus the optical fiber tip exhibits a crossover with wavelength from passive to active photon tip (see below).

13.3. NEAR-FIELD SCANNING OPTICAL MICROSCOPY

The conventional optical microscope has been widely used in chemistry and biology. Most of the development of NFO techniques has centered on the NFO application in microscopy. Different modes of operation and signal detection systems have been designed and developed. NSOM has gained increasing attention from both biomedical researchers and people in the physical science community.

13.3.1. Transmission and Fluorescence NSOM

The two most important modes of traditional optical microscopy are transmission and luminescence. For NFO's chemical and biological applications, these will probably continue to be the most important modes, especially if one includes near-field scanning optical spectroscopy (see below). The setup of NSOM is shown schematically in Figure 13.4. The advantages of using NSOM to study biological systems include the ability to acquire many types of information simultaneously. There has been a great effort to construct an NSOM system utilizing shear-force feedback to do high-resolution fluorescence microscopy while also obtaining a topographic image of the sample (37, 38). This flexible imaging system has been built on top of an inverted optical microscope that allows easy viewing and positioning, as well as efficient fluorescence collection. This system has been used to study biological systems including fluorescently labeled DNA and cell membranes (37), as shown in Figure 13.5. NSOM has also been used to generate fluorescence images of cytoskeletal actin within fixed mouse fibroblast cells (39). Comparison with other microscopic methods indicates a transverse resolution well beyond that of confocal microscopy, and contrast far more revealing than in force microscopy. Effects unique to the near field are shown to be involved in the excitation of fluorescence, yet the resulting images remain readily interpretable. As an initial demonstration of its utility, the technique is used to analyze the actin-based cytoskeletal structure between stress fibers and in cellular protrusions formed in the process of wound healing (39).

The same tip can be scanned over the same sample with an alternation of the contrast mechanisms (e.g., fluorescence and shear force), yielding images with a high degree of fidelity, as well as additional information (this is like adding the natural color to a three-dimensional topographic map). The best

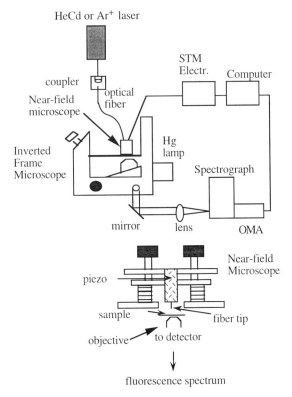

Figure 13.4. Schematic of experimental apparatus used for NSOM.

resolution to date has been claimed (5) to be about 12 nm (with 514 nm light). presumably this was achieved with a 20 nm diameter aperture. A signal of 50 nW has been claimed for an 80 nm aperture (5). Also, NSOM has been successfully applied in single-molecule localization, detection, and studies (see below) (9, 10, 40–44).

13.3.2.   Collection Mode NSOM

Various reflection and collection mode NSOM techniques have been devised (5, 7, 8, 45). One of the most elegant designs is a combined collection-feedback method (45–47). The sample is put on top of a prism, as shown in Figure 13.6, which is illuminated in a total internal reflection mode. A very thin quartz tube (collection probe) with a tip at its end scans over the surface of the

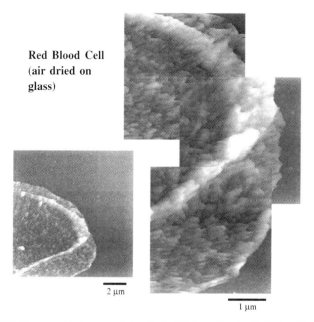

Red Blood Cell (air dried on glass)

2 μm

1 μm

Figure 13.5. Topographic images of a lysed, air-dried red blood cell obtained by NSOM.

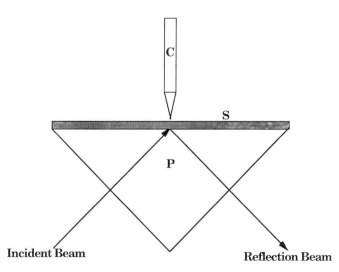

C

S

P

Incident Beam

Reflection Beam

Figure 13.6. Schematic drawing of an evanescent light collection mode. Key: C, collection probe; P, prism; S, sample.

prism. In the absence of a sample, the collector tip will collect light only if it touches the prism or is slightly above it (at about $\lambda/20$). As is well known (34), the light penetrates slightly from the prism into the air above it. The collection decreases exponentially with the gap size (this is the classical analog of quantum-mechanical tunneling through a barrier). With the addition of a sample to the prism, the gap decreases and more light is passed into the collector, depending on the local optical density of the sample. If the sample is homogeneous, then the local optical density is determined by the local thickness.

This method has excellent sensitivity in the vertical (z) direction—a resolution of about 2 Å. However, in the horizontal (x, y) directions the resolution is only about 0.1 μm. Horizontal resolution depends on the size of the tip. These tips have been produced by chemical etching; smaller tips could be produced by other methods mentioned earlier. We note that this is a zero-background method, with all of its advantages. On the other hand, the sample has to be extremely thin and quite transparent. It is also easy to confuse sample thickness variation with optical density variation. Alternatively, the sample may fluoresce or contain luminescent tags. This changes the contrast method and may improve the sensitivity. This method has also been used for spatially resolved fluorescence and Raman spectroscopy, with a spatial resolution of about 0.1 μm. Historically, the biggest advantage of this technique has been its inclusion of feedback via the intensity of the light leaking into the tip. Other feedback methods are described below.

Another type of NSOM is scanning plasmon near-field microscopy (SPNM) (48). A lateral resolution of 3 nm ($\lambda/200$) at an optical wavelength has been achieved. SPNM is based on the interaction of extended surface plasmons (SP), as introduced by Ritchie (49), with a sharp metal tip placed close to the surface of the object. In addition, this new technique may permit the detection and spectroscopic identification of single absorbed molecules. The origin of the surprisingly high resolution of SPNM is not clear.

13.3.3. Contrast in NSOM

There are several contrast methods, but not all of them are well understood: absorption, refractive index, reflection, and fluorescence (luminescence) (5, 8, 47, 50). One can also consider polarization and spectroscopy as separate modes of contrast. Furthermore, there is a large number of quantum effects, such as energy transfer, energy down conversion, and energy quenching (8, 28). The simplest optical contrast mechanisms in the near field regime (e.g., the refractive index) are not yet well understood (50) and are under intensive study. Actually, the microscopic quantum effects are better understood than the mesoscopic (near-field) optical interactions.

13.3.4. Sample Requirements

The most important consideration for sample preparation is sample rough-
ness. This is limited by the probe shape in the most obvious way (as for all
scanning probes). Sample thickness is an important factor for all transmission
(forward-scattering) modes of operation, but not for the reflectance (back-
scattering) and some "collection" modes. The near-field approach couples
optical resolution with distance from the probe; the higher the desired
resolution, the thinner the required sample. On the other hand, the contrast
mode (absorption, refractive index) may limit the thinness of the sample. Even
the fluorescence mode may be limited by the thinness of the sample, i.e., the
absorption cross section. However, this limitation can be overcome by increased intensity, by auxiliary fluorophores, or by quantum mechanisms
(energy transfer). Thus, the various luminescence modes appear to be the most
promising modes of forward-scattering near-field microscopy.

13.3.5. Luminescent Tags

In view of the above discussion, the selection of luminescent tags is as
important for NSOM as it is for traditional fluorescence microscopy. This is
an actively studied aspect of biology (12) and related fields. The tags are
usually tailor-made either for a "white" lamp (e.g., xenon) or for monochromatic sources (e.g., lasers), with the obvious use of spectral filters. In addition
to lamps or laser input, near-field microscopy sources include luminescent
probes, such as crystals, dye balls, or luminescent polymers, with broadband
but not white light characteristics (falling somewhere between lamps and
lasers). These nano-sources are located next to the sample; thus, no filter can be
inserted between them (only between the sample and the detector). Furthermore, the spectral absorption and photostability of the probes are major
factors — in addition, of course, to their chemical and/or biological selectivity.

13.3.6. Feedback and Shear Force Microscopy

All modern scanning-probe microscopies contain feedback as an essential
part. This keeps the probe close enough to the sample but prevents "crashing"
against it. A number of feedback methods have recently been applied to
NSOM. There are some new developments in the combination of NSOM with
STM and AFM (20–22, 51). For example, simultaneous STM and collection-
mode scanning NSOM images using tunneling regulation were obtained using
a gold-coated single optical fiber as both a tunneling and an NFO probe. The
method counts on the reliable STM distance regulation and the short separation between tip and sample (< 1 nm). With optimized fiber probes, both

STM and NSOM images show good resolution and image contrast. The first images were obtained in the collection mode with calibration samples of gratings of period down to 200 nm (20). Micropipettes have also been used for feedback (22), using STM feedback between the metallic coating of the micropipette and the sample. Obviously, this requires the sample to be an electrical conductor. In addition, the optical and electrical probing are centered away from each other (roughly by the radius of the probe). Bent micropipettes have also been used as combined AFM–NSOM probes to achieve submicrometer resolution (22, 32). Micropipettes have unique applicability as force probes in a variety of imaging conditions and a variety of scanned-tip microscopies. These probes are characterized in terms of the parameters that determine their force characteristics. Measurements are presented (22) showing that one can readily achieve force constants of 10 N/m

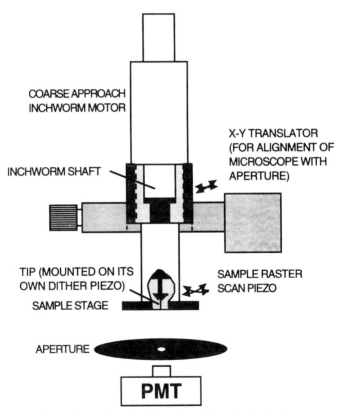

Figure 13.7. NSOM based on a lateral force feedback scheme.

and it is anticipated that a reduction in this force constant by 2 orders of magnitude can be achieved. Such probes can be produced simply with a variety of geometries that permit a wide range of force-imaging requirements to be met. Specifically, the galass micropipette probes are readily produced with apertures at the tip and can thus be applied to NSOM. This opens up the possibility of the long-awaited development of a universal feedback mechanism for NSOM. In addition, a combined AFM–NSOM probe (51) has been microfabricated. However, its resolution is low (5 μm).

The most notable development in NSOM feedback is shear force microscopy (52–55). As a distance regulation method, shear force microscopy has been developed to enhance the reliability, versatility, and ease of use of NSOM. The method relies on the detection of shear forces between the end of an NFO probe and the sample of interest. A schematic presentation is given in Figure 13.7. Here the NFO tip is dithered piezoelectrically. This lateral swinging slightly modulates the light reaching the detector. The modulation frequency is related to the resonating frequencies of the tip. As the tip comes within the (attractive) van der Waals force field of the sample, this frequency changes. A typical·dithering amplitude is 2 to 5 nm. The system can be used solely for distance regulation in NSOM, for simultaneous shear foce and near-field imaging, or for shear force microscopy alone. In the last case, uncoated optical fiber probes are found to yield images with consistently high resolution.

13.4. NEAR-FIELD SCANNING OPTICAL SPECTROSCOPY

Near-field scanning optical spectroscopy (NSOS) (5, 8, 56–59) is based on NSOM. It basically adds one more dimension, spectroscopy, to NSOM and can be used to obtain spectra of various nano-structures, such as subcellular structures and quantum wells. NSOS has inherited all the advantages of NSOM: its noninvasive nature, its ability to look at non-conducting and soft surfaces, and the addition of a spectral dimension. The ability to obtain spectroscopic information with nanometer-sized resolution makes NSOS very promising for a wide variety of biomedical and chemical research. Examples include the detection of fluorescent labels on biological samples and the isolation of local nanometer-sized heterogeneity in microscopic samples. Here we discuss the potential for addressing spectroscopy on a subwavelength scale. We have studied systems of tetracene and perylene doped in poly(methyl methacrylates) (PMMA), as well as microscopic crystals, to demonstrate that nanoscopic inhomogeneity can be detected in what might at first appear to be a homogeneous sample (56). The eventual goal is to obtain spectroscopic information with a spatial resolution of nanometer or even molecular sizes.

In NSOS, an optical probe with an emissive aperture that is subwavelength in size is positioned such that the sample is within the near-field region. With piezoelectric control of the fiber tip, the tip can be accurately positioned over a fluorescing region of the sample and a spectrum recorded. Excitation of the sample can be external, with detection through the fiber tip, or with the fiber tip itself and subsequent detection of the emitting photons. This means that it is not necessary for the sample to be of any particular thickness or opacity; however, it should have a relatively smooth surface. The optical probes used in NSOS are the same nanometer-sized optical fiber light sources used in NSOM. The experimental apparatus for measuring fluorescence spectra with high spatial resolution is very similar to that used in NSOM, shown in Figure 13.4 (59). The 442 nm line from an He:Cd laser or one of several lines from an argon–ion laser is coupled to an optical fiber with a high-precision coupler. The fiber tip is mounted in a hollow tube of piezoelectric material that is positioned by the usual STM control electronics. The sample (deposited on a glass slide) is mounted on the near-field microscope such that it is perpendicular to the exciting tip, and the entire apparatus rests on the base of the inverted-frame microscope with a reflected light fluorescence attachment. Excitation of the sample via the fiber tip generates fluorescence, which is collected by an objective, filtered (to remove laser light), and collimated before exiting the microscope. The fluorescence is then focused onto an optical multichannel analyzer (OMA). The data are then collected and analyzed on a computer.

13.4.1. Spectra of Heterogeneous Samples

Films of a 1.0 wt% mixture of tetracene in PMMA were prepared by evaporation from dichloromethane. The film thickness was approximately 200 to 300 μm, and it appeared optically clear. Thin films (< 10 μm) of 15 wt% perylene in PMMA were prepared by spin-coating a dichloromethane solution on a glass slide.

Tetracene-PMMA films examined under the fluorescence microscope show microaggregates of tetracene with an average size of ∼ 10 μm embedded in the polymer (56). The background fluorescence from the film appear greenish-yellow and is presumed to be from either isolated molecules or crystals smaller than those that can be resolved with the conventional optical microscope. What is surprising about the aggregates is that the fluorescence ranges in color from green to yellow to red. Thus, the macroscopic fluorescence spectrum obtained with Hg lamp excitation is very broad, containing contributions from the background and aggregates of all colors. With NSOS it is then easy to excite a specific aggregate and record its fluorescence spectrum, as shown in Figure 13.8.

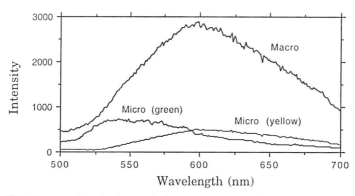

Figure 13.8. Macroscopic and microscopic fluorescence spectra of tetracene–PMMA film.

Similar results are obtained for microcrystals of perylene grown on a glass slide from benzene solution. Perylene is known to crystallize in two distinct forms (60), both having two molecules per unit cell. The α from crystallizes with the molecular planes overlapped and parallel to each other, Excitatoin of a molecule is conducive to the formation of excimers, which then produce broad fluorescence spectra with no vibrational fine structure (as would be expected from a repulsive ground state). In β perylene, the molecular planes are perpendicular, resulting in a spectrum that correlates with monomer emission. Under the microscope, α and β perylene are easily distinguished as bright green (β) and bright yellow (α) crystals. It is then easy to position an optical fiber tip directly above a microscopic crystal and record its spectrum. The left panel of Figure 13.9 shows the fluorescence spectra of rather large diameter (20–30 μm) single crystals of α and β perylene. The two crystals were a few microns apart, so their relative isolation produces as spectra that agree well with the literature spectra of large (> 1 cm) room-temperature single crystals (60). The right panel shown the fluorescence spectrum of a perylene crystal that was just observable under the microscope at 600X magnification. This would put its diameter at approximately 1 μm. Its color was not easily distinguished under the microscope, but with excitation via an optical fiber tip, it is easily identified as a perylene from its fluorescene spectrum.

NSOS has also been used to study (GaAs/AlGaAs) multiple quantum wells in the collection mode. The experiment has been carried out at low temperatures (150 K). Some information has been gained about the nanoscopic structures of quantum wells (38, 57, 58). Near-field microscopy/spectroscopy provides a means to access energies and homogeneous line widths for the individual eigenstates of these centers and thus opens up a rich area of physics involving quantum-resolved systems.

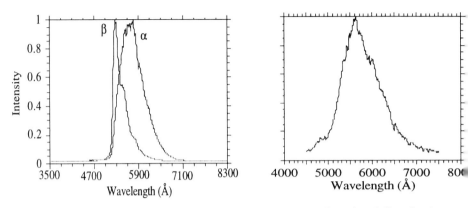

Figure 13.9. Fluorescence spectra of $\sim 30\,\mu m$ crystals of α and β perylene (left) and $\sim 1\,\mu m$ α perylene crystals (right).

13.4.2. Large-Probe Approach

In NSOS, the signal is divided and collected according to its corresponding wavelength. Often a strong enough fluorescence signal is not produced by an ultrasmall optical probe. One approach to overcome this difficulty is to use relative large probes (e.g., from 0.1 to 0.5 μm) but with a nanometer scanning distance at each step. At each step, one spectrum is collected and consecutive spectra are obtained by scanning the entire sample. With this approach, some interesting results have been obtained with the same perylene microcrystal sample. To demonstrate spatial resolution by monitoring spectral features, samples of spin-coated, perylene-doped PMMA films were studied. After spin-coating, these films were found to have regions of sharp color contrast easily observable under the microscope. That is, aggregates that fluoresce yellow are found next to blue-green fluorescent regions of the film that are presumbly due to the isolated molecules. Figure 13.10 shows the result of bringing a fiber optic tip within the near-field region of the sample and monitoring the fluorescence spectrum as the tip is scanned across such an interface. The spectra were obtained for a 0.5 μm tip that was scanned across a blue-yellow region toward a rather large aggregate ($\sim 5\,\mu m$) embedded in the film. Increments of only 400 Å were sufficient to measure a noticeable change in the spectrum. Note that the blue-yellow intensity does not change significantly, since it surrounds the aggregate and therefore contributes a constant background fluorescence.

That a measurable change in fluorescence spectra can be obtained with lateral movements of an exciting fiber tip that are 1 order of magnitude less

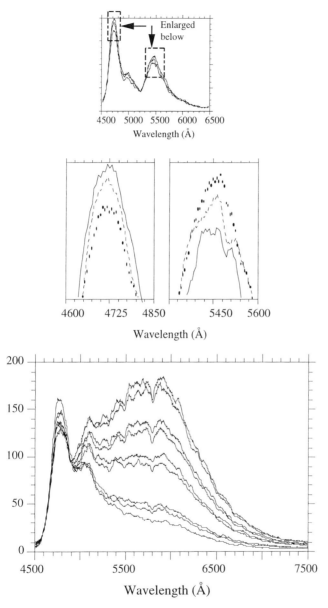

Figure 13.10. Fluorescence from a blue–yellow boundary in a perylene-doped PMMA spin-coated film (top). The peaks corresponding to blue and yellow emission are enlarged in the middle figure to illustrate the effect that scanning a fiber-optic tip across the interface has on the spectra. The size of the fiber tip was ~ 1 μm, and its lateral positions were separated by 1000 Å in successive spectra. The bottom figure shows similar results for a 0.5 μm tip positioned in increments of 400 Å.

than the diameter of the tip is highly encouraging in terms of our eventual goal of achieving molecular resolution. If different fluorescent species are separated by a distance of only 10 Å, then it is conceivable that a light source of only 100 Å will be sufficient to resolve this separation, provided that enough photons are available. Of course, interpretation of unknown heterogeneity will be difficult if the exciting tip is not comparable in size to the fluorescing region. If the tip is much larger than this region, then it would be difficult to quantify the number of such regions compared to their size within the near field. We have demonstrated the ability of NSOS to obtain high spatial resolution in a spectral dimension by studying different microscopic and nanoscopic samples, as well as quantum wells. In heterogeneous samples, it is possible to isolate heterogeneity on a nanometer-sized scale. Measurable changes in fluorescence spectra are obtained for lateral increments approximately one-tenth size of the aperture of the exciting fiber tip. This implies that it may be possible to differentiate heterogeneity at molecular sizes (~ 10 Å), with a light source 100 Å in diameter. Single-molecule detection is described below.

13.5. SINGLE-MOLECULE LOCALIZATION AND STUDY

Observing the dynamics of a single molecule represents the ultimate goal of analytical science. Can one image a single molecule *in vivo*? Can one see it wiggle or reactor break up? Can one simultaneously measure its energy dynamics? Can one monitor directly and in real time the ions or radicals released by a single ion gate or enzyme? Impressive evidence of progress has been provided by the successful applications of NFO in localizing and imaging single molecules (40). Several laboratories have succesfully demonstrated single-molecule localization and single-molecule studies by NFO (9, 10, 40–44).

Traditionally, molecular structure and dynamics have been observed by averaging techniques such as X-ray crystallography, electron diffraction, and various spectroscopies. On the other hand, electron microscopy and related method image single molecules at a heavy cost to their integrity—observing them in a vacuum and/or under highly perturbative conditions. Recent methods, such as STM and AFM, come closer to the ideal, but the molecules are still exposed to punishing electric fields or contact forces. These problems are particularly acute for the soft organic/biological molecules. In addition, the observation cannot be performed *in situ* or *in vivo* and rarely even *in vitro*. Furthermore, it is impossible or nearly unfeasible to observe the molecular dynamics. Near-field optical microscopy and spectroscopy (5, 8) are new tools providing hope for highly improved imaging at a relatively low cost to the sample (and the researcher).

Single-molecule studies represents the ultimate goal of analytical science (61–71). Observing the dynamics of a single molecule may have started with the direct patch-clamp assisted observation of single sodium gates, and in particular, the voltage jumps accompanying their opening and closing (61). Recent elegant single-molecule observations in the energy domain have been performed and reviewed (62–71). In this case, single molecules move around in the very high resolution laser spectral domain in samples of dilute mixed molecular crystals and polymers. Optical spectroscopy observations of single molecules have recently been made (65, 66). Here dye molecules are in a thin flow cell (65) or levitated microdroplet (66) and are fleetingly observed via laser fluorescence.

Using NFO to localize and detect single molecules has unique advantages. Actually, there is no need for a molecule-sized NFO probe for single-molecule localization and detection, as schematically shown in Figure 13.11 (see Color Plates). The probe size can be quite large compared to a one-molecule size. The report by Betzig and Chichester (40) defines the state of the art concerning optical imaging. Individual carbocyanine dye molecules in a submonolayer spread were imaged with NSOM. Molecules can be repeatedly detected and spatially localized (to about $\lambda/50$) with a sensitivity of at least 0.005 molecule/$(Hz)^{1/2}$ and the orientation of each molecule dipole can be determined. This information is exploited to map the electric field distribution in the near-field aperture with molecular spatial resolution. About two dozen isolated dye molecules are imaged within seconds, as shown in Figure 13.12. The molecular location is resolved within about 25 nm in the horizontal plane and 5 nm in the vertical plane. Furthermore, the much smaller molecular transition dipole is a point detector mapping out the electric field distribution of the near-field light source. In addition to imaging individual dye molecules, they obtained information on the orientation of these molecules (via polarization and transition dipole fitting). At the same time, they have been able to turn the tables, using a single molecule to map the electric field distribution in the vicinity of a nanometer light source.

The ability to observe the optical spectrum of a single molecule can provide insights into the interactions that distinguish one molecular environment from another (42). In this experiment, NSOS has been used to obtain the time-dependent emission spectrum of a single molecule in air at room temperature with a spatial resolution of about 100 nm. Single molecules of 1,1′-disectadecyl-3,3,3′,3′-tetramethylindocarbocyanine (diI) dispersed on PMMA have been examined. The spectra of individual molecules exhibit shifts of ± 8 nm relative to the average spectrum and are typically narrower, as is expected for spectral lines broadened inhomogeneously (i.e., by a distribution of molecular environments). There is also spectral variation of a width of up to 8 nm, some being as broad as the far-field many molecule spectrum. The emission spectra of some

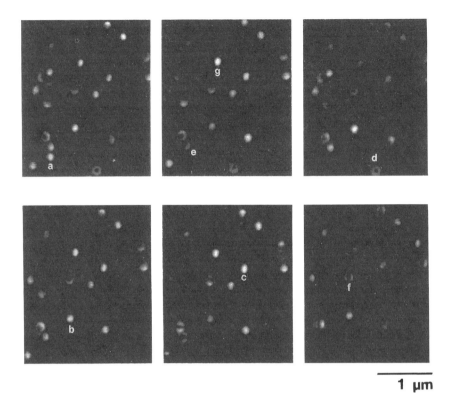

1 μm

Figure 13.12. Images of single molecules. Six sequential images of the same field of individual carbocyanine dye molecules detected by the NSOM fluorescence mode. Reprinted with permission from *Science*.

individual molecules exhibit time-dependent shifts of up to 10 nm. This variety in spectral position, width, shape, and time dependence can be understood within a model of inhomogeneous broadening in which there is a distribution of barrier heights to rearrangement of the molecular environment. Thus, properties of individual molecules can be understood thoroughly with NSOS. Single-molecule spectroscopy was also performed on a crystal of pentacene-doped *p*-terphenyl using a tip with a 60 nm diameter aperture at extremely low temperature (44). Individual molecules located a few hundreds nanometers from the tip had linewidths of 10–20 MHz at 1.8 K. Background fluorescence has been used as a useful distance-sensing signal, since it increased exponentially during approach with a characteristic length of about 100 nm. Single molecules closest to the tip were identified by Stark shifts.

The room-temperatuer dynamics of single molecules is of great interest. NFO has also been used to study the molecular dynamics of sulforhodamine 101 molecules dispersed on a glass surface at two different times (10). On the 10^{-2}–10^2 s timescale, intensity fluctuations in the emissions from single molecules are examined with polarization measurements, providing insight into their spectroscopic properties. The fluorescence lifetimes of single molecules are measured on a nanosecond timescale, and their excited-state energy transfer to the aluminum coating of the near-field probe is characterized. The feasibility of fluorescence lifetime imaging with single-molecule sensitivity, picosecond temporal resolution, and a spatial resolving power beyond the diffraction limit has been demonstrated.

NFO has also been applied in single-molecule photochemistry studies (11, 43). Fluorescence lifetimes of single rhodamine 6G (Rh6G) molecules on silica surfaces were measured with pulsed laser excitation, time-correlated single-photon counting, and NSOM (11). The fluorescence lifetime varies with the position of a molecule relative to a near-field probe. Qualitative features of lifetime decreases are consistent with molecular excited-state quenching effects near metal surfaces. The technique of NSOM provides a means of altering the environment of a single fluorescent molecule and its decay kinetics in a repeatable fashion. Ambrose et al. have also reported the photobleaching experiments of single molecules on a surface (43). Single molecules of Rh(6G) are detected on a silica surface under ambient conditions. Photobleaching of an individual molecule is revealed as an abrupt disappearance of fluorescence after seconds of constant excitation. Other reversible, upward, and downward changes in fluorescence are observed, possibly due to reorientation of molecules on the surface.

NFO has gained increasing attention in single-molecule studies. This will greatly facilitate the manipulation of single molecules in diagnostics and in biotechnology.

13.6. NEAR-FIELD NANOFABRICATION AND NANOMETER BIOCHEMICAL SENSORS

Optical microscopy and spectroscopy have often been utilized for chemical and biological analysis. For instance, a reagent is introduced into a cell and the color or spectrum of the cell changes, providing information regarding the pH or calcium content of the cell. More recently, fiber-optic chemical sensors (FOCS) have been introduced for such measurements (72). However, their spatial resolution has been limited by the size of the optical fiber, (e.g., 100 μm) (72). A spin-off from recent NFO technology has been the development of subwavelength FOCS (6, 19). The reduced size of the light sources, together with the enhanced molecular excitation cross-section (8, 41) and the good spectral and time resolution have enabled another unexpected development: rugged, ultraslim, ultrasensitive, and ultrafast, FOCS (6, 19) that require only attoliters of sample, zeptomoles (10^{-21} mol) of unknown, and milliseconds or microseconds of response time (6, 73). In addition, the nanometer sensors are small enough (e.g., 100 nm) to slip in to and out of a cell's membrane without any damage or leakage. Such biochemical sensors have been used to invesigate blood cells and, in particular, rat embryos (73). The sensor's chemical preparation (photopolymerization) is based on near-field nanofabrication (limiting the size of the produced probe) (19). In addition, the sensing occurs in the near-field regime of the optical excitation, greatly increasing the sensitivity per photon and per sensor molecule. This has decreased the volume needed for (nondistructive) analysis to well below a femtoliter (6, 19).

13.6.1. Photo-Nanofabrication

Using NFO, we have developed a novel nanofabrication technique: photo-nanofabrication, a new and controllable nanofabrication technology (6, 19, 74). It can produce nanometer-sized optical and exciton probes with or without a specific chemical or biological sensitivity. For probes with a specific chemical sensitivity, the probes are automatically FOCS. With NFO principle, photo-nanofabrication controls the size of the luminescent material grown at the top of a light transmitter, such as a micropipette or optical fiber tip, by photo-chemical reactions. These reactions are initiated and driven by an appropriate wevelength of light. The luminescent material is formed (synthesized) only in the presence of light and is "bonded" only to the area where light is emitted. The key to photo-nanofabrication is a near-field photochemical reaction in which the electromagnetic waves of the light sources are mapped by the photochemical process. Thus, the size of the luminescent probe is defined by the light-emitting aperture and is independent of the wavelength of the light used to promote the chemical reaction. The photochemical reaction occurs only in the near-field region (25, 30), where the photon flux and absorption cross section are the highest (see below).

To illustrate the principle of photo-nanofabrication, we here describe the near-field photopolymerization process by which submicrometer optical fiber pH sensors have been prepared. After silanization of the metal-coated fiber-optic tip, photopolymerization is controlled by the light emanating from the near-field light source. The size of the light source and the near-field evanescent photon profile control the size and shape of the immobilized, photoactive polymer (19). The pH sensors are prepared by incorporating the fluoresceinamine derivative acryloylfluorescein (FLAC) into an acrylamide-methylenebis(acrylamide) (BIS) copolymer that is attached covalently to

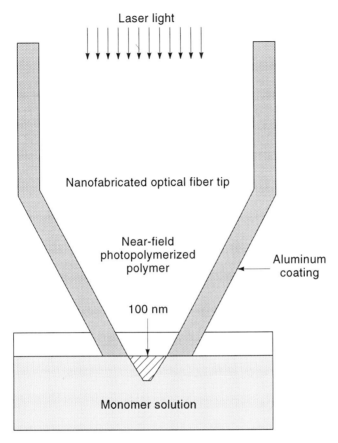

Laser light

Nanofabricated optical fiber tip

Near-field
photopolymerized
polymer

Aluminum
coating

100 nm

Monomer solution

Figure 13.13. Schematic drawing of near-field photopolymerization.

a silanized fiber tip surface by photopolymerization (19, 30). The polymerization process is schematically shown in Figure 13.13. The size of the polymer
grown on the aperture of the optical fiber tip is equal to or smaller than that of
the aperture. Figure 13.14 (see Color Plates) shows two photographs: one of
a subwavelength sensor and the other of the light probe from which the sensor
is nanofabricated. The greenish light is the fluorescence from the intracellular
fluorescence dye–embedded polymer covalently bonded to the activated fiber
surface. By using multistep photochemical synthesis, we have further miniaturized optical probes to sizes much smaller than the sizes of the original light
conductors. The ultimate goal of our photo-nanofabrication technique is to

produce optical, exciton, and sensor probes with molecular size for NSOM, NSOS, MEM, and FOCS by controllable molecular engineering.

13.6.2.   Multidye Photopolymerization

There is great potential for measuring two or more parameters with one miniaturized probe in both chemical and biological analysis (75–77). Multidye doped polymer and multistep photopolymerization have been tested with the fluorescent dye doped-polymer approach (77). By using a multidye solution for photochemical synthesis, we have prepared multifunctional optic probes of micrometer to submicrometer size. These probes emit multiwavelength photons and thus have multiple sensitivity potential providing either internal calibration or the scheme to build supertip sensors (8) based on energy transfer and other processes. There are two different ways to prepare a multidye sensor. The first is to use two or more dye molecules with polymerizable functional groups on photochemical promotion (19). Thus, a crosslinked copolymer will be synthesized, and all the dyes are covalently bonded to the surface of the light probe. The second way is to use a dye-doped polymer (77) in which the dyes are either bonded covalently to the probe or trapped inside the polymer. In preliminary experiments, we used a double-dye system, such as rhodamine B (RhB) with FLAC; or Calcium Green (a calcium-sensitive dye from Molecular Probes, Inc., Eugene, Oregon) with FLAC to prepare sensors and optical nanoprobes.

In Figure 13.15 shows the spectra of a few RhB/FLAC polymer probes prepared with different RhB concentrations in the monomer solutions. In the five emission spectra for RhB/FLAC polymer fiber tips, it can be seen that even with a very low concentration of RhB (10^{-7} M), there is still significant RhB emission in the spectrum. Compared to the pure FLAC polymer spectrum (on the left), there are always some red shifts. The higher the concentration of RhB, the larger the red shift, and the RhB emission becomes more and more dominant in the spectra. The disappearance of the FLAC spectral peak clearly indicates *energy transfer* from FLAC to RhB since the laser excitation of 488 nm is the absorption maximum for FLAC. This demonstrates that we can incorporate two dyes into the copolymer and thus may be able to create multisensitivity, miniaturized optical sensor probes.

13.6.3.   Multistep Photopolymerization

By using multistep photochemical synthesis (25, 30), we have further miniaturized optical probes to sizes much smaller than the sizes of the original light conductors. We developed a multistep near-field photopolymerization to fabricate smaller probes. In the above-described RhB inside

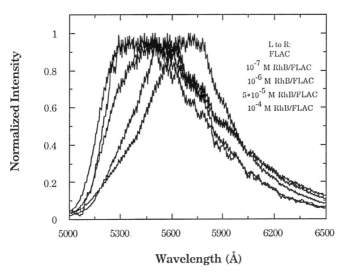

Figure 13.15. Fluorescence spectra of an RhB-doped FLAC multidye optic probe at room temperature.

FLAC polymer system, what is important is the distribution of RhB inside the polymer on the top of the fiber tip. It is reasonable to assume that RhB is homogeneously distributed inside the polymer; thus, there is essentially no effective size reduction for the probe if RhB fluorescence is targeted. However, performing multistep polymerization will ensure a size reduction in preparing the doped polymer probe. We note that usually a cone-shaped polymer (as shown in Figure 13.16) on the fiber tip is obtained in near-field photo-nanofabrication, which makes the miniaturization possible (30).

The basic principle of the multistep nanofabrication process is illustrated in Figure 13.17. We have used the combination of FLAC and RhB as an example of multistep photopolymerization. At the beginning of the process, only FLAC monomer solution is used for near-field photopolymerization. By controlling the reaction time, one should be able to know how much polymer is grown on a fiber tip. When the polymer is grown to a certain thickness, the RhB solution is added to the polymerization solution. We keep disturbance of the polymerization process to a minimum. Then the polymerization process on the fiber tip is continued as usual. Thus, we obtain a polymer tip with RhB only at the cone-shaped tip (6, 30). In multistep photochemical synthesis, the location of the active center is controlled, occurring only at the tip of the probe. We have thus demonstrated the principle of miniaturization of doped polymer probes by the multistep nanofabrication process.

Figure 13.16. SEM micrograph of a submicrometer optical fiber pH sensor.

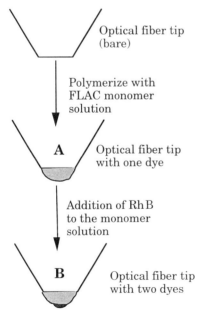

Optical fiber tip
(bare)

Polymerize with
FLAC monomer
solution

A Optical fiber tip
with one dye

Addition of Rh B
to the monomer
solution

B Optical fiber tip
with two dyes

Figure 13.17. Schematic drawing of multistep
photo-nanofabrication with RhB and FLAC.

In multistep photochemical synthesis, the distribution of the designated dye species (active center) can be controlled to occur only at the tip of the probe. Therefore, the probe made this way should have the following advantages: (1) RhB is permanently trapped in the polymer; (2) size reduction is realized by time control; (3) RhB stays on the top surface of the probe; (4) by increasing the concentration of FLAC in the monomer solution, we can make an efficient donor–acceptor energy transfer system in which FLAC is the energy donor site and RhB is the acceptor. We have successfully demonstrated the principle of miniaturization of light probes by both multidye probes and multistep nanofabrication processes. These procedures should be useful in the preparation of subwavelength FOCS.

13.6.4.   Subwavelength Biochemical Sensors

Photo-nanofabrication has enabled us to prepare subwavelength optical probes. One of the most successful applications of photo-nanofabrication to date is the preparation of submicrometer FOCS. We have taken SEM micrographs of the submicrometer optical fiber pH sensors, which clearly show that the sensor is smaller than the original light conductor and appears to become even smaller as the polymer grows (see Figure 13.16). For fabrication of the submicrometer sensor, the polymerization reaction temperature, polymerization time, and laser coupling efficiency had to be optimized to grow submicrometer-sized polymers on the activated fiber tip surfaces. We produced many different sizes, chemical sensors of down to about 0.1 μm. Figure 13.18 shows the pH sensitivity of the miniaturized sensor in the physiological range. This sensor has very high fluorescence intensity even though the sensors used here are extremely small. One major reason is the very high excitation efficiency in the near-field range (19).

13.6.5.   Data Acquisition and Internal Calibration Methodology

We have designed and constructed a new signal detection apparatus for smaller diameter sensors. It is based on an inverted-frame fluorescence Olympus microscope, and signal is collected by the microscope objective. In this apparatus, the microscope is connected with either a spectrometer and a photomultiplier tube (PMT) or an optical multichannel analyzer (OMA). This new apparatus increases collection efficiency by 1 or more orders of magnitude compared to back collection (through the same fiber) (72). The reagent is excited by the incident beam, and the fluorescence light together with the light of the incident beam is collected by an objective lens. The higher the numerical aperture of the objective used for collection, the stronger the signal collected. The fluorescence is transmitted by a dichroic mirror and

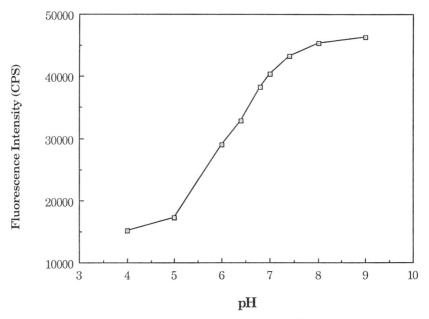

Figure 13.18. pH response of a submicrometer optical fiber pH sensor.

subsequently directed either into a PMT with a bandpass filter (540 nm) or into the OMA for analysis. The submicrometer fiber sensor is mounted on a three-way XYZ translational stage, with X, Y resolutions of 0.5 μm and Z resolution of 0.07 μm. The entire setup is placed on a laser air table for vibration isolation. All samples are mounted on the viewing stage of the microscope for observation before data collection. An argon ion laser beam is used for excitation of the dye polymer. A fluorescence spectrum of the dye polymer of the sensor tip is recorded for pH data analysis.

During the development of submicrometer optical fiber sensors, we noticed that the calibration of these sensors could be complicated and that reproducibility could be a problem for some samples if absolute light intensity measurements are used for pH determinations (25). Also, because these optical fiber sensors are very small, there is only a small amount of fluorophore for sensing. Thus, any significant photobleaching would reduce the accuracy of the measurement. Similary, the geometry for signal collection would be critical if absolute intensity measurements were used. These problems have been overcome by a newly developed internal calibration technique (6, 25).

To enhance the working ability of the miniaturized optical fiber sensors, we used several internal calibration methods to quantify pH. Our method is based on the fluorescence intensity ratio at different wavelengths of the same emission spectrum for a single dye. It is highly effective for small sensors, especially when dye species absorption differences are also utilized. Because various ratios can be obtained by selecting the intensities at different wavelengths of the same spectrum, this approach gives more than a double check for a single experiment. The change in the ratio per pH unit can be increased up to 10 times if two different excitation sources are used. Thus, it greatly enhances the sensitivity and accuracy of our measurments and greatly improves the working ability of these miniaturized sensors in biological samples. The following is an example of how pH is calculated with the internal calibration techniques.

For each solution, a fluorescence spectrum of the sensor is recorded by OMA system. The typical exposure time for taking the fluorescence spectrum is 0.5–10 s for submicrometer sensors. For each spectrum, the usuful ratios are obtained at three different wavelengths from a typical sensor we have made. The ratio of the intensities at two different wavelengths can be utilized to quantify pH. More ratios can be obtained if necessary. For a FLAC-based pH sensor, the ratios at 540/490 and 540/610 nm are large enough for a sensor to measure pH sensitively in the physiological range. The selection of a wavelength for the intensity ratio calculation is critical. Only certain wavelength combinations give accurate calibration and sensitive measurements of pH. It is also worth noting that different sensors have slightly different behaviors for their intensity ratios, probably due to the sensor geometry or dye distribution. Figure 13.19 gives a calibration curve for pH based on internal calibration methods (73).

13.6.6. Sensor Response Times, Sensitivity, and Stability

The miniaturization of the sensor results in very fast response times. The size of the sensor is between 0.1 and 0.5 μm, and no mechanical confinement is used. Thus, the analytes have immediate access to the dye on the sensor tip. This gives our sensors the shortest response times of any reported optical fiber sensors. The response times are well below 50 ms (Figure 13.20) (6).

The submicrometer pH sensor has a very low detection limit. The smallest volume that gives definite pH measurements has only a few thousand hydrogen ions (at pH 8) (19). The submicrometer pH sensor also has reasonable optical stability even when high laser power is used. We have established that removal of oxygen and/or efficient stirring significantly reduce photobleaching. Furthermore, the bleaching is almost imperceptible at the low laser power used under standard operating conditions. With the above performance values and considering the high fluorescence intensity recorded (19), we believe that it should be easy to reduce the sensor to tens of nanometers.

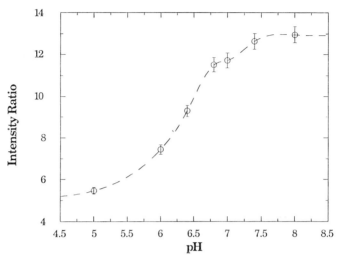

Figure 13.19. Miniaturized pH sensor internal calibration curve. The fluorescence intensity ratio is calculated as I 540 nm/I 610 nm from the sensor fluorescence spectra. The sensor is excited by a 488 nm Ar$^+$ laser beam at room temperature.

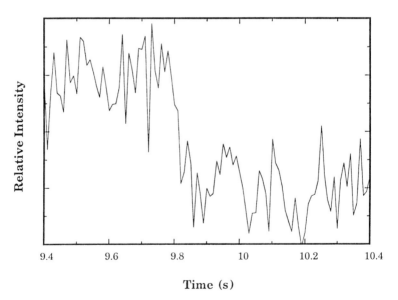

Figure 13.20. Submicrometer sensor response time measurement.

Compared to the state-of-the-art performance of optical fiber sensor (72), a thousandfold miniaturization of immobilized optical fiber sensors, a billion-fold sample reduction, and at least a hundredfold shorter response time have been achieved simultaneously by combining nanofabricated optical fiber tips with near-field photopolymerization. In addition, the submicrometer sensors have improved the detection limits by a factor for a billion (19).

13.7. *IN VIVO* MONITORING OF INTRACELLULAR SPECIES

The ability to control cellular homeostasis throughout development is impor-tant for the timing and regulation of critical events in metabolism, differenti-ation, and cell growth. The mechanisms underlying these processes in develop-ment have, however, not been clearly elucidated and are difficult to assess due to the small size of the organism and the rapidly changing cellular environ-ment. Many traditional methods for determination of cellular, biochemical, and molecular functions have, of necessity, ben invasive, requiring destruction of the cell or organism. Microoptical probes and ultrasmall microelectrodes, as well as other methodologies, have recently been developed to study minute functional biological units *in vivo* and *in vitro* (72–84), including such applica-tions as noninvasive monitoring of biochemical events in periportal and pericentral hepatocytes in the intact liver and from the visceral yolk sac epithelium of intact, viable rat conceptuses. Measurement of pH by conven-tional methods has been done by sensors far from the biological sample following uptake of a chemical indicator substrate or following tissue disrup-tion. One limitation of these conventional techniques has been the need to introduce into the cells or tissues chemicals that could interfere with normal function or, by themselves, alter the very aspects of homeostasis being monitored. The development of probes small enough to be used in extremely small, fragile, and dynamic biological systems, such as early embryos and single cells, will circumvent some of these limitations. The use of fiber optic chemical biosensors for both environmental and biological applications has been growing (72, 73).

Conventional fiber-optic sensors currently in use have tip diameters larger than 100 μm and are thus inappropriate for single-cell applications. In addi-tion, response times have been relatively long, typically seconds to minutes, making determinations of rapid, real-time responses impractical. The above-mentioned ultrasmall optical fiber sensors are capable of spatially resolved measurements in small organisms and single cells. These sensors are capable of very rapid (millisecond) monitoring of chemical and biological reactions. The working range of optical fiber sensors includes probes with tip dimensions in the submicrometer to about 10 μm range. The feasibility of the miniaturized optical fiber sensor has been tested with unknown chemical and biological samples. The biological single units tested by the submicrometer sensor probes are single and living blood cells, frog cells, and rat conceptuses. Most notable are the *in vivo* static and dynamic pH measurements on single and living rat conceptuses (78, 79).

13.7.1. Insertion of a Probe into a Single Biological Unit

There is a common understanding that if the probe is smaller than one tenth of the biological sample, insertion of the probe into the biological unit is considered noninvasive or minimally invasive. *In vivo* experiments then are possible. Our probe is well below the one tenth limit for all the samples we tested. As shown n Figure 13.21 (see Color Plantes) single cell analysis is feasible. The probe can be inserted into different regions of a single cell with the help of a micromanipulator. Therefore, *in vivo* subcellular structure analysis becomes possible.

We have successfully applied miniaturized optical fiber sensors in rat conceptus studies. The rat whole embryo culture system used in these experiments was developed and modified (83). Explanted, cultured, viable, gestational day 10–12 (GD 10–12) rat conceptuses, shown in Figure 13.22 (see Color Plates), are carefully placed in a customized perfusion chamber, which is positioned on the stage of an inverted fluorescence microscope. This apparatus maintains the viability of GD 10–12 rat conceptuses in serum-free medium for over an hour of monitoring. The miniaturized fiber optic biosensors are inserted noninvasively into the extraembryonic space of the rat conceptus, suspended in the customized perfusion chamber, causing little damage to or leakage from the surrounding visceral yolk sac (VYS). The probe is first mounted on a translational stage such as a micromanipulator. Then it is gently directed toward the cell membrane to punch a small hole in the cell membrane. A schematic of how the maniaturized fiber optic chemical sensor is inserted and positioned in the extraembryonic fluid (EEF) space of the conceptus is shown in Figure 13.22. We experienced some difficulties in punching the biological membrane, especially for blood cells smaller than rat embryos. But after careful localization and punching, we were also able to insert the probe into a single biological cell.

13.7.2. Static pH Measurement of Rat Conceptuses

Intracellular and extracellular pH or other species are differentially regulated throughout the course of mammalian embryogenesis. These fluctuations play an important role in the control of cell proliferation, metabolic regulation, and differentiation. Significant changes in the pH of the intracellular and intraconceptual milieus are also important in determining the relative distribution of chemicals from mother to embryo, especially during the teratogen-sensitive period of early organogenesis and prior to the establishment of a functional placenta. The extent to which chemical, environmental, and physiological factors influence the regulation of pH and lead to alterations of normal development is not well known.

The major objective of the development of ultramicrofiber-optic sensors is for biological applications in single cells. The ultramicrofiber-optic pH senor can discriminate pH changes in the physiological pH range. Preliminary experiments, using HBSS without the embryo present, verified the ability of the ultramicrofiber-optic sensor to discriminate less than 0.1 pH unit changes in the pH range of 6.6–8.6. Next, GD 10, 11, and 12 conceptuses were analyzed for differences in pH as a function of advancing gestational age. EEF pH measurements, determined by using the fiber-optic sensor, show values of 7.50–7.56 in the 10–16 somite, GD 10 rat conceptus, and pH values of 7.24–7.27 in the 32–34 somite, GD 12 conceptus. These data, as shown in Table 13.2, agree well with the mouse data presented for a conceptus of the same relative stage of development, based on somite number (80–82). Comparison of GD 12 rat data using the same reference sources (82) shows that our fiber-optic pH sensor records pH in the EEF at a level approximately 0.20 pH units lower than that reported. The discrepancies may be due to the differences between EEF alone and the pH of the whole conceptus or to real differences in the tissues themselves. In evaluating these discrepancies, it is important to note that our measurements are direct and can be accurately compared to other

Table 13.2. Static pH Measurements of Whole Rat Conceptuses: Comparison of pH Measurements in the Extraembryonic Fluid (EEF) in Whole Rat Conceptuses as Determined by the Ultramicrofiberoptic pH Sensor with Those Using a Radiochemical Method (82) in Whole Conceptal Homogenates

Day of Gestation (Rat)	Somites	pH (Fiberoptic)	Day of Gestation (Rat)	Somites	pH (Radiochemical)
Day 10(1)	10–16	7.56 ± 0.08	Day 10	n.a.	n.a.
Day 10(2)	10–16	7.50 ± 0.05			
Day 12(1)	32–34	7.27 ± 0.07	Day 11.5	27–30	7.47 ± 0.03*
Day 12(2)	32–34	7.24 ± 0.05	Day 12	33–37	7.44 ± 0.04*

Note: These data show a consistent level of EEF acidification with ascending gestational age in the rat conceptus.

n.a.: Data not available.

* Data obtained from Dean et al. (82) using the relative distribution of $[^{14}C]$-DMO to calculate intracellular pH.

determinations, regardless of size, age, or relative fluid volume of the cell or organism being monitored. The current application of our method does not require pools of conceptuses for determinations and can utilize *single live* conceptuses for the pH measurements. Traditional methods are based on relative concentration differences between maternal and embryonic tissues that occur as a radiolabeled weak acid diffuses preferentially into compartments of high relative pH. These methods also fail to account for difference in weak acid metabolism and altered distribution that may occur with time and between species. The use of a single embryo has numerous advantages over pooled embryos, in that it is possible to maintain structural and functional integrity while monitoring static and dynamic pH changes in the EEF.The conceptus also serves as its own control, can be accurately characterized as to developmental stage, can be monitored spatially and temporally in real time, and can be returned to the whole embryo culture to evaluate other relevant endpoints later in development.

13.7.3. Real-Time Monitoring during Environmental Changes and Direct Chemical Exposure

The miniaturization of this fiber-optical sensor has not occurred at the expense of response time, current sensors are able to respond to events of less than 50 ms (6). We have made use of this feature in determining the dynamic response of a rat conceptus to alterations in its *in vitro* environment. A perifusion system for dynamic experiments has been built (Figure 13.23). Experiments were conducted to ascertain the ability of the GD 12 rat conceptus to respond to changes in tis environment and to selectd chemical agents. Three different experimental protocols were used. First, the perifusate bathing the embryo was saturated with 100% nitrogen to create a condition of hypoxia. The measured pH over time did *not* vary significantly from the initial control readings. Even though hypoxia does not immediately alter hydrogen ion concentrations in the conceptus, as determined by measuring EEF pH, it does not preclude the possibility of intracellular changes taking place within cells of the embryo that do not affect the equilibrium between EEF and intracellular fluid compartments. Second, the conceptual response was determined by monitoring EEF for pH when the perifusate, containing HBSS, was altered over a pH range of 6.6–8.6. Again, the measured pH in EEF did *not* result in any significant variations from control levels. Finally, the dynamic effects of a thiol oxidant, diamide [used to oxidize intracellular glutathione (GSH) to glutathione disulfide (GSSG) and induce oxidative stress in the embryo], was evaluated (83). A 500 µM solution of diamide in the perifusate (HBSS, pH 7.4) results in an initial rapid decrease in pH over the first 30s, followed by a slower downward trend thereafter. An absolute decrease of about 0.3 pH units occurs

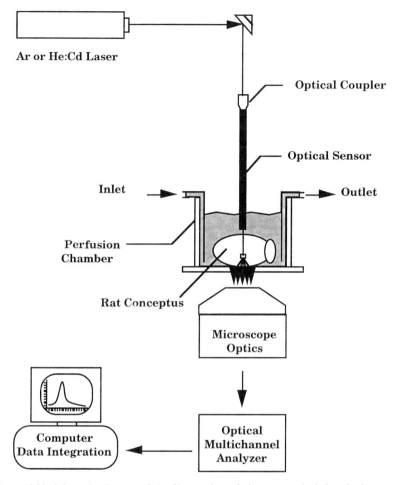

Figure 13.23. Schematic diagram of the fiber optic perfusion system depicting the intact rat conceptus in the customized monitoring chamber located on the stage of an inverted fluorescence microscope. The emission signal originating from the tip of the pH sensor inserted into the EEF compartment is detected through the microscope optics, directed to an OMA, and integrated using a microcomputer. This apparatus is able to maintain the viability of GD 10–12 rat conceptuses in a serum-free medium (HBSS) for over 1 h of monitoring.

within 3 min, as shown in Figure 13.24. A decrease in the pH_i of this magnitude has been shown to have significant effects on the activity of a number of enzymes that are critical for normal growth and development (84).

The ultramicrofiber-optic pH sensor can discriminate pH changes of less than a tenth of a pH unit over the range of 1 pH unit above and below the

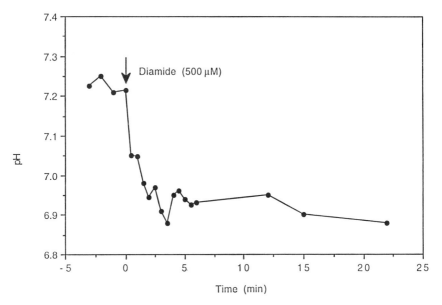

Figure 13.24. Drastic alterations is the conceptal environment, such as hypoxia and wide variations in extraconceptal pH, do not result in acute alterations of pH in the EEF. The thiol oxidant diamide, however, causes significant, rapid acidification of pH in the EEF within 30 s of exposure. The ultramicrofiber-optic sensor is capable of spatially resolving and monitoring dynamic changes in pH in viable intact rat conceptuses during organogenesis.

physiological pH of 7.4. The pH measurements were obtained in real time on a single, intact, viable rat conceptus under conditions of environmental change and direct chemical exposure. The insertion of the ultrasmall sensor through the VYS into the EEF appeared to cause no damage to or leakage from the involved tissues. This demonstrates the advantage of an essentially noninvasive approach compared to conventional means, which require the disruption of large numbers of conceptuses to obtain less sensitive and indirect measurements. Conceptuses of ascending developmental age undergo acidification of the EEF. The ability of ultramicrofiberoptic sensors to measure pH changes in real time in the intact rat conceptus demonstrates their potential applications for dynamic analysis in small multicellular organisms. Working sensor probe dimensions and response characteristics also make this approach feasible for use in single cells. The application of this novel technology to studies of developmental regulation, pharmacokinetics, toxicology, and physiology will provide valuable spatial and temporal information not heretofore available using conventional techniques.

13.8. ACTIVE LIGHT SOURCES AND MOLECULAR EXCITON MICROSCOPY

In NSOM, the size of the light source determines the resolution of the imaging, provided that it can be scanned in the near-field region. At nanometer size, the number of photons emitted from coated micropipettes or nanofabricated fiber tips is so small that it becomes very difficult to transmit, in the transmission mode NSOM, or excite a large enough number of fluorophores, in the fluorescence mode NSOM, such that their subsequent transmission or emission can be detected. Thus, it is impossible to detected a single molecule. Alternative light sources have been proposed and prepared for obtaining molecular resolution, most notably the molecular exciton source (4, 11, 30, 85). Such a light source incorporates a nanometer-sized organic crystal at the end of a pulled micropipette with an inner diameter of a few nanometers. Excitation of the crystal then results in the creations fo exciton whose travel through the crystal is limited only by the crystal's size, which potentially could be of a molecular dimension. Excitation of a sample would then proceed through an energy transfer mechanism or other near-field operations, which may be many orders of magnitude more efficient than emission and subsequent reabsorption by the sample. Therefore, we could accomplish NSOM with molecular resolution and sensitivity.

13.8.1. Active Light Sources

In NFO, the requirements of smallness and intensity are in direct conflict. All these passive light sources (2,3) are typically apertures letting light through, and when the size of the aperture is significantly below that of a wavelength, most of the light will be diffracted or reflected back rather than transmitted (8, 86, 71). For instance, for a hole in a metal plate for apertures of 500 Å or less, the intensity goes down superexponentially with aperture. This occurs because the metal surrounding the hole has to be at least about 500 Å thick (87) to be opaque enough to define a hole. The metal with the greatest opacity in the visible region is aluminum for which $d = 500$ Å (87) when the wavelength is 5000 Å. With a given minimal thickness, the emerging light intensity I is described by (86):

$$I \sim r^6 \exp(-L/r) \tag{1}$$

where L is the length (thickness) of the aperture. Reducing r by 1 order of magnitude reduces I by about 20 orders of magnitude. This assumes that the wavelength λ is much larger than the aperture:

$$\lambda \gg r \tag{2}$$

Equation (1) is valid only when Eq. (2) is valid. The length (thickness) L has to be significantly larger than the penetration length l of the radiation into the metal. The shortest known penetration length (for visible light) is a couple of hundred angstroms (in aluminum). Therefore, the smallest practical value for L is 500 Å (using aluminum).

Recently, three methods have been suggested for overcoming the above-mentioned handicap, i.e., the severe loss of light in subwavelength cavities.

1. *Transforming the photons into physically smaller energy quanta, such as excitons (88), followed by the retransformation of these energy quanta into photons* (*evanescent photons*). While there are significant inherent losses in such a process (see below), the dependence on aperture size is significantly less steep, i.e., linear with the cross section (8):

$$I \sim r^2 \tag{3}$$

   Therefore, effectively, there is either no loss or less severe loss of brightness with reduced size.

2. *Overcoming Eq. (1) by overcoming Eq. (2).* In practice, this means an effective reduction of λ. It is well known that within dielectric materials, λ is reduced from the vacuum (or air) value by a factor equal to the refractive index n. One can thus choose appropriate transparent materials with high n. The resultant reduction in λ effectively pushes down the diffraction limit—a well-known trick (e.g., the use of lenses with oil rather than air). In general, n has a value of about 1.3–1.5. However, using materials at the edge of their absorption wavelengths (or appropriately shifting the wavelength to this absorption edge) may effectively increase n by 1 order of magnitude.

3. *Using the waveguide (or "coax")* principle. The use of appropriately designed optical interfaces creates an "optical fiber" in the sub-wavelength regime of evanescent waves (8,45).

In a more general approach, the generation of photons in an *active* light source is separated from the problem of transmitting the photons through a cavity. In principle, the light source may generate light from electricity, heat, chemical processes, nuclear processes, excitons, or from light of a different wavelength (8, 87). This opens the way to a large number of potential schemes. It may even end the monotonic relation between size and brightness (see above).

Here is a more specific example of analysis with micropipettes. The optical imaging resolution limitation derives from the less than ideal characteristics of the aperture. There are no propagating electromagnetic modes in a sub-

wavelength cylindrical metallic waveguide such as a metal-coated micropipette. The least attenuated mode for a round aperture has been found to be
the TE_{11} mode, for which the energy decays at a rate of approximately

$$E_1 = E_0 \exp(-3.62\,L/r) \tag{4}$$

where L is the length of the aperture formed by the metal coating and r is the
radius (8). With a sufficiently rapid tapering of the micropipette, however,
this evanescent region can be kept short enough to obtain a fairly large
throughput of light. What ultimately limits the resolution is the finite
conductivity of the metallic coating around the micropipette. The electromagnetic wave penetrates the coating and decays witvn it at a finite rate as well,
given by (8)

$$E_2 = E_0 \exp(-d/c) \tag{5}$$

Where d is the depth of the penetration and c is the extinction length of the
metal (87). When the attenuation due to the waveguide effect exceeds the
attenuation in the metal, the contrast between the aperture and the surrounding medium becomes insufficient for superresolution applications. Thsu E_1
must be larger than E_2 to have an effective subwavelength light source.

Using molecular excitons, Lieberman et al. (4) proposed an active light
source for NSOM and MEM. The basic process is the conversion of a passive
light source into an active light and exciton source. The energy-packaging
abilities of certain materials are used to circumvent the boundary problem of
the edge of the aperture. According to our fundamental understanding of
energy propagation in materials, excitation can be confined to molecular and
atomic dimensions under appropriate conditions (88, 89). Using this property
of materials, one can develop a subwavelength light or exciton source by
growing a suitable crystal within the subwavelength confines of a micropipette. Then energy can be guided directly to the aperture at the tip instead
of being allowed to propagate freely in the form of an electromagnetic wave.
Such a material can be excited through an electrical or radiative process to
produce an abundance of excitons, allowing light to be effectively propagated
through the bottleneck created by the subwavelength dimensions of the tip
near the aperture. The excitons can be generated directly at the tip or within
the bulk of the material and allowed to diffuse to the tip via an excitonic
(electric dipole, Förster) transfer (89). In either case, in a suitable material these
excitons will then undergo radiative decay, producing a tiny source of light at
the very tip of the micropipette. The excitonic throughput is basically independent of the wavelength and is a linear function of the cross-sectional area of the
aperture (8).

As described above, in past schemes intensity was severely limited by the requirement to reduce the size of the light source. While this situation has been somewhat alleviated with the use of *active* sources (compared to "empty" holes), other problems have arisen. One such concern is the durability of the active material. For instance, the anthracene crystals used in the first exciton light source (4) deteriorated significantly within minutes [resulting in long-time amplification factors (25) of about 3 compared to several orders of magnitude initially]. Photochemistry, thermochemistry, and other processes are the obvious culprits. For instance, it is well known that in the presence of oxygen and ultraviolet light, anthracene is oxidized to anthraquinone (89), and other substances. Obviously, one can either avoid the oxygen, reduce the temperature, or replace the material with a more stable one. The problem is thus transformed into one of molecular engineering, a common occurrence with modern optics as well as other materials.

13.8.2. Preparation of Active Light Sources

Active light sources are those light and excitonic sources created by lumines-cent materials. There are two major techniques in the preparation of active subwavelength optical and exciton probes, as shown in Figure 13.25 (see Color Plates). The first one uses optical fiber probes and bonds luminescent materials covalently to the top surfaces of probes by photo-nanofabrication (19, 30, 74); the second one uses micropipettes or nanofabricated optical fiber tips to hold crystals or doped polymers at their tips (4, 28–33, 85). The nanofabricated optical fiber probes have been used successfully for the preparation of active light and exciton sources. For example, we have described the submicrometer optical fiber pH sensors prepared by incorporating fluoresceinamine into an acrylamide copolymer attached covalently to an activated fiber tip surface by near-field photopolymerization (6, 19). The same technique can be used to prepare other subwavelength active light and exciton sources.

Nanometer crystal light and exciton sources are prepared with organic and inorganic crystals or molecularly doped polymers grown inside the tip of a micropipette from a solution. We have used a large variety of crystals and doped polymers. The luminescent materials involved are anthracene, perylene, 9,10-diphenylanthracene (DPA), 9,10-dimethylanthracene (DMA), pyrene, fluorescein and its derivatives, rhodamine series dye, various aminoan-thracenes, tetracene, DCM (25), BASF dyes (25), rubrene, dendrimers (90, 91), uranyl compounds, CsCl, zinc sulfide (ZnS), and cadmium sulfide (CdS), as well as dye-doped polymers. These active probes can be excited by different laser lines from UV to visible. The techniques used in crystal growth inside the micropipette are crystal growth from solution, from melt, from vapor, from chemical reactions (25), and from other sources. The details of preparation of

Table 13.3. Active Light Source Throughput Amplification

Material Used for Active Light Source	Amplification Factor
Anthracene	3.2
1-Aminoanthracene	5.4
Perylene	6.8
DCM	2.9

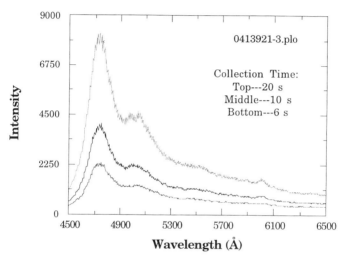

Figure 13.26. Spectra of a nanofabricated optical fiber perylene tip. Perylene is coated on a fiber tip and excited by a 4416 Å laser line.

nanometer crystal light sources were described previously (25, 30). The physical sizes of the active light or exciton probes are as small as 0.05 μm and are mostly in the range of 0.1–0.5 μm. The effective size of these active light or exciton sources has not been exactly determined. The light amplification factors of these active light sources are listed in Table 13.3 (25). Tiny polymer aggregates for microscopic studies have also been prepared. These light sources have subwavelength dimensions and could be used in a wide variety of fields. Perylene tip spectra are shown in Figure 13.26.

13.8.3. Molecular Exciton Microscopy

The concept of active light sources enables a totally new mode of NSOM, based not on the blocking or absorption of photons but rather on quenching directly the energy quanta that otherwise would have produced photons

(8, 11, 92). Even though MEM has not been fully implemented and well developed (25), it holds the promise of single-molecule spatial resolution with biochemical sensitivity. For instance, a thin, localized gold film (or cluster) can quench an excitation (or exciton) that would have been the precursor of photons. Furthermore, a single atom or molecule on the sample could quench (i.e., by energy transfer) the excitations located at the tip of the light source. For simplicity, we assume that the active part of the light source is a single atom, molecule, or crystalline site, serving as the "tip of the tip." This quenching energy transfer from the excitation source's active part (donor) to the sample's active part (acceptor) may or may not qualify technically as an NSOM technique. Currently, however, it is the best hope for single-atom or single-molecule resolution and sensitivity. The technique basically is a form of quantum optics microscopy. It has been called molecular exciton microscopy (MEM) (92).

MEM is conceptually quite similar to STM (17). Its basic working principle is shown schematically in Figure 13.27. The excitons "tunnel" from the tip to the sample. However, there is no driving voltage or field. Rather, it is the energy transfer matrix element that controls transfer efficiency. Its unusual matrix elements provide the highest sensitivity to distance, higher than that of STM and comparable to that of AFM. In addition, the most striking result of this direct energy transfer is its untrahigh sensitivity to isolated or single molecular chromophores. Quantum optics energy transfer is highly efficiency within the range of the "Förster radius." Thus, a single excitation could be "absorbed" by the sample acceptor. In contrast, based on the Beer–Lambert law (89), about a billion photons are needed to excite a single accepter in the absence of other acceptors. Furthermore, as the distance range is limited to about 10 nm for direct energy transfer, MEM is as much a near-field technique as STM or AFM, i.e., very sensitive in the single-digit nanometer range and much less sensitive beyond 10 nm. However, in combination with conventional NSOM, the range can be extended to about 200 nm. Thus, MEM is

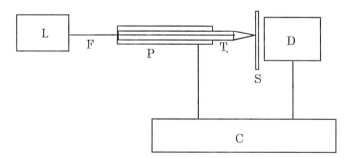

Figure 13.27. Schematic drawing of MEM. Key: L, light source; F, optical fiber; P, pizeoelectric tube; T, scanning exciton tip; S, sample; D, lightdetector; C, computer-controlled force feedback system.

a technique that is able to "zoom in" from macroscopic to nanoscopic distances. Obviously, such zooming in enhances the speed of operation. It also allows for a much more universal range of samples, from metal spheres and clusters to soft, *in vivo* biological units. In addition, MEM can use fluorophores, metal clusters, and so on to enhance contrast, sensitivity, and resolution with the help of NSOM. It can also be used in conjunction with lateral force feedback in the same way as NSOM. We should emphasize the MEM is still in its infancy. More development will be needed before practical applications become feasible. Some of the exciton light source results do lend credence to our MEM concept.

13.9. SOURCE–SAMPLE INTERACTION MODE

In our application of the NSOM technique to biological, chemical, and physical problems, it is apparent that the image contrast generated in the near field is not always a simple convolution of the optical properties of the tip and sample. The effect of tip–sample interactions can often be seen in NSOM data. These interactions are a consequence of the proximity of the probe and sample, a feature shared by all scanned-tip microscopies. Indeed, the basis of most scanned-tip microscopies is isolation of the effect of a given tip-sample interaction and exploitation of its dependence on some property of the sample to obtain a two-dimensional image. This is where NSOM differs from the other scanned-tip microscopies and the reason for its lack of any built-in feedback. In an ideal NSOM, the intensity distribution at the exit of the NSOM tip is constant and their is no tip–sample interaction. The image obtained is then a linear convolution of the point spread function of the tip and the optical properties of the sample. As NSOM has been applied to specific samples, it has become apparent that actual NSOM is not the ideal NSOM mentioned above, but rather, a hybrid of the ideal NSOM and various nonideal image contrast mechanisms rooted in tip–sample interactions.

In traditional optical, X-ray, and electron microscopy, the *source* first produces photons or electrons, which then interact with the sample. Obviously, this is not the case for scanning probe microscopies. Similarly, when a near-field *active* light source is close (much less than λ) to the sample, source–sample interactions arise. Several examples will now be presented.

13.9.1. Excitation Transfer Interactions

A molecular nano-crystal light source (say, anthracene) will transfer its energy to a chromophore molecule (say, rhodamine) or the sample (say, Langmuir–Blodgett film). Thus, the light source becomes an energy *donor* and the sample molecule an energy *aceptor*. The result is a fluorescence emission typical of the sample. Superficially, this may appear to be no different from the ordinary radiative process in which photons are first emitted from the source and then absorbed by the sample, causing it to fluoresce. However, in reality this is a *nonradiative* (Förester–Dexter energy transfer) process (88, 89). It may be much more efficient than the radiative process, and it will exhibit very different quantitative behavior. For instance, the dependence of fluorescence on the distance from the probe is much steeper (fourth to sixth power) compared to a very weak distance dependence for the radiative process (depending on geometry). Also, light polarization dependence may differ in the two cases.

Energy trasnfer experiments were conducted on dilute, shallow solutions of RhB covered with lipid monolayers and on monomolecular Langmuir–Blodgett (LB) films (93) of arachidic acid Di–O (4:1). Using both micropipettes and FOCS tips, we prepared a series of exciton probes for energy transfer studies. Excitons produced in the nanometer crystal source are the energy donor site for a probe-to-sample Förster transfer (89) or the spin-orbit interaction site for a probe-to-sample Kasha effect (94). To demonstrate the feasibility of the above ideas, we used an anthracene or DPA crystal tip and a monolayer LB film containing a dye molecule at low concentration. From a distance of approximately 0.5μm, the DPA tip is just a greenish-blue light

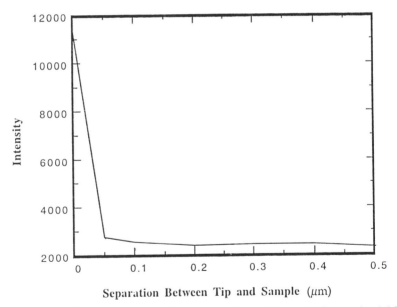

Separation Between Tip and Sample (μm)

Figure 13.28. Preliminary energy transfer experiment by a DPA crystal tip scanning 1:4 Di–O/arachidic acid LB film.

source. This emitted blue light is absorbed by the dye containing film, resulting in a yellow-orange dye fluorescence. This absorption is fairly inefficient and is barely dependent on the Z distance (crystal tip to LB film), as the number of dye molecules in the beam is practically independent of Z ($Z < 0.5\,\mu m$). However, once the DPA tip practically touches the LB film (coming within the Förster radius of about 5 nm), the mechanism of energy (exciton) transfer dominates the LB film excitation process and much stronger yellow-orange fluorescence is observed. This preliminary result is plotted in Figure 13.28. The large enhancement of acceptor fluorescence is attributed to nonradiative direct excitation (energy transfer), i.e., direct source–sample coupling, which is highly sensitive to z. In contrast, long-range indirect coupling via photons depends only weakly on z and is trivially accounted for by considerations of geometrical optics. These results lend credence to our theoretical modeling of nanometer or subnanometer optical and exciton probes.

13.9.2. Excitation Quenching and Transformation Interactions

A simple example consists of a molecular light source (e.g., anthracene) and a metallic sample (e.g., silver). At subwavelength ($< \lambda$) distances, the source's photon flux is modulated and even quenched by the sample—the Kuhn effect (88, 89). Again, this quenching effect can be considered a special case of the energy transfer mechanism, on without radiative reemission.

A special case of quenching is the Kasha effect (94). A heavy atom (atomic mass > 40) in the *sample* quenches the *source* emission. However, as a result, the source may emit light at a longer wavelength (i.e., phosphorescence rather than fluorescence). This transformation effect is based on interatomic spin-orbit coupling whereby one atom (the heavy one) has a high degree of spin-orbit coupling (relativistic effect) and the other atom (the light one) has very little of it on its own. The effect is "internal" when both atoms belong to the same molecule and is "external" when they are not. The effect is empirically observed spectroscopically as a change in the intensity or lifetime of an absorption or emission (89). For instance, a molecule such as anthracene has a fluorescence quantum efficiency of near unity because the first excited singlet state cannot transfer its energy, to the lower-lying first excited triplet state due to spin selection rules. However, on induced spin-orbit coupling, such singlet–transfer becomes possible both internally (with the aid of vibrational quanta that absorb the extra energy) and radiatively. Empirically, one observes quenching of the fluorescence (both intensity and lifetime are reduced). These effects have been observed only when both molecules are neighbors (or collide) inside the same phase (e.g., in liquid solution) (89). The interaction occurs at extremely short range (e.g., 5 Å or less). Our experiment employed this effect at the interface of two distinct phases, i.e., the scanning supertip and the acceptor

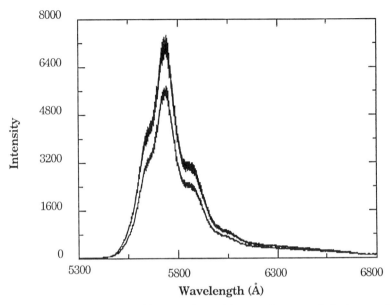

Figure 13.29. Perylene tip quenching experiment. Top: perylene crystal tip immersed in NaCl/water, bottom: perylene crystal tip immersed in NaI/water.

sample, where certain functional groups are chemically substituted for by heavy atoms such as Hg or I.

Our experiment involved a solid film of perylene in contact with a liquid solution (water) containing heavy atoms (NaI). We observed significant fluorescence quenching effects for two dituations: (1) the perylene film is at the bottom of an optical fiber tip that is scanned toward contact with the water solution (and back out again); (2) the perylene is a thin solid film floating on the water solution. As seen in Figure 13.29, the quenching is significant. We emphasize that blank experiments contain water solutions with NaCl (which show no quenching). Furthermore, as soon as the optical fiber tip (with the perylene "supertip") is pulled out of the water solution, the fluorescence returns to its old value. This eliminates the possibility that the perylene was dissolved, chipped off, or reacted chemically with the iodide (or formed some permanent van der Waals complex with it). This is the first observation of an interfacial Kasha effect. The interfacial effect occurs in the near-field zone and illustrates the principle of MEM. We believe that this source–sample Kasha effect presents one of the most sensitive and highest resolution methods combined. We have thus suggested (95) its use for *in situ* gene sequencing: the MEM scans a long DNA strand, after heavy-atom substitution of only one kind of base (say A), and detects these bases individually while scanning along the immobilized strand.

13.9.3. Single-Molecule Sensitivity

The goal of MEM is to achieve single-molecule resolution with single-molecule sensitivity. Elementary considerations (88, 89) show that nearly a billion photons of the right frequency must hit a highly absorbing molecule (oscillator strength near unity) to cause a single excitation (on the average). Assuming that this molecule is also an excellent light emitter (quantum efficiency near unity) and that the detector system requires about 100–1000 emitted photons/s, one needs about 10^{11}–10^{12} photons/s to emanate from the light source. This is now just possible with an 0.1 μm optical fiber (see above). However, a tough problem still remains—to filter out the billion times stronger excitation light. This calls for very demanding spectral and/or time and/or other optical filtration.

An alternative method of single-molecule excitation has been utilized by nature's photosynthesis (8,96). An antenna made of hundreds of dye molecules absorbs the light and transmits the excitation to the desired "active center." The transmission is done via excitation transfer (exciton transport), which usually is just a multistep Förster energy transfer—from one antenna molecule to the next and eventually from the nearest antenna molecule to the *active center* (acceptor) molecule.

An antenna of a million molecules needs only about 1000 photons to be excited (once). This excitation may be transmitted to the acceptor molecule with 90% efficiency under favorable conditions (88). Thus, the excitation sensitivity has been improved by 5 orders of magnitude. Furthermore, the problem of background discrimination (filtration of exciting light) has been reduced by a factor of a million. Even under less favorable antenna transfer conditions, the improvement is striking.

In practice, our working scheme is given by Figure 13.30. The antenna is a molecular crystallite or aggregate (say, anthracene). One of its molecules (active center) is closest to the single sample molecule (say, tetracene or perylene). This distance is 10–20 Å [much smaller than the Förster radius (89)]. The antenna (say, a million molecules) transfers excitations to the *active center*, which in turn transfers its energy to the single-sample molecule. Overall, only 10^4–10^6 photons are now required for a single excitation of the sample molecule, i.e., 10^3–10^5 time less than with direct excitation.

Here we switch terminology. The antenna tip is an *exciton tip*. First, the excitons are trapped by the active center. Then the exciton "tunnels" from the tip's active center to the sample molecule, causing the latter to emit a photon. Only about one exciton is needed for this single-molecule excitation. However, 10^3–10^5 photons are needed to produce that exciton. Thus, the exciton approach, exemplifying a direct source—sample interaction, significantly improves single-molecule sensitivity.

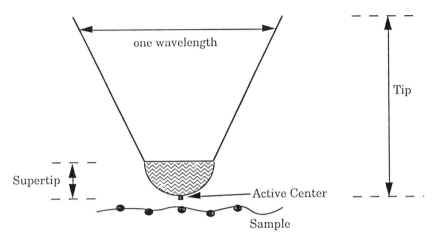

Figure 13.30. Schematic drawing of a tip, a supertip, and an active center. A fiber-optic tapered tip, a crystallite antenna supertip, and a single-molecule active center. The sample shows acceptor molecules or moieties.

13.9.4. Single-Molecule Spatial Resolution

In STM, a single atom at the tip is the active center. This is not achieved by atomic precision fabrication but by "natural selection"; the tip can be produced with ordinary scissors, and the electrons find a defect site of atomic size. If the same principle of defect site selection applies to the molecular exciton tip, then molecular spatial resolution has been achieved. Such resolution is the aim of present work in the field. Specialized molecular engineering approaches with supermolecules are also being investigated (90, 91).

13.9.5. Enhanced Analytical Sensitivity by Active Light Sources

As mentioned above, nearly a billion photons of the right frequency must hit a highly absorbing molecule to cause a single excitation in ordinary absorption and excitation (8, 89). With an extremely close active light source, the excitation of a single molecule may need only one exciton (8, 97). This opens the way for single-molecule detection and localization with our active light sources. It may also lead to new optical analysis techniques with a much improved detection limit or sensitivity.

"Detection limit" and "sensitivty" (98) are terms frequently used in analytical chemistry. The term "detection limit" here describes the lowest concentration or weight detected in chemical analysis. An active light and exciton source should have a higher detection limit than a passive light source. We emphasize that the active light source actually is an exciton source. When the exciton source is stated, first, it means that the optical probe produces excitons. Second, the excitons produced will transfer their energy to acceptor molecules through nonradiative means such as Förster energy transfer (89). Following is an example illustrating the principle of detection limit enhancement.

As shown in Figure 13.31, there are two kinds of light sources with two different concentrations of samples (dye molecules in a liquid solution). One of the optical light sources is a passive optical probe, such as an empty micropipette tip or an optical fiber tip, while the other is an active light source (i.e., an

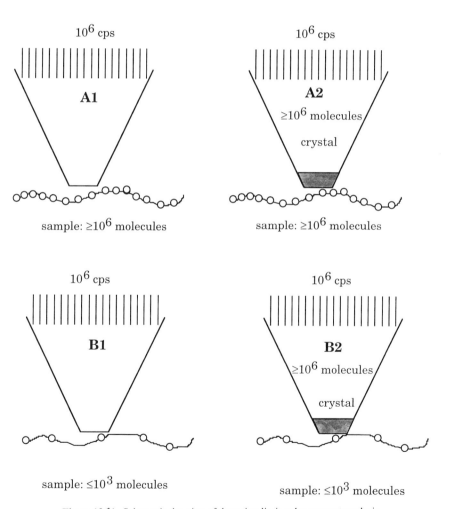

Figure 13.31. Schematic drawing of detection limit enhancement analysis.

exciton source such as anthracene or DPA crystal light sources). The samples are acceptor molecules that can be excited by both light sources. Suppose that the sample and the probe are within Förster energy transfer distance and the fluorescence quantum efficiency of the sample molecule is 1; then the detection limit will depend on the properties of the light sources and the concentrations of the samples. Suppose that there are 10^6 photons per second (pps) from the incident light beam in both probes. In cases A1 and B1, all the photons will excite sample molecules directly, while in cases A2 and B2, photons are first absorbed by the crystals deposited at the nano-optical probe tip. As we discussed above, excitons are produced, and their energy is directly transferred to acceptor molecules in the sample if such acceptor molecules exist within Förster energy transfer distance.

Theoretically, the number of fluorescence photons produced from the sample in each situation can be estimated (25). There are three assumptions. First, in far-field optics, about 1 billion photons are needed to excite one highly absorbent isolated molecule. Second, when a molecule is within the Förster radius of an exciton, the molecular excitation probability is on the order of unity. Third, the quantum yield of the sample molecule is assumed to be near unity. The follwoing definitions will be used:

C: number of molecules in the crystal light source
E: number of excitons proudced in the crystal light source
F: number of fluorescence photons produced in sample molecules
I: number of photons in the incident light beam
Q: excitation probability
S: number of molecules in a sample

In the A1 case, an order of magnitude estimation gives

$$D_{A1} = I_{A1} \times Q_{A1} \times S_{A1} = 10^6 \times 10^{-9} \times 10^6 \times 10^3 \quad \text{(pps)} \qquad (6)$$

In the A2 case, excitons are first produced inside the crystal light source:

$$E_{A2} = I_{A2} \times Q_{A2} \times C_{A2} = 10^6 \times 10^{-9} \times 10^6 = 10^3 \quad \text{(excitions per second)} \qquad (7)$$

Since only 10^3 excitons are produced, even if there are more than 10^3 acceptor molecules in the sample, the maximal number of fluorescence photons will only be 10^3 (pps):

$$F_{A2} = E_{A2} \times Q_{A2} = 10^3 \times 1 = 10^3 \quad \text{(cps)} \qquad (8)$$

Now we can define a ratio of fluorescence photons (R) to compare the fluorescence intensities in each situation; R is defined as the ratio of the fluorescence intensity of the sample acceptor excited by an active light source divided by that excited by a passive light source. Thus the ratio of the fluorescence intensities in the above two situations is

$$R_A = F_{A2}/F_{A1} = 1 \tag{9}$$

In these two cases where there are large numbers of sample molecules, there is no detection limit enhancement and no need for it. The efficient excitation of sample molecules by an active light source is not important due to the large number of acceptor molecules present in the sample. However, this is not the ideal situation to illustrate the enhancement of the detection limit. In case B we have a much lower concentration of sample molecules. The concentration is 1000 times less than that in case A, i.e., only 1000 molecules in the sample. The same calculations are shown here

In the B1 case:

$$F_{B1} = I_{B1} \times Q_{B1} \times S_{B1} = 10^6 \times 10^{-9} \times 10^3 = 1 \quad \text{(pps)} \tag{10}$$

In the B2 case, again, excitons are first produced inside the crystal light source:

$$E_{B2} = I_{B2} \times Q_{B2} \times C_{B2} = 10^6 \times 10^{-9} \times 10^6 = 10^3 \quad \text{(excitons per second)} \tag{11}$$

The fluorescence photons produced from the sample should be

$$F_{B2} = E_{B2} \times Q_{B2} = 10^3 \times 1 = 10^3 \quad \text{(pps)} \tag{12}$$

Now the two fluorescence intensities are different. The ratio is

$$R_B = F_{B2}/F_{B1} = 10^3 \tag{13}$$

This illustrates that by using an exciton source, one can achieve 1000 times better detection limit in analyzing very low concentration samples. This provides a novel approach for single-molecule detection, as well as for other fluorescence detection techniques aimed at extremely small or diluted samples.

We did preliminary detection limit experiments with different donor–acceptor systems. The key to a successful demonstration is that a large fraction of the excitons produced in the exciton sources must be within Förster energy transfer distance of the sample molecules. There are a few ways to construct such an exciton source (25, 28, 30). The first way is to have a thin layer of donor molecules, such as four layers of anthracene mleceules, which is about 40 Å thick. The second way is to produce the above-discussed funnel effect (88), i.e., an active center (8). The third way is to let the crystal tip touch the acceptor molecules in the sample. In all three cases, the energy quanta will be transferred from the exciton source to acceptor molecules in the sample.

The principle of detection limit enhancement has been demonstrated by active light sources as well as organic thin films. Active light sources such as anthracene, DPA, and perylene crystal tips were used as energy donors to transfer their energy to Rhodamine 6G (R6G) in aqueous solutions of different concentrations, from 10^{-3} to 10^{-7} M. Then 355 nm laser light was used to excite both R6G and the active light sources. The expected enhancement in detection limit was observed. For example, a DPA crystal light source was scanned down toward a 10^{-6} M R6G aqueous solution. The R6G fluorescence intensity was increased up to 15 times over a passive light source with the same photon output. It is thus clear that most of the light is limited from excitons at the surface of the crystal.

We also used floating organic films to illustrate the principle of detection limit enhancement. Anthracene crystal film was used as the exciton source, and R6G was used as the acceptor in aquous solution. The donor–acceptor system was excited with a bare optical fiber. A.thin layer of anthracene was made to float on the R6G aqueous solution by adding 1 drop of 0.025 M anthracene-benzene solution to the R6G. The fluorescence spectrum of R6G was recorded and compared with that obtained from the same concentration of R6G by direct excitation with the same light-transmitting optical fiber. As shown in Table 13.4, there is a large enhancement of R6G fluorescence intensity.

Table 13.4. Fluorescence Intensity Enhancement of R6G

Excitation Means	Sample	Integrated Fluorescence Intensity (Arbitrary Units)	Enhancement of Fluorescence Intensity
Optical fiber tip with 355 nm	R6G (10^{-6} M)	74,500	
Optical fiber tip with 355 nm	One drop of 0.025 M anthracene floating on R6G (10^{-6} M)	1,845,000	24.8

More than one factor may be responsible for the large enhancement in fluorescence intensity. One of the most obvious factors is the difference in excitation efficiency of R6G at 355 nm and in the anthracene fluorescence range. Anthracene has a wide fluorescence spectrum. We have taken an excitation spectrum of R6G. The difference between the excitation efficiency at 355 nm and at 450 nm is only 2.4 times, while Table 13.4 gives a fluorescence intensity difference of 24.8 times. This leaves a factor of at least 10 for the real near-field (excitonic) enhancement. Thus, it is fair to say that enhancement of fluorescence intensity by active light sources will result in a large enhancement in the detection limit for chemical analysis with optical detection.

13.10. DEVELOPMENT OF OPTICAL AND EXCITONIC SUPERTIPS

Even smaller probes are required and are in the making. In principle, an exciton light source can be as small as a single molecule or atom. At the same time, its position and scanning have to be defined in space as precisely as those of an STM tip (a randomly flying atom does not qualify). Existing designs for optical *supertips* (8) are based on the same principle as the green plant photosynthetic system (96). A submicometer antenna collects the photons by absorption and transfers the excitation energy to a single active center. From there the energy is either (1) radiated as a photon or (2) transferred to the sample in an energy transfer process (Förster–Dexter) (89). In either case, the result is generally affected by the nearby sample molecule: (1) The radiated excitation may be effected, for example, by intermolecular spin-orbit coupling (Kasha effect) (94). (2) The energy transfer results in a fluorescence or phosphorescence typical of the sample molecule. In the latter case, only virtual photons are produced by the supertip; this gives an excitation transfer tip (exciton tip), and only sample luminescence is detected. The world's largest ordered molecules, "dendrimers" (90, 91) , have been utilized or synthesized for this purpose (see below). Such a single molecule exhibits a 125 Å antenna with an active center of 10 Å or less. So far, only tips with aggreagtes of sample molecules have been utilized.

There are many methods and reasons for making supertips. For example, in principle, the optical fiber tip can be treated chemically to produce specific supertips for a variety of purposes: (1) wavelength shifters, such as crystallites that fluoresce to the red of the tip emission; (2) time "extenders" [some as (1)] utilizing prompt or delayed fluorescence or even phosphorescence; (3) highly sensitive opto-chemical nanosensors; (4) energy transfer supertips; and (5) heavy-atom sensors. Supertip development is the key to MEM, which relies on quantun optics mechanisms such as Förster energy transfer (89) or the Kasha effect (94). These interactions occur at the interface of the tip (its active center) and the sample (which are quantum mechanically coupled). For the highest resolution, this active center consists of a single molecule, or molecular cluster, that does the imaging. This molecule is then the energy donor site for the Förster energy transfer or the spin-orbit interaction site for the Kasha effect. Figure 13.30 shows the relation between the tip, the supertip, and the active center. As much as the supertip is part of the tip, the active center is part of the supertip. We note that a completely analogous situation is found in atomic force microscopy, where the active center is the force contact site at the tip of the supertip. In MEM the active center must be optically excited repeatedly. The design of the system of tips is thus geared to the need to supply the active center with plenty of excitation quanta. Supertips have been constructed on both micropipette tips and optical fiber tips. We will now describe several approaches to the development of supertips.

13.10.1.   Nanocrystal Designer Probes

One approach employs exciton-conducting crystallites that absorb the light of the fiber-optic tip and convert it into excitons, which ultimately produce photons again. We have successfully grown such crystals of perylene, diphenyl-anthracene, and so on, on the fiber tip. Alternatively, the crystallite is grown at the tip of a micropipette and the fiber-optic tip is pushed deep into the pipette, very close to the crystal tip. The crystallite acts as an antenna (compare this with the photosynthetic antenna) that channels the excitons to an active center that acts as an exciton trap (30). This trap collects excitation from as far away as 500–1000 Å. The active center is the key to *molecular engineering.*

The single impurity molecule (the "supertrap") creates a host "funnel" around it (89). This funnel consists of host crystal molecules perturbed by the impurity ("guest") molecules. The closer the host molecule is to the trap, the lower its excitation energy. The molecules in the funnel act as exciton traps, catching the excitation from the host crystal and passing it deeper and deeper (in energy) to the deepest of them all, the supertrap. For MEM the guest molecule is deposited on the surface of a molecular microcrystal and thus creates an energy funnel at its apex (active center). The best microcrystals are grown on the tip of a micropipette. This tip is excited *internally* by a fiber-optic tip. It gives as much intensity as with epiluminescence, i.e., the external excitation method (29). We have observed acceptable levels of luminescence intensities from a few molecules of RhB embedded in the surface of a DPA crystal, as shown in Figure 13.32. From the spectra, it is clear that even when dipping into very dilute RhB solutions (10^{-7} M), the emission of a tiny RhB active center is still detectable in an optical multichannel analyzer (OMA) apparatus. This preliminary work clearly demonstrates that supertips can be

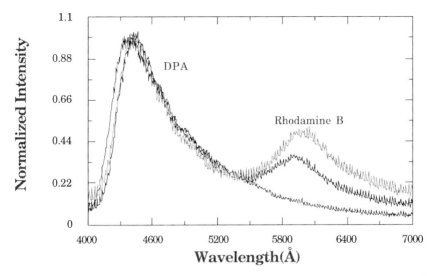

Figure 13.32. Fluorescence spectrum of active centers (RhB) and a supertip nanocrystal (diphenylanthracene). Note that an optical filter easily separates the two spectra.

prepared by dipping nanometer crystal tips in dye solutions. The crystal tips absorb the incident light, create excitons, and emit light again (with their typical fluorescence spectrum) or transfer energy to the active center and emit lower energy fluorescence. These tips are exciton supertips. By optimizing the selection of host-crystal guest–suppertrap pairs, we have been able to prepare supertips with high intensity and good stability (30, 97).

13.10.2.   Polymer Matrix Supertip

Another approach involves a polymeric matrix attached to the optical fiber tip by spatially controlled photopolymerization. The polymer is a copolymer consisting of acrylamide $N,N,$-methylenebisacrylamide (BIS), and appropriate dye monomer groups (19). This polymer supertip acts as an antenna, even though a less efficient one compared to a molecular crystal. However, with the large photon flux emanating from the tip, this lower efficiency should suffice. At the very tip of this supertip, the active center is attached by physical and chemical methods. The dye molecule on the top surface of the polymer is produced by photochemical reaction from a layer of precursor molecules deposited on the polymer tip by dipping it into a solution. The highest probability of a photochemical reaction is at the center of the tip, making it likely that the first and only active molecule is produced at this center. Several parameters can be used to control this operation: depth and duration of dipping, precursor concentration, intensity and/or exposure time, and photochemical reaction light intensity. By using a multidye solution for photochemical synthesis, we have prepared extremely small multifunctional optic and excitonic probes. These probes emit multiwavelength photons or produce excitons of different energy levels. By using multistep photochemical synthesis, we have miniaturized optical and exciton probes to sizes much smaller than those of the original light conductors. Multistep near-field photopolymerization (77) has been very effective in fabricating extremely small optical and excitonic supertips (41).

13.10.3. Single-Dendrimer Design

We have also designed a supertip made of a single symmetric macromolecule by using newly developed dendrimer supermolecules (90, 91). The so-called *starburst* phenylacetylene dendrimers include the largest structurally ordered molecule synthesized so far (D-127; see Figure 13.33). These fractal, tree-like supermolecules have spatially localized eigenfunctions and, in particular, localized electronic states. Furthermore, in the so-called SYNDROME family of dendrimers (90), simple theory leads to selectively lower excitation energies at the central locus of the ordered macromolecule, with energies increasing toward the rim. This is borne out experimentally by the vibronic spectra of the entire family of molecules (D-2 to D-127), with full internal consistency in the observed electronic energies (red shifts), vibrational quanta, Franck–Condon factors, overall transition moments, and picosecond spectral diffusion. The architecture of this series of dendrimers is controlled by organic synthetic methods. For example, the overall shape of a D-127 molecule is bowl-like, with a molecular size of about 125 Å. It can thus act as both an optical and a force active center. The large "rim" is bound to the tip by cumulative van der Waals bonding or covalent bonds. D-127 may be used in supertip preparation in two ways. The first is to make supertips in the range of 100 Å. In this case, D-127 is an energy transfer acceptor. It traps most of the energy quanta from the bulk of the tip, as shown below. D-127 is the light-emitting active center. The supertip prepared this way follows this scheme:

$$\text{tip} \rightarrow \text{supertip} \rightarrow \text{active center} \rightarrow \text{sample}$$

The second method of supertip preparation achieves approximately 10 Å resolution. To further demonstrate our energy funnel model (see above), we synthesized partial dendrimeric wedges with (and without) an excitation acceptor, a perylene derivative pendant, at the locus. As expected, the energy transfer from the large antenna ("tree canopy") to the small acceptor center was dramatic. The presence of the antenna (39) phenyl groups) increased the yield of the yellow perylenic emission by 3 orders of magnitude for a given excitation wavelength. This molecule is an ordered supermolecule transducer of absorbed radiation (STAR) (99), by analogy to the primary excitation, energy–collecting antennas of some natural photosynthetic systems. Overall, such a photonic subwavelength nano-lens may play a role in developing *molecular excitonics*, including luminescent optical nanoprobes, scanning exciton tunneling microscopy, and nanometer-scale fiber-optic chemical and biochemical sensors. To synthesize such a STAR molecule, the nanoarchitecture process is modified to prepare a D-127 molecule with a supertrap in the center. D-127 is synthesized with phenylacetylene (90). The synthesizing route has been modified to prepare similar dendrimers containing higher aromatic rings. A single such substituted group acts as a supertrap (8,89), collecting most of the excitation. This intramolecular exciton supertrap plays the role of the active center. It may act as an exciton donor, transferring excitation to an acceptor on the sample, and as the smallest possible light source. The modified dendrimer has a supertrap in the center with a size of about 10 Å. In such a supertip case, the supertrap center has different optical and excitonic properties from its surrounding molecules. The center can be used as a scanning tip that defines the scanning resolution. The supertip prepared this way follows the following scheme:

$$tip \rightarrow supertip \rightarrow antenna \rightarrow active\ center \rightarrow sample$$

The STAR molecule is designed not only for accelerated energy transfer but also for backward "spillover" of redundant energy in the pendant group. For instance, if a second excitation arrives while the first is still there, the two excitations are expected to "fuse" (88, 89), and the resulting higher energy excitation will transfer backward into the dendrimeric antenna state. Having many more nonradiative energy decay channels, it will be reduced from an S_n to an S_1 excitation. Then the S_1 excitation will flow again into the pendant supertrap. The nano-STAR represents a new class of "designer" molecules tailor-made for single-molecule light and exciton sources (8). Its large size (125 Å for D-127), stability, and efficiency will allow it to be used as a supertip for optical nanoprobes and nanosensors.

Supertips can double as NSOM and as MEM tips. The superresolution imaging of biologically interesting species relies on the dual function. The MEM operatin itself involves two different mechanisms: (1) active energy transfer (Förster–Dexter) (88, 89), in which the supertip is the exciton (energy) donor and the sample is the acceptor, and (2) passive interaction (sensor mode), exemplified by the Kasha effect (94), in which the sample either quenches the supertip's exciton or transforms it from a singlet to a triplet.

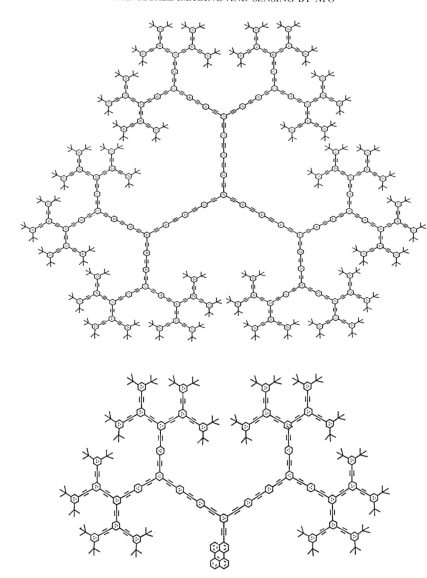

Figure 13.33. Dendrimer molecule (top) and nano-star molecule (bottom).

13.11. NEAR-FIELD OPTICAL CROSS SECTIONS

NFO has begun to gain recognition in several fields, including photofabrication, biomedical research, computer memory, microscopy, spectroscopy, and biochemical sensing. A special point of both theoretical and practical interest is the increased photoexcitation cross sections of optically active molecules in the near-field range. This factor is important for both the photofabrication process and the utilization of such probes. For example, NFO has been applied in two different ways in biochemical sensing: (1) for the nanofabrication of optical fiber biochemical sensors and (2) for the operation of these subwavelength sensors. In the first case, by applying NFO, we successfully demonstrated a new concept of near-field photochemical synthesis in which the dimension of the produced sensor is determined solely by the size of the light source. The synthesis maps the electromagnetic far-field or near-field profile of the light source. The second use of NFO occurs during the operation of these sensors. In NFO, the molecular absorption probability is much higher than that in far-field optics, where about 1 billion photons are needed to excite one isolated molecule (8). This near-field effect is important for both the photo-nanofabrication process and the utilization of such optical probes in biochemical analysis.

Near-field excitation (8, 25) is a new concept introduced for the excitation of subwavelength optical and excitonic probes. Effects unique to the near field are shown to be involved in the excitation by a subwavelength light source (6, 8, 19, 25, 39). There are three major reasons to introduce this new concept. First, geometrically, NFO encompasses a very small space; thus, only molecules inside the near-field region will be excited efficiently. The molecules to the left and right of the near-field light source will not be excited at all. This is in contrast to an evanescent wave spreading over a large region. Second, in the near-field region, the light intensity flux is much higher than that in the far-field region of the same light source. Thus, the excitation will be much stronger. Third, we believe that in the near-field region, the molecular absorption probability is much higher than in the far-field region (see above). Moreover, in the special case in which a molecule is within the Förester radius of an active near-resonance light source (8, 25), the molecular excitation probability is near unity. In the near-field case, the probability should be between unity and that in far-field optical excitation. From a quantum mechanics point of view, the electromagnetic wavefunction of light in the near-field region has become constrained. This may imply that the overlap between the molecular wavefunction and this constrained wavefunction should be significantly better. Thus, a higher excitation probability is expected. A classical analogy is the effect of the refractive index on the Einstein absorption coefficient. The absorption coefficient increases with λ^{-3}, where λ is the real wavelength. We assume that the effects of "photon squeezing" by the refractive index and by the near-field conditions are comparable. This leads to a greatly enhanced cross section. The promise of near-field photon sources for the nanofabrication and monitoring of electronic devices and optical memories is obviously of much current interest. So is the use of optical nanoprobes for microscopy, spectroscopy, and biochemical sensing (5, 6).

In addition, the mechanism of light–matter interaction may be different in the far- and near-field regimes (100), leading to different spectral selection rules and, in particular, to an enhanced cross section of light absorption (and thus fluorescence) (100). These phenomena are an extra bonus for near-field detection (19). Finally, good optical fiber tips do *not* affect light polarization (40,101) or even cross-correlation; ultrafast (80) fs nanometer light sources have been demonstrated (101).

13.12. SUMMARY

Different subwavelength light and exciton probes have been prepared by micropipettes and nanofabricated optical fiber tips. A variety of nanofabrication techniques have been developed to design and prepare these nano-optic and excitonic probes. These probes have been applied successfully in NSOM, NSOS, NFO biochemical sensing, and MEM. The diffraction limit has been overcome, enabling one to study various samples of great interest to biomedical and materials sciences with NFO. By using different microscopic and nanoscopic samples, spectroscopic studies have been carried out on heterogeneous samples and quantum well strucutres. The potential of NSOS has been demonstrated as a technique for obtaining nanometer-sized spatial resolution in a spectral dimension. A new near-field nanotechnology, photo nanofabrication, has been developed to prepare nanometer light and exciton probes. *In vivo* dynamic and static measurements of biochemical parameters have been carried out on individual, viable, intact rat conceptuses during the period of organogenesis or external environmental changes. The ability of the subwavelength sensors to measure biochemical changes in real time in the intact rat conceptus demonstrates their great potential appications for dynamic analysis in small, multicellular organisms and single cells. Supertips have been prepared, and their development has been discussed. The large variety of nanoprobes should lead to new applications in NFO, as well as in other microscopic analyses in biomedical and materials science. NFO has matured to the stage where chemists, biologists, and medical researchers can use it for studies where progress has been limited by available technologies.

We expect that NFO will be further developed and that NSOM, NSOS, and FOCS will become conventional tools in scientific research and development. The advances toward nanometer-resolved microscopy, spectroscopy, and biochemical sensing promise to push chemical and biological analysis much closer to one of its ultimate goals—the noninvasive detection and manipulation of single molecules, radicals, or ions, as well as the determinatino of a molecule's precise coordinates and the characterization of its structural conformation, its internal dynamics, and its energetics, all as a function of time and environmental perturbations.

There are still many technical problems to be solved—from understanding the contrast mechanism to controlling photobleaching (a standard problem in fluorescence microscopy). However, the future looks bright. Reversible bleaching (8) could be the basis for the highest density optical memories (with the "bit" occupying only a single molecule). More realistic in the near term would be pixels on the order of 100 Å. The necessary high scanning speeds are presently limited by probe intensity, but we expect much higher photon outputs to become possible. Indeed, subwavelength probes have already been turned into high-flux lasers (102). Analytical chemistry is being driven to the extreme of single dye molecule imaging and absolute ion or enzyme detection limits of zeptomoles or less (103, 104). Complex polymeric samples may finally be characterized on the molecular level by nanospectroscopy. Most important, it is possible that biosamples, including living cells, can be imaged down to the molecular level and analyzed spectroscopically or by chemical sensors. The intracellular molecular dynamics of organogenesis, metabolism, splitting, and chemical damage could be followed *in vivo* and in *real time*. At the same time, DNA could be sequenced *in situ* (95) or even repaired by the right probe at the right location. Further, molecular photonic process will be much better understood (105). The manipulation of matter at the atomic and molecular levels will also be facilitated with NFO (106). While much of this scenario might sound like science fiction, some or the above-mentioned achievements in NFO must have sounded like science fiction only a few years ago.

ACKNOWLEDGMENTS

We thank our colleagues who are partners in many of the described enterprises: Duane Birnbaum, Craig Harris, John Langmore, Greg Merrit, Jeff Moore, Eric Munson, Brad Orr, Steve Parus, Zhong-You Shi, Michael Shortreed, Steve Smith, Bjorn Thorsrud, and Zhifu Xu. Financial support came from U.S. Department of Energy Grant DE-FG02-90ER61085. The excitonic light source research and development was supported by National Science Foundation Grant DMR-9410709.

REFERENCES

1. Abbé, E. (1873). *Arch. Mikrosk. Anat.* **9**, 413.

2. Lewis, A., Isaacson, M., Muray, A., and Harootunian, A. (1984). Development of a 500 Å spatial resolution light microscope *Ultramicroscopy* **13**, 227.

3. Pohl, D. W., Denk, W., and Lanz, M. (1984) Optical stethoscopy: Image recording with resolution 1/20. *Appl. Phys. Lett.* **44**, 651.

4. Lieberman, K., Harush, S., Lewis, A., and Kopelman, R. (1990). A light source smaller than the optical wavelength. *Science* **247**, 59.

5. Betzig, E., Trautman, J. K., Harris, T. D., Weiner, J. S., and Kostelak, R. L. (1991). Breaking the diffraction barrier: Optical microscopy on a nanometric scale. *Science* **251**, 1468; Betzig, E., and Trautman, J. K. (1992). Near-field optics: Microscopy, spectroscopy, and surface modification beyond the diffraction limit. *Science* **257**, 189.

6. Tan, W., Shi, Z.-Y., Smith, S., Birnbaum, D., and Kopelman, R. (1992) Submicrometer intracellular chemical optical fiber sensors. *Science* **258**, 778–781.

7. Pohl, D. W. (1990). In *Advances in Optical and Electron Microscopy* (C. J. R. Sheppard and T. Mulvey, Eds.), p. 243. Academic Press, London.

8. Kopelman, R., and Tan, W. (1993). In *Spectroscopic and Microscopic Imaging of the Chemical State* (M. D. Morris, Ed.), p. 227. Dekker, New York.

9. Ambrose, W. P., Goodwin, P. M., Martin, J. C., and Keller, R. W. (1994). Alterations of single molecule fluorescence lifetimes in near field optical microscopy. *Science* **265**, 364–367.

10. Xie, X. S, and Dunn, R. C. (1994) *Science* **265**, 361–364.

11. Kopelman, R., Lewis, A., and Lieberman, K. (1989). Exciton microscopy and scanning optical nanoscopy. In *X-Ray Microimaging for the Life Sciences* (D. Attwood and B. Barton, Eds.), p. 166. Lawrence Berkeley Laboratory Berkeley, CA.

12. Herman, B., and Jacobson K. (1989). *Optical Micoscopy for Biology*. Wiley-Liss, New York.

13. Synge, E. H. (1928). *Philos. Mag.* [6] **6**, 356.

14. O'Keefe, J. A. (1956). Resolving power of visible light. *J. Opt. Soc. Am.* **46**, 359.

15. Ash, E. A. (1972). Super-resolution aperture scanning microscope. *Nature (London)* **237**, 510.

16. Morris, M. D. (1993). *Spectroscopic and Microscopic Imaging of the Chemical State*. Dekker, New York.

17. Binnig, G., Rohrer, H., Gerber, C. and Weibel, E. (1982). Surface studies by scanning tunneling microscopy. *Phys. Rev. Lett.* **49**, 57.

18. Schineir, J., Sonnenfeld, R., Hansma, P. K., and Tersoff, J. (1986). *Phys. Rev. C: Condens. Matter B* **34**, 4979.

19. Tan, W., Shi, Z.-Y., and Kopelman, R. (1992). The development of submicron optical fiber chemical sensor. *Anal. Chem.* **64**(21), 2985.

20. Garcia-Parajo, M., Cambril, E., and Chen, Y. (1994). Simultaneous scanning tunneling microscope and collection mode scanning near-field optical microscope using gold coated optical fiber probes *Appl. Phys. Lett.* **65**(12), 1498–1500.

21. Moers, M. H. P., Tack, R. G., van Hulst, N. F., and Bolger, B. (1993). A combined near-field optical and force microscope. *Scanning Microsc.* **7**(3), 789–792.

22. Shalom, S., Lieberman, K., Lewis, A., and Cohen, S. R. (1992). A micropipette force probe suitable for near field scanning optical microscopy. *Rev. Sci. Instrum.* **63**(9), 4061–4065.

23. Sakmann, B., and Neher, E. (eds.) (1983). *Single-Channel Recording.* Plenum, New York.

24. Thomas, R. C. (1978). *Ion-Sensitive Intracellular Microelectrodes.* Academic Press, New York.

25. Tan, W. (1993). *Ph. D. Thesis.* University of Michigan, Ann Arbor.

26. Brown, K. T., and Flaming, D. G. (1986). *Advanced Micropipette Techniques for Cell Physiology.* Wiley, New York.

27. Vogelmann, T., Martin, G., Chen, G., and Buttry, D. (1992). *Adv. Bot. Res.* **18**, 255.

28. Tan, W., and Kopelman, R. (1993). In *Dynamics in Small Confining Systems* (J. M. Drake et al., Eds.), pp. 287, 290. Materials Research Society, Pittsburgh, PA.

29. Lewis, A., and Lieberman, K. (1991). *Nature (London)* **354**, 214.

30. Tan, W., Shi, Z., Smith, S., and Kopelman, R. (1994). Photonanofabrication and optical nanoprobes. *Mol. Cryst. Liq. Cryst. Sci. Technol., Sect. A)* **252–253**, 535–549.

31. Kopelman, R., Lewis, A., Lieberman, K., and Tan, W. (1991). Evanescent luminescence and nanometer-size light source. *J. Lumin.* **48/49**, 87.

32. Lewis, A., and Lieberman, K., (1991) The optical near field and analytical chemistry. *Anal. Chem* **63**, 625A.

33. Kopelman, R., Smith, S., Tan, W., Zenobi, R., Liberman, K., and Lewis, A. (1992). Spectral analysis of surfaces at subwavelength resolution. *Proc. SPIE—Int. Soc. Opt. Eng.* **1637**, 33.

34. Born, M., and Wolf, E. (1959). *Principles of Optics* Pergamon, London.

35. Agranovich, V. M., and Galanin. M. D. (1982). *Electronic Excitation Energy Transfer on Condensed Matter.* North-Holland Publ., Amsterdam.

36. Aavikso, J., Freiberg, A., Lipmaa, J., and Reinot, T. (1987). Times resolved studies of exciton polaritons. *J. Lumin.* **37**, 313.

37. Monson, E., Merritt, G., Smith, S., Langmore, J. P., and Kopelman, R. (1995). Implementation of an NSOM system for fluorescence microscopy. *Ultramicroscopy* **57**, 257–262.

38. Smith, S., Monson, E., Merritt, G., Tan, W., Birnbaum, D., Shi, S. Y., Thorsurud, B. A., Harris, C., Grahn. H. T. et al. (1993). Tip/sample interactions. Contrast and near-field microscopy of biological and solid state samples. *Proc. SPIE—Int. Soc. Opt. Eng.* **1858**, 81–92.

39. Betzig, E., Chichester, R. J.; Lanni, F., and Taylor, D. L. (1993). Near-field fluorescence imaging of cytoskeletal actin. *Bioimaging* **1**(3), 129–135.

40. Betzig, E., and Chichester, R. J. (1993). Single molecules observed by near-field scanning optical microscopy. *Science* **262**, 1422–1425.

41. Kopelman, R., and Tan, W. (1993). Near-field optics: Imaging single molecules. *Science* **262**, 1382–1384.

42. Trautman, J. K., Macklin, J. J., Brus, L. E., and Betzig, E. (1994). Near-field spectroscopy of single molecules at room temperature. *Nature (London)* **369**, 40–42.

43. Ambrose, W. P., Goodwin, P. M., Martin, J. C., and Keller R. A. (1994). Single molecule detection and photochemistry on a surface using near field optical excitation. *Phys. Rev. Lett.* **72**(1), 160–163.

44. Moerner, W. E., Plakhotnik, T., Irngartinger, T., Wild, U. P., Pohl, D. W., and Hecht, B. (1994). Near-field optical spectroscopy of individual molecules in solids *Phys. Rev. Lett.* **73**(20), 2764–2767.

45. Reddick, R. C., Warmack, R. J., Chilcott, D. W., Sharp, S. L., and Ferrell, T. L. (1990). Photon scanning tunneling microscopy. *Rev. Sci. Instrum.* **61**, 3669.

46. Trautman, J. K., Betzig, E., Weiner, J. S., DiGiovani, D. J., Harris, T. D., Hellman, F., and Gyorgy, E. M. (1992). Image contrast in near-field optics. *J. Appl. Phys.* **71**, 465.

47. Heinzelmann, H., and Pohl, D. W. (1994). Scanning near field optical microscopy. *Appl. Phys.* **A59**(2), 89–101.

48. Specht, M., Pedarnig, J. D., Heckel, W. M., and Hansch, T. W. (1992). Scanning plasmon near-field microscope. *Phys. Rev. Lett.* **68**(4), 476–479.

49. Ritchie, R. H. (1957). *Phys. Rev.* **106**, 874.

50. Trautman, J. K., Betzig, E., Weiner, J. S., DiGiovani, D. J., Harris, T. D., Hellman, F., and Görgy, E. M. (1992). Image contrast in near-field optics. *J. Appl. Phys.* **71**, 465.

51. Quate, C. F. (1986). Vacuum tunneling: A new technique for microscopy. *Phys. Today*, August, p. 26.

52. Moers, M. H. P., Tack, R.G., Van Hulst, N. F., and Bolger B. (1994). Photon Scanning tunneling microscope in combination with a force microscope. *J. Appl. Phys.* **75**(3), 1254–1257.

53. Betzig, E., Finn, P. L., and Weiner, J. S. (1992). Combined shear force and near field scanning optical microscopy. *Appl. Phys. Lett.* **60**(20), 2484–2486.

54. Toledo-Crow, R., Chen, Y., and Vaez-Iravani, M. (1992). An atomic force regulated near-field scanning optical microscope. *Proc. SPIE—Int. Soc. Opt. Eng.* **1639**, 44.

55. van Hulst, N. F., Moers, M. H. P., Noordman, O. F. J., Faulkner, T., Segerink, F. B., van dert Werf, K. O., de Grooth, B. G., and Bolger, B. (1992). Operation of a scanning near-field optical microscope in reflection in combination with a scanning force microscope. *Proc. SPIE—Int. Soc. Opt. Eng.* **1639**, 36.

56. Kook, S. K., and Kopelman, R. (1992). Microfluorescence and microstructure of tetracene aggregates in poly(methyl methacrylate). *J. Phys. Chem.* **96**(26), 10672–10676.

57. Hess, H. F., Betzig, E., Harris, T. D., Pfeiffer, L. N., and West, K. W. (1994). Near-field spectroscopy of the quantum constituents of a luminescent system. *Science* **264**, 1740–1745.

58. Grober, R. D., Harris, T. D., Trautman, J. K., Betzig, E., Wegscheider, W., Pfeiffer, L., and West, K. (1994). Optical spectroscopy of a GaAs/AlGaAs quantum wire structure using near-field scanning optical microscopy. *Appl. Phys. Lett.* **64**(11), 1421–1423.

59. Birnbaum, D., Kook, S. K., and Kopelman, R. (1993). Near-field scanning optical spectroscopy: Spatially resolved spectra of microscrystals and nanoaggregates in doped polymers. *J. Phys. Chem.* **97**(13), 3091–3094.

60. Tanaka, J. (1963). *Bull. Chem. Soc. Jpn.* **36**, 1237.

61. Hille, B. (1984). *Ionic Channels in Excitable Membranes.* Sinaur Assoc. Sunderland, MA. (1994).

62. Moerner, W. E. (1994). *Science* **265**, 46–53; Moerner, W. E., and Kador, L. (1989). *Phys. Rev. Lett.* **62**, 2535; Kador, L., Horne, D. E., and Moerner, W. E. (1990) *J. Phys. Chem.* **94**, 1237; Basche, Th., and Moerner, W. E. (1993). *Angew. Chem., Int. Ed. Engl.* **32**, 457.

63. Orrit, M., and Bernard, J. (1990). *Phys. Rev. Lett.* **65**, 2716; Orrit, M., Bernard, J., and Personov, R. I. (1993). *J. Phys. Chem.* **97**, 10256–10268.

64. Smith, R. D., Cheng, X., Bruce, J. E., Hofstadler, S. A., and Anderson, G. A (1994). *Nature (London)* **369**, 137–139; Barnes, M. D., Ng, K. C., Whitten, W. B., and Ramsey, J. M (1993). *Anal. Chem.* **65**, 2360–2365.

65. Shera, E. B., Seitzinger, N. K., Davis, L. M., Keller, R. A., and Soper, S. A. (1990). *Chem. Phys. Lett.* **1744**, 553.

66. Whitten, W. B., Ramsey, J. M., Arnold, S., and Bronk, B. V. (1991). *Anal. Chem.* **63**, 1027.

67. Berndt. R., Gaish, R., Gimzewski, J. K., Reihl, B. Schlittler, R. R., Schneider, W. D., and Tschudy, M. (1993). *Science* **262**, 1425–1427; Ishikawa, M., Hirano, K., Hayakawa, T., Hosoi, S., and Brenner, S. (1994), *Jpn. J. Appl. Phys.* **33**, 1571–1576.

68. Kopelman, R., Tan, W., Lewis, A., and Lieberman, K. (1991). Scanning exciton microscopy and single molecule resolution and detection. *Proc. SPIE—Int. Soc. Opt. Eng.* **1435**, 96–101.

69. Lee, Y.-H., Maus, R. G., Smith B. W., and Winefordner, J. D. (1994) Laser-Induced fluorescence detection of a single molecule in a capillary. *Anal. Chem.* **66**(23), 4142–4149.

70. Nie, S., Chiu, D. T., and Zare, R. N. (1994) Probing individual molecules with confocal fluorescence microscopy. *Science* **266**, 1018–1021.

71. Fan, R.-R., and Bard, A. J. (1995). Electrochemical detection of single molecules *Science* **267**, 871–874.

72. Seitz, W. R. (1988). Chemical sensors based on immobilized indicators and fiber optics. *CRC Crit. Rev. Anal. Chem.* **19**, 135.

73. Tan, W., Shi, Z., Thorsrud, B. A., Harris, C., and Kopelman, R. (1994). Near-field fiber optic chemical sensors and biological applications, *Proc. SPIE—Int. Soc. Opt. Eng.* **2068**, 59–68.

74. Kopelman, R., Tan, W., and Shi, Z. (1994). Micro optical fiber light source and sensor and its method of fabrication. *PCT Int. Appl.* **WO 94 06040 A1 94 0317**, 1–30.

75. Bright, G. R., Whitker, J. F., Haugland, R. P., and Taylor, D. L. (1989). Heterogeneity of the changes in cytoplasmic pH upon serum stimulation of quiescent fibroblasts. *J. Cell. Physiol.* **141**, 410–419.

76. Wolfbeis, O. S., Weis, L., Leiner, M. J. P., and Ziegler, W. E. (1988). Fiber optic fluorosensor for oxygen and carbon dioxide. *Anal. Chem.* **60**, 2028–2030.

77. Tan, W., Shi, Z., and Kopelman, R. (1995). Miniaturized fiber optic chemical sensors with fluorescent dye doped polymers. *Sensors Actuators* **B 28**, 157–163.

78. Harris, C., Juchau, M. R., and Mirkes, P. E (1991). *Teratology* **43**, 229–239.

79. Freeman, S. J., Coakley, M., and Brown, N. A. (1987). In *Biochemical Toxicology: A Practical Approach* (K. Snell., and B. Mullock, Eds.) 83–107. IRL Press, Washington DC.

80. Scott, W. J., Jr., Duggan, C. A., Schreiner, C. M., Collins, M. D., and Nau, H. (1987). In *Approaches to Elucidate Mechanisms in Teratogenesis* (F. Weisch, Ed.), pp. 99–107. Hemisphere, New York.

81. Collins, M. D., Duggan, C. A., Schreiner, C. M., and Scott, W. J. Jr. (1989). *Am. J. Physiol.* **257**, R542-R549.

82. Dean, G. E., Fishkes, H., Nelson, P. J., and Rudnick, G. (1984). The hydrogen ion-pumping adenosine triphosphatase of platelet dense granule membrane. Differences from F1F0- and phosphoenzyme-type ATPases. *J. Biol. Chem.* **259**, 9569–9574.

83. Hiranruengchok, R., and Harris, C. (1993) Glutathione oxidation and embryotoxicity elicited by diamide in the developing rat conceptus in vitro. *Toxicol Appl. Pharmacol.* **120**, 62–71.

84. Waddell, W., and Butler T. (1959). *J. Clin. Invest.* **38**, 720–729.

85. Kopelman R., and Lewis, A. (1989). A nanometer dimension optical device with microimaging and nanoillumination capabilities. *U.S. Pat.* 07/380,099.

86. McDonald, A. (1972). Electric and magnetic coupling through small apertures in shield walls of any thickness. *IEEE Trans. Microwave Theory Tech.* **MTT-20**, 698.

87. Lewis, A., Betzig, E., Harootunian, A., Isaacson, M., and Kratschmer, E., (1988). Near-field imaging of fluorescence: In *Spectroscopic Membrane Probes* (L. M. Loew, Ed.), Vol. 2, p. 81. CRC Press, Boca Raton, FL.

88. Francis, A. H., and Kopelman, R. (1986). Excitation dynamics in molecular solids. In *Topics in Applied Physics, Laser Spectroscopy of Solids* (W. M. Yen and P. M. Selzer, Eds.), 2nd ed., Springer-Verlag, Berlin.

89. Pope, M., and Swenberg, E. (1982). *Electronic Processes in Organic Crystals.* Oxford Univ. Press, New York.

90. Xu, Z., and Moore, J. S. (1993). *Angrew. Chem.* **32**, 1354.

91. Xu. Z., Shi, Z.-Y., Tan, W., Kopelman, R., and Moore, J. S. (1993). *Polym. Prep., Am. Chem. Soc., Div. Polym. Chem.* **33**(1), 130–131.

92. Kopelman, R., Lewis, A., and Lieberman, K. (1989). Molecular exciton microscopy. *Biophys. J.* **55**, 450a.

93. Roberts, G. (Ed.) (1990). *Langumuir-Blodgett Films.* Plenum, New York.

94. Kasha, M. (1952). *J. Chem. Phys.* **20**, 71.

95. Kopelman, R., Langmore, J., Orr, B., Shi, Z.-Y., Smith, S., Tan, W. and Makarov, V. (1991). Scanning molecular exciton microscopy: A new approach to gene sequencing, *Human Genome Program Reports (1991–1992)*, p. 130. U. S. Department of Energy, Washington, D.C.

96. Fauman, E. B., and Kopelman. R. (1989). Excitons in molecular aggregates and a hypothesis on the sodium channel gating *Mol. Cell. Biophys.* **6**, 47.

97. Tan, W., Shi Z., and Kopelman, R. (1995). Molecular supertips for near-field optics. *Ultramicroscopy* (in press).

98. Yeung, E. S. (1994). Chemical analysis of single cells. *Acc. Chem. Res.* **27**, 409–414.

99. Shortreed, M., Shi, Z.-Y., Tan, W., Xu, Z., Moore, J., and Kopelman, R. (1996). *Science* (in press).

100. Kopelman, R, Tan, W., and Birnbaum, D. (1994). Subwavelength spectroscopy, exciton supertips and mesoscopic light-matter interactions. *J. Lumin.* **58**(1–6), 380–387.

101. Smith, S., Orr, B., Kopelman, R., and Noires, T. (1995). 100 Fs/100 nm near-field probes. *Ultramicroscopy*, **57**, 173–175.

102. Betzig, E., Grubb, S. G., Chichester, R. J., DiGiovanni, D. J., and Weiner, J. S. (1993). Fiber laser probe for near-field scanning optical microsocpy *Appl. Phys. Lett.* **63**(26), 3550–3552.

103. Yeung, E. S. (1995) *Adv. Chromatogr.* **35**, 1–51.

104. Tan, W., and Yeung, E. S. (1995). Simultaneous determination of enzyme activity and enzyme quantity in single human erythrocytes. *Anal. Biochem.* **226**(1), 74–79.

105. Kopelman, R. (1991). Exciton microscopy and reaction kinetics in restricted spaces. In *Physical and Chemical Mechanisms in Molecular Radiation Biology* (W. A. Glass and M. Varma, Eds.), Plenum Press, New York.

106. Avouris, P. (1995) Manipulation of matter at the atomic and molecular levels. *Acc. Chem. Res.* **28**, 95–102.

INDEX

477